LECTURES ON SOLID STATE PHYSICS

BY

GEORG BUSCH

Professor of Physics
Swiss Federal Institute of Technology,
Zurich, Switzerland

AND

HORST SCHADE

David Sarnoff Research Center,
RCA Laboratories, Princeton, N.J., USA

TRANSLATED BY

FERDINAND CAP

TRANSLATION EDITOR

D. TER HAAR

PERGAMON PRESS

OXFORD • NEW YORK • TORONTO
SYDNEY · PARIS · FRANKFURT

U. K.	Pergamon Press Ltd., Headington Hill Hall, Oxford OX3 0BW, England
U. S. A.	Pergamon Press Inc., Maxwell House, Fairview Park, Elmsford, New York 10523, U.S.A.
CANADA	Pergamon of Canada Ltd., P.O. Box 9600, Don Mills M3C 2T9, Ontario, Canada
AUSTRALIA	Pergamon Press (Aust.) Pty. Ltd., 19a Boundary Street, Rushcutters Bay, N.S.W. 2011, Australia
FRANCE	Pergamon Press SARL, 24 rue des Ecoles, 75240 Paris, Cedex 05, France
WEST GERMANY	Pergamon Press GmbH, 6242 Kronberg/Taunus, Pferdstrasse 1, Frankfurt-am-Main, West Germany

First English edition 1976

Library of Congress Cataloging in Publication Data

Busch, Georg, 1908–
Lectures on solid state physics.

(International series in natural philosophy; v. 79)
Translation of *Vorlesungen über Festkörperphysik*.
Includes bibliographies and index.
1. Solids. I. Schade, Horst, 1932– joint author II. Title.

QC176.B8713 1975   530.4'1   74–32180

First published in the German language as
Vorlesungen über Festkörperphysik
By: Birkhäuser Verlag, Basel 1973.

ISBN 0–08–016894–9

Printed in Hungary

Contents

Tables

Introduction

THIS book is based on the material of my lectures on solid state physics which I have been giving annually for some years; the lectures are mainly intended for students at the Department of Mathematics and Physics at the Swiss Federal Institute of Technology, Zürich.

Following repeatedly expressed requests from students and my own desire to arrange the lectures more freely, the idea grew to contain the essential topics in a book. My former assistant and co-worker, Dr. Horst Schade, took over the difficult and laborious task of writing down the manuscript for this book. My long-standing co-operation with Dr. Schade in teaching as well as in research had assured me beforehand that the presentation of the subject-matter would be completely in accordance with my own point of view. I did not assist in the compilation of the text, which due to lack of time would not have been possible for me. The merit and the responsibility for the contents of the book should therefore go to Dr. Schade whom I wish to thank herewith most cordially for his devoted work.

Zürich, October 1972 G. BUSCH

Preface

THE *Lectures on Solid State Physics* represent a selection from various branches of solid state physics. The choice of topics was primarily guided not by its scientific and practical importance but by the existing personal interests of the authors. We are aware that this book does not deal with all the aspects of the physics of the solid state, and some very important topics such as dielectrics and superconductivity are missing, or have only been described briefly. The maim aim of the lectures, as well as of this book, is the development of basic physical ideas, which lead to an understanding of phenomena and effects. Thus the assumptions and approach essential for the interpretation by models have, as far as possible, been clearly explained and established. Results of computation are always compared with experimental data, not only in an endeavour to show the limits of the validity of the models, but also to acquaint the reader with the order of magnitude of the observable effects. Many cross-references in the text are intended to point out important connections and common properties. This is to meet the often-expressed opinion that physics consists of many unrelated single facts and effects. The use of elementary mathematical methods is deliberate, because experience has shown again and again that difficulties in understanding lie not in the formalism but in the physical thinking process. This book is written for physics graduates, but should also be of interest to electrical engineers, chemists and metallurgists.

We wish to thank many colleagues and friends who helped us to produce this book. Our very special thanks go to PD Dr. S. Yuan who not only critically read the manuscript, but also made many valuable suggestions. We are also indebted to him for the compilations of the Periodic System and Periodic Tables of the Physical Properties of the Elements, and Physical Constants and Conversion Factors, as well as for several tables and the literature sources in the text.

We are most grateful to Dr. F. Hulliger for his careful proof-reading and his useful hints. Our further thanks go to PD Dr. P. Junod who let us have details of his lectures on magnetism, also to Prof. Dr. W. Baltensperger, Prof. Dr. A. H. Madjid, Dr. A. Menth and Prof. Dr. P. Wyder, who have read through parts of the manuscript.

Dr. Y. Baer, Dr. E. Bucher, Dr. F. Hulliger and Dr. C. Palmy contributed to the compilation of the Periodic Tables of the Physical Properties of the Elements, and Prof. Dr. A. Berg assisted with the production of the Fraunhofer diffraction patterns of crystalline, paracrystalline and amorphous structures (Figs. 1–3). Our thanks go also to them.

In addition we wish to thank Mr. L. Scherrer who has revised many illustrations with great care and patience, and Mrs. C. Winkler who typed parts of the manuscript.

The publishers have been most generous in meeting our requests and were most reliable in arranging the completion of the book. This we very much appreciate. We would also like to thank the Zentenarfonds der Eidgenössischen Technischen Hochschule Zürich for a contribution towards the printing costs.

Zürich and Princeton, N.J. G. BUSCH
October 1972 H. SCHADE

Preface to the English Edition

SINCE the German version of this book has been well received, we are encouraged to offer its English translation to a much larger readership. The title *Lectures on Solid State Physics* refers mainly to the origin of this book and is not meant to restrict the purpose of it to be merely a textbook for students. The present book should serve as a well-founded basis, not as a replacement, however, for lectures on solid state physics and, more generally, as a reference book for students as well as for professionals in the fields of solid state physics and chemistry, electrical engineering, and materials science.

As an addition to the English version, we have included a series of problems. Their solutions can be worked out entirely on the basis of the subject-matter contained in this book. Many of our co-workers and friends have participated in the formulation of these problems in recent years. In particular, we would like to acknowledge the great support by PD Dr. S. Yuan; we are very grateful to him also for his experienced advice on the selection of the problems presented here.

We would like to express our thanks to F. Cap for his knowledgeable translation of the main text, and to the editor D. ter Haar and the publisher for their support.

Zürich and Princeton, N.J.
July 1975

G. BUSCH
H. SCHADE

A. Characteristic Features of the Structure of Solids

A SOLID represents a particular form of aggregation of condensed matter. Whereas gases and usually also liquids occupy a given volume only if this volume is bounded by walls, solids have a geometrical shape independent of the presence of limiting walls; just like their volume, their shape is an invariant property, if pressure and temperature are constant and no external forces act on them. The solids are classified, according to their microscopic structure, as crystalline, paracrystalline, and amorphous.

A crystalline solid is characterized by a three-dimensional array of building blocks (atoms, ions, or molecules). The spatial position of these particles is fixed by three non-coplanar vectors (cf. p. 5) which span a lattice in space. This lattice is built up out of identical unit cells which generally have the form of parallelepipeds. It is a consequence of the structural periodicity that each lattice point has the same number of neighbours, i.e. the same coordination number. In the solid state most elements and simple chemical compounds have, as a rule, a crystalline structure.

In the case of paracrystalline solids the structure represents a distorted space lattice which also consists of unit cells. While the unit cells of a crystal are identical parallelepipeds, the lattice of a paracrystal consists of different distorted parallelepipeds with volumes which fluctuate around a statistical mean. Thus the lattice vectors are different for different cells, both in magnitude and in direction. Just as in the lattice of crystalline structures, each lattice point in a paracrystalline solid has the same coordination number. Paracrystalline structures are mainly observed in the case of macromolecular substances, e.g. high-molecular-weight polymers, cellulose, or proteins, whose building blocks themselves can be complex molecules. It should be mentioned that the paracrystalline structure is not only observed in the solid state, it may also occur in the liquid state; thus, for example, molten metals have been proved to be paracrystalline. In addition to this, we have to mention the so-called liquid crystals, which represent certain phases of organic materials, in which the molecules are arranged according to some ordering principle (nematic, cholesteric, and smectic structures).

In crystalline and paracrystalline solids the probability of finding a neighbouring particle when starting from a given particle depends on the azimuthal angle (so-called non-spherically symmetric particle distribution). In amorphous

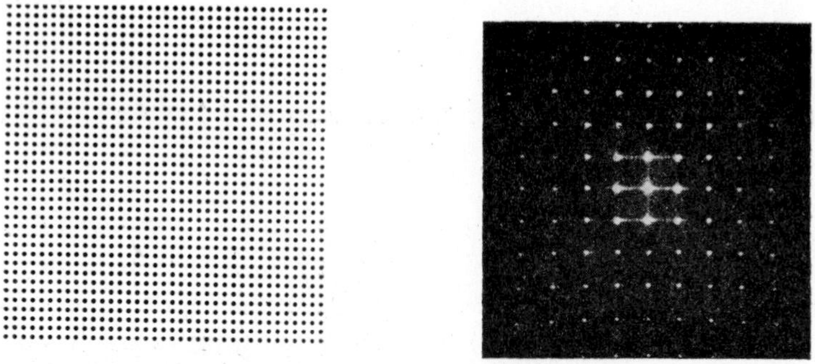

FIG. 1. Model and diffraction pattern of a crystalline structure.

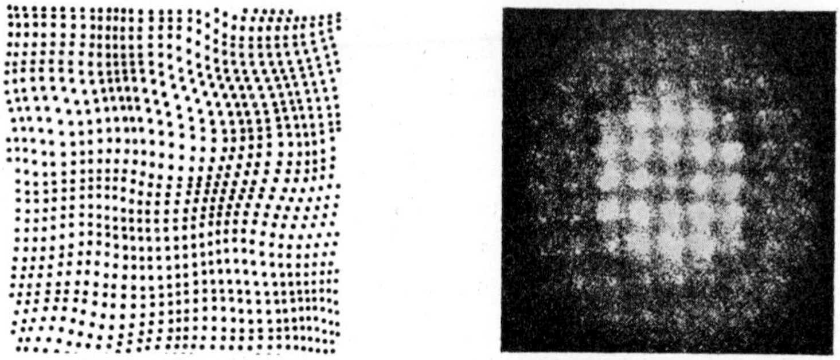

FIG. 2. Model and diffraction pattern of a paracrystalline structure.

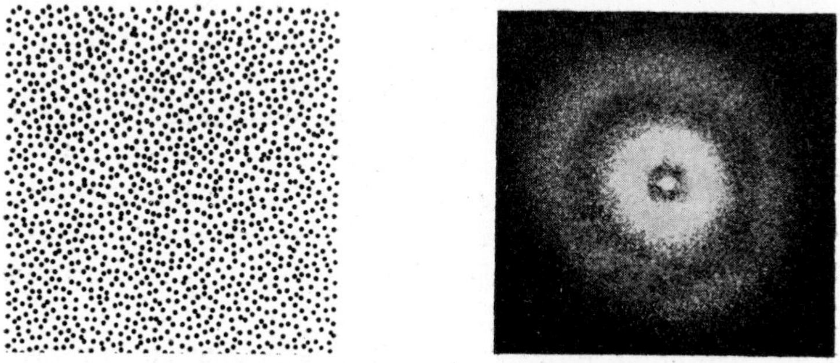

FIG. 3. Model and diffraction pattern of an amorphous structure.

solids, however, the particle distribution is spherically symmetric.

Disregarding the possibility of a so-called short-range order, the arrangement of the particles in amorphous structures is random. Accordingly, the particles have no well-defined coordination number and it is impossible to attribute a definite space lattice to amorphous structures. Amorphous substances are, for example, glass, ceramic materials, gels, and organic polymers. Also layers of solids precipitated at low temperatures are often amorphous. Substances which normally are crystalline can be obtained in an amorphous form when solidified very rapidly from the liquid state or when exposed to high-energy radiation.

In order to determine the structure of matter it is generally possible to use X-ray scattering (cf. pp. 16 ff.) or electron, neutron, or proton scattering. The structural parameters of a substance manifest themselves in the diffraction pattern; this can be demonstrated with the help of light diffraction patterns of two-dimensional model structures such as are shown in Figs. 1–3 (a crystalline, a paracrystalline, and an amorphous structure and the Fraunhofer diffraction patterns obtained with them).

The crystalline structure is represented by a square lattice (cf. Fig. 1). The diffraction centres are arranged in strict periodicity so that the scattered waves possess strict phase relations. The diffraction pattern is the result of a summation of the corresponding scattering *amplitudes* and displays well-defined "reflections".

The amorphous solid, however, is represented as an array of identical, randomly distributed scattering centres (cf. Fig. 3). The scattered waves emitted by the individual centres have no strict phase relations so that, on the average, no interference will be observed. The diffraction pattern obtained will be the same as that of a single scattering centre with the only exception that the intensity of the diffraction pattern is almost equal to the sum of the *scattering intensities* of all scattering centres.

In paracrystalline structures each scattering centre has on the average the same environment (cf. Fig. 2). Definite phase relations exist approximately only for small diffraction angles for which the lattice distortion has the least influence. Thus, for small diffraction angles, the pattern shows broadened "reflections". At large angles the definite phase relations lose their validity and the "reflections" vanish in the diffuse background. Thus the diffraction pattern of a paracrystalline structure displays both the characteristics of crystals and those of amorphous structures.

After this brief introduction to the structure of solids and its fundamental characteristics we shall consider in the following only *crystalline* solids and their physical properties.

References

Textbooks and Review Articles

AZAROFF, L. V., and BROPHY, J. J., *Electronic Processes in Materials* (McGraw-Hill, New York 1963).

BLAKEMORE, J. S., *Solid State Physics* (Saunders, Philadelphia 1969).

BORN, M., and GÖPPERT-MAYER, M., *Dynamische Gittertheorie der Kristalle*, in *Handbuch der Physik*, Editors H. GEIGER and K. SCHEEL, Vol. 24/2 (Springer, Berlin 1933).

BOWEN, H. J. M., *Properties of Solids and Their Atomic Structures* (McGraw-Hill, New York 1967).

DEKKER, A. J., *Solid State Physics* (Prentice-Hall, Englewood Cliffs, N.J., 1957).

HANNAY, N. B., *Solid State Chemistry* (Prentice-Hall, New York 1967).

HAUG, A., *Theoretical Solid State Physics*, Vol. 1 (Pergamon, Oxford 1971).

HELLWEGE, K. H., *Einführung in die Festkörperphysik* (*Introduction to Solid State Physics*), 3 volumes (Springer, Berlin 1968 [Heidelberger Taschenbücher Bd. 33–35]).

KITTEL, C., *Introduction to Solid State Physics*, 3rd ed. (Wiley, New York 1967).

KITTEL, C., *Quantum Theory of Solids* (Wiley, New York 1963).

MCKELVEY, J. P., *Solid State and Semiconductor Physics* (Harper and Row, New York 1966).

NYE, J. F., *Physical Properties of Crystals* (Oxford University Press, London 1957).

PEIERLS, R. E., *Quantum Theory of Solids* (Oxford University Press, London 1955).

SACHS, M., *Solid State Theory* (McGraw-Hill, New York 1963).

SEITZ, F., *Modern Theory of Solids* (McGraw-Hill, New York 1940).

SINNOT, M. J., *The Solid State for Engineers* (Wiley, New York 1958).

SLATER, J. C., *Quantum Theory of Matter*, 2nd ed. (McGraw-Hill, New York 1968).

SLATER, J. C., *Quantum Theory of Molecules and Solids*, Vol. 3 (McGraw-Hill, New York 1967).

SMITH, R. A., *Wave Mechanics of Crystalline Solids*, 2nd ed. (Chapman and Hall, London 1969).

SOMMERFELD, A., and BETHE, H., *Elektronentheorie der Metalle* in *Handbuch der Physik*, Editors H. GEIGER and K. SCHEEL, Vol. 24/2 (Springer, Berlin 1933). Reprint in Heidelberger Taschenbücher, Vol. 19 (Springer, Berlin 1967).

VOIGT, W., *Lehrbuch der Kristallphysik* (Teubner, Leipzig 1928).

WANG, S., *Solid State Electronics* (McGraw-Hill, New York 1966).

WANNIER, G. H., *Elements of Solid State Theory* (Cambridge University Press, London 1959).

WERT, C. A., and THOMSON, R. M., *Physics of Solids*, 2nd ed. (McGraw-Hill, New York 1970).

ZIMAN, J. M., *Principles of the Theory of Solids* (Cambridge University Press, London 1964).

Progress Series

REISS, H. (Editor), *Progress in Solid State Chemistry* (Pergamon, Oxford 1964 ff.).

SAUTER, F. (Editor), continued by MADELUNG, O., *Festkörperprobleme (Solid State Problems)* (Vieweg, Braunschweig 1962 ff.).

SEITZ, F., TURNBULL, D. and EHRENREICH, H. (Editors), *Solid State Physics* (Academic Press, New York 1955 ff.).

WOLFE, R. (Editor), *Applied Solid State Science* (Academic Press, New York 1969 ff.).

B. Interference Effects in Crystals

I. Geometrical Properties of Perfect Crystals

1. TRANSLATION OPERATIONS IN CRYSTALS

A crystal consists of a three-dimensionally periodic array of atoms; the periodicity in the solid state structure is of fundamental importance in solid state physics as will be shown in the following.

Geometrically, the periodic structure of the crystals is represented by a space lattice. A space lattice is obtained when points are subject to infinitely often repeated translations in three non-coplanar directions; it is therefore called a translation lattice.

When *all* lattice points are determined by the definition of three non-coplanar vectors $\boldsymbol{a}_1$, $\boldsymbol{a}_2$, $\boldsymbol{a}_3$ (translation vectors), the result is a simple primitive translation lattice. Such lattices in which each point has precisely the same environment are also called point lattices. When an arbitrary lattice point is chosen as the origin of the coordinate system (cf. Fig. 4) all other lattice points P_n are determined by the position vectors (lattice vectors)

$$\boldsymbol{r} = n_1\boldsymbol{a}_1 + n_2\boldsymbol{a}_2 + n_3\boldsymbol{a}_3 \,; \tag{B. 1}$$

n_1, n_2, n_3 are integers.

The three vectors $\boldsymbol{a}_1$, $\boldsymbol{a}_2$, $\boldsymbol{a}_3$ span a volume which is called a simple primitive unit cell. It contains a single lattice point; each of the corner points is common

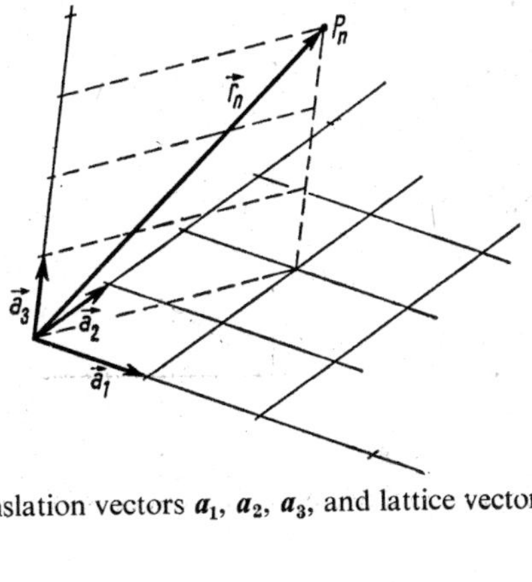

FIG. 4. Translation vectors $\boldsymbol{a}_1$, $\boldsymbol{a}_2$, $\boldsymbol{a}_3$, and lattice vector $\boldsymbol{r}_n$.

to the eight unit cells meeting at it; with respect to the unit cell a corner point is counted as $\frac{1}{8}$. The directions of $\boldsymbol{a}_1$, $\boldsymbol{a}_2$, $\boldsymbol{a}_3$ are the crystallographic axes, their magnitudes are the unit periods (identity periods) along the axes.

When special relations linking the three vectors are taken into account (cf. Table 1) one obtains seven different sets of these three vectors (i.e. seven simple primitive lattices). As the three-dimensional periodic structure of the crystals is determined by the directions and lengths of the three vectors $\boldsymbol{a}_1$, $\boldsymbol{a}_2$, $\boldsymbol{a}_3$, there exist seven systems of crystals.

TABLE 1. The crystal systems

Crystal system	Crystal axes	Angles made by the axes	Examples
1) Triclinic	$a_1 \neq a_2 \neq a_3$	$\alpha \neq \beta \neq \gamma \neq 90°$	$CuSO_4 \cdot 5H_2O$, $K_2Cr_2O_7$
2) Monoclinic	$a_1 \neq a_2 \neq a_3$	$\alpha = \gamma = 90°, \beta \neq 90°$	α-Se, Na_2CO_3
3) Ortho-rhombic	$a_1 \neq a_2 \neq a_3$	$\alpha = \beta = \gamma = 90°$	Ga, $AgNO_3$
4) Tetragonal	$a_1 = a_2 \neq a_3$	$\alpha = \beta = \gamma = 90°$	β-Sn, KH_2PO_4
5) Hexagonal	$a_1 = a_2 \neq a_3$	$\alpha = \beta = 90°, \gamma = 120°$	Zn, SiO_2
6) Rhombohedral	$a_1 = a_2 = a_3$	$\alpha = \beta = \gamma \neq 90°$	As, $CaCO_3$
7) Cubic	$a_1 = a_2 = a_3$	$\alpha = \beta = \gamma = 90°$	Cu, NaCl

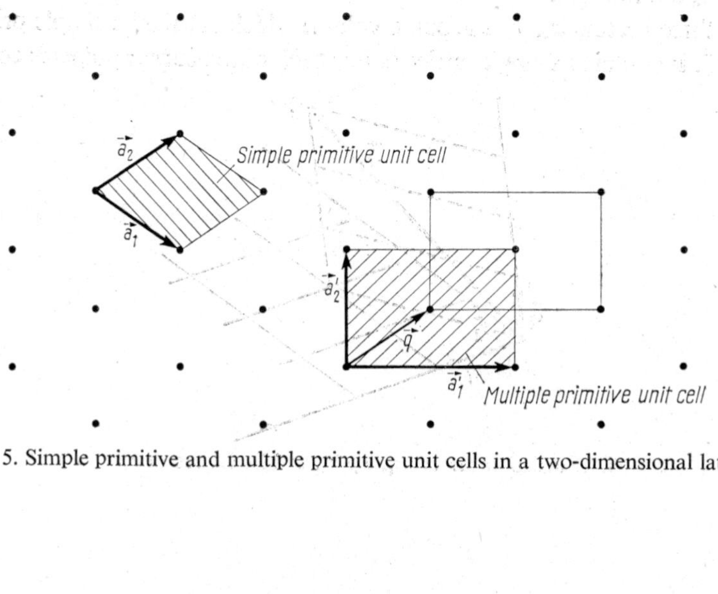

FIG. 5. Simple primitive and multiple primitive unit cells in a two-dimensional lattice.

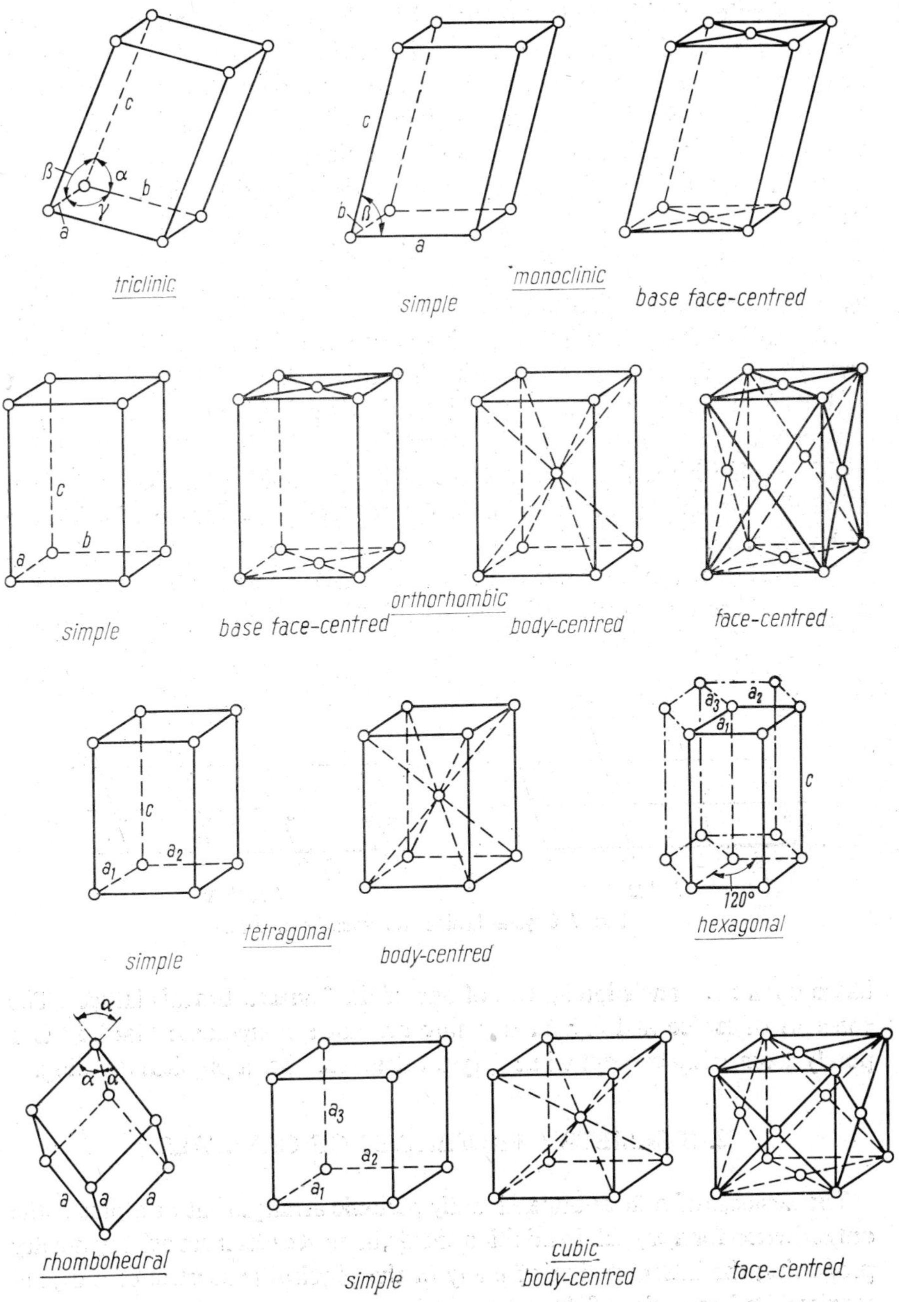

FIG. 6. The fourteen Bravais lattices.

The simple primitive unit cell formed by the vectors $\boldsymbol{a}_1$, $\boldsymbol{a}_2$, $\boldsymbol{a}_3$, however, does not always display all the symmetries of the given translation lattice. In many cases these symmetries are clearly recognized only when a larger unit is considered, namely the so-called multiple primitive unit cell formed by the vectors $\boldsymbol{a}_1'$, $\boldsymbol{a}_2'$, $\boldsymbol{a}_3'$; it contains more than one lattice point (cf. Fig. 5). The lattice points in the multiple primitive unit cell are determined by the basis vectors $\boldsymbol{q}_i$, i.e.,

$$\boldsymbol{q}_i = u_i\boldsymbol{a}_1' + v_i\boldsymbol{a}_2' + w_i\boldsymbol{a}_3' . \qquad (B.\ 2)$$

u_i, v_i, w_i are basis coordinates (cf. p. 35).

We can imagine a multiple primitive lattice as being formed by a parallel arrangement of congruent, simple primitive lattices, one inside the other so that each lattice point belongs to one of these simple primitive lattices.

According to Bravais there exist seven simple primitive lattices and seven multiple primitive lattices, i.e. fourteen different primitive translation lattices, also called Bravais lattices (cf. Fig. 6). The notations used for these lattices are given on p. 10.

A crystal structure is obtained when identical atoms, ions, or groups of atoms are assigned to equivalent lattice points (cf. Fig. 7). Each crystal structure is

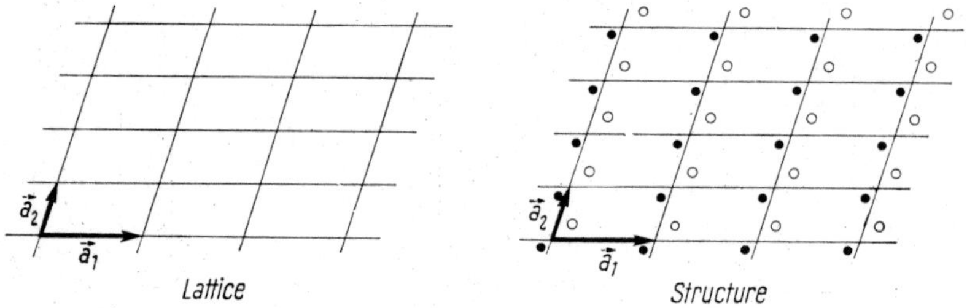

FIG. 7. Crystal lattice and crystal structure.

based upon the translation system of one of the fourteen Bravais lattices. The concepts of lattice and crystal structure are not synonymous: "lattice" is a purely geometrical concept while "crystal structure" has a physical meaning.

2. SYMMETRY PROPERTIES OF CRYSTALS

The existence of a three-dimensionally periodic arrangement of atoms is the only criterion for a crystal. In addition to this, in most cases there exist symmetry properties; the interpretation of many of the physical properties of a crystal requires the knowledge of these symmetries.

A crystal is said to have symmetry properties when its structural units are identical in different directions. It is then possible, by means of so-called repetition or symmetry operations, to bring the crystal into self-coincidence. In order to perform these operations we have to make use of the following symmetry elements which are arranged in the crystal in a definite way:

1. *Rotational axes.* Only 1-, 2-, 3-, 4-, and 6-fold rotation axes are compatible with the translation properties of the fourteen Bravais lattices. The multiplicity X of the rotational axis tells how often a rotation operation about the angle $2\pi/X$ must be performed before the crystal is brought back to its original position.

2. *Inversion centres.* A crystal possesses an inversion centre (a symmetry centre) if the structure is brought into self-coincidence by the operation $\boldsymbol{r} \to -\boldsymbol{r}$, where $\boldsymbol{r}$ characterizes the position of an arbitrary point in the crystal with respect to the centre.

3. *Rotation–inversion axes.* The operation by which self-coincidence is achieved is a rotation followed by an inversion.

3a. *Mirror planes.* The symmetry operation of a two-fold rotation–inversion axis is identical with the reflection in a plane perpendicular to the rotation axis.

4. *Screw axes.* The self-coincidence operation consists of a rotation about the angle $2\pi/X$ ($X = 1, 2, 3, 4, 6$) combined with a translation in the direction of the rotation axis by a certain fraction of the identity period in the direction of the axis. Screw axes have a definite sense which, for example, is the cause of the different rotation of the polarization plane of linearly polarized light in certain crystals.

5. *Glide planes.* The symmetry operation in the glide plane consists of a reflection and a translation by a certain fraction of an identity period parallel to the plane.

Within the framework of group theory, symmetry operations in which at least one point remains unchanged are described by point groups. There exists a total of thirty-two different crystallographic point groups, which are based on the symmetry elements of (1)–(3), as well as combinations of these symmetry elements. Each crystallographic point group determines a certain symmetry of the crystal. According to the thirty-two point groups we have thirty-two classes of crystals (point symmetry classes).

As already mentioned, a crystal structure is obtained when a certain arrangement of atoms (basis) is attributed to each lattice point of a Bravais lattice. When the basis consists of a single atom whose centre of mass coincides with the lattice point (e.g. the crystals of the elements Cu, Ag, Au) the crystal possesses the highest point group symmetry possible for the crystal system considered. Generally, however, the basis is a complex arrangement of atoms and itself possesses one of the thirty-two point group symmetries, which also determines (and generally reduces) the crystal symmetry. Here the symmetry elements (4) and (5) may appear in addition; their symmetry operations also comprise translations of the order of magnitude of elementary periods.

When we consider a crystal structure in which all symmetry elements are denoted by the corresponding symbols, we obtain a periodic arrangement of symmetry elements which is called a space group. A space group is characterized by the Bravais lattice as well as by the type and position of the symmetry elements in the unit cell. Based on group-theoretical principles, Schoenflies (1891) and Fedorov (1892) defined 230 possible space groups. A crystal structure is sufficiently defined by knowing its space group. A space group can, however, contain different crystal structures; for instance, Cu-type and NaCl-type structures (see Table 2) belong to the same space group.

3. NOTATION OF CRYSTAL STRUCTURES

The symbols used for Bravais lattices indicate whether the translation lattice is simple or multiple primitive:

- *P* a simple primitive lattice,
- *I* a body-centred lattice,
- *F* a uniformly face-centred lattice,
- *C*, *A*, *B* a basis-face-centred lattice; *C*, *A*, *B* denote the face in which the axes (a_1, a_2; a_2, a_3; a_3, a_1) lie,
- *R* a rhombohedral lattice.

Two systems are used to symbolize the crystal classes: the system given by Schoenflies and the one following Hermann and Mauguin, which has been internationally accepted for the identification of space groups ("International symbols"). Here we shall not enter into details as regards the fundamentals and arguments for the various symbol systems; the following compilation may be taken as an aid in understanding the notation used in the literature.

SCHOENFLIES SYMBOLS

- C_j *j*-fold rotational axis ($j = 1, 2, 3, 4, 6$),
- S_j *j*-fold rotation–inversion axis ($j = 1, 2, 3, 4, 6$),
- D_j *j* two-fold rotational axes perpendicular to a principal rotational axis ($j = 2, 3, 4, 6$),

$C_i \equiv S_1$ inversion centre,
$C_s \equiv S_2$ single symmetry plane,
T four three-fold and three two-fold rotation axes in a tetrahedral arrangement,
O four three-fold and three four-fold rotation axes in an octahedral arrangement.

Additional indices give the position of a symmetry plane with respect to the axes of rotation:

h "horizontal" = perpendicular to the rotation axis,
v "vertical" = parallel to the rotation axis,
d "diagonal" = parallel to the principal rotation axis, bisecting the angles between the two-fold rotation axes.

INTERNATIONAL SYMBOLS

The symbols describing the crystal classes comprise information for at most three geometrically preferred directions which are parallel to the symmetry axes or the normals to the symmetry planes.

The symbols for the space groups first give a symbol for the translation system (*P*, *I*, *F*, *C*, *B*, *A*, *R*) which is followed by the other symmetry elements (in the same sequence as for the crystal classes):

X X-fold rotational axis,
$\bar{X}$ X-fold rotation–inversion axis,
$m(\equiv \bar{2})$ mirror plane,
$\bar{1}$ inversion centre,
$\frac{X}{m}$, X/m X-fold rotation axis normal to mirror plane,
Xm X-fold rotation axis parallel to mirror plane,
$X2$ X-fold principal rotational axis normal to two-fold rotational axis (axes),
$\bar{X}2$ X-fold rotation–inversion axis normal to two-fold rotation axis (axes),
$\bar{X}m$ X-fold rotation–inversion axis parallel to mirror plane(s),
$\frac{X}{m}m$, X/mm X-fold rotation axis normal to mirror plane and parallel to mirror plane(s).

4. SIMPLE CRYSTAL STRUCTURES

In Table 2 are compiled the essential properties of several crystal structures. Elements, alloys, and compounds crystallizing in these structures are given in

TABLE 2. Several simple structures of elements (*A*), alloys and compounds (*B*, *C*)

Symbol†	Structural type	Number of atoms per unit cell	Position of atoms in unit cell (basis coordinates)	Number of nearest neighbours (coordination number)	Distance of nearest neighbours
A 1	face-centred cubic structure (Cu-type)	4	$000;\ \frac{1}{2}\frac{1}{2}0;\ \frac{1}{2}0\frac{1}{2};\ 0\frac{1}{2}\frac{1}{2}$	12	$\frac{a}{\sqrt{2}}$
A 2	body-centred cubic structure (W-type)	2	$000;\ \frac{1}{2}\frac{1}{2}\frac{1}{2}$	8	$\frac{\sqrt{3}\,a}{2}$
A 3	Hexagonal close packed structure (Mg-type)	2	$\frac{2}{3}\frac{1}{3}0$ $\frac{1}{3}\frac{2}{3}\frac{1}{2}$	12	$\sqrt{\frac{a^2}{3}+\frac{c^2}{4}}$
A 4	diamond structure	8	$000;\ \frac{1}{2}\frac{1}{2}0;\ \frac{1}{2}0\frac{1}{2};\ 0\frac{1}{2}\frac{1}{2}$ $\frac{1}{4}\frac{1}{4}\frac{1}{4};\ \frac{3}{4}\frac{3}{4}\frac{1}{4};\ \frac{3}{4}\frac{1}{4}\frac{3}{4};\ \frac{1}{4}\frac{3}{4}\frac{3}{4}$	4	$\frac{\sqrt{3}\,a}{4}$
B 1	NaCl-structure	(Na): 4 (Cl): 4	(Na): $000;\ \frac{1}{2}\frac{1}{2}0;\ \frac{1}{2}0\frac{1}{2};\ 0\frac{1}{2}\frac{1}{2}$ (Cl): $\frac{1}{2}\frac{1}{2}\frac{1}{2};\ 00\frac{1}{2};\ 0\frac{1}{2}0;\ \frac{1}{2}00$	6	$\frac{a}{2}$ (Na—Cl)
B 2	CsCl-structure	(Cs): 1 (Cl): 1	(Cs): 000 (Cl): $\frac{1}{2}\frac{1}{2}\frac{1}{2}$	8	$\frac{\sqrt{3}\,a}{2}$ (Cs—Cl)
B 3	zinc blende structure (sphalerite ZnS type)	(Zn): 4 (S): 4	(Zn): $000;\ \frac{1}{2}\frac{1}{2}0;\ \frac{1}{2}0\frac{1}{2};\ 0\frac{1}{2}\frac{1}{2}$ (S): $\frac{1}{4}\frac{1}{4}\frac{1}{4};\ \frac{3}{4}\frac{3}{4}\frac{1}{4};\ \frac{3}{4}\frac{1}{4}\frac{3}{4};\ \frac{1}{4}\frac{3}{4}\frac{3}{4}$	4	$\frac{\sqrt{3}\,a}{4}$ (Zn—S)
C 1	fluorite structure (CaF_2 type) and anti-fluorite structure	(Ca): 4 (F): 8	(Ca): $000;\ \frac{1}{2}\frac{1}{2}0;\ \frac{1}{2}0\frac{1}{2};\ 0\frac{1}{2}\frac{1}{2}$ (F): $\frac{1}{4}\frac{1}{4}\frac{1}{4};\ \frac{3}{4}\frac{3}{4}\frac{1}{4};\ \frac{3}{4}\frac{1}{4}\frac{3}{4};\ \frac{1}{4}\frac{3}{4}\frac{3}{4}$ (F): $\frac{3}{4}\frac{3}{4}\frac{3}{4};\ \frac{1}{4}\frac{1}{4}\frac{3}{4};\ \frac{1}{4}\frac{3}{4}\frac{1}{4};\ \frac{3}{4}\frac{1}{4}\frac{1}{4}$	(Ca): 8 (F): 4	$\frac{\sqrt{3}\,a}{4}$ (Ca—F)

† After P. P. Ewald and C. Hermann, *Struktur-Bericht*, Z. f. Kristallographie, Supplementary Volume 1931 (Akademische Verlagsgesellschaft, Leipzig).

Table 3. Many substances exist in different modifications at one and the same temperature (polytypes). Thus, for example, ZnS may crystallize in the zinc-blende or the wurtzite structures. The polytypical character is particularly well developed in the case of the hexagonal α-SiC structures; the numerous polytypes are the consequence of regular changes in the packing sequence (cf. p. 112).

Alloys always have a more or less extended region of homogeneity and thus exist in non-stoichiometric ratios (cf. pp. 231 ff.). But also many alloys, in particular numerous oxides, are not always strictly stoichiometric (cf. pp. 142 ff.).

TABLE 3. Structures and lattice constants of elements, alloys, and compounds. The data are given for room temperature, unless the temperatures are given [after Landolt-Börnstein, Vol. 1/4 (Springer, Berlin 1955), pp. 81–89; W. B. Pearson, *Handbook of Lattice Spacings and Structure of Metals*, Vol. 2 (Pergamon, Oxford 1967), pp. 79 ff.; G. V. Samsonov (Ed.), *Handbook of the Physicochemical Properties of the Elements* (Plenum, New York 1968), pp. 110–123; A. Taylor and B. J. Kagle, *Crystallographic Data on Metal and Alloy Structures* (Dover, New York 1963), pp. 254 ff.; R. W. G. Wyckoff, *Crystal Structures*, 2nd ed. (Interscience, New York 1963–1969), mainly Vol. 1]

Face-centred cubic structure (A 1)

Substance	*a* [Å]	Substance	*a* [Å]	Substance	*a* [Å]
Ca	5.59	Rh	3.80	Cu	3.61
α-Sr	6.08	Ir	3.84	Ag	4.09
				Au	4.08
γ-Ce	5.16	Ni	3.52		
Yb	5.49	Pd	3.89	Al	4.05
		Pt	3.92		
				Pb	4.95
Ac	5.31				
α-Th	5.08				
				Ne	4.46 (4°K)
				Ar	5.31 (4°K)
				Kr	5.71 (92°K)
				Xe	6.25 (88°K)

Body-centred cubic structure (A 2)

Substance	*a* [Å]	Substance	*a* [Å]	Substance	*a* [Å]
Li	3.51	Ba	5.02	α-Cr	2.88
Na	4.29			Mo	3.15
K	5.23 (5°K)	Eu	4.58	α-W	3.17
Rb	5.63				
Cs	6.08 (173°K)	V	3.03	α-Fe	2.87
		Nb	3.30		
		Ta	3.30		

(*Continued overleaf*)

TABLE 3 *(Cont.)*

Hexagonal close packed structure (A 3)

Substance	*a* [Å]	*c* [Å]	Substance	*a* [Å]	*c* [Å]
He	3.58	5.84 (1.45°K, 37 atm)	α-Ti	2.95	4.68
			α-Zr	3.23	5.15
			α-Hf	3.19	5.05
Be	2.29	3.58			
Mg	3.21	5.21	Re	2.76	4.46
α-Sc	3.31	5.27	Ru	2.71	4.28
Y	3.65	5.73	Os	2.74	4.32
α-La	3.77	2·6.08†	α-Co	2.51	4.07
α-Pr	3.67	2·5.94†			
α-Nd	3.66	2·5.80†	Zn	2.66	4.95
α-Gd	3.64	5.78	Cd	2.98	5.62
α-Tb	3.60	5.69			
α-Dy	3.59	5.65	α-Tl	3.46	5.52
α-Ho	3.58	5.62			
α-Er	3.56	5.59			
α-Tm	3.54	5.55			
α-Lu	3.50	5.55			

† The factor 2 is due to the fact that, strictly speaking, α-La, α-Pr, and α-Nd exist in the structure of the double hexagonal close packing.

Diamond structure (A 4)

Substance	*a* [Å]	Substance	*a* [Å]
C (diamond)	3.57	Ge	5.66
Si	5.43	α-Sn (gray tin)	6.50 (286°K)

NaCl structure (B 1)

Substance	*a* [Å]	Substance	*a* [Å]	Substance	*a* [Å]
LiH	4.08	PbS	5.94	PrN	5.16
		PbSe	6.12	PrP	5.87
LiF	4.02	PbTe	6.46	PrAs	6.01
LiCl	5.13			PrSb	6.37
LiBr	5.49	CaO	4.80	PrBi	6.46
LiJ	6.00	CaS	5.68		
		CaSe	5.91	DyN	4.89
NaF	4.62	CaTe	6.35	DyP	5.65
NaCl	5.63			DyAs	5.79
NaBr	5.96	TiO	4.18	DySb	6.16
NaJ	6.46	VO	4.08	DyBi	6.25
KF	5.33	MnO	4.44	UC	4.96
KCl	6.28	FeO	4.31	UN	4.89

(Continued on facing page)

TABLE 3 *(Cont.)*

NaCl structure (B 1) (Cont.)

Substance	a [Å]	Substance	a [Å]	Substance	a [Å]
KBr	6.59	CoO	4.26	UO	4.93
KJ	7.05	NiO	4.19 (548°K)		
				NpO	5.01
RbF	5.63	EuO	5.14	PuO	4.96
RbCl	6.58	EuS	5.97	AmO	5.05
RbBr	6.85	EuSe	6.19		
RbJ	7.34	EuTe	6.60		
CsF	6.01				
AgF	4.92				
AgCl	5.54				
AgBr	5.77				

CsCl structure (B 2)

Substance	a [Å]	Substance	a [Å]	Substance	a [Å]
				β'-CuZn (β-brass)	2.95
CsCl	4.11	BeCo	2.61		
CsBr	4.29	BeNi	2.62	β'-AgZn	3.16
CsJ	4.56	BePd	2.82	β'-AgCd	3.33
		γ-BeCu	2.70	AgLa	3.78
TlCl	3.84			AgCe	3.74
TlBr	3.97	MgLa	3.96	AgGd	3.65
TlJ	4.20	MgCe	3.91	AgDy	3.61
		MgPr	3.89	AgLu	3.54
β-LiAg	3.17	MgAg	3.31		
LiHg	3.29	MgTl	3.64	β-AlCo	2.86
LiTl	3.43			β-AlNi	2.89
				AlOs	3.00

Zinc blende structure (B 3)

Substance	a [Å]	Substance	a [Å]	Substance	a [Å]
CuF	4.26	ZnS	5.42	AlP	5.47
CuCl	5.41	ZnSe	5.66	AlAs	5.66
CuBr	5.68	ZnTe	6.09	AlSb	6.14
CuJ	6.05				
		CdS	5.83	GaP	5.45
AgJ	6.47	CdSe	6.05	GaAs	5.65
		CdTe	6.48	GaSb	6.10
BeS	4.89				
BeSe	5.14	HgS	5.85	InP	5.87
BeTe	5.63	HgSe	6.08	InAs	6.06
		HgTe	6.43	InSb	6.48
				β-SiC	4.36

(Continued overleaf)

TABLE 3 (*Cont.*)

Fluorite and anti-fluorite structure† *(C 1)*

Substance	a [Å]	Substance	a [Å]	Substance	a [Å]
LaH_2	5.67	UN_2	5.32	Li_2O	4.62
CeH_2	5.58	UO_2	5.47	Li_2S	5.72
PrH_2	5.51	NpO_2	5.43	Li_2Se	6.02
DyH_2	5.20	PuO_2	5.40	Li_2Te	6.52
		AmO_2	5.38	Na_2O	5.56
NbH_2	4.55	CmO_2	5.37	K_2O	6.45
				Rb_2O	6.76
CaF_2	5.45	$PtAl_2$	5.91		
SrF_2	5.78	$PtGa_2$	5.92	Mg_2Si	6.35
BaF_2	6.19	$PtSn_2$	6.42	Mg_2Ge	6.39
				Mg_2Sn	6.76
CdF_2	5.40	$AuAl_2$	6.00	Mg_2Pb	6.81
HgF_2	5.54	$AuGa_2$	6.08		
PbF_2	5.93	$AuIn_2$	6.52	Ir_2P	5.54
				$IrSn_2$	6.34
EuF_2	5.80			$CoSi_2$	5.36
α-CeO_2	5.41				

† The fluorite and anti-fluorite structures differ in that the positions of the cations and anions are interchanged.

The unit cell of the cubic structures is a cube of volume a^3, and for the hexagonal structure *A 3* a rhombic prism of volume $\sqrt{3}\, ca^2/2$ (cf. Fig. 6). From the volume of the unit cell and the masses of the atoms contained in it a microscopic density can be calculated, the so-called X-ray density ϱ_X. Because of the various lattice defects (cf. p. 118) this density often differs from the density ϱ measured macroscopically. In this connection another useful quantity should be mentioned, namely the atomic volume which is defined by the quotient of atomic weight A and density ϱ, i.e., $V_A = A/\varrho$.

II. X-ray Diffraction by Crystals

The geometrical properties of the crystals are based on the requirement of a three-dimensional periodicity in the crystal structure. The symmetry properties can be derived theoretically, and some of them are immediately recognized from the shape of a bulk crystal.

While the basic theoretical investigations into crystal structure were finished at the end of the last century, the experimental verification was possible only in 1912 (Friedrich and Knipping), with the help of X-rays. These experiments proved not only the periodicity of the crystal structure but also the wave nature of the X-rays.

The analysis of crystal structures requires radiation of a wavelength which is of the order of the interatomic spacing in the crystal. Apart from electron and neutron rays one uses mainly X-rays in structural analyses because of their weak absorption.

The theory of the interaction of X-rays with a periodic structure is based on investigations by von Laue (1912). Under the influence of electromagnetic radiation the electrons of the atomic shell are excited to execute oscillations so that each atom in the crystal is the origin of an electromagnetic spherical wave of the frequency of the incident wave: the X-ray wave is diffracted. All diffracted waves are added according to the Huygens–Fresnel principle. In certain directions because of phase correlations the diffracted X-ray wave displays interference maxima which can be observed.

While the so-called geometrical theory yields information only on the conditions for interference, the dynamical theory yields additional information on the intensity and the shape of the interference maxima. The dynamical theory takes into account the thermal motion of the crystal elements. The temperature dependence of the intensity of X-ray reflections is determined by the Debye–Waller factor and is derived on pp. 72 ff.

The geometrical theory of interferences dealt with in the following applies not only to the diffraction of X-rays but also to that of electrons and neutrons.

1. VON LAUE'S GEOMETRICAL THEORY

We shall describe the diffraction of X-rays in a triclinic crystal. The translation system of the crystal is assumed to correspond to a simple primitive triclinic lattice whose lattice points are occupied by identical atoms. The geometrical theory is based on the following:

1. Strict periodicity of the crystal structure; all atoms are at rest, i.e. the absolute temperature T is equal to zero.
2. Multiple scattering is neglected; the atoms scatter only the incident wave.
3. The phase differences of the diffracted wave at the point of observation are uniquely determined by the lattice geometry (hence the name "geometrical theory").

Let us first calculate the amplitude of an X-ray wave diffracted from a *single* centre; thereafter we shall consider diffraction by *many* centres.

(a) *DIFFRACTION FROM A* SINGLE *CENTRE*

The scattering centre P and the point of observation B for the scattered X-ray wave are determined by the position vectors $\boldsymbol{r}$ and $\boldsymbol{R}$ (cf. Fig. 8). A plane X-ray wave of angular frequency ω is incident in the direction of the wave vector $\boldsymbol{k}_0$.

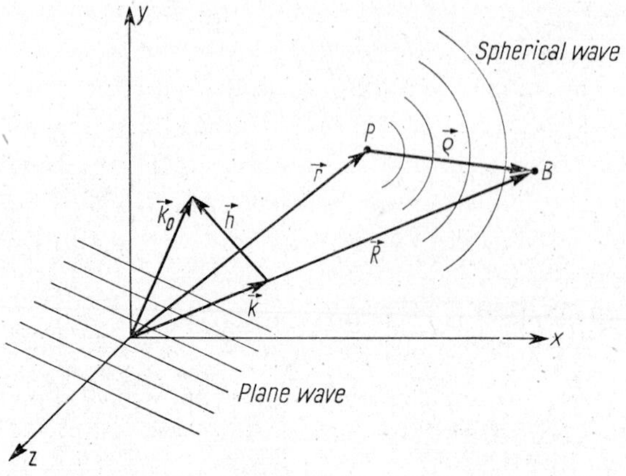

FIG. 8. Scattering of a plane wave by a centre *P*.

Its amplitude A_p at the scattering centre P is at the time t given by

$$A_P = ae^{-i\omega t}e^{i(\boldsymbol{k}_0 \cdot \boldsymbol{r})}. \tag{B. 3}$$

Assume the fraction ψ of the amplitude to be scattered at P as a coherent spherical wave; its amplitude at the point of observation, B, is given by

$$A_B = \psi \frac{A_P}{\varrho} e^{i(\boldsymbol{k} \cdot \boldsymbol{\varrho})} \tag{B. 4}$$

where

$$|\boldsymbol{k}| = |\boldsymbol{k}_0| = \frac{2\pi}{\lambda}, \tag{B. 5}$$

with ψ the scattering factor, that is the fraction of the scattered amplitude of the incident wave, $\boldsymbol{\rho}$ the position vector from the scattering centre P to the observation point B, $\boldsymbol{k}$ the wave vector in the direction of observation, c the velocity of light, and λ the X-ray wavelength.

In the case $R \gg r$ the vectors $\boldsymbol{R}$ and $\boldsymbol{\rho}$ are almost parallel and we have

$$\varrho \approx R - \frac{1}{k}(\boldsymbol{k} \cdot \boldsymbol{r}) \approx R. \tag{B. 6}$$

Thus we obtain from eqn. (B. 4)

$$A_B = \frac{\psi}{R} a e^{-i\omega t} e^{i(\boldsymbol{k}_0 . \boldsymbol{r})} e^{i(\boldsymbol{k}.[\boldsymbol{R}-\boldsymbol{r}])}$$

$$= \frac{a}{R} e^{-i\omega t} e^{i(\boldsymbol{k}.\boldsymbol{R})} \psi e^{i([\boldsymbol{k}_0-\boldsymbol{k}].\boldsymbol{r})}. \tag{B. 7}$$

With

$$A_0 = \frac{a}{R} e^{-i\omega t} e^{i(\boldsymbol{k}.\boldsymbol{R})} \tag{B. 8}$$

and

$$\boldsymbol{h} = \boldsymbol{k}_0 - \boldsymbol{k} \tag{B. 9}$$

the scattering amplitude due to a single scattering center P, observed at a remote point B, is given by

$$A_B = A_0 \psi e^{i(\boldsymbol{h}.\boldsymbol{r})}. \tag{B. 10}$$

(b) *DIFFRACTION FROM* MANY *CENTRES*

Let us assume the scattering centres to be identical atoms with a scattering factor ψ located at the lattice points of a triclinic Bravais lattice. The position vectors of the scattering centres are the lattice vectors given by the three translation vectors $\boldsymbol{a}_1$, $\boldsymbol{a}_2$, $\boldsymbol{a}_3$:

$$\boldsymbol{r}_n = n_1\boldsymbol{a}_1 + n_2\boldsymbol{a}_2 + n_3\boldsymbol{a}_3. \tag{B. 11}$$

All diffracted waves interfere with one another, and the total scattering amplitude observed at the remote point B is obtained as the sum of the scattered waves emitted by all scattering centres:

$$A_{\text{tot}} = \sum_{n_1, n_2, n_3} A_B = A_0 \sum_{n_1, n_2, n_3} \psi e^{i(\boldsymbol{h}.[n_1\boldsymbol{a}_1 + n_2\boldsymbol{a}_2 + n_3\boldsymbol{a}_3])}$$

$$= A_0 \psi \sum_{n_1, n_2, n_3} e^{i\{n_1(\boldsymbol{a}_1.\boldsymbol{h}_1) + n_2(\boldsymbol{a}_2.\boldsymbol{h}_2) + n_3(\boldsymbol{a}_3.\boldsymbol{h}_3)\}}. \tag{B. 12}$$

It is physically interesting to ask for the conditions of a maximum coherent scattering amplitude at point B, since these conditions determine the intensity peaks observed in the experiment, and conclusions can be drawn as to the structure of the crystal under investigation. The scattering amplitude A_{tot} will be highest when at point B all individual scattered waves are in phase, i.e.,

$$n_1(\boldsymbol{a}_1 . \boldsymbol{h}) + n_2(\boldsymbol{a}_2 . \boldsymbol{h}) + n_3(\boldsymbol{a}_3 . \boldsymbol{h}) = 2\pi p, \tag{B. 13}$$

where p is an integer.

In the case of a lattice given by the vectors $\boldsymbol{a}_1$, $\boldsymbol{a}_2$, $\boldsymbol{a}_3$ eqn. (B. 13) represents the conditions for the vector $\boldsymbol{h}$ which defines the direction of observation with respect to the direction of incidence of a monochromatic X-ray wave. Equivalent to (B.13) are the three von Laue equations

$$(\boldsymbol{a}_1 . \boldsymbol{h}) = 2\pi h_1, \quad (\boldsymbol{a}_2 . \boldsymbol{h}) = 2\pi h_2, \quad (\boldsymbol{a}_3 . \boldsymbol{h}) = 2\pi h_3, \tag{B. 14}$$

where h_1, h_2, h_3 are integers.

The values of $\boldsymbol{h}$, for which the three von Laue equations are *simultaneously* satisfied, determine the directions of observation for the largest scattering amplitudes; the value of $\boldsymbol{h}$ must satisfy the condition that its projections on the directions of $\boldsymbol{a}_1$, $\boldsymbol{a}_2$, $\boldsymbol{a}_3$ multiplied by the magnitudes a_1, a_2, a_3 are integral multiples of 2π ($\boldsymbol{h}$ has the dimension of a reciprocal length).

The evaluation of the von Laue equations becomes clear when we introduce the important concept of the reciprocal lattice; it is determined by the vectors *reciprocal* to $\boldsymbol{a}_1$, $\boldsymbol{a}_2$, $\boldsymbol{a}_3$.

By way of a lemma we define the reciprocal vectors.

Lemma: Resolve a given vector $\boldsymbol{w}$ into components in the three given directions $\boldsymbol{a}_1$, $\boldsymbol{a}_2$, $\boldsymbol{a}_3$, so that

$$\boldsymbol{w} = x_1\boldsymbol{a}_1 + x_2\boldsymbol{a}_2 + x_3\boldsymbol{a}_3 . \tag{B. 15}$$

Find the values x_1, x_2, x_3.

A solution to this problem is immediately obtained when the scalar product of the vector $\boldsymbol{w}$ with the vector $[\boldsymbol{a}_2 \wedge \boldsymbol{a}_3]$, and $[\boldsymbol{a}_3 \wedge \boldsymbol{a}_1]$, and $[\boldsymbol{a}_1 \wedge \boldsymbol{a}_2]$, respectively, is formed:

$$(\boldsymbol{w} . [\boldsymbol{a}_2 \wedge \boldsymbol{a}_3]) = x_1(\boldsymbol{a}_1 . [\boldsymbol{a}_2 \wedge \boldsymbol{a}_3]), \tag{B. 16}$$

i.e.

$$x_1 = \left(\boldsymbol{w} . \frac{[\boldsymbol{a}_2 \wedge \boldsymbol{a}_3]}{(\boldsymbol{a}_1 . [\boldsymbol{a}_2 \wedge \boldsymbol{a}_3])}\right) = \left(\boldsymbol{w} . \frac{[\boldsymbol{a}_2 \wedge \boldsymbol{a}_3]}{V_{UC}}\right). \tag{B. 17}$$

Similarly we obtain

$$x_2 = \left(\boldsymbol{w} . \frac{[\boldsymbol{a}_3 \wedge \boldsymbol{a}_1]}{(\boldsymbol{a}_2 . [\boldsymbol{a}_3 \wedge \boldsymbol{a}_1])}\right) = \left(\boldsymbol{w} . \frac{[\boldsymbol{a}_3 \wedge \boldsymbol{a}_1]}{V_{UC}}\right), \tag{B. 18}$$

$$x_3 = \left(\boldsymbol{w} . \frac{[\boldsymbol{a}_1 \wedge \boldsymbol{a}_2]}{(\boldsymbol{a}_3 . [\boldsymbol{a}_1 \wedge \boldsymbol{a}_2])}\right) = \left(\boldsymbol{w} . \frac{[\boldsymbol{a}_1 \wedge \boldsymbol{a}_2]}{V_{UC}}\right). \tag{B. 19}$$

$V_{UC} = (\boldsymbol{a}_1 . [\boldsymbol{a}_2 \wedge \boldsymbol{a}_3])$ is the unit cell volume.

We define the vectors reciprocal to $\boldsymbol{a}_1$, $\boldsymbol{a}_2$, $\boldsymbol{a}_3$ as follows:

$$\boldsymbol{b}_1 = 2\pi \frac{[\boldsymbol{a}_2 \wedge \boldsymbol{a}_3]}{V_{UC}}, \quad \boldsymbol{b}_2 = 2\pi \frac{[\boldsymbol{a}_3 \wedge \boldsymbol{a}_1]}{V_{UC}}, \quad \boldsymbol{b}_3 = 2\pi \frac{[\boldsymbol{a}_1 \wedge \boldsymbol{a}_2]}{V_{UC}}. \tag{B. 20}$$

The factor 2π is convenient for the definition of the Brillouin zones; in momentum space they define the interference conditions governing electromagnetic waves as well as particle waves in periodic structures (cf. pp. 39, 67, and 209). The reciprocal vectors $\boldsymbol{b}_1$, $\boldsymbol{b}_2$, $\boldsymbol{b}_3$ are perpendicular to the vectors $\boldsymbol{a}_2$ and $\boldsymbol{a}_3$, $\boldsymbol{a}_3$ and $\boldsymbol{a}_1$, $\boldsymbol{a}_1$ and $\boldsymbol{a}_2$, respectively. They have the dimension of a reciprocal length, their magnitudes are proportional to $1/a_1$, $1/a_2$, $1/a_3$. The reciprocal vectors $\boldsymbol{b}_1$, $\boldsymbol{b}_2$, $\boldsymbol{b}_3$ are the translation vectors of the so-called reciprocal lattice.

Generally we have

$$(\boldsymbol{a}_i . \boldsymbol{b}_j) = 2\pi\delta_{ij}; \quad \delta_{ij} = 1 \quad \text{for} \quad i = j, \quad \delta_{ij} = 0 \quad \text{for} \quad i \neq j. \tag{B. 21}$$

The required decomposition of the vector $\boldsymbol{w}$ is thus

$$\boldsymbol{w} = \frac{1}{2\pi}[(\boldsymbol{w}.\boldsymbol{b}_1)\boldsymbol{a}_1 + (\boldsymbol{w}.\boldsymbol{b}_2)\boldsymbol{a}_2 + (\boldsymbol{w}.\boldsymbol{b}_3)\boldsymbol{a}_3]. \tag{B. 22}$$

The result obtained for this lemma will be used to determine the X-ray interference maxima. In the same way as the vector $\boldsymbol{w}$ the vector $\boldsymbol{h}$ is now decomposed into components parallel to the translation vectors $\boldsymbol{b}_1$, $\boldsymbol{b}_2$, $\boldsymbol{b}_3$ of the reciprocal lattice

$$\boldsymbol{h} = \frac{1}{2\pi}[(\boldsymbol{h}.\boldsymbol{a}_1)\boldsymbol{b}_1 + (\boldsymbol{h}.\boldsymbol{a}_2)\boldsymbol{b}_2 + (\boldsymbol{h}.\boldsymbol{a}_3)\boldsymbol{b}_3]. \tag{B. 23}$$

Using von Laue's equations (B. 14) we obtain

$$\boldsymbol{h} = h_1\boldsymbol{b}_1 + h_2\boldsymbol{b}_2 + h_3\boldsymbol{b}_3. \tag{B. 24}$$

This means the following: When $\boldsymbol{h}$ is a lattice vector of the reciprocal lattice, the von Laue equations are satisfied. The scattering amplitude A_{tot} at a remote point of observation, B, will then and only then be maximal. The position of

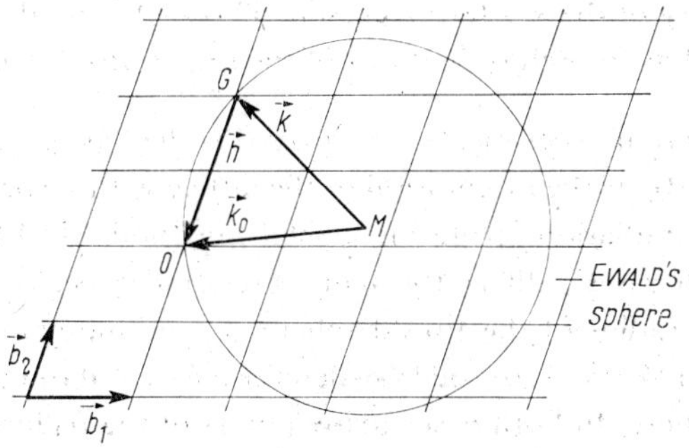

FIG. 9. Ewald's construction.

each point in the reciprocal lattice, referred to an arbitrary lattice point as the origin, will then be fixed by the triple set h_1, h_2, h_3.

The positions of the X-ray interference peaks can be obtained by means of a simple geometrical construction suggested by Ewald, from the maximum conditions of eqn. (B. 24) (cf. Fig. 9). For the given triclinic lattice the reciprocal lattice is constructed which is defined by the three translation vectors $\boldsymbol{b}_1$, $\boldsymbol{b}_2$, $\boldsymbol{b}_3$. The direction of incidence of X-rays of wavelength λ is also assumed to be given and is plotted in the reciprocal lattice: from an arbitrary lattice point O of the reciprocal lattice we draw the vector $-\boldsymbol{k}_0(|\boldsymbol{k}_0| = 2\pi/\lambda)$. The end point of this vector determines a point M which, in general, is no lattice point of the reciprocal lattice since the direction and the wavelength of the incident X-ray are arbitrary. The vector $+\boldsymbol{k}_0$ points from M to O. With M as centre we draw a sphere of radius $|\boldsymbol{k}_0| = 2\pi/\lambda$; it is called Ewald's sphere.

If lattice points G lie on this sphere, the directions are determined at which intensity peaks will arise: from G to O we have the lattice vector $\boldsymbol{h}$ which satisfies the von Laue condition of eqn. (B. 14). The direction in which the observed intensity is maximum is the direction of the wave vector $\boldsymbol{k}$ of the scattered X-ray wave; according to eqn. (B. 9), this vector is defined by $\boldsymbol{k}_0$ and $\boldsymbol{h}$, and it points from M to G.

The construction according to Ewald immediately yields the following results:

1. The primary ray is always present since the point O will always lie on the sphere. In this case the lattice vector $\boldsymbol{h} = 0$, i.e. the direction of incidence ($\boldsymbol{k}_0$) and the direction of observation ($\boldsymbol{k}$) coincide.
2. For monochromatic radiation single crystals yield only a few interference peaks or none, since in the general case no points of the reciprocal lattice lie on the surface of the sphere.
3. If the direction of $\boldsymbol{k}_0$ of the incident wave coincides with the direction of a preferred crystallographic axis, the interference pattern will display the symmetry of the scattering crystal. This fact is used in structural analyses as well as for determinations of the orientation of a crystal.

In X-ray diffraction experiments it is thus either the wavelength which is varied (von Laue method), or the orientation of the lattice with respect to the incident monochromatic radiation (rotating-crystal method, Debye–Scherrer–Hull method). In both cases sufficiently many interference peaks are obtained; in the first case the radius of the Ewald sphere is continuously varied, in the second the reciprocal lattice is rotated in a definite manner about the point O on the surface of the sphere. In both cases many points of the reciprocal lattice come to lie on the sphere and give rise to interference maxima.

2. EXPERIMENTAL METHODS

(a) *X-RAY SPECTRA*

X-rays are produced when fast electrons hit matter. The emission of the radiation is determined by the following two processes:

1. The electrons are slowed down in the electric field of the nuclei and lose part of their kinetic energy which is emitted in the form of electromagnetic radiation. The spectrum of this so-called bremsstrahlung is continuous (white X-rays) above a sharp short-wavelength edge λ_{min}, which is given by the electrons' accelerating potential U:

$$\lambda_{min} = \frac{hc}{eU}, \tag{B. 25}$$

where h is Planck's constant, c is the velocity of light, and e is the unit charge.

2. The electrons ionize the atoms of the anticathode. In this process electrons are knocked out from the inner shells, and electrons of the outer shells, emitting the characteristic radiation, jump to the vacant places in the inner shells. Corresponding to the electron transitions this radiation has a line spectrum which is characteristic of the material of the anticathode and is superimposed on the bremsstrahlung spectrum. The so-called K-series in the X-ray spectrum

TABLE 4. Wavelengths of characteristic X-ray emission and the corresponding absorption edges of several common materials for anticathodes and the suitable β-filter materials and their K-absorption edges [after the *International Tables for X-ray Crystallography*, Vol. 3 (Kynoch Press, Birmingham 1962), pp. 60, 69, 71]

Anti-cathode material	*K*-series: Characteristic spectral lines λ [Å]				Absorption edge λ_{edge} [Å]	β-filter (λ_{edge} [Å])
	$K\alpha_2$	$K\alpha_1$	$K\beta_1$	$K\beta_2$		
Cr	2.2935	2.2896	2.0848	—	2.0701	V (2.2690)
Fe	1.9399	1.9360	1.7565	—	1.7433	Mn (1.8964)
Co	1.7928	1.7889	1.6208	—	1.6081	Fe (1.7433)
Ni	1.6617	1.6578	1.5001	1.4886	1.4880	Co (1.6081)
Cu	1.5443	1.5405	1.3922	1.3810	1.3804	Ni (1.4880)
Mo	0.7135	0.7093	0.6323	0.6210	0.6198	Zr (0.6888)
Ag	0.5638	0.5594	0.4970	0.4870	0.4858	Rh (0.5338)
	L-series					
	$L\alpha_2$	$L\alpha_1$	$L\beta_1$	$L\beta_2$		
W	1.4847	1.4764	1.2818	1.2446	1.2155	Cu (1.3804)

appears when an L, M, N, ... electron drops to a vacant place in the K-shell ($K\alpha$-lines result from transitions from the L-shell to the K-shell, $K\beta$-lines from M to K transitions). In structural analyses the $K\alpha$ radiation is used in most cases. The $K\beta$ radiation (shorter wavelength) can be separated from the $K\alpha$ radiation by suitable β-filters. The anticathode material may be copper, molybdenum, or tungsten. The wavelengths of the characteristic X-ray emission of several substances are compiled in Table 4.

The usual X-ray tubes emit in the spectral range between about 0.2 and 3 Å (cf. Fig. 10). The accelerating voltage of the electrons is of the order of 10 kV.

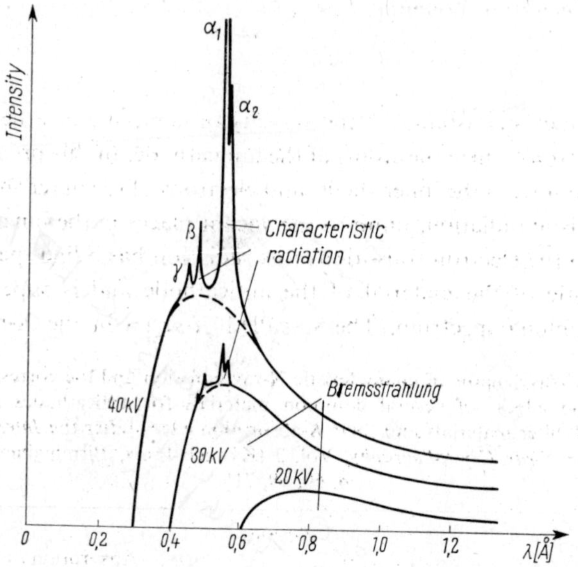

FIG. 10. Intensity distribution of an X-ray tube for various accelerating voltages (according to L. V. Azaroff, *Elements of X-ray Crystallography*, McGraw-Hill, New York 1968).

In the following we shall give a brief description of X-ray diffraction methods. The various types of diffraction patterns are immediately understood with the help of Ewald's construction. As to the quantitative analysis of interference patterns we refer the reader to the relevant literature.

(b) *THE VON LAUE METHOD*

The von Laue method is specially suited for determinations of the crystal orientation. A single crystal is exposed to white X-ray radiation (cf. Fig. 11),

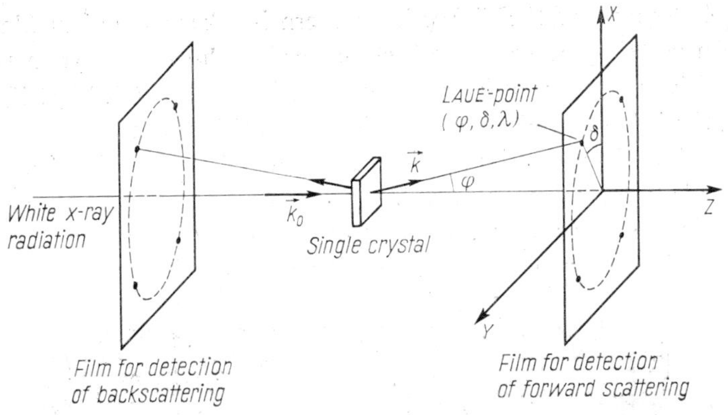

FIG. 11. The von Laue method.

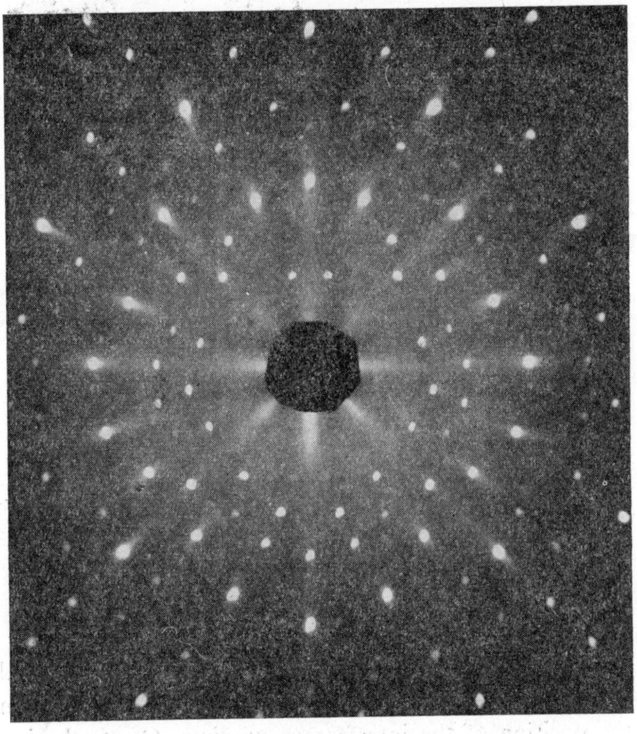

FIG. 12. The von Laue diffraction pattern of a cubic crystal.

and a photograph of the diffraction pattern is taken on a plane film. It consists of a certain arrangement of points which shows the symmetry of the crystal with respect to its orientation to the incident X-ray (cf. Figs. 12 and 13).

With the help of Ewald's construction the positions and wavelengths of the interference peaks are easily predicted. This may be shown for the diffraction by a simple primitive cubic crystal.

The translation vectors $\boldsymbol{b}_1$, $\boldsymbol{b}_2$, $\boldsymbol{b}_3$ of the reciprocal lattice are perpendicular to one another, their magnitudes are $2\pi/a$ ($a = a_1 = a_2 = a_3$). The wave vector $\boldsymbol{k}_0$ of the incident X-ray is assumed to be parallel to one of the translation vectors and to make the angle φ with $\boldsymbol{k}$ in the case of an interference maximum. The plane in which $\boldsymbol{k}_0$ and $\boldsymbol{k}$ lie is assumed to make an angle of δ with the plane containing $\boldsymbol{k}_0$ and a translation vector perpendicular to $\boldsymbol{k}_0$. Knowing φ and δ, we know the position of the interference maxima. On the basis of Ewald's construction and simple geometrical considerations the following relations are

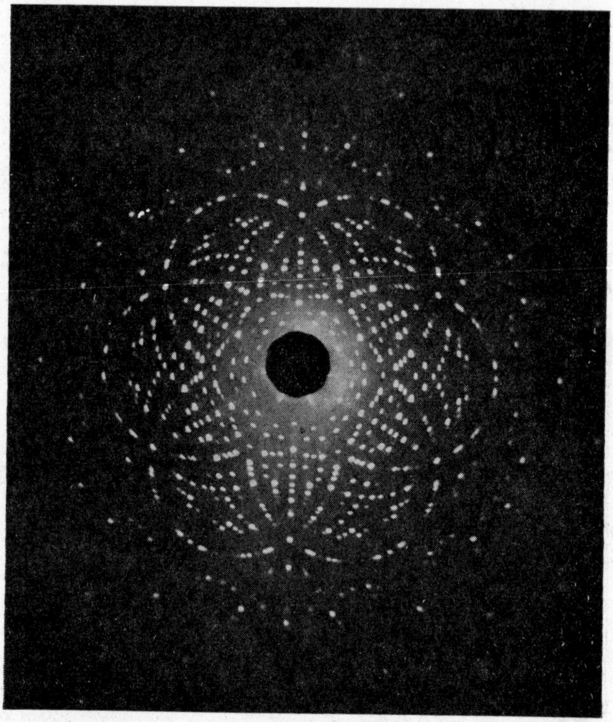

FIG. 13. The von Laue diffraction pattern of a hexagonal crystal (barium–platinate–titanate).

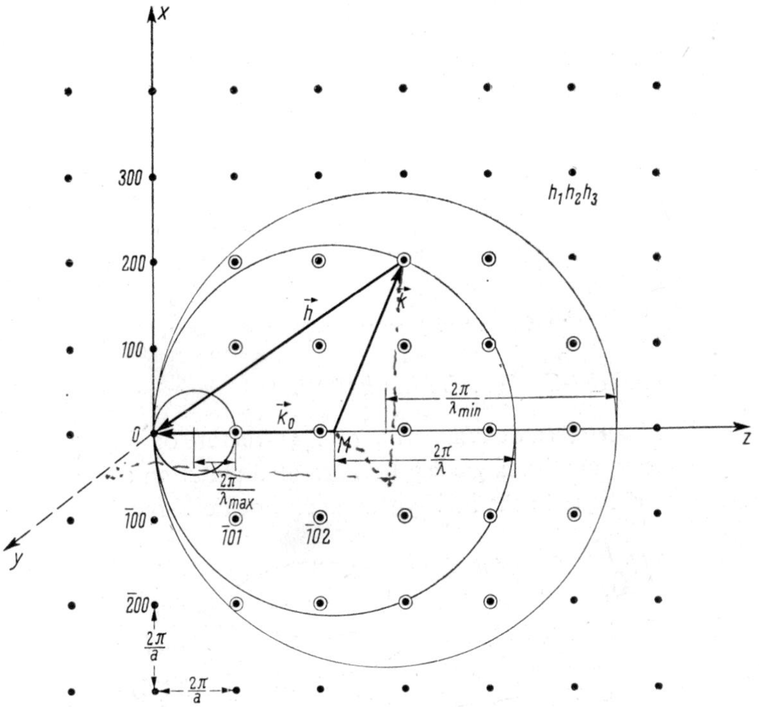

FIG. 14. Ewald's construction of von Laue diffraction patterns (Only the circled dots indicate diffraction maxima.)

obtained (Fig. 11 and Fig. 14):

$$\tan \delta = \frac{h_2}{h_1}, \tag{B. 26}$$

$$\tan \alpha = \frac{h_3}{(h_1^2+h_2^2)^{1/2}}, \tag{B. 27}$$

$$2\alpha = \varphi, \tag{B. 28}$$

$$\cos \varphi = \frac{1-\tan^2 \alpha}{1+\tan^2 \alpha} = \frac{h_1^2+h_2^2-h_3^2}{h_1^2+h_2^2+h_3^2}, \tag{B. 29}$$

$$\sin \varphi = \frac{(h_1^2+h_2^2)^{1/2} 2\pi/a}{2\pi/\lambda} = \frac{\lambda}{a}(h_1^2+h_2^2)^{1/2}. \tag{B. 30}$$

When we eliminate φ from eqns. (B. 29) and (B. 30), we obtain the wavelengths λ for which interference maxima are observed:

$$\lambda = \frac{|h_3|}{h_1^2+h_2^2+h_3^2} 2a. \tag{B. 31}$$

Only the points h_1, h_2, h_3 of the reciprocal lattice which lie outside the Ewald sphere of radius $2\pi/\lambda_{max}$ and inside the sphere of radius $2\pi/\lambda_{min}$ (and the points lying on the surfaces of the spheres) determine the interference maxima (von Laue spots). The minimum wavelength λ_{min} is given by eqn. (B. 25). For λ_{max} we have

$$\lambda_{max} = 2a. \qquad \text{(B. 32)}$$

All wavelengths $\lambda > \lambda_{max}$ yield Ewald spheres with radii smaller than $2\pi/2a$. Apart from the origin, no other lattice point will lie on such spheres, and therefore no interference maxima are obtained for $\lambda > \lambda_{max}$.

(c) *THE ROTATING-CRYSTAL METHOD*

This method permits a determination of the structure of single crystal samples. During exposure the single crystal is rotated about an axis which is perpendicular or oblique to the incident monochromatic X-ray. The interference

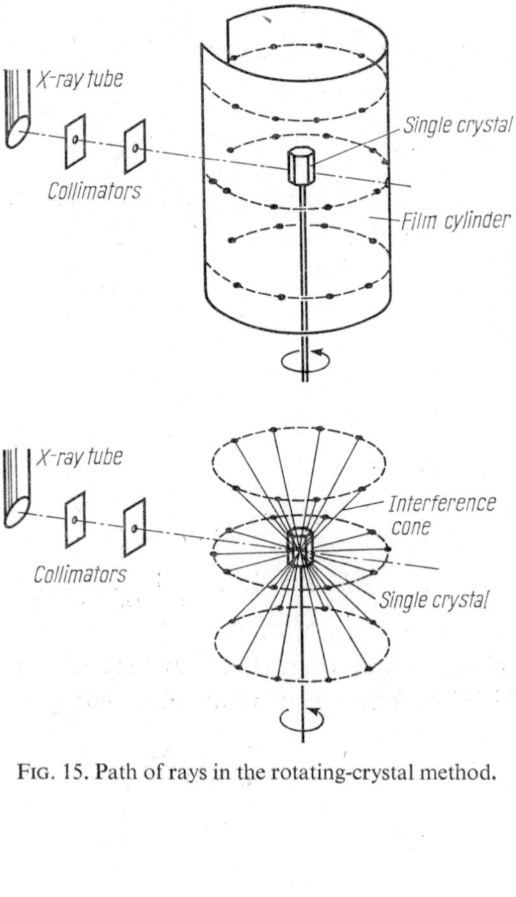

FIG. 15. Path of rays in the rotating-crystal method.

maxima are recorded on a film which, in general, forms a cylinder around the rotation axis (cf. Fig. 15).

Corresponding to the rotation of the crystal the reciprocal lattice rotates about a straight line parallel to the rotation axis while the Ewald sphere remains unchanged. The total number of possible interferences is equal to the number of points of the reciprocal lattice which lie inside a toroid and on its surface. The interference maxima appear on the film as spots arranged along

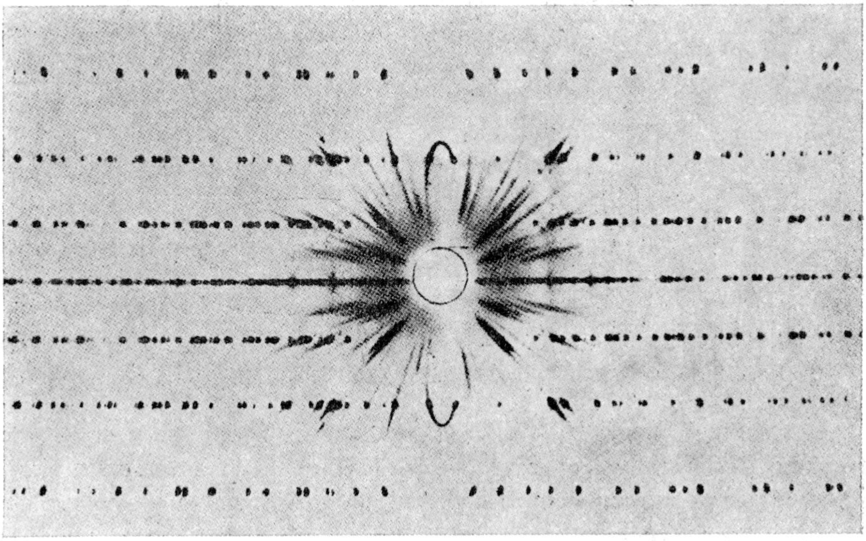

FIG. 16. Photograph taken by the rotating-crystal method.

lines perpendicular to the rotation axis, which are called layer lines (cf. Fig. 16). The interference spots of a layer line pertain to the points of the reciprocal lattice on a plane perpendicular to the rotation axis.

(d) *THE POWDER METHOD* (according to Debye–Scherrer, or to Hull)

Unlike the first two methods, for which single crystals were necessary, this method is based on the use of a powder consisting of many small crystallites; this powder is exposed to monochromatic X-rays. As in the case of the rotating-crystal method, the diffraction pattern is taken on a film mounted with cylindrical geometry concentric around the sample; the incident X-ray is perpendicular to the cylinder axis (cf. Fig. 17).

From Ewald's construction it is obvious that in this case the diffraction pattern will consist of individual lines (cf. Fig. 24). To illustrate this, we assume the Ewald sphere to be fixed and admit every possible rotation of the reciprocal lattice about the origin O, corresponding to all possible orientations of the various small crystallites. Each lattice point h_1,h_2,h_3 will then move on a sphere

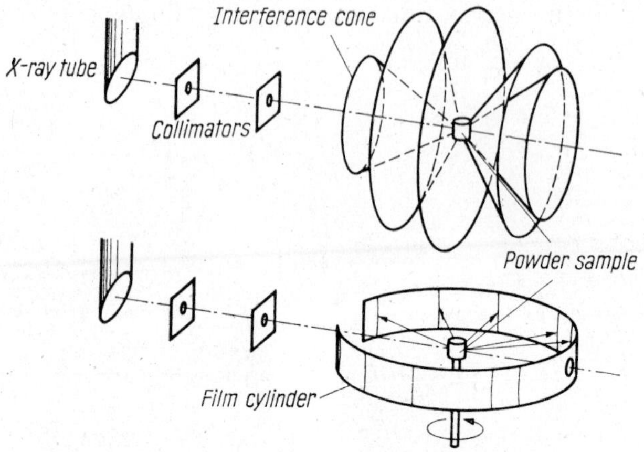

FIG. 17. Path of rays in the powder method (Debye–Scherrer–Hull method).

around O. When this sphere intersects with Ewald's sphere, the curve of intersection will be a latitudinal circle with $\boldsymbol{k}_0$ as axis. All points on such a circle yield interference spots.

3. BRAGG'S INTERPRETATION OF X-RAY DIFFRACTION

Every plane containing three non-colinear lattice points is called a crystal plane. The orientation of parallel crystal planes is characterized by the Miller indices which are determined as follows (see Fig. 18):

1. We determine the coordinate frame formed by the crystal axes $\boldsymbol{a}_1$, $\boldsymbol{a}_2$, $\boldsymbol{a}_3$, the intercepts of the three axes with the crystal plane, and give their coordinates n_1, n_2, n_3 in terms of a_1, a_2, a_3.
2. The reciprocals $1/n_1$, $1/n_2$, $1/n_3$ of these three axial sections are then multiplied by a factor p which is chosen such that the products are irreducible integers. The values thus obtained (p/n_1, p/n_2, p/n_3) are the Miller indices h, k, l.

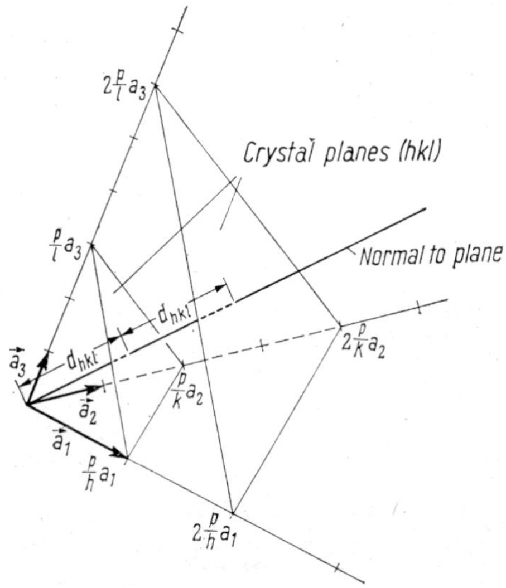

FIG. 18. Determination of the crystal planes (hkl) by the Miller indices.

Every set of parallel crystal planes is thus "indexed", i.e. characterized by the indices (hkl). When a plane to be indexed intersects an axis on the negative side of the origin, the corresponding index is marked by placing a minus sign above it (e.g. ($h\bar{k}l$)). A set of planes possessing the same symmetry properties is characterized by $\{hkl\}$; for example, in the cubic lattice the group $\{1\;1\;1\}$ comprises the following eight sets of planes: $(1\;1\;1)$, $(\bar{1}\;\bar{1}\;\bar{1})$, $(1\;\bar{1}\;1)$, $(\bar{1}\;1\;\bar{1})$, $(\bar{1}\;\bar{1}\;1)$, $(1\;1\;\bar{1})$, $(\bar{1}\;1\;1)$, $(1\;\bar{1}\;\bar{1})$.

Directions in a crystal are likewise identified by three figures written as $[uvw]$; the integral numbers u, v, w are irreducible and have the same ratios as the components of a vector in the desired direction referred to the axes $\boldsymbol{a}_1$, $\boldsymbol{a}_2$, $\boldsymbol{a}_3$. Components in negative directions of the crystal axes are again marked by a minus sign above the corresponding figure (e.g. $[\bar{u}vw]$). Equivalent directions in a crystal are denoted by $\langle uvw \rangle$; for example, in a cubic crystal the face diagonals $\langle 1\;1\;0 \rangle$ comprise the following twelve directions: $\langle 1\;1\;0 \rangle$, $\langle 1\;\bar{1}\;0 \rangle$, $\langle \bar{1}\;1\;0 \rangle$, $\langle \bar{1}\;\bar{1}\;0 \rangle$, $\langle 1\;0\;1 \rangle$, $\langle \bar{1}\;0\;1 \rangle$, $\langle 1\;0\;\bar{1} \rangle$, $\langle \bar{1}\;0\;\bar{1} \rangle$, $\langle 0\;1\;1 \rangle$, $\langle 0\;\bar{1}\;1 \rangle$, $\langle 0\;1\;\bar{1} \rangle$, $\langle 0\;\bar{1}\;\bar{1} \rangle$.

Especially in cubic crystals the Miller indices define both planes and directions in the spatial lattice: the direction $[hkl]$ is perpendicular to the plane (hkl). This does not hold true for arbitrary crystal systems.

The interplanar spacing d_{hkl} is given by the direction cosines α_1, α_2, α_3 of the normals to the planes, the Miller indices h, k, l and the factor p, which again must be chosen so that the numbers p/h, p/k, p/l are integral and irreducible as to their ratios (p/h, p/k, p/l are the axial sections cut off by the crystal planes)

$$d_{hkl} = \frac{p}{h} a_1 \cos \alpha_1 = \frac{p}{k} a_2 \cos \alpha_2 = \frac{p}{l} a_3 \cos \alpha_3 . \qquad \text{(B. 33)}$$

From the viewpoint of Bragg's theory, X-ray interferences are due to reflections from the sets of crystal planes. Since X-rays are insignificantly absorbed in crystals, the incident beam is reflected from many mutually parallel crystal planes (cf. Fig. 19).

When the radiation reflected from the individual planes has phase differences of one or several wavelengths, reflection maxima appear, the so-called X-ray

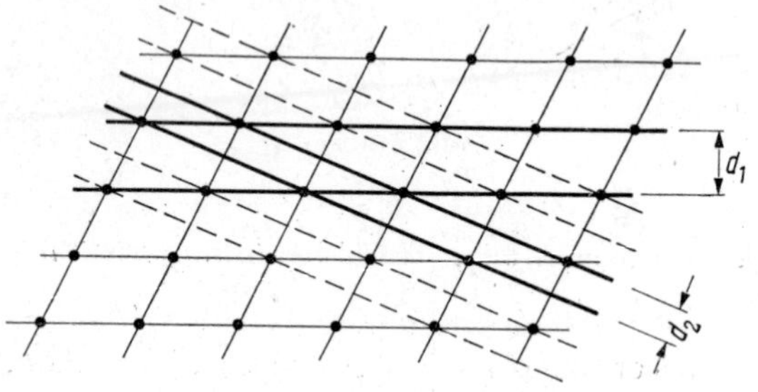

FIG. 19. Sets of crystal planes in a two-dimensional lattice.

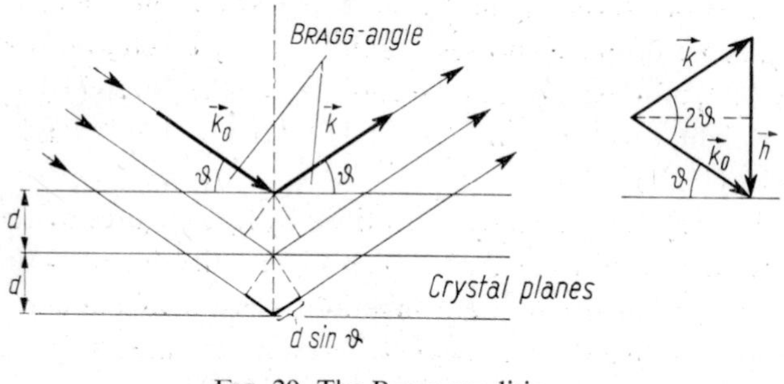

FIG. 20. The Bragg condition.

reflection spots. The corresponding condition is obvious from Fig. 20 and is known as the Bragg condition:

$$2d \sin \vartheta = n\lambda, \qquad \text{(B. 34)}$$

d is the interplanar spacing, ϑ is the glancing angle (Bragg angle), and n is an integer.

This shows us that the wavelength for Bragg reflection from parallel crystal planes of separation d can be at most

$$\lambda_{\max} = 2d. \qquad \text{(B. 35)}$$

For a special case this conclusion had been deduced directly from Ewald's construction [cf. eqn. (B. 32)]; however, it can be derived from it also generally.

Bragg's condition is based only on the periodicity of the crystal structure and describes the geometry of the X-ray interferences. For this equation the arrangement of the atoms (scattering centres) in the crystal planes is insignificant; it influences only the intensity of the various X-ray reflection spots.

The von Laue equations (B. 14) and Bragg's equation (B. 34) are completely equivalent. The directions of incidence and observation ($\boldsymbol{k}_0$ and $\boldsymbol{k}$) make an angle 2ϑ. By virtue of eqns. (B. 5) and (B. 9) the magnitude of the vector $\boldsymbol{h}$ is then given by

$$|\boldsymbol{h}| = 2\frac{2\pi}{\lambda}\sin\vartheta. \tag{B. 36}$$

With this relation the von Laue equations (B. 12) can be written in the form

$$\begin{aligned}
(\boldsymbol{a}_1.\boldsymbol{h}) &= a_1\frac{4\pi}{\lambda}\sin\vartheta\cos\alpha_1 = 2\pi h_1 = 2\pi mh,\\
(\boldsymbol{a}_2.\boldsymbol{h}) &= a_2\frac{4\pi}{\lambda}\sin\vartheta\cos\alpha_2 = 2\pi h_2 = 2\pi mk,\\
(\boldsymbol{a}_3.\boldsymbol{h}) &= a_3\frac{4\pi}{\lambda}\sin\vartheta\cos\alpha_3 = 2\pi h_3 = 2\pi ml,
\end{aligned} \tag{B. 37}$$

where $\cos\alpha_1$, $\cos\alpha_2$, $\cos\alpha_3$ are the direction cosines of the vector $\boldsymbol{h}$ with respect to $\boldsymbol{a}_1$, $\boldsymbol{a}_2$, $\boldsymbol{a}_3$; h, k, l are irreducible integers, and m is a common factor of h_1, h_2, h_3.

The direction cosines of the vector $\boldsymbol{h}$ satisfy the proportionalities

$$\cos\alpha_1 \propto \frac{h}{a_1}, \qquad \cos\alpha_2 \propto \frac{k}{a_2}, \qquad \cos\alpha_3 \propto \frac{l}{a_3}. \tag{B. 38}$$

On the other hand, the direction cosines of the normal to a set of parallel crystal planes (hkl) satisfy the very same relations [cf. eqn. (B. 33)].

Thus the von Laue equations are always satisfied when the vector $\boldsymbol{h}$ is perpendicular to a set of crystal planes. This means that the X-ray interference spots are due to the reflection of the incident X-ray from the crystal planes.

When we substitute the distance d_{hkl} of two neighbouring planes (B. 33) in eqn. (B 37), we obtain Bragg's equation

$$2d_{hkl}\sin\vartheta = n\lambda, \tag{B. 39}$$

where $n = mp$, an integer, is the "order of reflection".

From the equivalence of the von Laue equations and Bragg's equation it follows that each point *mh, mk, ml* of the reciprocal lattice that lies on Ewald's sphere corresponds to a set of planes (*hkl*) in the original lattice, from which the X-rays are reflected.

According to Ewald's construction the von Laue and/or Bragg equations are always satisfied when the two equations (B. 5) and (B. 9) apply. Correspondingly, interference spots are observed if the wave vectors of the incident and the diffracted waves differ by a reciprocal-lattice vector. In addition, we must assume the energies of the incident and diffracted waves to be the same, i.e. the absolute magnitudes of the vectors $\boldsymbol{k}_0$ and $\boldsymbol{k}$ must be equal. From these two conditions it follows that

$$2(\boldsymbol{k}\,.\,\boldsymbol{h})+|\boldsymbol{h}|^2 = 0 \qquad \text{(B. 40)}$$

or (cf. Fig. 20)

$$|\boldsymbol{k}|\,\sin\vartheta = \frac{|\boldsymbol{h}|}{2}\,. \qquad \text{(B. 41)}$$

These relations are identical with Bragg's reflection condition. It can be shown that the interplanar spacing d in the original lattice is proportional to the absolute magnitude $|\boldsymbol{h}|$ of a reciprocal-lattice vector perpendicular to these planes, i.e.

$$d = \frac{2\pi n}{|\boldsymbol{h}|}\,. \qquad \text{(B. 42)}$$

With this relation and the relation $|\boldsymbol{k}| = 2\pi/\lambda$ we obtain Bragg's reflection condition from eqn. (B. 41).

4. THE STRUCTURE FACTOR

The conditions for the appearance of X-ray interference spots have been obtained from a consideration of the diffraction of an X-ray wave in a structure which is described by a Bravais lattice of identical atoms. Bragg's equation, or the von Laue equations, determine all reflections possible with such simple structures; however, they do not yield information on the intensities of the reflected waves.

From eqn. (B. 12) we have seen that the intensity of the diffracted X-ray wave depends on the atomic scattering factor ψ. The scattering factor accounts for the fact that, compared with the wavelength of the X-ray radiation, the scattering centre has finite dimensions. The scattered wave consists of the algebraic sum of the components of electromagnetic radiation emitted by the individual shell electrons. The value of the scattering factor is proportional to the number of shell electrons, i.e. to the atomic number of the element considered, and depends on the X-ray wavelength and the glancing angle ϑ.

For structures which are derived from a multiple primitive lattice or a lattice with a basis, the intensities are determined essentially by the so-called structure factor, which also contains the atomic scattering factor ψ. The structure factor accounts for the kind of atoms as well as their geometrical arrangement in the unit cell. In the following we shall show that the structure factor and thus

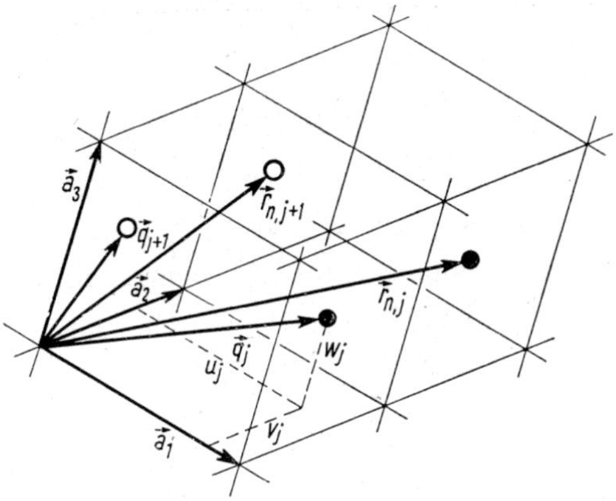

FIG. 21. Lattice with basis.

the intensity of certain reflections can be equal to zero. Structures derived from a lattice with basis have generally fewer reflections than structures with the corresponding simple primitive lattices.

The positions of atoms in a lattice with basis are determined (cf. Fig. 21) by

$$\boldsymbol{r}_{nj} = n_1\boldsymbol{a}_1+n_2\boldsymbol{a}_2+n_3\boldsymbol{a}_3+\boldsymbol{q}_j\,, \tag{B. 43}$$

where $\boldsymbol{q}_j$ is the basis vector defining the position of the jth atom in the unit cell through the basis coordinates u_j, v_j, w_j:

$$\boldsymbol{q}_j = u_j\boldsymbol{a}_1+v_j\boldsymbol{a}_2+w_j\boldsymbol{a}_3\,. \tag{B. 44}$$

Now we obtain for the total scattering amplitude of X-ray radiation [cf. eqn. (B. 12)]

$$\begin{aligned} A_{\text{tot}} &= A_0 \sum_{n_1, n_2, n_3, j} \psi_j e^{i(\boldsymbol{r}_{nj}.\boldsymbol{h})} \\ &= A_0 \sum_j \psi_j e^{i(\boldsymbol{q}_j.\boldsymbol{h})} \sum_{n_1, n_2, n_3} e^{i[n_1(\boldsymbol{a}_1.\boldsymbol{h})+n_2(\boldsymbol{a}_2.\boldsymbol{h})+n_3(\boldsymbol{a}_3.\boldsymbol{h})]}\,. \end{aligned} \tag{B. 45}$$

The sum over n_1, n_2, n_3 reaches the highest values when the von Laue equations are satisfied (cf. pp. 20 ff.). The sum over j is called the structure factor S:

$$S = \sum_j \psi_j e^{i(q_j.h)} = \sum_j \psi_j e^{i[u_j(a_1.h)+v_j(a_2.h)+w_j(a_3.h)]}. \tag{B. 46}$$

With the help of the von Laue equations (B. 14) we obtain the structure factor for the interference maximum $(h_1h_2h_3)$:

$$S = \sum \psi_j e^{2\pi i(u_jh_1+v_jh_2+w_jh_3)}. \tag{B. 47}$$

In this way a certain value of the structure factor is attributed to each point of the reciprocal lattice.

THE STRUCTURE FACTORS OF THE BODY-CENTRED CUBIC LATTICE AND OF THE FACE-CENTRED CUBIC LATTICE

Assume all lattice points occupied by identical atoms. The unit cell of the *body-centred* cubic lattice contains two lattice points with the following coordinates u_j, v_j, w_j (cf. Table 2 and Fig. 22 a).

j	u_j	v_j	w_j
1	0	0	0
2	$\frac{1}{2}$	$\frac{1}{2}$	$\frac{1}{2}$

Thus the structure factor S_{bcc} is given by

$$S_{bcc} = \psi[1+e^{\pi i(h_1+h_2+h_3)}]. \tag{B. 48}$$

Hence we obtain

$$S_{bcc} = 2\psi \quad \text{for} \quad h_1+h_2+h_3 = 2p,$$
$$S_{bcc} = 0 \quad \text{for} \quad h_1+h_2+h_3 = 2p+1,$$
$$p = 0, 1, 2 \ldots$$

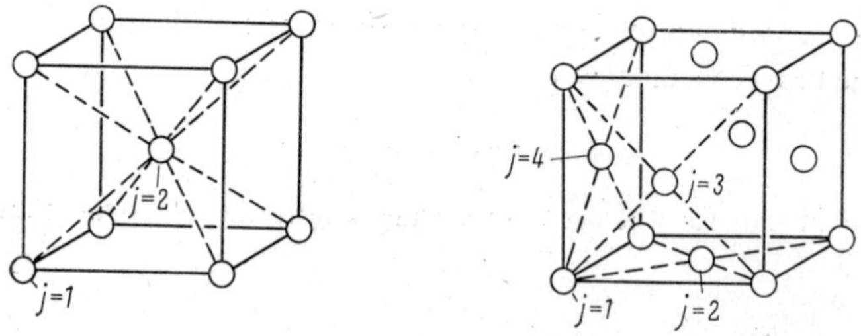

(a) body-centred cubic lattice (b) face-centred cubic lattice

FIG. 22. Cubic lattice with basis.

All reflections from crystal planes $(h_1h_2h_3)$ become equal to zero (extinction) when $h_1+h_2+h_3$ is odd.

The unit cell of a *face-centred* cubic lattice contains four lattice points with the following coordinates u_j, v_j, w_j (cf. Table 2 and Fig. 22 b):

j	u_j	v_j	w_j
1	0	0	0
2	$\frac{1}{2}$	$\frac{1}{2}$	0
3	$\frac{1}{2}$	0	$\frac{1}{2}$
4	0	$\frac{1}{2}$	$\frac{1}{2}$

Thus the structure factor S_{fcc} is given by

$$S_{\text{fcc}} = \psi[1+e^{i\pi(h_1+h_2)}+e^{i\pi(h_1+h_3)}+e^{i\pi(h_2+h_3)}]. \tag{B. 49}$$

Hence we obtain

$S_{\text{fcc}} = 4\psi$ when the numbers h_1, h_2, h_3 are all odd or all even (e.g. 111, 113, 220);

$S_{\text{fcc}} = 0$ when the numbers h_1, h_2, h_3 are mixed even–odd (e.g. 100, 110, 112).

Finally, by way of example of the NaCl structure, we shall consider the case of a structure for which the lattice points in the unit cell are occupied by atoms of different atomic number. The lattice of the NaCl crystal consists of two face-centred cubic lattices of one and the same type of atoms each, which are displaced with respect to one another in the three directions of the cube edges by $a/2$ (cf. Fig. 23). The basis of this lattice comprises eight lattice points, four occupied by atoms of type A and four occupied by atoms of type B.

j	u_j	v_j	w_j	
1	0	0	0	ψ_A
2	$\frac{1}{2}$	$\frac{1}{2}$	0	
3	$\frac{1}{2}$	0	$\frac{1}{2}$	
4	0	$\frac{1}{2}$	$\frac{1}{2}$	
5	$\frac{1}{2}$	$\frac{1}{2}$	$\frac{1}{2}$	ψ_B
6	$\frac{1}{2}$	0	0	
7	0	$\frac{1}{2}$	0	
8	0	0	$\frac{1}{2}$	

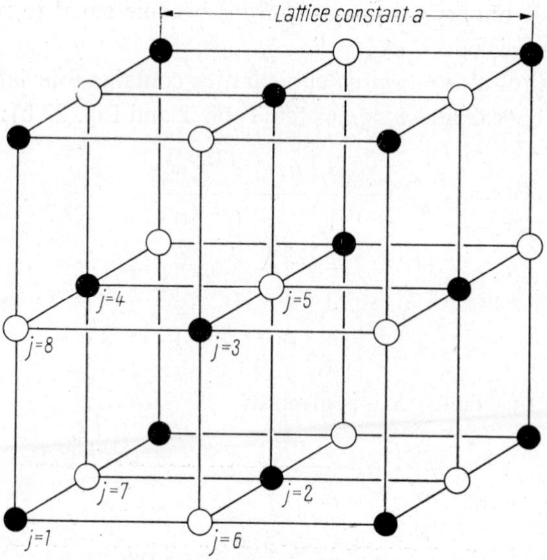

FIG. 23. NaCl structure; the basis of this cubic lattice with lattice constant a comprises eight lattice points.

For the structure factor we obtain

$$\begin{aligned} S_{\text{NaCl}} &= \psi_A[1+e^{i\pi(h_1+h_2)}+e^{i\pi(h_1+h_3)}+e^{i\pi(h_2+h_3)}] \\ &\quad +\psi_B[e^{i\pi(h_1+h_2+h_3)}+e^{i\pi h_1}+e^{i\pi h_2}+e^{i\pi h_3}] \\ &= [\psi_A+\psi_B e^{i\pi(h_1+h_2+h_3)}]\,[1+e^{i\pi(h_1+h_2)}+e^{i\pi(h_1+h_3)}+e^{i\pi(h_2+h_3)}]. \end{aligned} \quad \text{(B. 50)}$$

Hence it follows that

$S_{\text{NaCl}} = 0$, if the numbers h_1, h_2, h_3 are mixed even–odd;

$S_{\text{NaCl}} = 4(\psi_A+\psi_B)$, if the numbers h_1, h_2, h_3 are all even;

$S_{\text{NaCl}} = 4(\psi_A-\psi_B)$, if the numbers h_1, h_2, h_3 are all odd.

When we compare, for example, the X-ray diffraction patterns of NaCl with those of KCl (which has the same structure), we can observe a clear difference (cf. Fig. 24): the reflections $h_1h_2h_3$ for which all h_i are odd are lacking in the KCl pattern. Accordingly the atomic scattering factors and thus the determining numbers of shell electrons of potassium and chlorine must be equal. This is only possible if the scattering centres are ions, namely the *ions* K^+

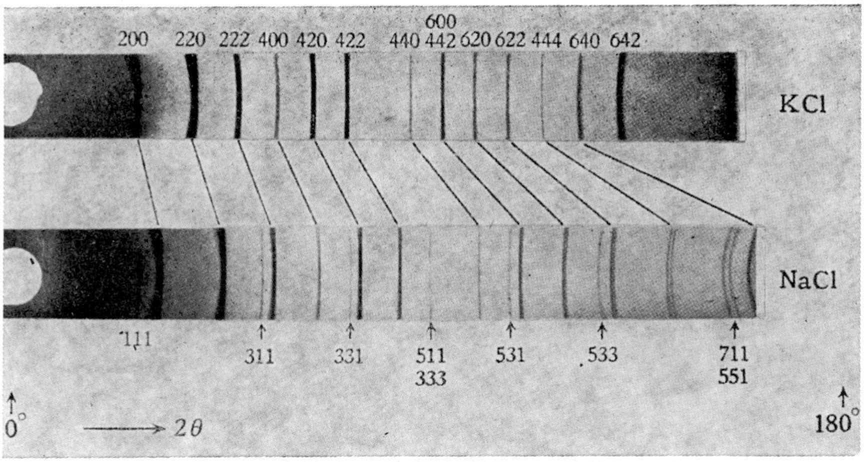

FIG. 24. X-ray diffraction pattern of KCl and NaCl taken by the powder method (from L. Bragg, *The Crystalline State*, Vol. 1, Bell, London 1955).

and Cl^-, since

$$\psi_A \propto Z_{K^+} = Z_K - 1 = 19 - 1 = 18,$$
$$\psi_B \propto Z_{Cl^-} = Z_{Cl} + 1 = 17 + 1 = 18.$$

This is a direct experimental proof of the fact that KCl is an ionic crystal.

The structural elements of the alkali halide crystals are always *ions* of a metal and a halogen, i.e. independent of the lattice type, the alkali halides are ionic crystals. CsCl, CsBr, and CsI have a body-centred cubic structure (CsCl structure), that is, the metal ions and the halogen ions form primitive cubic lattices which are displaced with respect to one another by $a/2$ in the directions of the cube edges. All other alkali halides crystallize in the NaCl type.

III. Brillouin Zones

Having introduced the concept of the reciprocal lattice, we have arrived at a simple interpretation of the von Laue equations. An X-ray wave of wavelength λ, with the direction of incidence of $\boldsymbol{k}_0$, is scattered with maximum intensity in the direction of $\boldsymbol{k}$ when the vector $\boldsymbol{h} = \boldsymbol{k}_0 - \boldsymbol{k}$ is a lattice vector of the reciprocal lattice.

It is possible to identify each lattice vector of the reciprocal lattice with an $\boldsymbol{h}$ vector. Then the mid-normal plane to $\boldsymbol{h}$ will be the locus of all arrow points of the vectors $\boldsymbol{k}_0$ and $\boldsymbol{k}$ which start from the end points of this $\boldsymbol{h}$ vector. Denoting all wave vectors for which the diffracted X-ray wave has maximum intensity by $\boldsymbol{k}_L$, we can write the equation for the mid-normal plane to the lattice vector $\boldsymbol{h}$ as follows:

$$(\boldsymbol{k}_L \,.\, \boldsymbol{h}) = \frac{|\boldsymbol{h}|^2}{2}. \tag{B. 51}$$

When we determine all mid-normal planes to the vectors $\boldsymbol{h}$ which connect an arbitrary lattice point with its nearest neighbours of the reciprocal lattice, these planes will enclose a volume of size $V_{BZ} = (\boldsymbol{b}_1 \cdot [\boldsymbol{b}_2 \wedge \boldsymbol{b}_3])$ which is called the first Brillouin zone ($\boldsymbol{b}_1$, $\boldsymbol{b}_2$, and $\boldsymbol{b}_3$ are the translation vectors of the reciprocal

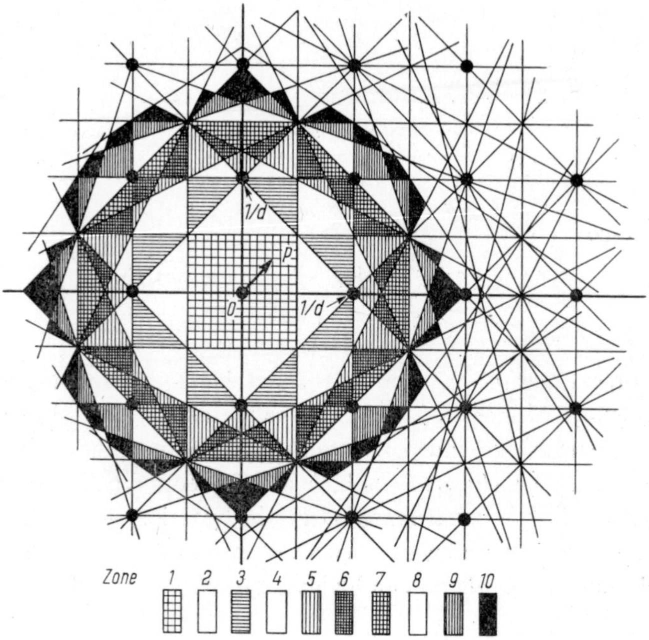

FIG. 25. Brillouin zones for a two-dimensional square lattice (from L. Brillouin, *Wave Propagation in Periodic Structures*, McGraw-Hill, New York 1946).

lattice). In the same way the nth Brillouin zone is obtained as the volume (with the same size) which lies outside the $(n-1)$th zone. For the case of two dimensions, Fig. 25 shows the division of the reciprocal lattice into Brillouin zones; here a Brillouin zone is the area formed by the mid-normals to the vectors connecting a lattice point with its nearest neighbours.

The Brillouin zone surfaces are defined by the von Laue equations, i.e. Bragg reflection can take place when the point of the vector $\boldsymbol{k}_0$ (beginning at the origin) lies on the surface of a Brillouin zone. We shall show later that the electrons in a solid display their wave character so that also electron waves can undergo Bragg reflections. The subdivision of the reciprocal space into zones is of fundamental importance for the investigation of electron motion in periodic structures (cf. p. 210).

The shape of the Brillouin zones depends solely on the geometrical properties of the Bravais lattice upon which the crystal structure is based, and is independent of the chemical composition and the number of atoms in the unit cell. For a given lattice all Brillouin zones have the same volume V_{BZ}, and it is always possible, with the help of translations, to show the exact congruency between the first Brillouin zone and all higher Brillouin zones (cf. Fig. 25 for the two-dimensional case).

References

Structure

BUERGER, M. J., *Elementary Crystallography: An Introduction to the Fundamental Geometrical Features of Crystals* (Wiley, New York 1956).
HUME-ROTHERY, W., and RAYNOR, G. V., *The Structure of Metals and Alloys* (Institute of Metals, London 1956).
KLEBER, W., *Einführung in die Kristallographie* (Verlag Technik, Berlin 1963).
NIGGLI, P., *Lehrbuch der Mineralogie und Kristallchemie*, 2 vols. (Bornträger, Berlin 1942).
NIGGLI, P., *Grundlagen der Stereochemie* (Birkhäuser, Basel 1945).
PAULING, L., *The Nature of the Chemical Bond*, 3rd ed. (Cornell University Press, Ithaca 1960).
WELLS, A. F., *Structural Inorganic Chemistry* (Oxford University Press, London 1962).
WINKLER, H. G. F., *Struktur und Eigenschaften der Kristalle*, 2nd ed. (Springer, Berlin 1955).

Structural Analysis

ARNDT, U. W., and WILLIS, B. T. M., *Single Crystal Diffractometry* (Cambridge University Press, London 1966).
AZAROFF, L. V., and BUERGER, M. J., *The Powder Method in X-ray Crystallography* (McGraw-Hill, New York 1958).
BACON, G. E., *Neutron Diffraction* (Oxford University Press, London 1960).
BRAGG, W. L., *The Crystalline State*, Vol. 1 (Bell, London 1955).
BUERGER, M. J., *X-ray Crystallography* (Wiley, New York 1942).
BUERGER, M. J., *Crystal Structure Analysis* (Wiley, New York 1960).
GUINIER, A., *Théorie et technique de la radiocristallographie* (Dunod, Paris 1956).
VON LAUE, M., *Materiewellen und ihre Interferenzen*, 2nd ed. (Akademische Verlagsgesellschaft, Leipzig 1948).
VON LAUE, M., *Röntgenstrahl-Interferenzen* (Akademische Verlagsgesellschaft, Frankfurt 1960).

B. INTERFERENCE EFFECTS IN CRYSTALS

Various Review Articles in:

Handbuch der Physik, FLÜGGE, S. (Editor) (Springer, Berlin 1955 ff.),
Vol. 7/1, 2: *Crystal Physics* I, II (1955, 1958),
Vol. 30: *X-rays* (1957),
Vol. 32: *Structural Research* (1957),
Vol. 33: *Particle Optics* (1956).

Reference Books

LANDOLT-BÖRNSTEIN, *Zahlenwerte und Funktionen*, Vol. 1/4: *Kristalle* (Springer, Berlin 1955).

LONSDALE, K., *et al.* (Editor), *International Tables for X-ray Crystallography* (Kynoch, Birmingham 1952 ff.),
Vol. 1: *Symmetry Groups* (1952),
Vol. 2: *Mathematical Tables* (1959),
Vol. 3: *Physical and Chemical Tables* (1962).

PEARSON, W. B., *A Handbook of Lattice Spacings and Structures of Metals and Alloys*, 2 vols. (Pergamon, Oxford 1958, 1967).

EWALD, P. P., and HERMANN, C. (Editors), *Struktur-Bericht*, 7 vols. (Akademische Verlagsgesellschaft, Leipzig 1913–1939).

WILSON, A. J. C./PEARSON, W. B. (Editors of "The International Union of Crystallography"), *Structure Reports for 1940–1960* (Oosthoek, Utrecht 1956–1968).

WYCKOFF, R. W. G., *Crystal Structures*, 6 vols., 2nd ed. (Interscience, New York 1963–1969).

C. Lattice Dynamics

STRICTLY speaking, crystals possess the geometrical properties described in the foregoing chapter only if the absolute temperature is equal to zero and the external pressure is hydrostatic. In the following we shall consider the solid-state properties which depend immediately on the crystal's internal energy. We are mainly interested in the problem of the variation of this energy under the influence of temperature and pressure. Relations describing this influence can be obtained on the basis of thermodynamics; in addition data on the microscopic structure of the solid are required in some cases, e.g. the derivation of the equation of state, see pp. 101 ff.

I. Thermodynamic Fundamentals

Let us consider a solid of volume V which, at the temperature T, is under the influence of the hydrostatic pressure p. Within the framework of thermodynamics, the compressibility is defined as

$$\varkappa = -\frac{1}{V}\left(\frac{\partial V}{\partial p}\right)_T \tag{C. 1}$$

and the volume expansion coefficient as

$$\alpha = \frac{1}{V}\left(\frac{\partial V}{\partial T}\right)_p . \tag{C. 2}$$

It is also possible, without knowing details of the structure of the solid, to calculate the compression work and the *difference* of the specific heats at constant pressure and at constant volume.

The compression work. If the hydrostatic pressure is increased and the temperature is kept constant, the initial volume V_0 of a solid is reduced to V; for this the compression work A_c is required. The necessary pressure increase is obtained from the compressibility equation (C. 1),

$$dp = -\frac{1}{\varkappa}\frac{dV}{V_0} . \tag{C. 3}$$

Assume the volume V_0 to correspond to pressure $p = 0$. The pressure increase will then be equal to the pressure applied,

$$p = -\frac{1}{\varkappa}\,\frac{V-V_0}{V_0}, \tag{C. 4}$$

and for the compression work we have

$$A_c^{\swarrow} = -\int_{V_0}^{V} p\,dV = \frac{1}{\varkappa V_0}\int_{V_0}^{V} (V-V_0)\,dV,$$

$$A_c^{\swarrow} = \frac{1}{2\varkappa}\,\frac{(V-V_0)^2}{V_0}. \tag{C. 5}$$

The difference of specific heats. Thermodynamically we have in general

$$c_p - c_V = -T\left(\frac{\partial V}{\partial T}\right)_p^2 \left(\frac{\partial p}{\partial V}\right)_T \tag{C. 6}$$

and hence, with (C. 1) and (C. 2), we find

$$c_p - c_V = \frac{\alpha^2 VT}{\varkappa}. \tag{C. 7}$$

Whereas the measurement of specific heats is always carried out at constant pressure, the theories of specific heats are based on the assumption of constant

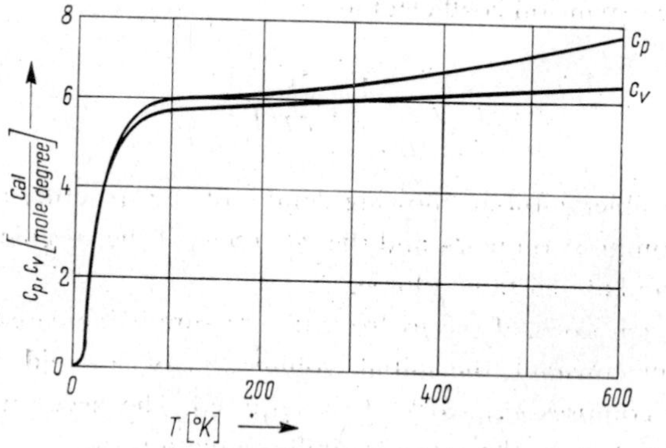

FIG. 26. Specific heat of lead at constant pressure and at constant volume as a function of temperature (according to F. Seitz, *Modern Theory of Solids*, McGraw-Hill, New York 1940).

volume (cf. pp. 53 ff.). The general relation (C. 7) is therefore important for comparisons between experimental and theoretical values of specific heats. The difference between c_p and c_V is vanishing at $T = 0°K$ but increases linearly with increasing T and, particularly at higher temperatures, it cannot be neglected (cf. Fig. 26).

II. The Internal Energy

The total energy of a closed system is called the intrinsic or internal energy U. The internal energy U of a solid consists of the kinetic and potential energy of the structural elements (atoms, ions, molecules), the lattice defects, and the electrons.

The internal energy is divided into the following contributions:

1. The lattice energy U_l depends on the forces acting between the structural elements. As to its numerical value it is equal to the energy necessary to disintegrate the crystal of volume V into free atoms, ions, or molecules. "Free" particles do not interact, their mutual distances are infinitely large so that their potential energy can be set equal to zero. Thus the lattice energy is negative; it depends only on the volume of the crystal, that is, $U_l(V)$, and is, because of the weak temperature dependence of the volume, only indirectly dependent on the temperature.

2. The energy of the lattice vibrations $U_v(V, T)$ is—apart from the zero-point energy (see p. 59)—given by the increment of energy of the crystal when it is heated for a given volume V from absolute zero to the temperature T. This vibrational energy depends on both the volume and the temperature and consists of the kinetic energy of the vibrating structural elements and their potential energy with respect to their rest positions.

3. Further contributions ΔU are, for example, the energies of the lattice defects (see p. 118), of the electron gas (see p. 169), or of the spin waves.

In the framework of the present chapter we only treat the first two contributions to the internal energy and we base our considerations on the following model of an ideal crystal: the atoms (ions) are on average localized at lattice sites through mutual forces, but they show a temperature motion in the form of spatial vibrations around their rest positions. We thus have for the internal energy

$$U = U_l(V) + U_v(V, T). \tag{C. 8}$$

The contributions are calculated on the basis of the models by Mie (lattice energy) and Einstein or Debye (energy of the lattice vibrations).

1. THE LATTICE ENERGY

Experimentally it is found that forces are necessary in order to deform solids. The resultant forces in the solid are due to the forces between the structural elements of the lattice. It has been suggested by Mie to assume that the attractive as well as the repulsive forces between two atoms (or ions) can be derived from a

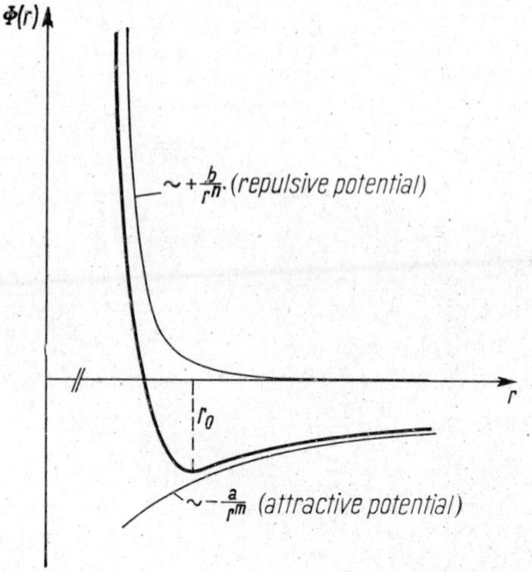

FIG. 27. Potential energy between two atoms as a function of their relative distance apart.

potential which is inversely proportional to a power of the distance; so the potential energy of two atoms at distance r (cf. Fig. 27) is given by

$$\phi(r) = -\frac{a}{r^m} + \frac{b}{r^n}, \tag{C. 9}$$

where a, b, m, n are constants.

The first term refers to attraction, the second to repulsion. The equilibrium distance r_0 results from the condition of minimum potential energy, i.e.

$$\frac{d\phi}{dr} = 0. \tag{C. 10}$$

It must be assumed that at large distances the repulsive forces are necessarily weaker than the attractive forces so that the particles can condense into a solid state. At small distances, however, the repulsive forces must predominate, otherwise stable equilibrium would be impossible. Therefore $n > m$. This is observed, for example, in the uniaxial deformation of a solid: uniaxial compression requires greater forces than uniaxial expansion, when a certain relative change of length is to be achieved.

The total lattice energy of a crystal of volume V is obtained by means of summation over the potential energies between all pairs of lattice particles, i.e.

$$U_l = \tfrac{1}{2} \sum_i \sum_{\substack{j \\ i \neq j}} \phi_{ij}(r_{ij}). \tag{C. 11}$$

The factor $\frac{1}{2}$ accounts for the fact that the energy contribution of each pair is taken only once. Substitution of eqn. (C. 9) in eqn. (C. 11) yields

$$U_l = -\frac{1}{2} \sum_i \sum_{\substack{j \\ i \neq j}} \frac{a_{ij}}{(r_{ij})^m} + \frac{1}{2} \sum_i \sum_{\substack{j \\ i \neq j}} \frac{b_{ij}}{(r_{ij})^n}. \tag{C. 12}$$

If the distances r_{ij} between the lattice particles is referred to the distance of the nearest neighbours, which is defined by the number of lattice particles in the crystal volume V, we obtain

$$r_{ij} = c_{ij} r_1 \tag{C. 13}$$

where

$$r_1^3 = \frac{V}{2N_P}\,; \tag{C. 14}$$

N_P is the number of pairs of atoms or ions in the volume V (the number of molecules in ionic crystals).

Using eqn. (C. 13), we can rewrite (C. 12)

$$U_l = -\frac{A'}{r_1^m} + \frac{B'}{r_1^n} \tag{C. 15}$$

where

$$A' = \sum_i \sum_{\substack{j \\ i \neq j}} \frac{a_{ij}}{2c_{ij}^m} \quad \text{and} \quad B' = \sum_i \sum_{\substack{j \\ i \neq j}} \frac{b_{ij}}{2c_{ij}^n}. \tag{C. 16}$$

Substitution of eqn. (C. 14) in eqn. (C. 15) yields

$$U_l = -\frac{A}{V^{m/3}} + \frac{B}{V^{n/3}} \tag{C. 17}$$

where

$$A = (2N_P)^{m/3} A' \quad \text{and} \quad B = (2N_P)^{n/3} B'. \tag{C. 18}$$

We see that the constants A and B depend on the number of lattice particles in the crystal volume, on their spatial arrangement (i.e. on the crystal structure), and on the powers m and n.

The crystal has such a structure that the resulting force acting on each structural element is equal to zero, i.e. the atoms are in positions of equilibrium. Thermodynamic equilibrium is reached when the free energy F of the crystal is lowest. The free energy is defined by

$$F = U - TS, \tag{C. 19}$$

where S is the entropy.

At $T = 0$ and $p = 0$ the free energy is equal to the lattice energy. The condition of equilibrium will then read

$$\left(\frac{\partial U_l}{\partial V}\right)_{T=0} = 0 \tag{C. 20}$$

and yield the relation

$$\frac{m}{3} \frac{A}{V_0^{(m/3)+1}} = \frac{n}{3} \frac{B}{V_0^{(n/3)+1}}, \tag{C. 21}$$

V_0 being the equilibrium volume. Using this relation we can eliminate, for example, the constant B in eqn. (C. 17) and obtain the equilibrium lattice energy:

$$U_{l0} = -\frac{A}{V_0^{m/3}}\left(1 - \frac{m}{n}\right). \tag{C. 22}$$

So far we have not made any assumptions as to the nature of the lattice forces. The power law by Mie was first nothing more than a hypothesis in order to discuss the thermoelastic properties of solids within the framework of a theory. The constants A, m, n can only be determined empirically, when no assumptions on the nature of the forces (i.e. chemical bonding) are introduced.

According to the type of the forces acting between the lattice particles, we distinguish essentially four different types of chemical bond. For each type of bond the lattice energy must be considered separately. Of all types of bonds that of the heteropolar (ionic) bond is known best. So we restrict ourselves to a consideration of the lattice energy in ionic crystals.

(a) *THE LATTICE ENERGY OF IONIC CRYSTALS*

The main part of the binding forces in ionic crystals is of electrostatic character. The Coulomb energy u_{coul} for two point charges Ze at the distance r

amounts to

$$u_{\text{coul}} = -\frac{1}{4\pi\varepsilon_0}\frac{Z^2e^2}{r}, \tag{C. 23}$$

where ε_0 is the vacuum permittivity.

The total Coulomb energy U_{coul} of an ionic crystal is obtained by summing over all pairs of ions. It consists of an attractive and a repulsive component between the ions of opposite and equal charge, respectively.

Hence results a negative contribution to the lattice energy which represents the cohesion of ionic crystals.

Summation is carried out according to a method by Madelung and Ewald: assuming the ions to represent point charges on lattice points, it yields the expression

$$U_{\text{coul}} = -\alpha_M N_P \frac{Z^2e^2}{4\pi\varepsilon_0}\frac{1}{r_1}, \tag{C. 24}$$

where α_M is the Madelung constant, N_P is the number of ion pairs, Z is the ion valence, e is the elementary charge, and r_1 is the nearest neighbour separation.

The Madelung constant α_M is defined by

$$\alpha_M = r_1 \sum_j (\pm)\frac{1}{r_j}, \tag{C. 25}$$

where r_j is the distance between the reference ion and the jth neighbouring ion, $+$ means *different* signs of the charges at distance r_j and $-$ means *equal* signs. α_M depends only on the type of the lattice and is independent of the lattice dimensions and the kind of ions. In the case of complex structures the Madelung constant is derived from eqn. (C. 24) instead of (C. 25). If $Z_{\text{cation}} \neq Z_{\text{anion}}$, N_P is the number of molecules and Z is the largest common factor

TABLE 5. Madelung constants α_M for various types of structure (after M. P. Tosi, *Solid State Physics* **16**, 1 [1964], Table 1)

Structural type	Examples	α_M
Rocksalt NaCl	AgBr, EuS	1.748
Cesium chloride CsCl	CsBr, TlBr	1.763
Zinc blende ZnS	CuCl, GaAs	1.638
Wurtzite ZnS	ZnO, GaN	1.641
Fluorite CaF_2	LaH_2, UO_2	5.039
Rutile TiO_2	CrO_2, MnF_2	4.816
Corundum Al_2O_3	V_2O_3, Cr_2O_3	25.031
Perovskite $CaTiO_3$	$BaTiO_3$, KJO_3	12.377

of Z_{cation} and Z_{anion}; for example, $Z = 2$ for TiO_2 and UO_2, $Z = 1$ for CaF_2 and Al_2O_3. The values of α_M are compiled for various structural types in Table 5.

The attractive potential U_{coul} of the lattice energy is therefore proportional to $1/r_1$, i.e. the corresponding power is $m = 1$ [cf. eqn. (C.22)]. The lattice energy is then obtained as

$$U_{l0} = -\frac{A}{V_0^{1/3}}\left(1-\frac{1}{n}\right), \tag{C. 26}$$

and using eqn. (C. 24)

$$U_{l0} = -\alpha_M N_P \frac{Z^2e^2}{4\pi\varepsilon_0}\frac{1}{r_1}\left(1-\frac{1}{n}\right). \tag{C. 27}$$

Thus the lattice energy of an ionic crystal is known if we know the value of n of the repulsive potential. The value of n can be determined from the compressibility as will be shown in the following.

The lattice energy is a function of the volume. Since the relative change of volume caused by a variation of the external hydrostatic pressure p is small, the lattice energy can be given by Taylor series at $V = V_0$:

$$U_l(V) = U_{l0}+\left(\frac{\partial U_l}{\partial V}\right)_{V_0}\Delta V+\frac{1}{2}\left(\frac{\partial^2 U_l}{\partial V^2}\right)_{V_0}(\Delta V)^2+\ \dots \tag{C. 28}$$

At $V = V_0$ the lattice energy U_l is minimum [cf. eqn. (C. 20)], and therefore

$$\left(\frac{\partial U_l}{\partial V}\right)_{V_0} = 0 \tag{C. 29}$$

so that we have to a first approximation

$$U_l(V) = U_{l0}+\frac{1}{2}\left(\frac{\partial^2 U_l}{\partial V^2}\right)_{V_0}(\Delta V)^2. \tag{C. 30}$$

The second term must be equal to the compression work at $T = 0°K$; with (C. 5) we arrive at

$$\left(\frac{\partial^2 U_l}{\partial V^2}\right)_{V_0} = \frac{1}{\varkappa V_0}. \tag{C. 31}$$

By twice differentiating Mie's assumption (C. 17) and using for the equilibrium condition (C. 21) we obtain a relation for the compressibility which is independent of the character of the chemical bonding:

$$\frac{1}{\varkappa} = \frac{m(n-m)}{9}\frac{A}{V_0^{(m/3)+1}}. \tag{C. 32}$$

For ion crystals $m = 1$ and according to (C. 27) we obtain

$$A = \alpha_M N_P V_0^{1/3} \frac{Z^2 e^2}{4\pi\varepsilon_0} \frac{1}{r_1} \qquad \text{(C. 33)}$$

and thus

$$\frac{1}{\varkappa} = \frac{n-1}{9} \frac{\alpha_M Z^2 e^2}{4\pi\varepsilon_0} \frac{1}{r_1} \frac{N_P}{V_0}. \qquad \text{(C. 34)}$$

When we denote by n_P the number of ion pairs in a volume r_1^3 we obtain

$$\frac{1}{\varkappa} = \frac{n-1}{9} \frac{\alpha_M Z^2 e^2}{4\pi\varepsilon_0} \frac{n_P}{r_1^4}. \qquad \text{(C. 35)}$$

A measurement of $\varkappa$ therefore permits a determination of the repulsion exponent n which is of the order of 10 for many ionic crystals (cf. Table 6).

TABLE 6. Repulsion exponents n and lattice energy U_l (in kcal/mole) of alkali halides. U_l (calc.) was calculated from electrostatic attraction and repulsion (C. 20), U_l(exp.) was measured in experiments with the Born–Haber cyclic process; all values refer to normal pressure and room temperature (after M. P. Tosi, *Solid State Physics* **16,** 1 [1964], Tables 8, 11, 12)

	F	Cl	Br	I	
	6.2	7.3	7.7	7.0	n
Li	239.3	194.0	183.3	165.7	$-U_l$ (calc.)
	242.3	198.9	189.8	177.7	$-U_l$ (exp.)
	6.4	8.4	8.3	8.0	n
Na	210.0	180.9	170.5	157.1	$-U_l$ (calc.)
	214.4	182.6	173.6	163.2	$-U_l$ (exp.)
	7.4	8.6	9.1	9.2	n
K	187.1	162.9	157.0	147.1	$-U_l$ (calc.)
	189.8	165.8	158.5	149.9	$-U_l$ (exp.)
	8.1	9.5	9.5	10.1	n
Rb	181.0	158.8	151.5	143.8	$-U_l$ (calc.)
	181.4	159.3	152.6	144.9	$-U_l$ (exp.)
	10.2	10.6	10.5	11.1	n
Cs	176.4	152.0	146.1	138.6	$-U_l$ (calc.)
	172.5	155.4	149.4	142.4	$-U_l$ (exp.)

The high values of $n > m$ cause repulsion mainly between the nearest-neighbour ions (in spite of their opposite charges!). This repulsion drops rapidly as the ion separation increases.

Compressibility and repulsion exponent are inversely proportional to one another. In many cases the high value of n supports the assumption that the ions are rigid, hard spheres ($n \to \infty$) which are arranged in the lattice in such a way that the nearest neighbours touch each other. The lattice constant is then determined by the ion radii (cf. Table 7).

TABLE 7. Ion radii and lattice constants of alkali halides with NaCl structure. The lattice constant of AB is almost equal to twice the sum of the ion radii of A^+ and B^- [after C. Kittel, *Introduction to Solid State Physics*, 2nd ed. (Wiley, New York 1956), pp. 80, 82]

		F^-	Cl^-	Br^-	I^-
	Ion radii [Å]	1.36	1.81	1.95	2.16
		Lattice constants [Å]			
Li^+	0.68	4.02	5.13	4.49	6.00
Na^+	0.98	4.62	5.63	5.96	6.46
K^+	1.33	5.33	6.28	6.59	7.05
Rb^+	1.47	5.63	6.58	6.85	7.34

The repulsion of ions at small separations is based on the fact that the charge distribution inside the ionic sphere becomes important and neighbouring electron clouds interact with one another. The repulsive force is, in principle, due to the Pauli principle; quantum-mechanical methods permit a calculation of the repulsion without making use of empirical quantities.

Knowing n we can calculate the lattice energy U_l of an ion crystal; the main part ($\approx 90\%$) of this energy consists of the electrostatic potential of attraction [cf. eqn. (C. 27) and Table 6].

(b) *THE BORN–HABER CYCLIC PROCESS*

It is in general not possible to determine the lattice energy by direct measurements since, when energy is supplied to the ion crystal, it disintegrates into neutral molecules or atoms and not into mutually independent ions. The lattice energy, however, is defined as the energy necessary to decompose a crystal (e.g. NaI) into its structural elements (Na^+ and I^- ions). This energy can be determined indirectly from thermochemical data, with the help of the Born-

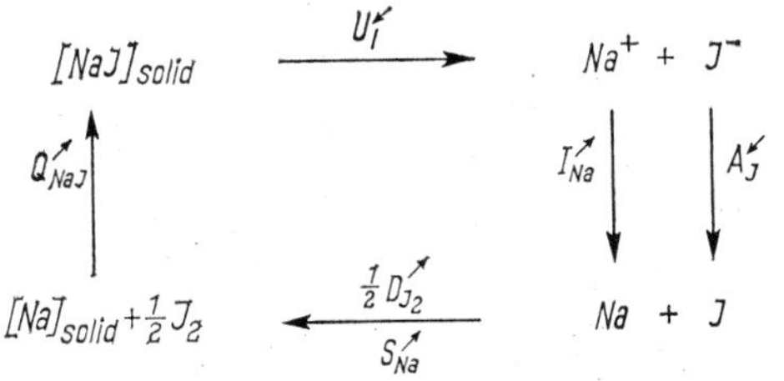

FIG. 28. The Born–Haber cycle.

Haber cyclic process. This makes it possible to check the value of U_{l0} in eqn. (C. 27). Considering the case of the NaI crystal as an example, we shall describe the individual partial processes of such a cyclic process (cf. Fig. 28). The algebraic sum of all energies involved in this process is equal to zero.

1. A disintegration of the NaI crystal into its structural elements, the Na^+ and I^- ions, requires the lattice energy $U_{l0}^{\swarrow}$.
2. Na^+ passes over to the gaseous state, giving off the ionization energy $I_{Na}^{\nearrow}$, I^- takes the form of atomic I by absorption of the energy $A_I^{\swarrow}$ which is equal to the electron affinity of I.
3. The gaseous Na sublimates under release of the sublimation heat $S_{Na}^{\nearrow}$, forming solid, metallic Na; atomic I passes over to its molecular form, $\frac{1}{2}I_2$, under release of the dissociation energy $\frac{1}{2}D_{I_2}^{\nearrow}$.
4. The solid Na and the gaseous $\frac{1}{2}I_2$ join to form crystalline NaI under release of the production heat $Q_{NaI}^{\nearrow}$.

From these energies the lattice energy can be calculated:

$$U_{l0} = Q_{NaI} + S_{Na} + \tfrac{1}{2} D_{I_2} + I_{Na} - A_I\,. \tag{C. 36}$$

The quantities on the right-hand side of eqn. (C. 36) are known in general, the lattice energies calculated from them agree with those obtained from eqn. (C. 27) and also with those calculated from quantum-theoretical considerations.

2. THE ENERGY OF THE LATTICE VIBRATIONS

Let us consider a crystal containing N lattice particles which vibrate in three directions about their equilibrium positions (lattice sites). In order to calculate the energy of these vibrations, we consider the crystal to be a system of $3N$ linear harmonic oscillators. The total energy of lattice vibrations is obtained by adding the energy contributions of the linear oscillators:

$$U_v = \sum_{i=1}^{3N} \bar{u}(\omega_i, T). \tag{C. 37}$$

The mean energy $\bar{u}(\omega_i, T)$ of an oscillator of the frequency ω_i at the temperature T amounts to

$$\bar{u}(\omega_i, T) = \frac{\hbar\omega_i}{2} + \frac{\hbar\omega_i}{\exp\left(\frac{\hbar\omega_i}{kT}\right) - 1}, \tag{C. 38}$$

where $\hbar = h/2\pi$ is Dirac's constant and k is the Boltzmann constant.

It is now necessary to determine all possible values of ω_i, the so-called eigenfrequency spectrum.

(a) *EINSTEIN'S THEORY*

The simplest model for the vibration spectrum is due to Einstein: the spectrum is assumed to consist of the single frequency ω_E, and the lattice vibration energy according to eqn. (C. 37) is obtained as

$$U_v = U_v(V, T) = 3N\left[\frac{\hbar\omega_E}{2} + \frac{\hbar\omega_E}{\exp\left(\frac{\hbar\omega_E}{kT}\right) - 1}\right]. \tag{C. 39}$$

U_v refers to a crystal of volume V, which contains N lattice particles. By means of $U_v(V, T)$ the specific heat of the lattice vibrations at constant volume of the crystal can be determined:

$$c_V \equiv \left(\frac{\partial U}{\partial T}\right)_V = \left(\frac{\partial U_v}{\partial T}\right)_V. \tag{C. 40}$$

With Einstein's assumption of a single eigenfrequency we have

$$c_V = 3Nk \frac{\left(\frac{\hbar\omega_E}{kT}\right)^2 \exp\left(\frac{\hbar\omega_E}{kT}\right)}{\left[\exp\left(\frac{\hbar\omega_E}{kT}\right) - 1\right]^2}. \tag{C. 41}$$

In the case of $\hbar\omega_E \ll kT$ the exponential functions of eqn. (C. 41) can be expanded and the Dulong–Petit law is obtained:

$$c_V = 3Nk. \tag{C. 42}$$

Thus, in the case of sufficiently high temperatures the mean energy of a linear oscillator is equal to the equipartition value kT, and the contribution to the specific heat is equal to k.

At lower temperatures the classical equipartition value of the mean energy must be replaced by the function $\bar{u}(\omega_E, T)$ [cf. eqn. (C. 38)] which is obtained from quantum-theoretical considerations. If $\hbar\omega_E \gg kT$ we obtain from eqn. (C. 41)

$$c_V = 3Nk\left(\frac{\hbar\omega_E}{kT}\right)^2 \exp\left(-\frac{\hbar\omega_E}{kT}\right). \tag{C. 43}$$

As $T \to 0$, also $c_V \to 0$; this is in agreement with the Third Law of thermodynamics and also agrees qualitatively with experimental experience. As regards the quantitative agreement, however, this is poor in the case of $\hbar\omega_E \gg kT$; the values of c_V obtained experimentally (at low temperatures $c_p \approx c_V$, cf. p. 44) do not satisfy the exponential law of eqn. (C. 43).

(b) *DEBYE'S THEORY*

In spite of the simple assumptions it is based upon, the specific heat theory by Debye is in relatively good agreement with experiments. According to Debye the eigenfrequency spectrum is calculated on the basis of the phenomenological theory of elasticity of continuous media. The discontinuous model of the crystal is replaced by a continuum, i.e. by an isotropic, homogeneous, elastic medium which is subdivided into cubic cells of volume L^3. The deviation from the

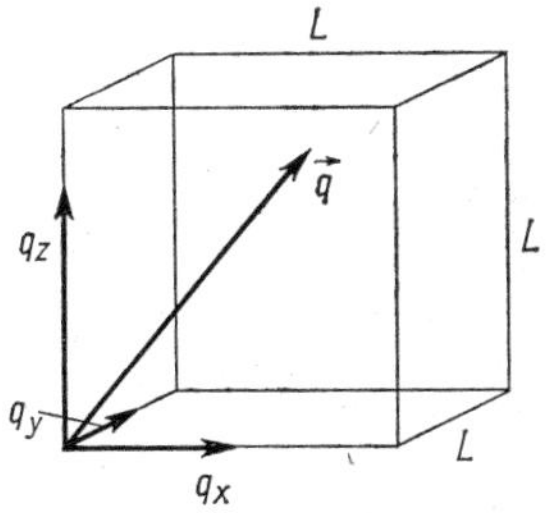

FIG. 29. Wave vector $\boldsymbol{q}$ in a cubic elementary cell of volume $V = L^3$.

mean density of the medium is denoted by $S(x, y, z, t)$. The problem is reduced to a determination of the possible modes and eigenfrequencies in the cell L^3 (cf. Fig. 29). This eigenvalue problem requires a solution to the wave equation

$$\frac{\partial^2 S}{\partial t^2} = c^2 \nabla^2 S \tag{C. 44}$$

with suitable boundary conditions; c is the velocity of propagation of the perturbation S, in the present case it is the sound velocity in the medium considered. As the medium was assumed to be isotropic, it is independent of the direction. The wave equation (C. 44) is satisfied for plane waves

$$S(x, y, z, t) = \psi(x, y, z)e^{i\omega t} \tag{C. 45}$$

where

$$\psi(x, y, z) = Ae^{i(q_x x+q_y y+q_z z)} \tag{C. 46}$$

and

$$c = \frac{\omega}{q}; \tag{C. 47}$$

$\boldsymbol{q}$ and q are the wave vector and the wave number. We shall only consider stationary solutions and restrict ourselves to a discussion of the positional part $\psi(x, y, z)$.

The boundary conditions are obtained from the requirement for homogeneity of the crystal, i.e. all cells of volume L^3 are assumed to possess identical properties. The periodicity of the crystal can then be expressed by the so-called periodic boundary conditions

$$\psi(x+L, y, z) = \psi(x, y+L, z) = \psi(x, y, z+L) = \psi(x, y, z). \tag{C. 48}$$

With eqn. (C. 46) we hence obtain a set of possible q values

$$q_x = n_x\frac{2\pi}{L}, \quad q_y = n_y\frac{2\pi}{L}, \quad q_z = n_z\frac{2\pi}{L}, \tag{C. 49}$$

$$q^2 = q_x^2+q_y^2+q_z^2 = \left(\frac{2\pi}{L}\right)^2 n^2, \tag{C. 50}$$

where n_x, n_y, n_z are integers. By virtue of the periodic boundary conditions we thus obtain a discrete sequence of q values:

$$q = 0, \quad \pm\frac{2\pi}{L}1, \quad \pm\frac{2\pi}{L}\sqrt{2}, \quad \pm\frac{2\pi}{L}\sqrt{3}, \ldots$$

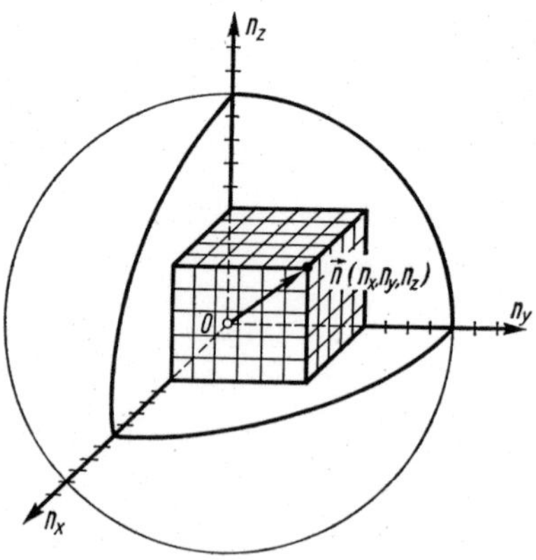

FIG. 30. Number space.

In a number space with the axes n_x, n_y, n_z there corresponds to each q value a representative point with integral coordinates n_x, n_y, n_z and a volume of unity (cf. Fig. 30). The number Z_q^0 of the q values between $q = 0$ and $q = q(n)$ is then given by the spherical volume (radius n) in the number space:

$$Z_q^0 = \frac{4\pi}{3} n^3 = \frac{L^3}{6\pi^2} q^3 . \qquad \text{(C. 51)}$$

The eigenvalue density, i.e. the number of q values between q and $q+dq$, will then be given by

$$\frac{dZ_q}{dq} = \frac{L^3}{2\pi^2} q^2 . \qquad \text{(C. 52)}$$

Corresponding to the three directions of oscillations, three modes are attributed to each wave number q; a longitudinal wave with the propagation velocity c_l and two transverse waves with the propagation velocity c_t. Using eqns. (C. 47) and (C. 50), we obtain the number of possible eigenfrequencies between 0 and ω:

$$Z_\omega^0 = \frac{L^3}{6\pi^2} \omega^3 \left(\frac{1}{c_l^3} + \frac{2}{c_t^3} \right); \qquad \text{(C. 53)}$$

from this we can derive the frequency density $\varrho(\omega)$, i.e. the number of possible frequencies between ω and $\omega + d\omega$:

$$\frac{dZ_\omega}{d\omega} = \varrho(\omega) = \frac{L^3}{2\pi^2} \omega^2 \left(\frac{1}{c_l^3} + \frac{2}{c_t^3} \right). \qquad \text{(C. 54)}$$

When we define a mean propagation velocity $\bar{c}$ of the elastic perturbation according to

$$\frac{3}{\bar{c}^3} = \frac{1}{c_l^3} + \frac{2}{c_t^3}, \qquad \text{(C. 55)}$$

we obtain for the frequency density

$$\varrho(\omega) = \frac{3}{2\pi^2} \frac{L^3}{\bar{c}^3} \omega^2 . \qquad \text{(C. 56)}$$

From this we see that the eigenfrequency spectrum (cf. Fig. 31) becomes denser and denser as ω increases, and as $\omega \to \infty$ it should become continuous: $\varrho(\omega \to \infty) = \infty$. It is, however, justified to assume an elastic continuum only in the case of wavelengths which are longer than the distance between the lattice particles, since the crystal actually consists of N lattice particles with $3N$ degrees of freedom. The total number of eigenoscillations of such a system is

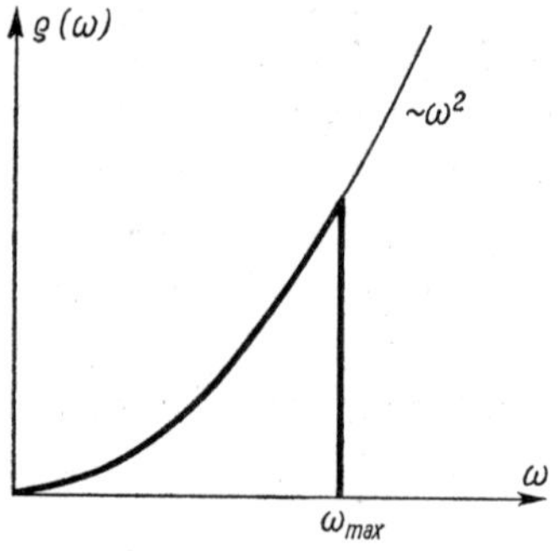

FIG. 31. Frequency density $\varrho(\omega)$ according to the Debye model.

equal to the number of degrees of freedom, i.e. $3N$. Instead of the infinitely many eigenoscillations of the continuum, Debye only takes into account the $3N$ lowest eigenfrequencies. Thus the spectrum is cut off at a certain maximum frequency ω_{max}, which is given by the following condition:

$$3N = \int_0^{\omega_{max}} dZ = \int_0^{\omega_{max}} \varrho(\omega)\, d\omega = \frac{1}{2\pi^2} \frac{L^3}{\bar{c}^3} \omega_{max}^3 . \tag{C. 57}$$

We can now rewrite eqn. (C. 56):

$$\varrho(\omega) = \frac{9N}{\omega_{max}^3} \omega^2 \quad \text{(valid when} \quad \omega \leqslant \omega_{max}). \tag{C. 58}$$

Knowing the number $\varrho(\omega)\, d\omega$ of eigenfrequencies in the frequency range ω to $\omega + d\omega$, we can determine the vibrational energy U_v. Each eigenoscillation has the energy of an oscillator of frequency ω. Summation over all oscillator energies is replaced by integration, because of the high frequency density. So we have instead of (C. 37)

$$U_v = \int_0^{\omega_{max}} \bar{u}(\omega)\, \varrho(\omega)\, d\omega \tag{C. 59}$$

and, after substituting (C. 38) and (C. 58),

$$U_v = \frac{9N}{\omega_{max}^3} \int_0^{\omega_{max}} \frac{\hbar\omega^3}{\exp\left(\frac{\hbar\omega}{kT}\right) - 1}\, d\omega + U_0 \tag{C. 60}$$

where

$$U_0 = \frac{9N\hbar}{2\omega_{max}^3} \int_0^{\omega_{max}} \omega^3 \, d\omega = \frac{9}{8} N\hbar\omega_{max} . \tag{C. 61}$$

The zero point energy U_0 is independent of temperature. Using the substitutions

$$\Theta = \frac{\hbar\omega_{max}}{k} \tag{C. 62}$$

and

$$x = \frac{\hbar\omega}{kT} \tag{C. 63}$$

eqn. (C. 60) becomes

$$\begin{aligned} U_v &= 9NkT\left(\frac{T}{\Theta}\right)^3 \int_0^{\Theta/T} \frac{x^3}{e^x - 1} dx + U_0 \\ &= 9NkT\left(\frac{T}{\Theta}\right)^3 D\left(\frac{\Theta}{T}\right) + U_0 . \end{aligned} \tag{C. 64}$$

The quantity $\Theta = \hbar\omega_{max}/k$ is called the characteristic temperature or Debye temperature. The function

$$D\left(\frac{\Theta}{T}\right) = \int_0^{\Theta/T} \frac{x^3}{e^x - 1} dx \tag{C. 65}$$

is called the Debye function; it has been tabulated.

Let us now calculate the vibrational energy and from it the specific heat of lattice vibrations for the limiting cases of high temperatures $T \gg \Theta$ and low temperatures $T \ll \Theta$ (cf. Fig. 32).

1. In the case of sufficiently high temperatures $T \gg \Theta$ the ratio $\hbar\omega/kT = x \ll 1$, and in the Debye function the denominator of the integrand can be expanded so that the vibrational energy is obtained as

$$\begin{aligned} U_v &\approx 9NkT\left(\frac{T}{\Theta}\right)^3 \int_0^{\Theta/T} \frac{x^3 \, dx}{1 + x + \frac{x^2}{2} + \ \dots \ - 1} + U_0 \\ &= 9NkT\left(\frac{T}{\Theta}\right)^3 \left[\frac{1}{3}\left(\frac{\Theta}{T}\right)^3 - \frac{1}{8}\left(\frac{\Theta}{T}\right)^4\right] + U_0 . \end{aligned} \tag{C. 66}$$

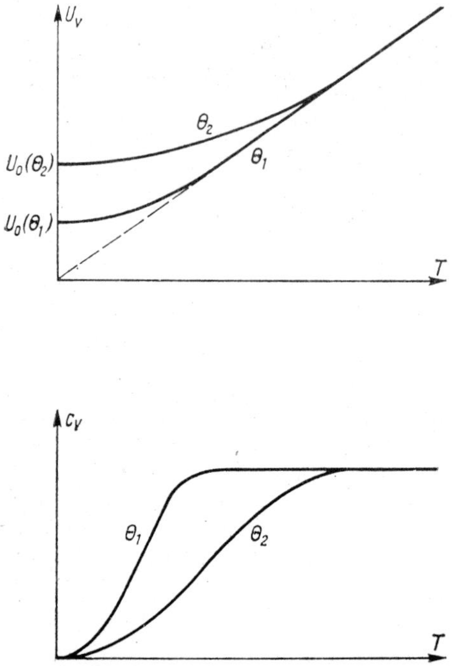

FIG. 32. Vibrational energy U_v and specific heat c_V as functions of temperature T for two characteristic temperatures Θ_1 and Θ_2 (for $\Theta_2 > \Theta_1$).

When we take (C. 61) into account as well as the definition of Θ, we obtain

$$U_v \approx 3NkT. \tag{C. 67}$$

Hence, with (C. 40), we obtain for the specific heat at high temperatures

$$c_V \approx 3Nk, \tag{C. 68}$$

i.e. we have again the Dulong–Petit value which was also derived for the Einstein model (cf. p. 54).

2. In the case of sufficiently low temperatures, $T \ll \Theta$, we have $\hbar\omega/kT = x \gg 1$. Since for high x values the integrand in the Debye function becomes very small, the upper limit of integration $\Theta/T \gg 1$ can be replaced by ∞. In this approximation the Debye function is equal to a numerical value and thus independent of Θ/T:

$$D\left(\frac{\Theta}{T}\right) \approx \int_0^\infty \frac{x^3}{e^x - 1}\, dx = \frac{\pi^4}{15}. \tag{C. 69}$$

Hence we obtain for the vibrational energy

$$U_v \approx \frac{9}{15}\pi^4 NkT\left(\frac{T}{\Theta}\right)^3 + U_0 \qquad \text{(C. 70)}$$

and for the specific heat at low temperatures

$$c_V \approx \frac{12}{5}\pi^4 Nk\left(\frac{T}{\Theta}\right)^3 \propto T^3 . \qquad \text{(C. 71)}$$

According to Debye's theory the temperature dependence of the specific heat can be represented by a universal curve which is common for all substances; this curve is obtained when the specific heat divided by the number N of lattice particles is plotted versus the reduced temperature T/Θ. When we compare, for example, the molar specific heats of Na and NaCl, we have for a given temperature

$$\frac{c_V}{N_A}(\text{Na}) = \frac{c_V}{2N_A}(\text{NaCl}), \qquad \text{(C. 72)}$$

as in one mole Na there are N_A lattice particles while in one mole NaCl there are $2N_A$ lattice particles.

(c) *COMPARISON OF DEBYE'S THEORY WITH EXPERIMENTS*

According to present-day information the T^3 law for low temperatures [eqn. (C. 71)] generally applies to temperatures $T < \Theta/12$. The values measured for

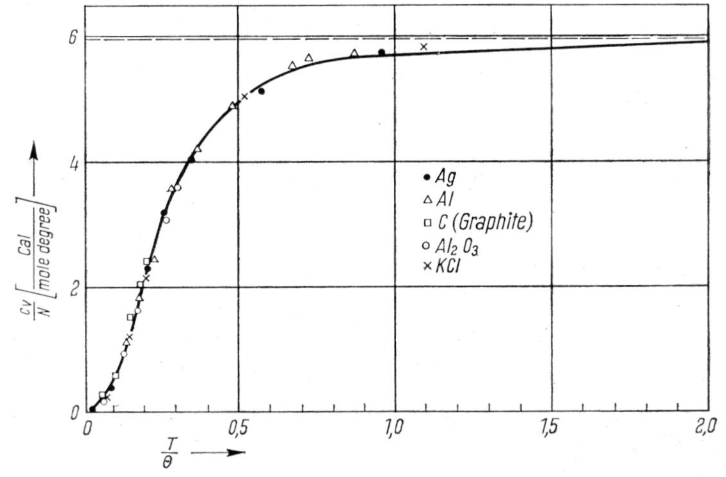

FIG. 33. Specific heat of various substances per number of lattice particles per mole as a function of the dimensionless temperature (according to F. Seitz, *Modern Theory of Solids*, McGraw-Hill, New York 1940).

a great number of substances (elements and compounds) lie on the universal curve c_V/N versus T/Θ, as is to be expected from theory (cf. Fig. 33). Figure 34 shows several curves c_V vs. T for elements with strongly different Debye temperatures. It should be mentioned that the Dulong–Petit saturation value is reached

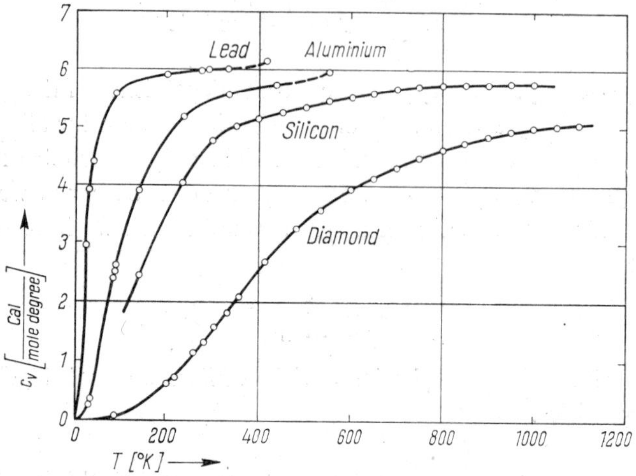

FIG. 34. Specific heat c_V versus temperature T for various elements (after F. K. Richtmyer and E. H. Kennard, *Introduction to Modern Physics*, McGraw-Hill, New York 1947).

at temperatures which are the lower, the lower the Debye temperature (cf. Fig. 32 and Table 8).

The values of the Debye temperature Θ_{th} calculated from measurements of the specific heats at temperature T are compiled in Table 8. Independent of Θ_{th} the Debye temperature Θ_{el} can be obtained from elasticity data. Using the relation $\Theta = \hbar\omega_{max}/k$, we can rewrite eqn. (C. 57) in the form

$$\Theta = (6\pi^2 n)^{1/3} \frac{\hbar}{k} \bar{c}, \tag{C. 73}$$

where $n = N/L^3$ is the number of lattice particles per unit volume.

The mean sound velocity $\bar{c}$ depends on the elastic constants. For an isotropic medium we have

$$\bar{c} = \left(\frac{E}{\varrho}\right)^{1/2}, \tag{C. 74}$$

TABLE 8. Debye temperatures of several elements and compounds at $T \approx \Theta/2$ (after E. S. R. Gopal, *Specific Heats at Low Temperatures*, Heywood, London 1966, p. 33)

Element	Θ [°K]	Element	Θ [°K]	Element	Θ [°K]	Element	Θ [°K]
Ac	100	Cu	310	La	130	Re	300
Ag	220	Dy	155	Li	420	Rh	350
Al	385	Er	165	Mg	330	Sb	140
Ar	90	Fe	460	Mn	420	Se	150
As	275	Ga (orth.)	240	Mo	375	Si	630
Au	180	Ga (tetrag.)	125	N	70	Sn (fcc)	240
B	1220	Gd	160	Na	150	Sn (tetrag.)	140
Be	940	Ge	370	Nb	265	Sr	170
Bi	120	H (para)	115	Nd	150	Ta	230
C (diamond)	2050	H (ortho)	105	Ne	60	Tb	175
C (graphite)	760	H (n–D_2)	95	Ni	440	Te	130
Ca	230	He	30	O	90	Th	140
Cd (hcp)	280	Hf	195	Os	250	Ti	355
Cd (bcc)	170	Hg	100	Pa	150	Tl	90
Ce	110	I	105	Pb	85	V	280
Cl	115	In	140	Pd	275	W	315
Co	440	Ir	290	Pr	120	Y	230
Cr	430	K	100	Pt	225	Zn	250
Cs	45	Kr	60	Rb	60	Zr	240

Compound	Θ [°K]	Compound	Θ [°K]	Compound	Θ [°K]	Compound	Θ [°K]
AgBr	140	$CrCl_3$	100	KCl	230	RbI	115
AgCl	180	Cr_2O_3	360	KI	195	SiO_2	
As_2O_3	140	Cu_3Au		LiF	680	(quartz)	255
As_2O_5	240	ordered	200	MgO	800	TiO_2	
BN	600	disordered	180	MoS_2	290	(rutile)	450
CaF_2	470	FeS_2 (cub.)	630	NaCl	280	ZnS (cub.)	260
$CrCl_2$	80	KBr	180	RbBr	130		

where E is the elastic modulus and ϱ is the density. Thus the Debye temperature is proportional to the root of the elastic modulus, i.e. for isotropic media

$$\Theta_{el} = (6\pi^2 n)^{1/3} \frac{\hbar}{k} \left(\frac{E}{\varrho}\right)^{1/2}. \qquad \text{(C. 75)}$$

The agreement between the values of Θ_{th} and Θ_{el} is generally good (cf. Table 9).

TABLE 9. Debye temperatures Θ_{th} and Θ_{el} at $T = 0°K$, obtained from measurements of the specific heat or the elastic modulus (after K. A. Gschneidner, Jr. *Solid State Physics* **16**, 275 (1964), Tab. 16)

Element	Θ_{th} [°K]	Θ_{el} [°K]
Cu	342	345
Ag	228	227
Au	165	162
Be	1160	1462
Mg	396	387
Zn	316	324
Cd	252	212
Al	423	428
In	109	111
C (diamond)	2240	2240
Si	647	649
Ge	378	375
Sn (white)	236	202
Pb	102	105
V	326	399
Ta	247	262
Mo	459	474
W	388	384
Fe	457	477
Ni	427	476
Pd	283	275

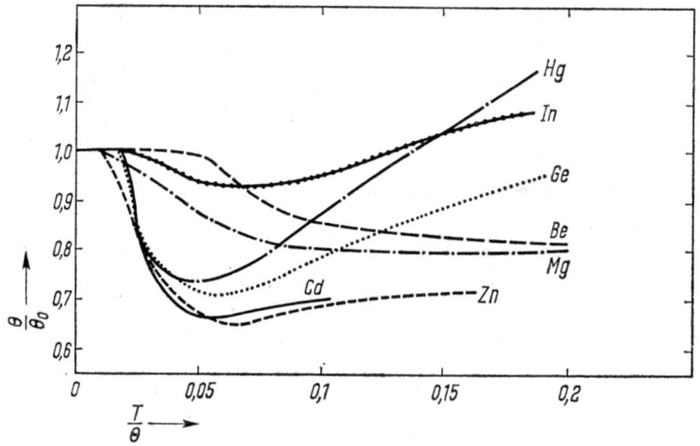

FIG. 35. Temperature dependence of Debye temperatures of various elements (from D. H. Parkinson, *Rep. Prog. Phys.* **21**, 226 (1958)).

If Debye's theory were strictly correct, the Θ_{th} value determined from c_V should be temperature-independent by definition. Deviations from the theory are mainly explained phenomenologically, assuming a temperature dependence of the Debye temperature $\Theta_{th}(T)$: the values of Θ_{th} determined from the measured values of $c_V(T)$ are indeed temperature dependent (cf. Fig. 35). The difference between the curves $\Theta_{th}(T)$ and Θ_{th} = const rarely exceeds 20% and is in many cases smaller than 10%.

As a whole Debye's theory is in satisfactory agreement with experiment. The residual discrepancies between theory and experiment are easily understood when the frequency spectrum assumed by Debye, $\varrho(\omega) \propto \omega^2$ [cf. eqn. (C. 58)], is compared with calculated or measured frequency spectra (cf. Figs. 53 and 56) which are not even approximately described by (C. 58). The physical cause of the $\Theta_{th}(T)$ dependence is not the temperature dependence of the propagation velocity $\bar{c}$, but the fact that the frequency density deviates strongly from $\varrho(\omega) \propto \omega^2$. With the help of known frequency spectra typical $\Theta_{th}(T)$ curves can be calculated within the framework of Debye's theory which agree with the curves obtained experimentally.

The general theory of lattice vibrations does not yield a simple relation between specific heat and temperature. It is therefore expedient and usual to maintain the Debye formalism and to give the temperature dependence of the Debye temperature, $\Theta_{th}(T)$.

(d) *PHONONS*

In both the Einstein and Debye theories as well as generally in lattice dynamics the energy of a state of vibration of frequency ω is set equal to the mean energy $\bar{u}(\omega)$ of a linear oscillator.

According to the possible values of oscillator energy,

$$u(\omega) = (n+\tfrac{1}{2})\hbar\omega, \tag{C. 76}$$

the vibrational energy of each state is quantized. By analogy with photons, the energy quanta of the electromagnetic field, the energy quanta of the lattice vibrations are called phonons. As they have zero spin, they obey Bose–Einstein statistics. The number $N(\omega, T)$ of phonons which have energy $\hbar\omega$ and are excited at temperature T is, if the phonon gas is in thermal equilibrium, given by

$$N(\omega, T) = \frac{1}{\exp\left(\dfrac{\hbar\omega}{kT}\right) - 1}. \tag{C. 77}$$

The vibrational energy U_v in a solid of temperature T is obtained by integrating over the energies of all excited phonons, i.e.

$$U_v = \int_0^{\omega_{\max}} N(\omega, T)\, \hbar\omega\, \varrho(\omega)\, d\omega. \tag{C. 78}$$

Except for the zero-point energy this relation is identical with (C. 59).

The concept of the phonons is expedient in descriptions of interactions between the lattice vibrations and excitations by electric or magnetic fields or radiation. Such interactions, such as phonon–phonon, photon–phonon, electron–phonon, and neutron–phonon interactions, are regarded as collision processes. For such collision processes it is a fundamental requirement that energy and momentum are conserved. The conservation condition may be referred to the colliding particles or to the whole crystal as a "rigid body" to which momentum can also be transferred (cf. p. 69).

Whereas photons propagate also in empty space (at the velocity of light, c), phonons can propagate only in a medium with the appropriate sound velocity c_s.

By analogy with the photon, the energy of a phonon is

$$u_P = \hbar\omega. \tag{C. 79}$$

Unlike the photon, however, the phonon has no momentum as it merely represents a mechanical state of vibration with zero average matter transport. None the less we *formally* define a phonon momentum

$$\boldsymbol{p}_P = \hbar\boldsymbol{q}, \tag{C. 80}$$

whose amount is given by the de Broglie relation

$$p_P = \frac{h}{\lambda} = \hbar q, \tag{C. 81}$$

where $\boldsymbol{q}$, q are the wave vector and wave number of the phonon and c_s is the sound velocity in the crystal.

In the framework of Debye's theory this momentum and the energy of a phonon are related by

$$u_P = \hbar\omega = p_P c_s = \hbar q c_s = \frac{h}{\lambda} c_s, \tag{C. 82}$$

where c_s is the sound velocity in the crystal.

This formal definition of a momentum proves reasonable as in "normal" collisions with phonons the momentum conservation law is satisfied when the phonons are assumed to possess a momentum $\hbar\boldsymbol{q}$. Strictly speaking, this momentum $\hbar \boldsymbol{q}$ is a crystal momentum (cf. p. 218) which is connected with the generation or annihilation of a phonon and which results in a motion of the crystal's centre of mass. For example, the scattering of a photon from a lattice atom can result in the generation (emission) or the annihilation (absorption) of a phonon (cf. Fig. 36). In the case of "normal" scattering processes the conservation laws of energy and momentum are valid:

$$u = u_0 \pm \hbar\omega \tag{C. 83}$$

and

$$\hbar \boldsymbol{k} = \hbar \boldsymbol{k}_0 \pm \hbar \boldsymbol{q}; \tag{C. 84}$$

u_0, u are the phonon energies before and after collision, $\boldsymbol{k}_0$, $\boldsymbol{k}$ are the corresponding wave vectors; the signs + and − refer to phonon absorption and emission, respectively.

Apart from the "normal" processes (N processes) there exist the so-called umklapp processes (U processes), which are due to the periodicity of the lattice. They occur if in the reciprocal lattice the end point of the vector $\boldsymbol{k}_0 \pm \boldsymbol{q}$ lies outside the first Brillouin zone (cf. p. 39). Since the wave vectors are defined only to within a lattice vector $\boldsymbol{h}$ of the reciprocal lattice (cf. p. 210), the wave vector reduced to the first Brillouin zone will after the collision be equal to

$$\boldsymbol{k} = \boldsymbol{k}_0 \pm \boldsymbol{q} - \boldsymbol{h}. \tag{C. 85}$$

A multiplication of this relation by $\hbar$ shows that the momentum law applies not only to the collision partners but also to the crystal as a whole. Physically, an umklapp process means the absorption or emission of a phonon with simultaneous Bragg reflection of the other collision partner (e.g. electron, photon, or neutron). The term "umklapp process" is due to the fact that the direction of the *reduced* vector $\boldsymbol{k}_0 \pm \boldsymbol{q} - \boldsymbol{h}$ is almost inverse to the direction of the corresponding vector $\boldsymbol{K}_0 \pm \boldsymbol{q}$ which has not been reduced to the first Brillouin zone (cf. Fig. 36). In particular at low temperatures the umklapp processes are essential for the establishment of thermal equilibrium in the crystal.

In the following we shall consider examples of photon–phonon interactions, such as the influence of lattice vibrations on X-ray interferences, that is, the thermal scattering of X-rays.

3. RECOIL-LESS EMISSION AND ABSORPTION

The geometrical conditions for X-ray interferences (cf. pp. 16 ff.) have been derived under the assumption of strict periodicity of the crystal structure which, disregarding the zero point motion, would be satisfied only at absolute zero.

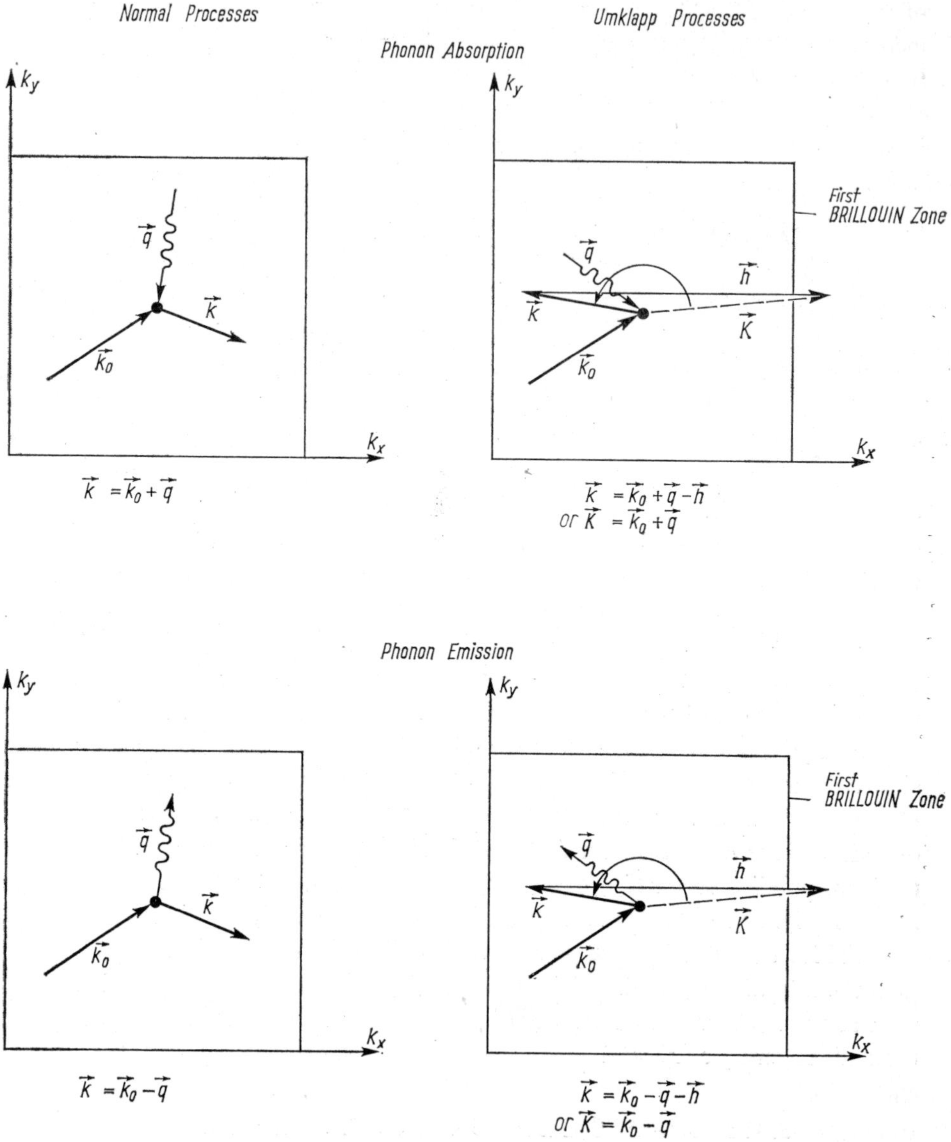

FIG. 36. Phonon absorption and phonon emission in the case of normal and umklapp processes.

The intensity of the interference spots, however, depends on the temperature of the crystal: the intensity of a reflection spot decreases as the temperature increases (cf. Fig. 37). Moreover, X-rays also undergo diffuse scattering in directions for which the Bragg condition is usually not satisfied. These effects are chiefly due to the lattice vibrations.

In order to treat the thermal scattering of X-rays we start from the laws of momentum and energy conservation for the collision process considered. In a

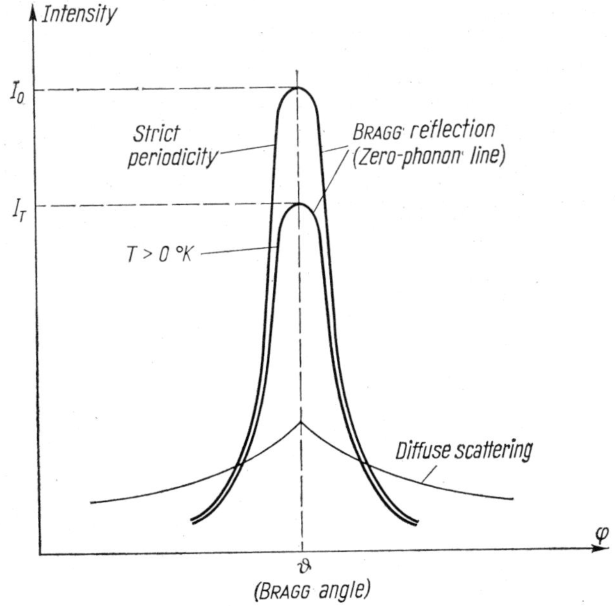

FIG. 37. Thermal scattering of X-rays. The scattered intensity is plotted as function of the observation angle and is given for a Bragg reflection from an exactly periodic lattice and from a lattice with lattice vibrations ($T > 0$°K), respectively. The background of the diffuse scattering processes is also indicated.

photon reflection the crystal of mass M absorbs momentum $\boldsymbol{p}$ which is given by

$$\hbar \boldsymbol{k}_0 = \hbar \boldsymbol{k} + \boldsymbol{p}. \tag{C. 86}$$

This equation means momentum conservation in the case of an unchanged state of eigenvibrations of the crystal.

It is easy to see that the energy transfer from the photon to the crystal of mass M will be negligibly small. The recoil energy E_r is given by

$$E_r = \frac{p^2}{2M}. \tag{C. 87}$$

The momentum transferred from a photon of energy u_X to the crystal can be at most equal to twice the photon momentum. Using de Broglie's relation, we thus obtain for the maximum recoil energy

$$E_r = 2\frac{u_X^2}{Mc^2}. \tag{C. 88}$$

For example, when an X-ray quantum of 10^4 eV hits a crystal of mass 1 g, the ratio of recoil energy to photon energy is of the order of 10^{-29}.

Thus in all practical cases the recoil energy will be insignificant and the X-rays are reflected without recoil. The photon energy remains virtually unchanged, i.e. the wavelength λ and the magnitudes of the wave vectors $\boldsymbol{k}_0$ and $\boldsymbol{k}$ (before and after scattering) are the same. Scattering is therefore elastic. Ewald's construction and Bragg's condition of reflection are based on this assumption; they are therefore justified. In the case of Bragg reflection scattering is elastic and coherent. When a photon incident at the Bragg angle ϑ is reflected, the crystal obtains the momentum

$$\boldsymbol{p} = \hbar\boldsymbol{h} \tag{C. 89}$$

where $\boldsymbol{h}$ is the corresponding lattice vector of the reciprocal lattice [cf. eqn. (B. 9) and Fig. 38].

Scattering processes which yield no interference spots but a diffuse background are called diffuse scattering. Such processes are incoherent and in most cases also inelastic. As the temperature rises lattice vibrations are excited in the crystal which influence the X-ray scattering in two different ways. On the one hand, photon–phonon interaction gives rise to inelastic diffuse scattering; on the other hand, the departure of the atom from their equilibrium position weakens the intensity of the Bragg reflections.

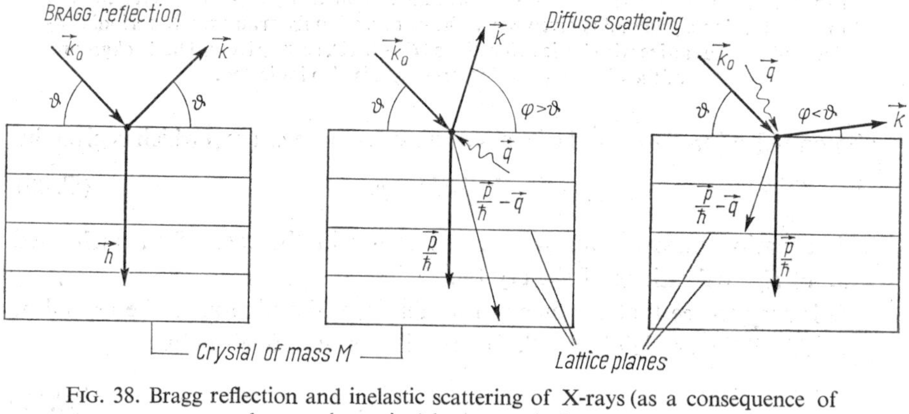

FIG. 38. Bragg reflection and inelastic scattering of X-rays (as a consequence of phonon absorption) by a crystal of mass M.

Inelastic scattering can also be considered to be a collision process in which the X-ray quantum is scattered, with simultaneous emission or absorption of one or several phonons, at an angle which is different from Bragg's angle. Although this is in principle an inelastic scattering process, the photon energy after the collision is virtually unchanged, i.e. the absolute magnitudes of the wave vectors $\boldsymbol{k}_0$ and $\boldsymbol{k}$ are equal to a good approximation. The photon energy is given by

$$u_X = \hbar \frac{2\pi c}{\lambda_X}. \tag{C. 90}$$

Similarly, the phonon energy [eqn. (C. 79)] is equal to

$$u_P = \hbar \frac{2\pi c_s}{\lambda_P}. \tag{C. 91}$$

The wavelengths λ_X of the X-rays and λ_P of the phonons of highest energies are equal as to the order of magnitude (cf. pp. 28 and 90). The ratio of photon and phonon energies will then be approximately equal to the ratio of the velocities of light and sound, that is, $u_X/u_P \approx c/c_s \approx 10^5$. In the collision between X-ray photon and phonon the energy of the photon remains virtually unchanged while its momentum is changed. It is precisely because the wavelengths λ_X and λ_P are of the same order of magnitude that the momentum exchange between photon and phonon is essential. The photon momentum is

$$p_X = \hbar \frac{2\pi}{\lambda_X} \tag{C. 92}$$

and the momentum attributed to the phonon (C. 81) is given by

$$p_P = \hbar \frac{2\pi}{\lambda_P}. \tag{C. 93}$$

The appearance of inelastic diffuse scattering can be understood with the help of the conservation laws of energy and momentum (cf. Fig. 38). As the photon energy remains virtually unchanged, we have

$$|\boldsymbol{k}_0| \approx |\boldsymbol{k}|. \tag{C. 94}$$

At the same time the momentum conservation law must be satisfied which, for the so-called one-phonon processes, is given by eqn. (C. 84) or (C. 85). One-phonon processes are processes in which a single phonon is created or annihilated. For the many-phonon processes (which are less probable) we have instead of eqn. (C. 85)

$$\boldsymbol{k} = \boldsymbol{k}_0 + \sum_i \boldsymbol{q}_i - \frac{\boldsymbol{p}}{\hbar}, \tag{C. 95}$$

where $\boldsymbol{q}_i$ are the wave vectors of the phonons involved in the scattering process. In the case of inelastic coherent scattering $\boldsymbol{p}$ is given by eqn. (C. 89). It is an essential fact that, owing to the action of the phonons, X-rays of arbitrary angle of incidence can be scattered in directions which are given not by Bragg's condition but by the modified conservation laws [(C. 83) and (C. 84), (C. 85), (C. 95)].

At temperatures $T > 0°K$ the X-ray quanta are scattered not only diffusely, but also according to Bragg's condition, i.e., with no phonons involved in the collision process, $\boldsymbol{q}_i = 0$. The Bragg reflections are therefore called zero-phonon processes and accordingly the Bragg reflections are given the name zero-phonon lines (cf. Fig. 37). The intensity of these zero-phonon lines is a function of the temperature; it will be calculated in the next section.

(a) *THE DEBYE–WALLER FACTOR*

The intensities of the Bragg reflections decrease as the temperature of the crystal increases. This is illustrated in Fig. 39. It must be stressed that the line width of the reflections is independent of the temperature. Thus the Bragg

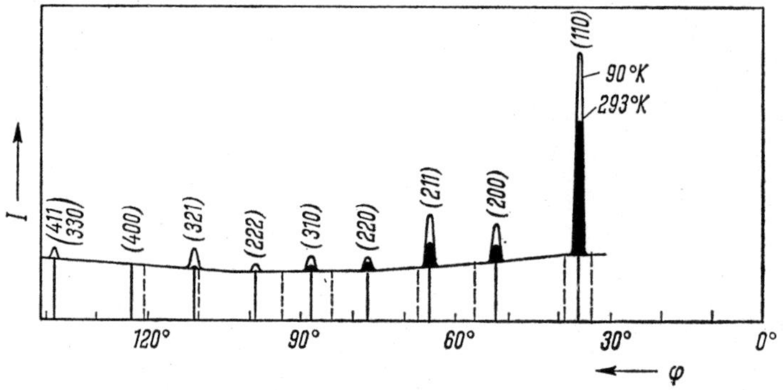

FIG. 39. Intensity of interference patterns after Debye–Scherrer for lithium at 90°K and 293°K, respectively [from G. Pankow, *Helv. Phys. Acta* **9**, 87 (1936)].

reflections remain sharp in spite of the lattice vibrations of the scattering atoms. The line width depends mainly on the nature of the crystal and the apparatus used.

The intensity decrease of the Bragg reflections with increasing temperature is described by the Debye–Waller factor

$$D(T) = \frac{I_T}{I_0}, \tag{C. 96}$$

where I_T is the scattered intensity for $T > 0°K$ and I_0 is the scattered intensity in the case of a strictly periodic crystal structure.

The Debye–Waller factor represents the probability that the X-ray quantum does not interact with phonons in the collision process (zero-phonon process), i.e., that it undergoes elastic and coherent scattering. The intensity of the diffuse (i.e., incoherent and inelastic) scattering (cf. p. 70) is compensated by the intensity drop of the Bragg reflections.

In the following we shall show that the Debye–Waller factor is due only to the displacement of the atoms from their equilibrium positions. Let us consider a simply primitive lattice; $\boldsymbol{\Delta r}_n$ denotes the displacement of the atoms of the nth unit cell from the position of rest. The displacements $\boldsymbol{\Delta r}_n$ are time-dependent and vary at about the phonon frequency which may reach an order of magnitude of 10^{13} sec^{-1}. Compared with the duration of the experiment these variations are rapid; they are slow, however, compared with the X-ray frequencies which lie in the range of 10^{19}–10^{20} sec^{-1}. X-ray scattering can therefore be assumed to take place by atoms "at rest" which are displaced from their equilibrium positions. The reflections observed represent mean values, averaged over large times, of diffractions from all possible configurations of atoms. We assume the same statistical laws to govern the displacements of each atom. At a given instant of time the totality of possible displacements of all N atoms of the crystal is equal to the totality of all possible displacements of a single atom at N different instants of time. This holds under the supposition that these instants of time are separated by instants which are large compared with the lattice vibration period. Under these assumptions the two means are equal, a fact which will be used in the following calculation of the Debye–Waller factor.

We start from the total scattering amplitude of X-rays which has been given on p. 35 for a crystal with a strictly periodic structure. For a simple primitive lattice (one of our initial assumptions) and for strict periodicity, the structure factor is identical with the scattering factor ψ of the atoms of the crystal considered; the total scattering amplitude [eqn. (B. 45)] is then given by

$$A_{\text{tot}} = A_0 \psi \sum_n e^{i(\boldsymbol{h} \cdot \boldsymbol{r}_n)} . \tag{C. 97}$$

Because of the lattice vibrations the atoms are displaced by $\boldsymbol{\Delta r}_n$ from their equilibrium positions and the lattice vectors must be replaced by (cf. Fig. 40)

$$\boldsymbol{r}_n(T) = \boldsymbol{r}_n + \boldsymbol{\Delta r}_n . \tag{C. 98}$$

The structure factor F_n of the nth unit cell is then different from the scattering factor ψ, i.e.

$$F_n = \psi S_n = \psi e^{i(\boldsymbol{h} \cdot \boldsymbol{\Delta r}_n)} . \tag{C. 99}$$

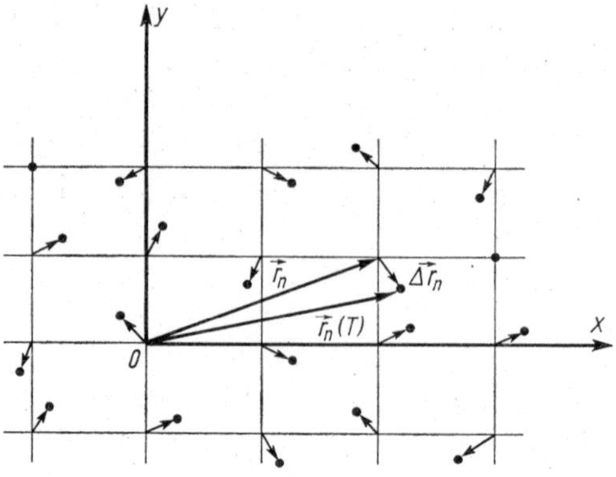

FIG. 40. Lattice vectors and displacements of the lattice atoms from their equilibrium position due to lattice vibrations.

For the total scattering amplitude and thus for the intensity

$$I_T = |A_{tot}(T)|^2 \tag{C. 100}$$

observed at $T>0°\text{K}$ the ensemble average $\overline{F_n}^{\text{ens}}$ of all structure factors is characteristic. Instead of eqn. (C. 97) we obtain

$$A_{tot}(T) = A_0\psi\overline{S_n}^{\text{ens}} \sum_n e^{i(\boldsymbol{k}.\boldsymbol{r}_n)} \tag{C. 101}$$

and, using eqns. (C. 96), (C. 100), and (C. 101), we find for the Debye–Waller factor

$$D(T) = \left(\overline{S_n}^{\text{ens}}\right)^2. \tag{C. 102}$$

As already mentioned, the ensemble average is equal to the time average of the displacements in the nth unit cell. An expansion of the exponential function in eqn. (C. 99) yields

$$\overline{S_n}^{\text{ens}} = 1 + i\overline{(\boldsymbol{h}.\boldsymbol{\Delta r})}^{\text{ens}} - \tfrac{1}{2}\overline{(\boldsymbol{h}.\boldsymbol{\Delta r}_n)^2}^{\text{ens}} + \ldots \,. \tag{C. 103}$$

The linear term vanishes as, averaged over time, the displacement of the atoms from their positions of rest is equal to zero, i.e.

$$\overline{(\boldsymbol{h}.\boldsymbol{\Delta r}_n)}^{\text{ens}} = \left(\boldsymbol{h}.\overline{\boldsymbol{\Delta r}_n}^{\text{ens}}\right) = 0. \tag{C. 104}$$

The quadratic term takes the form

$$\frac{1}{2}\overline{(\boldsymbol{h}.\boldsymbol{\Delta r}_n)^2}^{\text{ens}} = \frac{1}{2}h^2\frac{\overline{\Delta r_n^2}^{\text{ens}}}{3} \tag{C. 105}$$

when we assume the contributions of the displacement components $\boldsymbol{\Delta r}_n$ to be equal when averaged over time. When we set

$$\boldsymbol{\Delta r}_n = \boldsymbol{\Delta x} + \boldsymbol{\Delta y} + \boldsymbol{\Delta z} \tag{C. 106}$$

and

$$\boldsymbol{h} = \boldsymbol{h}_x + \boldsymbol{h}_y + \boldsymbol{h}_z\,, \tag{C. 107}$$

we have

$$\overline{(\boldsymbol{h}.\boldsymbol{\Delta r}_n)^2}^{\text{ens}} = h_x^2\,\overline{\Delta x^2} + h_y^2\,\overline{\Delta y^2} + h_z^2\,\overline{\Delta z^2}\,. \tag{C. 108}$$

We rearrive at (C. 105) using

$$\overline{\Delta x^2} = \overline{\Delta y^2} = \overline{\Delta z^2} = \tfrac{1}{3}\,\overline{\Delta r_n^2}\,. \tag{C. 109}$$

Equation (C. 103) can be rewritten as

$$\overline{S_n}^{\text{ens}} = 1 - \frac{h^2}{2}\,\frac{\overline{\Delta r_n^2}^{\text{ens}}}{3} + \ldots \tag{C. 110}$$

or

$$\overline{S_n}^{\text{ens}} = e^{-M} \tag{C. 111}$$

where

$$M = \frac{h^2}{2}\,\frac{\overline{\Delta r_n^2}^{\text{ens}}}{3}\,. \tag{C. 112}$$

Using Bragg's reflection condition (B. 41) we obtain

$$M = \frac{8\pi^2 \sin^2 \vartheta}{\lambda^2}\,\frac{\overline{\Delta r_n^2}^{\text{ens}}}{3}\,. \tag{C. 113}$$

In order to calculate the mean square displacement $\overline{\Delta r_n^2}^{\text{ens}}$ we make use of Debye's theory of specific heat (cf. pp. 55 ff.). For a state of an eigenvibration of frequency ω the atoms vibrate about their positions of rest with amplitudes given by

$$\boldsymbol{\Delta r}_{n,\,\omega} = \boldsymbol{a}_\omega \cos(\omega t). \tag{C. 114}$$

The direction of $\boldsymbol{a}_\omega$ characterizes the polarization of the mode. The time-averaged square of the amplitude $\boldsymbol{\Delta r}_{n,\,\omega}$ of the nth atom is given by

$$\overline{\Delta r_{n,\,\omega}^2}^{t} = \frac{a_\omega^2}{2}\,. \tag{C. 115}$$

As $3N$ eigenmodes are possible in the crystal, the time average of the total amplitude of the nth atom is given by an integral over the frequencies of these $3N$ modes:

$$\overline{\Delta r_n^2}^{t} = \int_0^{\omega_{\max}} \frac{a_\omega^2}{2}\,\varrho(\omega)\,d\omega \tag{C. 116}$$

with the frequency density according to Debye [eqn. (C. 58)]

$$\varrho(\omega)=\frac{9N}{\omega_{\max}^3}\,\omega^2\,. \tag{C. 117}$$

The maximum amplitude a_ω for an eigenmode of frequency ω is obtained by means of an energy balance. N atoms of mass m which oscillate with a frequency ω contribute

$$u(\omega)=Nm\omega^2\frac{a_\omega^2}{2} \tag{C. 118}$$

to the crystal's vibrational energy. This contribution is equal to the energy of the corresponding oscillator [cf. eqn. (C. 38)]:

$$u(\omega)=\frac{\hbar\omega}{2}+\frac{\hbar\omega}{\exp\left(\frac{\hbar\omega}{kT}\right)-1}\,. \tag{C. 119}$$

Hence we obtain for the maximum square of the amplitude

$$\frac{a_\omega^2}{2}=\frac{\hbar}{Nm\omega}\left[\frac{1}{2}+\frac{1}{\exp\left(\frac{\hbar\omega}{kT}\right)-1}\right]. \tag{C. 120}$$

Substitution of eqns. (C. 117) and (C. 120) in (C. 116) yields the mean square displacement under the assumption that the time average is equal to the ensemble average:

$$\overline{\Delta r_n^2}^{\,t}=\overline{\Delta r_n^2}^{\,\mathrm{ens}}\frac{9}{\omega_{\max}^3 m}\int\limits_0^{\omega_{\max}}\left[\frac{\hbar\omega}{2}+\frac{\hbar\omega}{\exp\left(\frac{\hbar\omega}{kT}\right)-1}\right]d\omega. \tag{C. 121}$$

With the substitutions (cf. p. 59)

$$\Theta=\frac{\hbar\omega_{\max}}{k} \tag{C. 122}$$

and

$$x=\frac{\hbar\omega}{kT} \tag{C. 123}$$

we obtain from eqn. (C. 121)

$$\overline{\Delta r_n^2}^{\,\mathrm{ens}}=\frac{9}{4}\,\frac{\hbar^2}{mk\Theta}\,P\left(\frac{T}{\Theta}\right) \tag{C. 124}$$

where the function $P(T/\Theta)$ is defined as

$$P\left(\frac{T}{\Theta}\right) = 1+4\left(\frac{T}{\Theta}\right)^2 \int_0^{\Theta/T} \frac{x\,dx}{e^x-1}. \qquad \text{(C. 125)}$$

After substituting eqns. (C. 111), (C. 113), and (C. 124) in (C. 102) we obtain for the Debye–Waller factor

$$D(T) = \exp(-2M) = \exp\left[-3\frac{(2\pi\hbar)^2}{mk\Theta}\frac{\sin^2\vartheta}{\lambda^2}P\left(\frac{T}{\Theta}\right)\right]. \qquad \text{(C. 126)}$$

This shows that the intensity drop of the Bragg reflections is strong if the mass of the scattering atoms is small and the Debye temperature is low. In this case the amplitudes of the lattice vibrations are large and occasionally such a lattice is called "soft". Table 10 shows that the attenuation of the reflections is the

TABLE 10. Debye temperatures, mean square displacements, and Debye–Waller factors of various substances at $T = 293$°K. The values given for the Debye–Waller factors and the exponents M do not refer to certain reflections $h_1h_2h_3$, but are extrema obtained for $\vartheta = 90°$ [cf. eqns. (C. 113) and (C. 126)]. They are calculated for Cu $K\alpha$ radiation ($\lambda = 1.54$ Å) [after A. Guinier, *X-Ray Diffraction* (Freeman, San Francisco 1963), p. 192]

Material	Θ [°K]	$\overline{\Delta r^2}$ [Å^2]	$D = e^{-2M}$	$\frac{M(0°K)}{M(293°K)}$
Lead	88	0.074	0.18	0.075
Silver	215	0.026	0.57	0.18
Tungsten	280	0.009	0.82	0.22
Copper	315	0.020	0.64	0.26
Aluminium	398	0.030	0.49	0.32
Iron	453	0.012	0.77	0.36
Diamond	1860	0.005	0.88	0.85

smaller, the higher the Debye temperature. In diamond at room temperature, for example, relatively few phonons are excited and thus the deviations from the positions of rest of the atoms are small and the Debye–Waller factor is large. The intensity drop of the reflections also depends on the X-ray wavelength λ and the glancing angle ϑ; the reflection intensity is the lower, the shorter the wavelength and the larger the glancing angle.

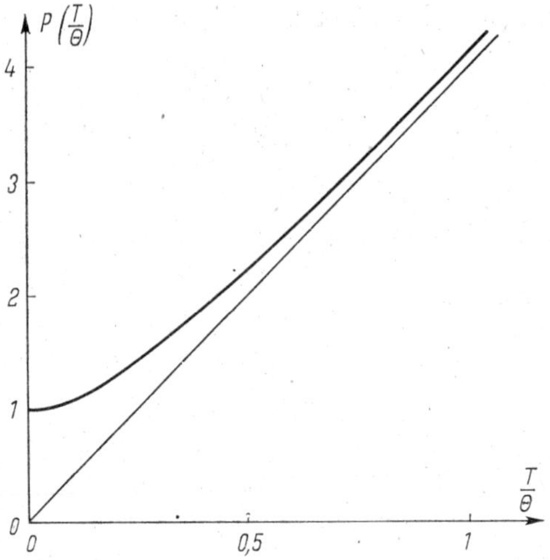

FIG. 41. The function $P(T/\Theta)$ for the computation of the temperature dependence of the Debye–Waller factor.

The temperature dependence of the Debye–Waller factor is contained in the function $P(T/\Theta)$ which will be discussed for two ranges of temperature (cf. Fig. 41):

1. If $T \gg \Theta$, similarly as in the case of the Debye function (C. 65), the integrals of (C. 125) can be expanded; we obtain

$$P\left(\frac{T}{\Theta}\right) \approx 4\frac{T}{\Theta}, \tag{C. 127}$$

i.e., at high temperatures the intensity of the Bragg reflections drops exponentially as the temperature is raised. Substitution of (C. 127) in (C. 126) yields

$$D(T) \approx \exp(-c_1 T). \tag{C. 128}$$

The temperature-independent constant c_1 is given by (C. 126).

2. If $T \ll \Theta$, the integral in (C. 125) is equal to a numerical value, i.e.

$$P\left(\frac{T}{\Theta}\right) \approx 1+4\frac{\pi^2}{6}\left(\frac{T}{\Theta}\right)^2. \tag{C. 129}$$

Then we can write the Debye–Waller factor in the form

$$D(T) = \exp(-c_0 - c_2 T^2). \tag{C. 130}$$

The constants c_0 and c_2 are again obtained from a comparison with (C. 126). The constant c_0 is due to the effect of the zero-point energy of the oscillators.

This means that even a reduction of temperature to absolute zero does not reduce the thermal motion of the atoms to nothing. Even at $T = 0°K$ the Debye–Waller factor is smaller than unity. However, eqn. (C. 126) and Fig. 41 show that the influence of temperature on the intensities of the reflections decreases with decreasing temperature. It must be mentioned that a measurement of the Debye–Waller factor makes it possible to determine the Debye temperatures. In general the Debye temperatures determined from X-ray data lie a little below the values obtained from measurements of the specific heats or elastic data (cf. pp. 61 ff.).

(b) *THE MÖSSBAUER EFFECT*

The Debye–Waller factor is fundamental for the observability of the Mössbauer effect. This effect, discovered in 1958 by Mössbauer, consists in the recoil-free emission and resonance absorption of γ quanta by nuclei bound in a solid. The Debye–Waller factor $D_\gamma(T)$ is here the fraction of nuclei in the crystal which emit or absorb γ quanta without recoil, that is, without interaction with phonons.

To understand the Mössbauer effect consider the following: a free nucleus at rest, of mass m_{nucl}, is assumed to drop from its excited state of energy E_a to the ground state E_g and to emit a γ quantum. In this process the nucleus obtains a recoil energy E_r:

$$E_r = \frac{E_0^2}{2m_{\text{nucl}}c^2} \tag{C. 131}$$

where

$$E_0 = E_a - E_g \gg E_r; \tag{C. 132}$$

E_0 is the nuclear excitation energy.

The energy of the γ quantum emitted is reduced by the recoil energy and thus equal to $E_0 - E_r$. Usually this energy is not sufficient to excite another identical nucleus at rest to the state E_a; its excitation requires the excitation energy plus the recoil energy, $E_0 + E_r$. The emission and absorption lines of free nuclei (e.g. in a gas) are therefore shifted with respect to one another (cf. Fig. 42). If the emission and absorption lines partially overlap owing to the spectral line widths, resonance absorption (resonance fluorescence of the nuclei) may arise, i.e., a fraction of the γ quanta emitted by the excited nuclei is absorbed by identical nuclei in the ground state so that these nuclei become excited. Nuclear motion which may be due to temperature, earlier emission or absorption processes, or external means (e.g. the motion of absorber or emitter in an ultracentrifuge) gives rise to Doppler effects. They result in a broadening or displace-

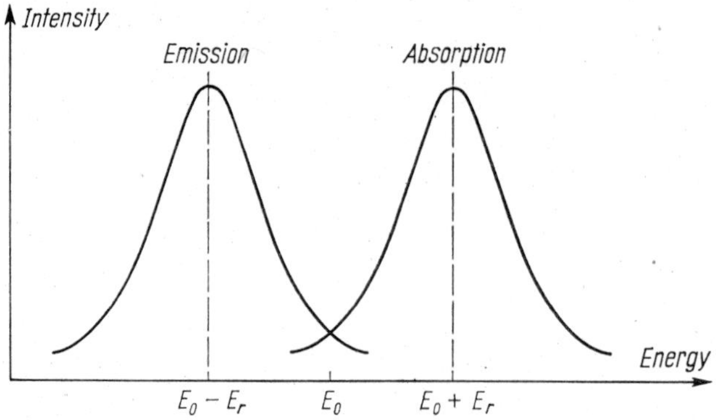

FIG. 42. Emission and absorption spectra for γ transitions of energy E_0 of free nuclei.

ment of the emission and absorption lines, which may lead to stronger overlapping and increased resonance absorption.

In contrast to the expected decrease of the Doppler effects and the resonance absorption with decreasing temperature, Mössbauer found that in a solid resonance absorption may increase strongly as the temperature drops (this was first observed with ^{191}Ir for the 129-keV γ transition). This is explained by the fact that the emitting and absorbing nuclei are bound in the crystal so that, under certain conditions, the recoil energy can be absorbed by the crystal as a whole. In eqn. (C. 131) the mass of the nucleus must then be replaced by the mass of the whole crystal so that the recoil energy E_r becomes negligibly small compared with the excitation energy E_0. In this case the γ quanta are emitted and absorbed with virtually no recoil. The emission and absorption lines lie at almost the same energy E_0 and the condition for resonance absorption is almost perfectly satisfied. The probability of recoil-free processes decreases as the temperature increases, and the recoil energy is then absorbed rather by the individual atoms; this is connected with the emission or absorption of phonons.

The emission and absorption spectra for γ-rays, which are the result of a certain transition between nuclear levels, will thus have two components: (1) A broad emission or absorption line below or above the excitation energy E_0. Shape and position of this line depend on the number of inelastic (recoil) interactions. (2) A very sharp line at the energy E_0 which is due to the recoil-free processes. This "recoil-free" line is also called the Mössbauer line.

The similarity of the effects connected with X-ray scattering in crystals and the emission and absorption of γ quanta in crystals is obvious. The inelastic processes which are due to interactions with phonons give rise to diffuse X-ray scattering, and to broad, mutually shifted emission and absorption lines of γ quanta. The recoil-free processes manifest themselves as Bragg reflection (cf. p. 72) or as Mössbauer effect, and yield the zero-phonon or Mössbauer lines. The temperature motion of the scattering centres, and of the emitting or absorbing atoms, results in an intensity decrease but not in a broadening of these lines. In both cases the relative intensity is determined by a Debye–Waller factor which gives the probability of recoil-free processes.

It is easy to derive the Debye–Waller factor for recoil-free emission or absorption of γ quanta from that for X-ray interferences. We must take into account that the lattice atom in its vibration will either emit a γ quantum or absorb one, while in X-ray scattering the X-ray quantum is absorbed and re-emitted. Accordingly the maximum recoil energies which can be transferred to the atom are in principle different for these two cases; in the case of emission or absorption of a γ quantum of energy E_0 [cf. eqn. (C. 131)] we have

$$E_{r\gamma} = \frac{E_0^2}{2m_{\text{nucl}}c^2} \qquad \text{(C. 133)}$$

and in the case of the scattering of an X-ray quantum of energy E_0 from an atom of mass m_A [cf. eqn. (C. 88)] we have

$$E_{rX} = 2\frac{E_0^2}{m_A c^2}. \qquad \text{(C. 134)}$$

In the case of X-ray interferences the recoil energy which can be transferred to the scattering atom is introduced into the Debye–Waller factor $D(T)$. If this energy is then replaced by the recoil energy $E_{r\gamma}$ transferred by the γ quantum, the Debye–Waller factor $D_\gamma(T)$ characteristic for the Mössbauer effect is obtained. We again restrict ourselves to the Debye approximation for the phonon spectrum. With the photon energy E_0 [cf. eqn. (C. 90) with $u_X = E_0$] we obtain from (C. 126)

$$D(T) = \exp\left[-3\frac{E_0^2}{m_A c^2}\,\frac{\sin^2\vartheta}{k\Theta}\,P\left(\frac{T}{\Theta}\right)\right]. \qquad \text{(C. 135)}$$

The photons of energy E_0 incident at the Bragg angle ϑ are scattered by an atom of mass m_A through an angle of 2ϑ (cf. Fig. 20). If the atom were free, the following recoil energy would be transferred in the scattering process:

$$E_{rX} = \frac{E_0^2}{m_A c^2}[1-\cos 2\vartheta] = 2\frac{E_0^2}{m_A c^2}\sin^2\vartheta. \qquad \text{(C. 136)}$$

When we substitute this value in eqn. (C. 135) we obtain

$$D(T) = \exp\left[-\frac{3}{2}\frac{E_{rX}}{k\Theta}P\left(\frac{T}{\Theta}\right)\right]. \tag{C. 137}$$

Replacing E_{rX} by $E_{r\gamma}$ [cf. eqn. (C. 133)], we obtain the Debye–Waller factor for recoil-free emission or absorption of γ quanta:

$$D_\gamma(T) = \exp\left[-\frac{3}{4}\frac{E_0^2}{m_{\text{nucl}}c^2k\Theta}P\left(\frac{T}{\Theta}\right)\right]. \tag{C. 138}$$

As to its amount the resonance absorption is determined by the product of two Debye–Waller factors; one determines the intensity of recoil-free emission, the other that of recoil-free absorption. These factors are also denoted the Lamb–Mössbauer factors. For identical source and absorber (cf. p. 83) of the same temperature the temperature dependence of resonance absorption is given by

$$f_M = \exp(-4M), \tag{C. 139}$$

where

$$M = \frac{3}{8}\frac{E_0^2}{m_{\text{nucl}}c^2k\Theta}P\left(\frac{T}{\Theta}\right). \tag{C. 140}$$

(α) *Experimental verification of the Mössbauer effect.* The verification of the Mössbauer effect is based on a measurement of resonance absorption of γ quanta. The Mössbauer lines appear in the γ-ray spectrum on the background of inelastic interactions if the Debye–Waller factor exceeds a value of about 10^{-2}. In the low-temperature approximation this condition [cf. eqn. C. 129)] is equivalent to the condition

$$\frac{E_0^2}{2m_{\text{nucl}}c^2} = E_r \lesssim 2k\Theta \tag{C. 141}$$

and limits the observation of the Mössbauer effect to low-energy ("soft") γ-quantum transitions, for which the recoil energy of the free nucleus does not exceed twice the maximum energy in the Debye spectrum. This means that at sufficiently low temperatures we may expect to observe a Mössbauer effect for relatively heavy nuclei, with γ transitions of energies up to about 10^2 keV. For the 129-keV transition in ^{191}Ir, for which recoil-free resonance absorption was first observed, the recoil energy $E_r = 0.046$ eV and the maximum Debye energy $k\Theta \approx 0.025$ eV, i.e. at sufficiently low temperatures the Debye–Waller factor is about 1%. It is a consequence of the condition $E_r \lesssim 2k\Theta$ that the Mössbauer effect has been studied only for a few nuclear transitions (mainly for transitions in ^{57}Fe, ^{119}Sn, ^{161}Dy, ^{169}Tm, ^{197}Au). ^{57}Fe, for example, has a γ

transition with $E_0 = 14.4$ keV; for this value eqn. (C. 138) yields even at room temperature a fraction of 70% for the recoil-free emission. The Mössbauer effect of the 14.4-keV transition in ^{57}Fe is thus particularly high; since in addition the resonance line is extremely sharp (line width $\Gamma = 5\times10^{-9}$ eV) this transition is suitable for investigations into many interesting problems of various fields of physics (cf. p. 85).

Equation (C. 141) is easy to understand when we take into consideration that, for sufficiently low recoil energies $E_r \ll k\Theta$, there are relatively few oscillators in the Debye frequency spectrum which may take up the recoil energy so that the probability of recoil-free processes is high. Larger recoil energies can be taken up by a relatively higher number of oscillators of higher energies, and accordingly the fraction of recoil-free processes drops as the γ-quantum energy increases.

The experimental arrangement for measurements of the Mössbauer effect (cf. Fig. 43) consists essentially of a source containing the γ-active isotope, an absorber which contains the same isotope in the ground state, and a γ-quantum

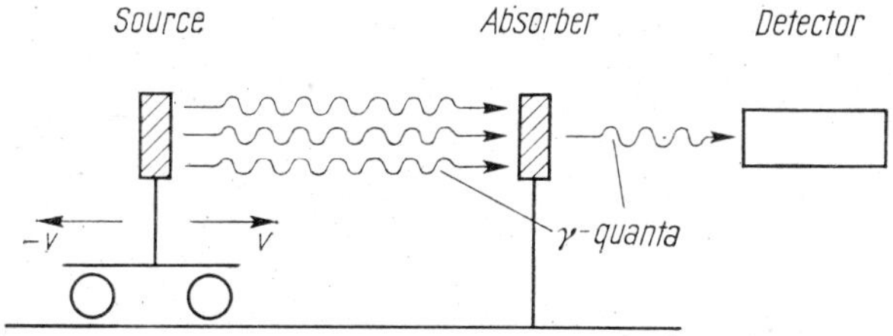

FIG. 43. Experimental arrangement for measurement of the Mössbauer effect.

detector. A moving mechanism guarantees a relative velocity v between source and absorber. The Doppler shift caused hereby is

$$\Delta E = \frac{v}{c} E_0. \tag{C. 142}$$

This Doppler shift of the emitted γ energy E_0 makes resonance absorption impossible so that more γ quanta pass through the absorber and are counted by the detector. Knowing the relative radiation intensity as a function of the relative velocity, we can determine the line widths, line splitting, or line shifts of certain γ transitions. If the isotopes in source and absorber are kept at exactly the same conditions, the radiation intensity (counting rate) is minimum for $v = 0$ (cf. Fig. 44). The counting rate in terms of the relative velocity is called the Mössbauer spectrum.

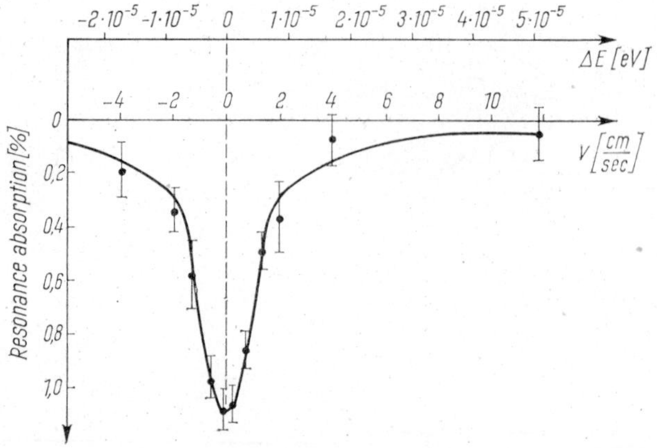

FIG. 44. Resonance absorption of the 129 keV γ radiation of ^{191}Ir as a function of the relative velocity between source and absorber and as a function of the Doppler shift, respectively [after R. L. Mössbauer, *Z. Naturforsch.* **14a**, 211 (1959)].

The velocities necessary for a detection of resonance absorption must be chosen according to the line width and line splitting which are to be observed. The natural line width Γ is linked with the mean lifetime τ of the excited state by the relation

$$\Gamma = \frac{\hbar}{\tau}. \tag{C. 143}$$

Typical lifetimes of excited states which reach the ground state in a γ decay lie between 10^{-7} and 10^{-10} sec; from eqn. (C. 143) we obtain typical values for the line widths between 10^{-8} and 10^{-5} eV. The Doppler shifts necessary for a detection of such lines must be of the same order of magnitude. So we obtain from eqn. (C. 142) typical relative velocities between 10^{-3} and 10^{+1} cm/sec, when we take into account that the γ energies E_0 are of the order of 10–100 keV.

As in Mössbauer experiments it is possible to observe the lines at almost their natural widths, the recoil-free resonance absorption of soft γ quanta represents a spectroscopical method of extremely high resolution E_0/Γ. This permits a detection of very small changes in energy of γ quanta, as, for example, in the case of hyperfine splitting of nuclear levels or the motion of a γ quantum in the gravitational field. The evaluation of hyperfine structure measurements becomes simpler when line splitting occurs either only in the source or only in the absorber. Often so-called single-line sources are used which permit a "scanning" of the hyperfine structure splitting of the absorber.

(β) *Applications of the Mössbauer effect.* The Mössbauer effect can be used to study various problems in general physics, nuclear physics, solid state physics, and chemistry. In the following we shall consider only two effects which can be studied with the help of the Mössbauer effect and which are important in solid-state physics: the nuclear Zeeman effect (magnetic hyperfine structure splitting) and the isomer shift (a kind of chemical shift).

When a nucleus possessing a magnetic moment of its own (cf. p. 432) is placed in a magnetic field, the degenerate levels are split up. This splitting appears in the Mössbauer spectrum if it exceeds the natural line width. In the case of ^{57}Fe, for example, a magnetic field applied to the nucleus causes a splitting of the ground state into two levels and of the excited state into four levels. The γ transitions between the two states are permitted if the magnetic quantum number M_I is changed by one at most. With this selection rule six different γ transitions are possible (cf. Fig. 45). Accordingly, six Mössbauer lines can be

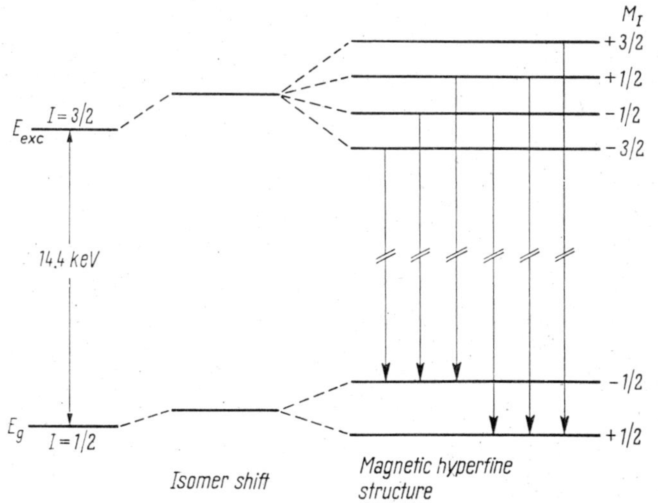

FIG. 45. Magnetic hyperfine structure of the ground state level E_g and of the excited 14.4 keV level E_{exc} in ^{57}Fe; the selection rule $|\Delta M_I| \leq 1$ for the magnetic quantum number M_I is valid for the γ transitions between the two levels (nuclear spin quantum number $I = \frac{3}{2}$ and $I = \frac{1}{2}$).

observed (cf. Fig. 46) provided that the splitting caused by the magnetic field is sufficient. In metallic ferromagnetic iron at room temperature the internal magnetic field at the position of the nucleus amounts to about 4×10^7 A/m. The hyperfine splitting caused by this field is of the order of 10^{-7} eV and thus exceeds the natural line width of the 14.4-keV γ transition in ^{57}Fe (5×10^{-9} eV) by a factor of about 30. This means that the energy resolution necessary for a detection of the line splitting must be at least 10^{11}.

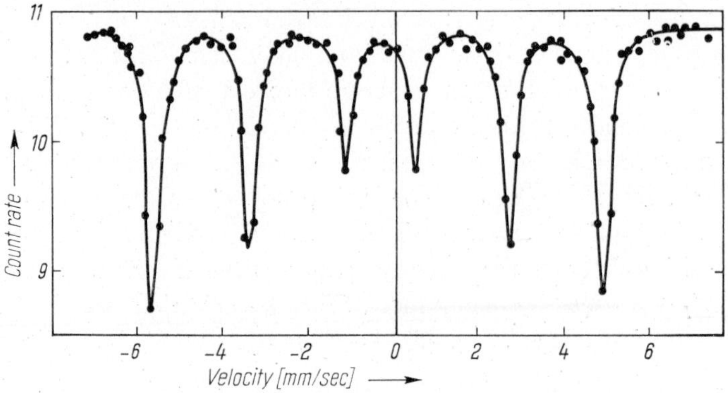

FIG. 46. Magnetic hyperfine structure of the Mössbauer lines of ^{57}Fe; obtained with a single line source (^{57}Co in Pt) and a ferromagnetic metallic Fe absorber. The six Mössbauer lines correspond to the six γ transitions which were shown in Fig. 45 [after W Kerler and W. Neuwirth, *Z. Phys.* **167,** 176 (1962)].

Besides the magnitude of the internal magnetic field we can also determine the sign when we measure the line splitting in its dependence on the external magnetic field. For metallic iron we arrive at the surprising result that the internal field is antiparallel to the magnetic moment of the atom and thus to the macroscopic magnetization. In general, the Mössbauer effect is an important method for determining magnitude and sign of the internal magnetic fields at the nucleus in ferromagnetic and antiferromagnetic substances.

Another hyperfine interaction is the isomer shift which is due to the electrostatic interaction between the charge distributions in the nucleus and the electronic shell. The interaction energy is proportional to the electron charge density at the nucleus and to a mean square radius of the nucleus; it results in a shift of the nuclear levels. The transition energy E_0 between two nuclear levels is changed if the mean square nuclear radius and thus the nuclear level shift depends on the state of excitation of the nucleus. The change in transition energy is called the isomer shift (cf. Fig. 47); it cannot be observed directly. In a Mössbauer experiment, however, it is possible to measure the difference between two isomer shifts if the Mössbauer nuclei in source and absorber are in two different chemical compounds; then the electron densities at the nucleus are different in source and absorber. The energy shift between recoil-free

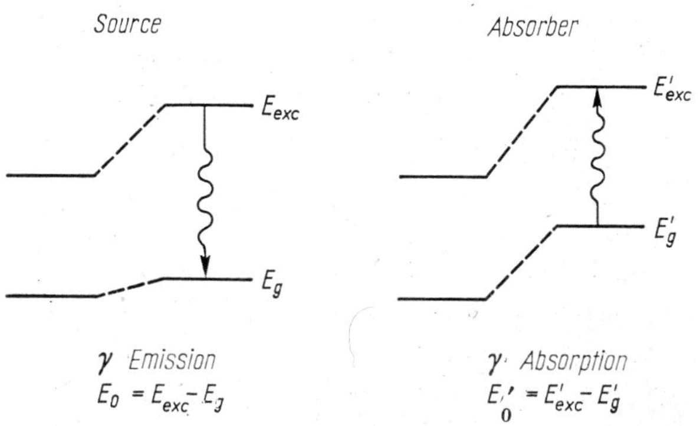

FIG. 47. Isomer shift of the nuclear levels of an isotope which is situated as source and absorber in two different chemical compounds. The energy difference for γ emission (E_0) and for γ absorption (E'_0) gives rise to a chemical shift δ. The left half in each of the figures shows the nuclear levels of the corresponding pointlike nucleus.

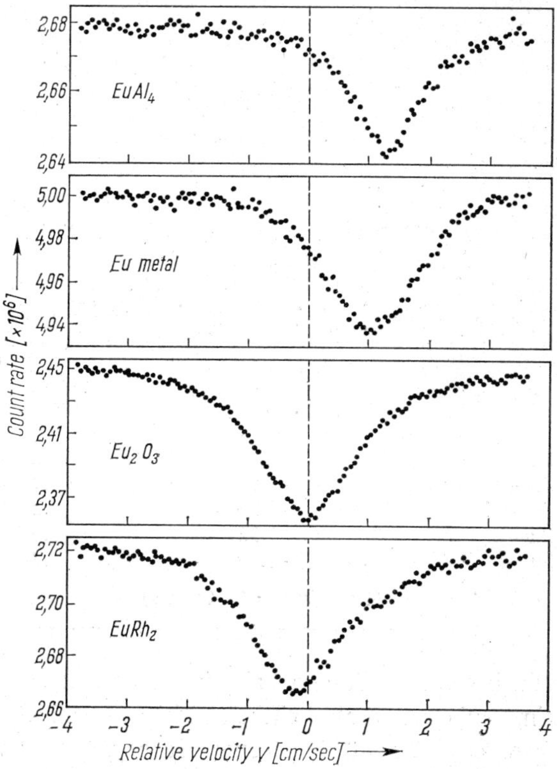

FIG. 48. Isomer shift for the 97 keV γ transition of ^{153}Eu in some europium compounds [after U. Atzmony *et al.*, *Phys. Rev.* **156**, 262 (1967)].

emission and absorption is proportional to the difference of the squares of the nuclear radii in the ground state and in the excited state, and proportional to the difference of the electron densities at the nuclei in the two compounds. The isomer shift manifests itself in a changed resonance condition: resonance will occur not at $v = 0$ but at $v = \delta \gtrless 0$ (cf. Fig. 48).

Because of its dependence on the electron density at the nucleus, the isomer shift yields valuable information in problems of the chemical bond in solids. Only s-electrons possess a finite probability $|\psi_s(0)|^2$ of being at the nucleus. In general, however, the electron density $|\psi_s(0)|^2$ will depend not only on the number of s-electrons, but also on the electrons of the other shells which partly screen the nuclear charge and thus diminish the nuclear attraction to the s-electrons. The dependence of the isomer shift on the chemical bond can be simply observed if the Mössbauer isotope has complete shells and if the chemical bond involves only the outer, incompletely filled shells. ^{119}Sn may serve as an example; it has a 24-keV γ transition. Here the isomer shift is determined by the five s-electrons; it is large if the electron affinity of the bonding partner is small. The difference of the electron affinity of two bonding partners can be expressed in terms of electro-negativity differences. Using the tin compounds as an example, we show in Table 11 that the isomer shift increases with the difference in electro-negativity referred to a constant source.

TABLE 11. Isomer shifts δ of several Sn compounds compared with β-Sn in the Mössbauer effect of the ^{119}Sn nuclei ($E_\gamma = 24$ keV) and differences $\Delta\chi$ of the electro-negativity of anion and cation [after O. C. Kistner, V. Jaccarino and L. R. Walker, in *The Mössbauer Effect*, edited by D. M. J. Compton and A. H. Schoen (Wiley, New York 1962), p. 264]

Compound	δ [mm/sec]	$\Delta\chi$
Sn—F	−2.77	2.2
Sn—O	−2.54	1.7
Sn—Cl	−2.26	1.2
Sn—Br	−1.32	1.0
Sn—J	−1.19	0.7

(1 mm/sec $\cong 8 \cdot 10^{-8}$ eV)

An isomer shift is generally observed also in measurements of the magnetic hyperfine splitting and quadrupole splitting, which we shall not discuss in this connection. In these cases the Mössbauer spectra are symmetric not with respect to $v = 0$ but with respect to δ, the velocity corresponding to the isomer shift (cf. Fig. 46).

4. MODELS OF A DISCRETE CRYSTAL

In their derivations of the spectrum of the lattice vibrations Einstein and Debye made no assumptions as to the atomic structure of crystals. They characterized the crystals only by means of the crystal volume (i.e., the number of oscillating lattice particles) and the macroscopically periodic structure (periodic boundary conditions, cf. p. 56).

Theories which are based on the equations of motion of the lattice particles yield more detailed information on the spectrum of the lattice vibrations. Born and von Karman have shown that the mathematically simple example of a one-dimensional lattice model may help us to understand the principal properties of the general three-dimensional lattice. In the following we restrict ourselves to a discussion of the lattice vibrations of a diatomic one-dimensional "crystal".

(a) *LINEAR DIATOMIC CRYSTAL MODEL*

Let us assume that the unit cell of the one-dimensional crystal contains one particle of mass M and one of mass m. The equilibrium positions of the particles have all the same separation, a, and are continuously numbered: $n = 0, 1, 2, 3, \ldots, \infty$. Thus the crystal consists of an infinitely long linear chain of links of mass M and mass m. The even numbered points $2n$ are the equilibrium positions of the masses M, the odd numbered points $2n-1$ are those of the masses m (cf. Fig. 49). The forces between the particles can be symbolized by identical springs characterized by springs with a spring constant f. The force pushing the particles back to their initial positions is assumed to depend only on the distance to their nearest neighbours. Let the deviations from the rest positions be ξ_{2n} for the masses M and ξ_{2n-1} for the masses m.

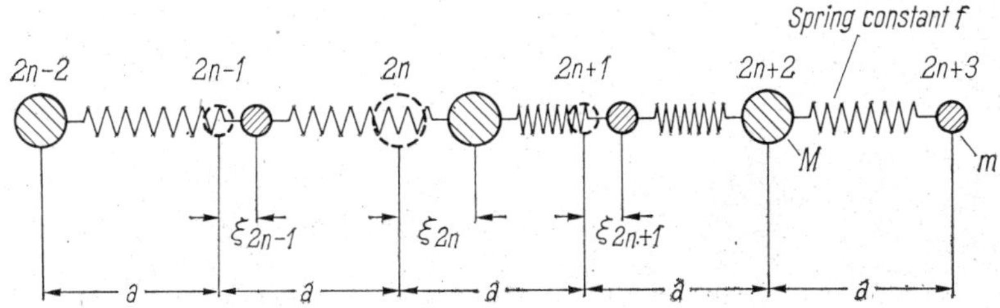

FIG. 49. Lattice vibrations in a one-dimensional two-atomic crystal.

Newton's equations of motion for the two species of particles are

$$\begin{aligned} M\ddot{\xi}_{2n} &= f(\xi_{2n+1}-\xi_{2n})-f(\xi_{2n}-\xi_{2n-1}), \\ m\ddot{\xi}_{2n-1} &= f(\xi_{2n}-\xi_{2n-1})-f(\xi_{2n-1}-\xi_{2n-2}). \end{aligned} \qquad \text{(C. 144)}$$

When we define the frequencies of free oscillations by

$$\omega_0^2 = \frac{f}{M} \quad \text{and} \quad \Omega_0^2 = \frac{f}{m}, \tag{C. 145}$$

we obtain

$$\begin{aligned} \ddot{\xi}_{2n} &= \omega_0^2(\xi_{2n+1}+\xi_{2n-1}-2\xi_{2n}),\\ \ddot{\xi}_{2n-1} &= \Omega_0^2(\xi_{2n}+\xi_{2n-2}-2\xi_{2n-1}). \end{aligned} \tag{C. 146}$$

Here we have a system of infinitely many coupled differential equations. In order to obtain the stationary solutions to this system, we choose functions periodic in space and time, similar to the harmonic waves in the elastic continuum [cf. eqns. (C. 45) and (C. 46)]:

$$\begin{aligned} \xi_{2n} &= Ae^{i\omega t}e^{iq2na} && \text{for the masses } M,\\ \xi_{2n-1} &= Be^{i\omega t}e^{iq(2n-1)a} && \text{for the masses } m. \end{aligned} \tag{C. 147}$$

A and B are time-independent maximum amplitudes of the masses M and m, $q = 2\pi/\lambda$ is the wave number, and na is a position coordinate (only the amplitudes at the lattice points are physically significant).

Substituting (C. 147) in the equations of motion (C. 146), we obtain two linear homogeneous equations for the amplitudes A and B:

$$\begin{aligned} -\omega^2 A &= \omega_0^2 B(e^{iqa}+e^{-iqa})-2\omega_0^2 A,\\ -\omega^2 B &= \Omega_0^2 A(e^{iqa}+e^{-iqa})-2\Omega_0^2 B. \end{aligned} \tag{C. 148}$$

Apart from the trivial solution $A = B = 0$ these equations have a single solution for A and B which is obtained by setting the coefficient determinant equal to zero:

$$\begin{vmatrix} 2\omega_0^2-\omega^2 & -\omega_0^2(e^{iqa}+e^{-iqa}) \\ -\Omega_0^2(e^{iqa}+e^{-iqa}) & 2\Omega_0^2-\omega^2 \end{vmatrix} = 0. \tag{C. 149}$$

Hence we obtain immediately the condition of solubility as a relation between ω and q:

$$\omega_\pm^2 = \omega_0^2+\Omega_0^2 \pm [(\omega_0^2+\Omega_0^2)^2-4\omega_0^2\Omega_0^2 \sin^2(qa)]^{1/2}, \tag{C. 150}$$

which represents a dispersion relation as $\omega/q \neq \partial\omega/\partial q \neq$ constant. [For elastic continua, however, $\omega/q = c =$ const., cf. eqn. (C. 47).]

The $\omega(q)$ relation (C. 150) is a periodic function of q and the range of q values can therefore be restricted to

$$0 \leqslant \bar{q} \leqslant \frac{\pi}{a}. \tag{C. 151}$$

The wave number, $\bar{q}$, limited in this way, is called a reduced wave number; the range given is the first Brillouin zone of the one-dimensional crystal. Moreover, $\omega(q)$ is symmetric with respect to the value $q = \pi/2a$. So we have to discuss (C. 150) only for the range $0 \leq \bar{q} \leq \pi/2a$.

For each q value there exist two values ω_+ and ω_- according to the two solutions for ω^2. The function $\omega(q)$ consists of two separate branches $\omega_+(q)$ and $\omega_-(q)$, which pertain to different limited frequency ranges. To demonstrate this we shall calculate the special $\omega_\pm$ values for $\bar{q} = 0$ and $\bar{q} = \pi/2a$:

$$\bar{q} = 0: \qquad \omega_+ = 2^{1/2}(\Omega_0^2+\omega_0^2)^{1/2}, \qquad \omega_- = 0; \tag{C. 152}$$

$$\bar{q} = \frac{\pi}{2a}: \qquad \omega_+ = 2^{1/2}\Omega_0, \qquad \omega_- = 2^{1/2}\omega_0. \tag{C. 153}$$

Here we assumed $M > m$, i.e., according to eqn. (C. 145), $\Omega_0 > \omega_0$.

For frequencies in the range $2^{1/2}\omega_0 < \omega < 2^{1/2}\Omega_0$ the dispersion relation is satisfied only for complex wave numbers q. This means that the amplitude is damped as far as the position coordinate is concerned. In this frequency range there exist no solutions in the form of harmonic waves; it is therefore also called the "forbidden frequency band". All frequencies above the maximum frequency $\omega_{max} = 2^{1/2}(\Omega_0^2+\omega_0^2)^{1/2}$ are likewise "forbidden".

The function $\omega(q)$ is shown in Fig. 50. The oscillations in the frequency range $0 < \omega < 2^{1/2}\omega_0$ are called acoustic or elastic oscillations ($\omega_-(q) \to$ acoustic branch), those in the frequency range $2^{1/2}\Omega_0 < \omega < 2^{1/2}(\Omega_0^2+\omega_0^2)^{1/2}$ are called optical oscillations ($\omega_+(q) \to$ optical branch). The oscillation processes which

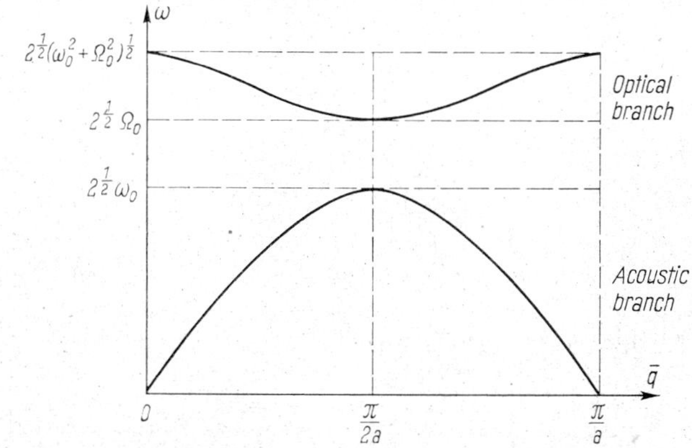

FIG. 50. Dispersion relation $\omega(q)$ of a one-dimensional two-atomic crystal.

give rise to these two branches are different because the amplitude ratios of neighbouring particles have different signs in the two frequency ranges. At any time t the amplitude ratio can be calculated from eqns. (C. 147) and (C. 148):

$$\frac{\xi_{2n}}{\xi_{2n-1}} = \frac{A}{B} e^{iqa}, \tag{C. 154}$$

where

$$\frac{A}{B} = \frac{2\omega_0^2}{2\omega_0^2-\omega^2} \cos(qa) = \frac{2\Omega_0^2-\omega^2}{2\Omega_0^2 \cos(qa)}. \tag{C. 155}$$

In the case of acoustic oscillations we have $2\omega_0^2-\omega^2 > 0$ and $2\Omega_0^2-\omega_0^2 > 0$ (see Fig. 50) and hence $\xi_{2n}/\xi_{2n-1} > 0$, that is, neighbouring particles swing in the same direction. In the approximation $q \ll 1/a$ ($\lambda \gg a$) the amplitude ratio becomes independent of the masses M and m: $\xi_{2n}/\xi_{2n-1} \to 1$. This means that the linear crystal vibrates as a whole, as a rigid system. In this approximation the amplitudes of neighbouring particles display only an infinitesimal difference. Adding the two equations of motion (C. 146) then yields the differential equation of the vibrating string, i.e., the one-dimensional wave equation of the elastic continuum [cf. eqn. (C. 44)]. In this approximation the ratio ω/q must pass over to a constant velocity value, as in the approximation of $\lambda \gg a$ it follows from eqn. (C. 150) for $\omega_-(q)$ that

$$\omega_-(q) = \left(\frac{2a}{m+M} fa\right)^{1/2} q = \text{const.}\, q. \tag{C. 156}$$

The oscillations with frequencies $0 < \omega_- < 2^{1/2}\omega_0$ are therefore called elastic or acoustic because they are identical with the elastic waves in a continuum for long (compared with the lattice constant) wavelengths.

For optical oscillations we have $2\omega_0^2-\omega^2 < 0$ and $2\Omega_0^2-\omega^2 < 0$, and hence $\xi_{2n}/\xi_{2n-1} < 0$, that is, neighbouring particles oscillate in opposite directions. In the approximation $q \ll 1/a$ ($\lambda \gg a$) we obtain for the amplitude ratio $\xi_{2n}/\xi_{2n-1} = -m/M$. The amplitudes are inversely proportional to the corresponding masses and the centre of mass of the unit cell remains at rest. If the masses possess electric charges of the same magnitude and opposite signs (linear model of an ionic crystal), the oscillations with a negative amplitude ratio give rise to dipole moments which vary in time with the frequency ω_+. Thus we have for each pair of ions an electrical oscillator as a source of electromagnetic radiation. On the other hand, electromagnetic radiation of frequency ω_+ is absorbed, as it excites particles of different mass and charge to execute dipole oscillations. The frequencies are of the order of magnitude of 10^{13} to

10^{14} sec^{-1} and correspond to infrared radiation. The general terminology "optical oscillations" is due to the interaction described above between the crystal and light.

Because of the conservation of momentum (cf. p. 67) light of a wavelength of the order of 50 μm will excite only optical oscillations with wave number $q \approx 0$, as $q = 2\pi/\lambda \ll 2\pi/a$. The corresponding frequency is given by

$$\omega_{\max} \approx 2^{1/2}(\Omega_0^2+\omega_0^2)^{1/2} = (2f)^{1/2}\left(\frac{1}{m}+\frac{1}{M}\right)^{1/2}. \tag{C. 157}$$

In ionic crystals this maximum frequency of optical lattice vibrations can be directly measured. It is determined from the frequency of the so-called residual radiation ("Reststrahlen"), an almost monochromatic radiation left over when white light is multiply reflected from the surfaces of an ionic crystal. We know from optics that strong absorption and strong reflection appear at almost the same frequencies: the reflection maximum is on the short-wavelength side near the corresponding absorption maximum. The position of the latter for the residual radiation yields essentially the maximum frequency of the optical lattice vibrations and thereby the spring constant f [cf. eqn. (C. 157)]. In Table 12 we have compiled wavelengths of residual-radiation absorption maxima for alkali halides.

TABLE 12. Wavelengths for the absorption maxima of residual radiation in alkali halides (after J. T. Houghton and S. D. Smith, *Infrared Physics* [Oxford University Press, 1966], p. 95)

λ [μm]	F	Cl	Br	I
Li	32.9	58.5	58.5	69.4
Na	40.6	61.0	73.9	85.6
K	52.1	70.8	84.7	99
Rb	62.5	84.1	112	134
Cs	78.6	101	135	163

(b) *LINEAR MONATOMIC CRYSTAL MODEL*

When we assume equal masses $m^* = m = M(\rightarrow \omega^* = \omega_0 = \Omega_0)$ the dispersion relation (C. 150) takes the following form:

$$\omega^+ = 2\omega^* \left|\cos\frac{qa}{2}\right|,$$
$$\omega^- = 2\omega^* \left|\sin\frac{qa}{2}\right|. \tag{C. 158}$$

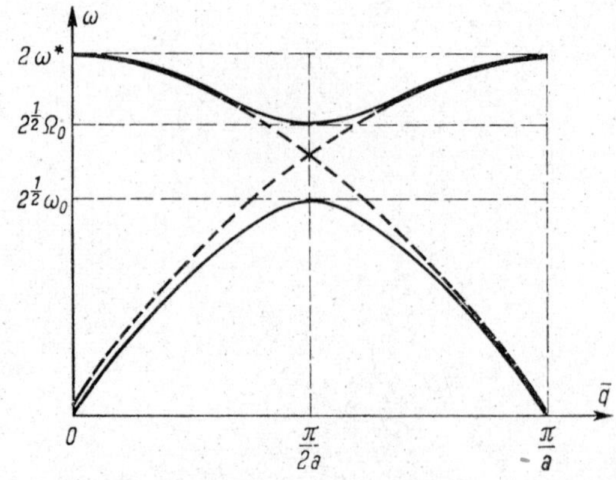

FIG. 51. Dispersion relations $\omega(q)$ for the one-atomic (dashed curves) and the two-atomic one-dimensional crystal.

Now the two branches $\omega_+(q)$ and $\omega_-(q)$ have the same limited frequency range $0 \leqslant \omega \leqslant 2\omega^*$. A "forbidden" frequency interval does therefore no longer exist between the two branches (cf. Fig. 51). The optical and the acoustic branches are connected as follows:

$$\omega_+(q): \quad 0 \leqslant \bar{q} \leqslant \frac{\pi}{2a}, \quad \text{optical oscillations,}$$

$$\frac{\pi}{2a} \leqslant \bar{q} \leqslant \frac{\pi}{a}, \quad \text{acoustic oscillations;}$$

$$\omega_-(q): \quad 0 \leqslant \bar{q} \leqslant \frac{\pi}{2a}, \quad \text{acoustic oscillations,}$$

$$\frac{\pi}{2a} \leqslant \bar{q} \leqslant \frac{\pi}{a}, \quad \text{optical oscillations.}$$

The branch of $\omega_+(q)$ is physically significant only if neighbouring masses have different signs of their electrical charges, i.e. if the unit cell contains two different particles. When we formulate the equations of motion for the linear chain of perfectly identical particles, we obtain only the dispersion relation $\omega_-(q)$. In this case the Brillouin zone is twice as large as for particles which are not perfectly identical.

Under the assumption that the masses m^*, m, and M are related by $m^* = 2mM/(m+M)$, we compare the dispersion relations of the monatomic and the diatomic lattice models in Fig. 51.

(c) *LATTICE VIBRATION SPECTRUM FOR THE LINEAR MODEL*

For an infinite crystal model the number of eigenoscillations is infinite in every interval $d\omega$ of allowed frequencies as all $\bar{q}$ values may occur. The solutions (C. 45) are subject to boundary conditions if the motion is assumed to be spatially periodic. Satisfaction of the boundary conditions yields a selection of finitely many $\bar{q}$ values and thus a finite number of eigenfrequencies. The spectrum of the oscillations informs us about the number of frequencies $\varrho(\omega)\, d\omega$ in the interval ω to $\omega + d\omega$.

The spatial period of the oscillations of the diatomic linear chain (lattice constant $2a$) is $2aN$, i.e., N unit cells form a fundamental region $L = 2aN$ (cf. p. 56). The periodic boundary conditions are the following:

$$\begin{aligned} \xi_{2n} &= \xi_{2(n+N)}\,, \\ \xi_{2n-1} &= \xi_{2(n+N)-1}\,. \end{aligned} \tag{C. 159}$$

Hence we obtain the selection of q values

$$q = n\frac{2\pi}{2aN} = n\frac{2\pi}{L} \tag{C. 160}$$

where $n = 0, 1, 2 \ldots N$.

The q values are distributed at equal separations $2\pi/L$. The number Z_q^0 of the q values from $q = 0 \ldots q$ is given by

$$Z_q^0 = \frac{L}{2\pi} q. \tag{C. 161}$$

In the Brillouin zone ($q = 0 \ldots 2\pi/2a$) the number of allowed q values is equal to N. The numbers of eigenvalues with wave numbers between q and $q+dq$ is then

$$dZ_q = \frac{L}{2\pi}\, dq = \frac{2aN}{2\pi}\, dq. \tag{C. 162}$$

Using the corresponding dispersion relation $\omega_{\pm}(q)$ we obtain the frequency density as a function of $\omega_{\pm}$:

$$dZ_\omega = \frac{dZ_q}{d\omega}\, d\omega = \varrho(\omega)\, d\omega \tag{C. 163}$$

with

$$\varrho(\omega) = \frac{L}{2\pi} \left| \frac{dq}{d\omega} \right|. \tag{C. 164}$$

In the case of equal masses, i.e., for a monatomic chain with lattice constant a, it is thus easy to calculate $\varrho(\omega)$. Let us calculate $\varrho(\omega)$ for the branch $\omega_-(q)$. We obtain from (C. 158)

$$\frac{d\omega}{dq} = \omega^* a \cos\frac{qa}{2} = a\left[\omega^{*2} - \left(\frac{\omega_-}{2}\right)^2\right]^{1/2} \tag{C. 165}$$

and from eqn. (C. 164)

$$\varrho(\omega) = \frac{N}{2\pi}\left[\omega^{*2} - \left(\frac{\omega_-}{2}\right)^2\right]^{-1/2}. \tag{C. 166}$$

The spectrum of eigenfrequencies **(Fig. 52)** is thus different from the Debye spectrum $\varrho(\omega) \propto \omega^2$ [cf. eqn. (C. 58)]. The function $\varrho(\omega)$ is real only for frequencies below the maximum frequency ($\omega < 2\omega^*$).

With the help of eqn. (C. 164) and Fig. 51 we can for the diatomic chain immediately describe the qualitative shape of the frequency spectrum $\varrho(\omega)$ (cf. Fig. 52). The maxima and minima of the branches $\omega_\pm(q)$ correspond to an

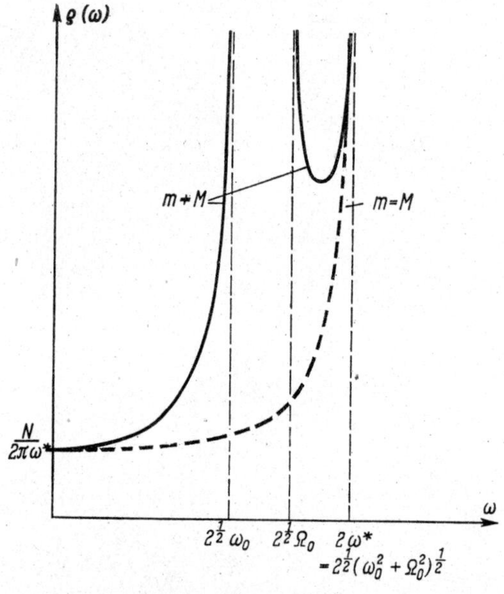

FIG. 52. Frequency spectrum for the one-atomic (dashed curve) and two-atomic one-dimensional crystal.

infinitely high frequency density. The spectrum consists of two parts which are separated by a "forbidden band". This is a typical property of the linear chain model.

(d) *DISPERSION RELATIONS AND LATTICE VIBRATION SPECTRA OF REAL CRYSTALS*

A calculation of the dispersion relations for a real crystal is based on the equation of motion of the lattice particles. Just as in the one-dimensional case we have z independent equations of motion if the unit cell contains z particles. The displacements ξ of the particles from their positions of rest are given by vectors. The vector components correspond to one longitudinal and two transverse waves. In this way we obtain $3z$ linear homogeneous equations which have a non-trivial solution if the corresponding determinant is equal to zero. From this condition we obtain an equation for ω^2 of degree $3z$ (cf. p. 90). Thus the dispersion relations comprise generally $3z$ frequency branches (3 acoustic and $3z-3$ optical branches); in concrete cases, of course, some of these branches may coincide.

As in eqn. (C. 164) we can use the dispersion relations $\omega(\boldsymbol{q})$ to calculate the frequency spectrum $\varrho(\omega)$. The frequency density of a branch reaches maxima for values of ω for which the derivative of the dispersion relation with respect to the wave number q vanishes. The total frequency density is obtained as the sum over the frequency densities of all dispersion branches. The spectrum does contain a maximum frequency but often no "forbidden" frequency bands. Figure 53 shows a calculated frequency spectrum for silver. The calculation was based on the assumption of central forces between the nearest and next-nearest neighbours; for the lattice constant and the three elastic constants experimental data were used.

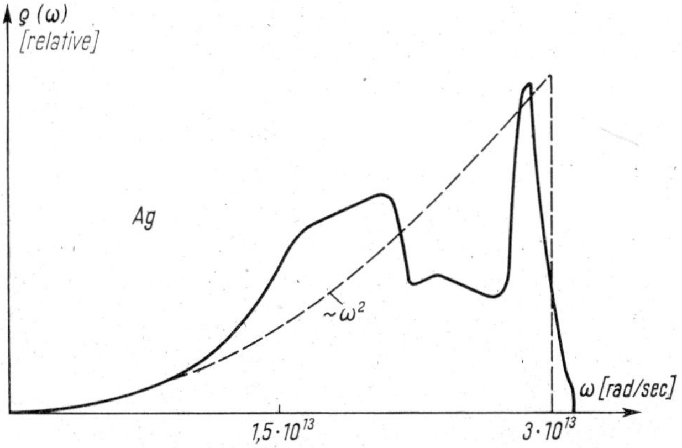

FIG. 53. Frequency spectrum of phonons in silver calculated under the assumption of central forces between the nearest and next-nearest neighbours and using three elastic constants and the lattice constant (from Landolt-Börnstein, vol. 1/4, Springer, Berlin 1955).

The results of the calculations of dispersion relations and frequency spectra of phonons are generally in very good agreement with experiment. A great number of experiments are based on an investigation of inelastic scattering processes which, accounting for the conservation laws of energy and momentum [cf. eqns. (C. 83) and (C. 84)], arise between phonons and photons or real particles, such as, for example, the inelastic scattering of neutrons, X-ray or light quanta. For such scattering experiments thermal neutrons are suited particularly well as their energy and momentum are of the same order of magnitude as the corresponding values which are essential in the production or annihilation of phonons. Therefore scattering results in a noticeable change in energy and momentum of the incident neutrons.

The neutrons are scattered from the nuclei in the crystal by virtue of the nuclear forces. Moreover, because of their spin, the neutrons undergo magnetic interaction with the magnetic moments of the electron shells and the nuclei. The neutron scattering cross-section is different not only for different elements but even for isotopes of one element. It consists of two components, a coherent and an incoherent one. Coherent scattering takes place at identical scattering centres if the scattered waves interfere with one another. Isotopes or scattering centres of different orientations of their spins give rise to incoherent scattering, i.e., the centres act independently of one another and the scattered waves cannot interfere.

Coherent and incoherent scattering can be an inelastic as well as an elastic process. While elastic scattering yields information on the crystal structures and, in particular, on the magnetic structures, inelastic scattering is suitable for lattice dynamics investigations. When we choose the de Broglie wavelength λ_n of the neutrons so large that the Bragg condition is no longer satisfied, i.e. $\lambda_n > 2d_{hkl}$ [cf. eqn. (B. 35)], neutron scattering will be mainly inelastic. The corresponding neutron energies are smaller than the thermal energy kT and we speak of "cold neutron scattering".

The measurement of phonon dispersion relations $\omega(\boldsymbol{q})$ requires a predominantly coherent inelastic scattering of the neutrons. The crystal is irradiated from different directions by monochromatic neutron beams and for each direction of incidence the energy and direction of the scattered neutrons are measured. With the help of the conservation laws the dispersion relations $\omega(\boldsymbol{q})$ of the phonons (cf. Fig. 54) can be derived from the measurements. From the experimentally derived dispersion relations we can then numerically calculate the frequency spectrum $\varrho(\omega)$ (cf. Fig. 55).

In the case of cubic crystals the frequency spectrum $\varrho(\omega)$ of the phonons can also be determined directly by measuring the energy distribution of the inco-

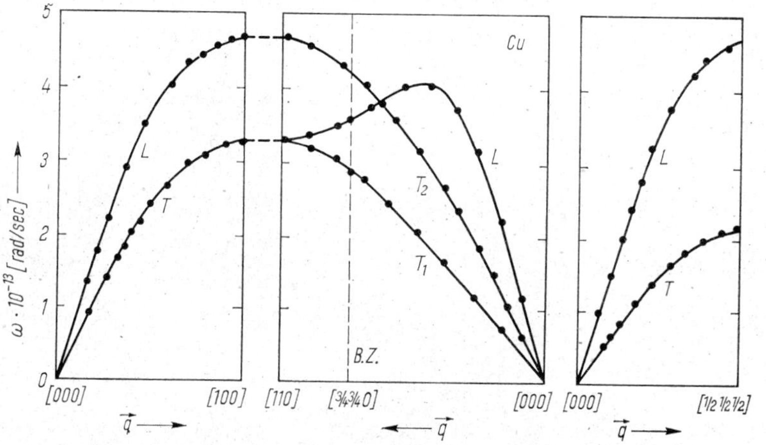

FIG. 54. Branches of the dispersion relation for phonons in copper, derived from inelastic coherent scattering of monochromatic neutrons [after E. C. Svensson, B. N. Brockhouse and J. M. Rowe, *Phys. Rev.* **155**, 619 (1967)].

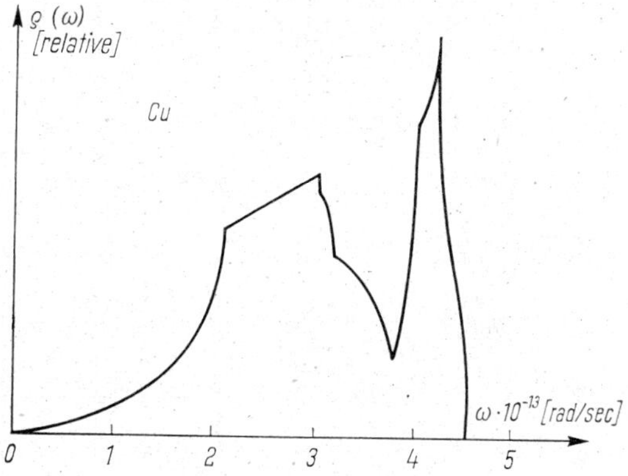

FIG. 55. Frequency spectrum of phonons in copper calculated from the dispersion curve of neutron scattering given in Fig. 54 [after E. C. Svensson, B. N. Brockhouse, and J. M. Rowe, *Phys. Rev.* **155**, 619 (1967)].

herently scattered neutrons. The scattering is then governed only by energy conservation. In the case of one-phonon processes the neutron energy change is given by [cf. eqn. (C. 83)]

$$\frac{\hbar^2}{2m_n}|k^2-k_0^2| = \hbar\omega. \qquad \text{(C. 167)}$$

If the incident neutrons are monochromatic, the number of the scattered neutrons of energy $\hbar^2k^2/2m_n$ is proportional to the number of phonons of energy

$\hbar\omega$. The energy spectrum of the scattered neutrons has therefore essentially the same shape as the phonon frequency spectrum $\varrho(\omega)$.

For cubic crystals with one atom per unit cell the incoherent differential scattering cross-section is, according to Plazcek and van Hove, directly proportional to the frequency density, i.e.,

$$\frac{d^2\sigma_{\text{incoh}}}{d\Omega\, dk} = \frac{S}{4\pi}\frac{2\hbar k^2}{m_{\text{nucl}}k_0}\exp(-2M)\frac{(\boldsymbol{k}-\boldsymbol{k}_0)^2}{|k^2-k_0^2|}\left[\frac{1}{\exp\left(\frac{\hbar\omega}{kT}\right)-1}+\frac{1}{2}(1\pm 1)\right]\varrho(\omega), \tag{C. 168}$$

where Ω is the solid angle, S is the incoherent scattering cross-section of the nucleus, m_{nucl} is the nuclear mass, $\exp(-2M)$ is the Debye–Waller factor; $+$ stands for phonon emission, $-$ for phonon absorption.

For a direct measurement of $\varrho(\omega)$ we can use crystals of substances which scatter electrons mainly incoherently. This is observed for only a few elements; as an example we give measurements for nickel and vanadium in Fig. 56.

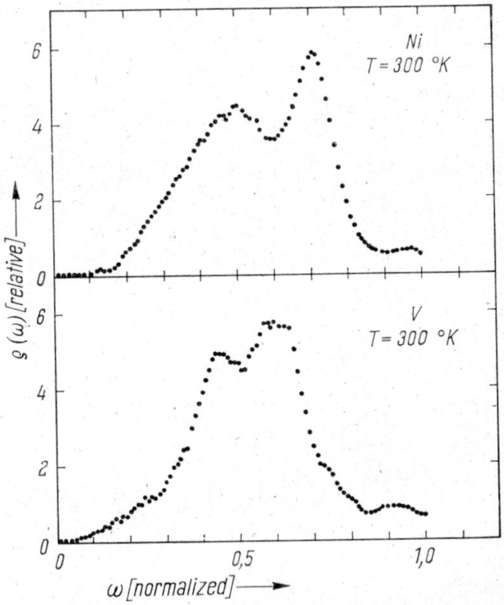

FIG. 56. Frequency spectrum of phonons for nickel and vanadium derived from inelastic incoherent neutron scattering (after B. Mozer, K. Otnes, and H. Palevsky, in *Lattice Dynamics*, edited by R. F. Wallis, Pergamon, Oxford 1965).

5. DEBYE'S EQUATION OF STATE FOR SOLIDS

The general equation of state of a solid is a relation between the mechanical stress tensor, the strain tensor, and the temperature. We restrict ourselves to hydrostatic pressures p. As in the case of liquids and gases, the equation of state will then give the relation between pressure p, volume V, and temperature T of the solid.

A derivation of the equation of state is expediently based on the definition of the free energy $F(V, T)$,

$$F = U - TS, \tag{C. 169}$$

where U is the internal energy and S is the entropy.

The equation of state follows immediately from the thermodynamic equation

$$p = -\left(\frac{\partial F}{\partial V}\right)_T, \tag{C. 170}$$

which represents the required relation between p, V, T and which is obtained as follows:

From the definition eqn. (C. 169) it follows that

$$dF = dU - T\,dS - S\,dT. \tag{C. 171}$$

With

$$dS = \frac{\delta Q}{T} \tag{C. 172}$$

and the first law of thermodynamics

$$dU = \delta Q + \delta A \tag{C. 173}$$

(δQ is the differential of added heat, δA that of the work performed on the solid), we can rewrite eqn. (C. 171) in the form

$$dF = \delta A - S\,dT. \tag{C. 174}$$

As we consider only variations of state in terms of p, V, T, we have [cf. eqn. (C. 5)]

$$\delta A = \delta A_c = -p\,dV. \tag{C. 175}$$

F is a state quantity and therefore dF is a total differential. Using eqn. (C. 175) we obtain from (C. 174)

$$p = -\left(\frac{\partial F}{\partial V}\right)_T \tag{C. 176}$$

and

$$S = -\left(\frac{\partial F}{\partial T}\right)_V. \tag{C. 177}$$

The internal energy of the solid is the result of various contributions (cf. p. 45) from which we consider only U_l and U_v. With eqn. (C. 8) we obtain an analogous decomposition of the free energy into a lattice contribution F_l and a vibrational contribution F_v:

$$F = U_l + U_v - TS = F_l + F_v . \qquad \text{(C. 178)}$$

Like U_l, F_l is practically independent of T; therefore

$$F_l = U_l \quad \text{and} \quad \left(\frac{\partial F}{\partial T}\right)_V = \left(\frac{\partial F_v}{\partial T}\right)_V . \qquad \text{(C. 179)}$$

Using eqns. (C. 177) and (C. 178) we can write

$$F_v = U_v - TS = U_v + T\left(\frac{\partial F_v}{\partial T}\right)_V \qquad \text{(C. 180)}$$

and

$$U_v = F_v - T\left(\frac{\partial F_v}{\partial T}\right)_V = \left(\frac{\partial\left(\frac{F_v}{T}\right)}{\partial\left(\frac{1}{T}\right)}\right)_V . \qquad \text{(C. 181)}$$

A substitution of eqn. (C. 178) in (C. 170) yields

$$p + \left(\frac{\partial U_l}{\partial V}\right)_T = -\left(\frac{\partial F_v}{\partial V}\right)_T . \qquad \text{(C. 182)}$$

The quotient $(\partial F_v/\partial V)_T$ can be determined only if we know the vibrational energy U_v. We shall use the Debye model (cf. pp. 55 ff.) according to which U_v has the form [cf. eqn. (C. 64)]:

$$U_v = Tf\left(\frac{\Theta}{T}\right). \qquad \text{(C. 183)}$$

It can be shown by means of eqn. (C. 181) that the free energy, too, has the form

$$F_v = Tg\left(\frac{\Theta}{T}\right). \qquad \text{(C. 184)}$$

The volume dependence of F_v is contained in the Debye temperature: a change in volume gives rise to a change of the eigenfrequencies ω_i and thus of the frequency density $\varrho(\omega)$. Since, however, the number $3N$ of oscillators remains constant, the frequency $\omega_{\max} = k\Theta/\hbar$ must change according to eqn. (C. 57).

Then eqn. (C. 182) can be written in the form

$$p + \left(\frac{\partial U_l}{\partial V}\right)_T = -T\frac{\partial g}{\partial \Theta}\frac{\partial \Theta}{\partial V} = -\frac{\partial g}{\partial\left(\frac{\Theta}{T}\right)}\frac{\partial \Theta}{\partial V} . \qquad \text{(C. 185)}$$

Using eqns. (C. 184) and (C. 181) we obtain

$$p+\left(\frac{\partial U_l}{\partial V}\right)_T = -\frac{\partial\left(\frac{F_v}{T}\right)}{\Theta\partial\left(\frac{1}{T}\right)}\frac{\partial\Theta}{\partial V} = -U_v\left(\frac{\frac{\partial\Theta}{\Theta}}{\frac{\partial V}{V}}\right)\frac{1}{V} \tag{C. 186}$$

and hence the Debye equation of state

$$\left[p+\left(\frac{\partial U_l}{\partial V}\right)_T\right]V = \gamma U_v \tag{C. 187}$$

with the Grüneisen parameter

$$\gamma = -\frac{d(\log\Theta)}{d(\log V)}. \tag{C. 188}$$

According to eqn. (C. 67) the vibrational energy for $T \gg \Theta$ is $U_v \approx 3NkT$. The equation of state is then similar to the van der Waals equation of real gases and the term $(\partial U_l/\partial V)_T$ is an analogue to the internal pressure.

If $p = 0$ and $T = 0$ ($\rightarrow U_v = 0$) the equation of state becomes identical with the condition for thermodynamic equilibrium [cf. eqn. (C. 20)]

$$\left(\frac{\partial U_l}{\partial V}\right)_T = 0. \tag{C. 189}$$

The Grüneisen parameter given by eqn. (C. 188) describes the relative change of the limiting frequency ω_{max} and the Debye temperature Θ in terms of the relative change in volume. According to this definition of γ we have the proportionality

$$\Theta \propto \frac{1}{V^\gamma} \qquad (\gamma > 0). \tag{C. 190}$$

A differentiation of the equation of state (C. 187) with respect to the temperature at constant volume yields a relation between γ, the compressibility $\varkappa$, the volume expansion α, and the specific heat c_V, from which the γ values can be determined experimentally:

$$V\left(\frac{\partial p}{\partial T}\right)_V = \gamma c_V. \tag{C. 191}$$

Using the thermodynamic relation

$$\left(\frac{\partial p}{\partial T}\right)_V = -\left(\frac{\partial p}{\partial V}\right)_T\left(\frac{\partial V}{\partial T}\right)_p \tag{C. 192}$$

and the definitions of $\varkappa$ and α [eqns. (C. 1) and (C. 2)] we obtain

$$\gamma = \frac{\alpha V}{\varkappa c_V}. \tag{C. 193}$$

The γ values of numerous elements and compounds are of the same order of magnitude, namely unity (cf. Table 13). This is in agreement with the experimental fact that expansion coefficient and compressibility, which vary for different substances over several orders of magnitude, are proportional to one another.

TABLE 13. Thermodynamic data of cubic crystals at room temperature (after A. Eucken, *Lehrbuch der chemischen Physik* [Akademische Verlagsgesellschaft, Leipzig 1944], p. 675)

	Cubic expansion coefficient $\alpha \cdot 10^6$ [deg^{-1}]	Compressibility $\varkappa \cdot 10^{12}$ [dyne^{-1} cm^2]	Molar heat c_V [erg/deg]	Molar volume V [cm^3]	Grüneisen constant $\gamma = \frac{\alpha V}{\varkappa c_V}$
Na	216	15.8	26.0	23.7	1.25
K	250	33	25.8	45.5	1.34
Cu	49.2	0.75	23.7	7.1	1.96
Ag	57	1.01	24.2	10.3	2.40
Al	67.8	1.37	22.8	10.0	2.17
C	2.9	0.16	5.66	3.42	1.10
Fe	33.6	0.6	24.8	8.1	1.60
Pt	26.7	0.38	24.5	9.2	2.54
NaCl	121	4.2	48.3	27.1	1.61
KCl	114	5.6	49.7	37.5	1.54
KBr	126	6.7	48.4	43.3	1.68
KI	128	8.6	48.7	53.2	2.12

It is easy to see that the Grüneisen parameter is related to the shape of the potential curve $U_l(V)$ which is determined by the coefficients m and n (cf. p. 46): the function $U_l(V)$ indicates a non-linear dependence of the force on the spacing of two lattice particles; the inclination of the force curve is a measure of the elastic modulus which is therefore dependent on the spacing, i.e. on the crystal volume, and which is given essentially by the second derivative of the lattice potential with respect to the volume. The variation in the elastic modulus with volume, that is, a deviation from Hooke's law, gives, according to eqn. (C. 75), a variation of Debye temperature with volume, and is thus, according to eqn. (C. 188), determining the value of γ. According to Grüneisen γ is simply related to the repulsion exponent n:

$$\gamma = \frac{n+2}{6}. \tag{C. 194}$$

The dependence on the attraction exponent m can be neglected if $n \gg m$; this condition is satisfied in many cases (cf. p. 51).

Thus the Grüneisen parameter depends only on the type of lattice forces and is virtually constant for crystals of the same structure. According to eqn. (C. 194) γ is independent of the temperature. This agrees with experiments according to which the ratio of expansion coefficient and specific heat is almost temperature-independent. The compressibility is similarly independent of temperature [cf. eqn. (C. 34)].

References

BAK, T. A. (Editor), *Phonons and Phonon Interactions* (Benjamin, New York 1964).

BARKER, J. A., *Lattice Theories of the Liquid State* (Pergamon, Oxford 1963).

BLACKMAN, M., *The Theory of the Specific Heat of Solids*, in *Reports on Progress in Physics*, Vol. 8 (Physical Society, London 1941).

BLACKMAN, M., *Specific Heat of Solids*, in *Handbuch der Physik*, Vol. 7/1 (Springer, Berlin 1955).

BORN, M., and HUANG, K., *Dynamical Theory of Crystal Lattices* (Oxford University Press, London 1954).

BRILLOUIN, L., *Wave Propagation in Periodic Structures* (Dover, New York 1953).

DELAUNEY, J., *Theory of Specific Heat and Lattice Vibrations*, in *Solid State Physics*, Vol. 2 (Academic Press, New York 1956).

EUCKEN, A., *Lehrbuch der chemischen Physik*, Vol. 2/2 (Akademische Verlagsgesellschaft, Leipzig 1944).

GOPAL, E. S. R., *Specific Heats at Low Temperatures* (Heywood, London 1966).

GSCHNEIDNER, K. A., JR., *Physical Properties and Interrelationships of Metallic and Semimetallic Elements*, in *Solid State Physics*, Vol. 16 (Academic Press, New York 1964).

KEESOM, P. H., and PEARLMAN, M., *Low Temperature Heat Capacity of Solids*, in *Handbuch der Physik*, Vol. 14/1 (Springer, Berlin 1956).

LEIBFRIED, G., *Gittertheorie der mechanischen und thermischen Eigenschaften der Kristalle*, in *Handbuch der Physik*, Vol. 7/1 (Springer, Berlin 1955).

MARADUDIN, A. A., MONTROLL, E. W., and WEISS, G. H., *Theory of Lattice Dynamics in the Harmonic Approximation*, in *Solid State Physics*, Suppl. 3 (Academic Press, New York 1963).

MITRA, S. S., *Vibration Spectra of Solids*, in *Solid State Physics*, Vol. 13 (Academic Press, New York 1962).

PARKINSON, D. H., *The Specific Heat of Metals at Low Temperature*, in *Reports on Progress in Physics*, Vol. 21 (Physical Society, London 1958).

SHAM, L. J. and ZIMAN, J. M., *The Electron–Phonon Interaction*, in *Solid State Physics*, Vol. 15 (Academic Press, New York 1963).

SLATER, J. C., *Introduction to Chemical Physics* (McGraw-Hill, New York 1939).

STEVENSON, R. W. H. (Editor), *Phonons* (Oliver and Boyd, Edinburgh 1966).

TOSI, M. P., *Cohesion of Ionic Solids in the Born Model*, in *Solid State Physics*, Vol. 16 (Academic Press, New York 1964).

VOGT, E., *Physikalische Eigenschaften der Metalle*, Vol. 1 (Akademische Verlagsgesellschaft, Leipzig 1958).

WALLIS, R. F. (Editor), *Lattice Dynamics* (Pergamon, Oxford 1965).

D. Imperfections

So far we have treated crystals as perfect periodic structures ("perfect crystals"). Real crystals contain time-averaged deviations from the ideal structure, and these fundamentally cannot be eliminated. All deviations from a strict three-dimensional periodicity of the crystal structure are called lattice defects or imperfections. Lattice defects which do not entail a loss of the so-called long-range order in the crystal are called "imperfections of the first kind". Such imperfections are, for example, the lattice vibrations (treated in the previous chapter); the thermal motion of the lattice particles represents a perturbation of the strictly periodic structure which, however, vanishes when averaged over time. The imperfections of the first kind also comprise the structural and chemical imperfections which may occur in numerous combinations and which will be dealt with in later sections. In the case of "imperfections of the second kind", which will not be considered here, the solid-state structure has no long-range order, i.e. the structural particles have no "ideal" positions and display statistical fluctuations in their positions relative to the nearest neighbours. This applies to paracrystalline and amorphous substances (cf. pp. 1 ff.).

Crystal imperfections are essential for an understanding of numerous phenomena, such as strength, plasticity, diffusion, ionic conduction, defect conduction in semiconductors, colour centres, which could in principle not be explained within the framework of the perfect crystal theory.

I. Structural Imperfections

Structural imperfections refer to stoichiometric crystals without impurities, and comprise all deviations from a perfectly regular arrangement of the crystal building blocks in the perfect lattice. According to the spatial dimensions of these imperfections we distinguish point defects (atomic imperfections), line defects, and planar defects.

1. POINT DEFECTS

We distinguish the following point defects (cf. Fig. 57):

(a) *Vacancies.* Regular lattice sites are vacant. Lattice particles have migrated to the surface and have left their sites unoccupied. These defects are also called Schottky defects.

(b) *Interstitials.* Lattice particles have moved from their regular sites, leaving vacancies behind, to interstitial positions in the lattice. The combination of an interstitial and a vacancy is called Frenkel defect.

(c) *Disorder in ordered alloys.* In compounds or ordered alloys lattice sites are occupied by "wrong" atoms.

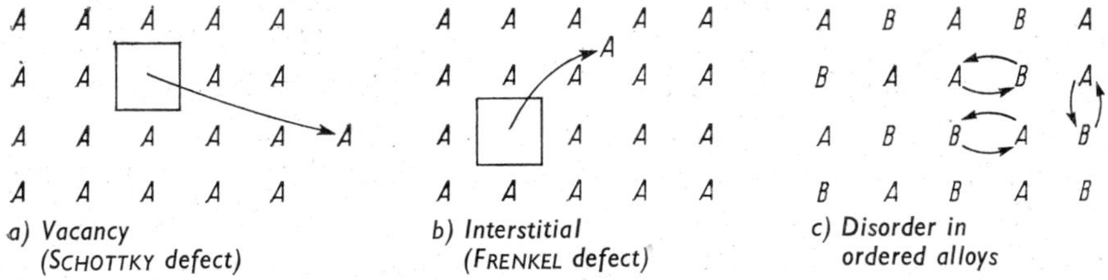

FIG. 57. Point defects.

2. LINE DEFECTS

Line defects are lattice imperfections along closed lines or open lines which end at the crystal surface. The crystal structure is disturbed around these lines in a volume whose radial extension is of the order of about one atomic separation. These one-dimensional lattice defects are called dislocations.

The geometry of a dislocation is generally complex; however, it can be regarded as a combination of two special types, the edge dislocation and the screw dislocation. They will be described in the following.

(a) *Edge dislocation* (cf. Fig. 58): a lattice plane which ends inside the crystal will strongly distort the lattice along its boundaries. The edge of the lattice "half" plane is the dislocation line. Near it the environment of the lattice particles is different from that in the undisturbed crystal and the lattice forces are therefore changed.

(b) *Screw dislocation* (cf. Figs. 59 and 60): the lattice planes perpendicular to a certain direction are degenerated into forming a single connected screw surface. The screw axis is the dislocation line, near it the lattice is strongly distorted. We may illustrate this by means of the following consideration.

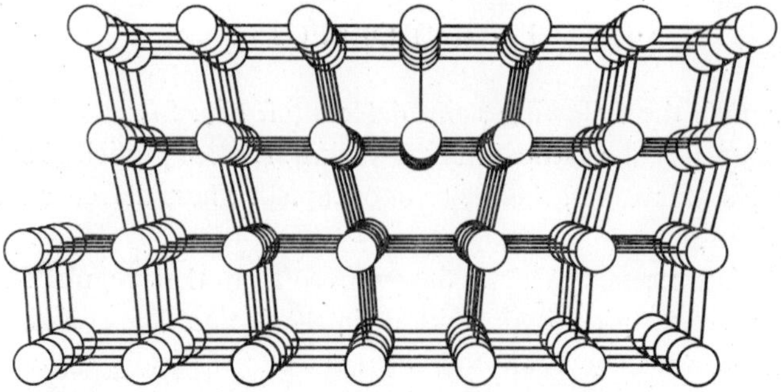

FIG. 58. Arrangement of atoms near an edge dislocation in a cubic crystal [from J. E. Goldman (editor), *The Science of Engineering Materials*, Wiley, New York 1957].

We cut a perfect crystal along a plane parallel to a set of lattice planes, in such a way that the end of the cut in the interior of the crystal is a straight line. The lattice planes on either side of the cut are then displaced with respect to each other in the direction of this straight line by a lattice vector (Burgers vector, see below) and rejoined. In this way we have distorted the lattice planes perpendicular to the straight line to form a screw surface. The straight line is the dislocation line (a displacement parallel to the cut in a direction perpendicular to the straight line would have resulted in an edge dislocation). A dislocation line is essentially different from a series of point defects arranged along a line and cannot be replaced by it.

In general the geometry of a dislocation is determined by the so-called Burgers vector which, as to magnitude and direction, gives the displacement of two parts of the crystal. The Burgers vector must be a lattice vector if the crystal

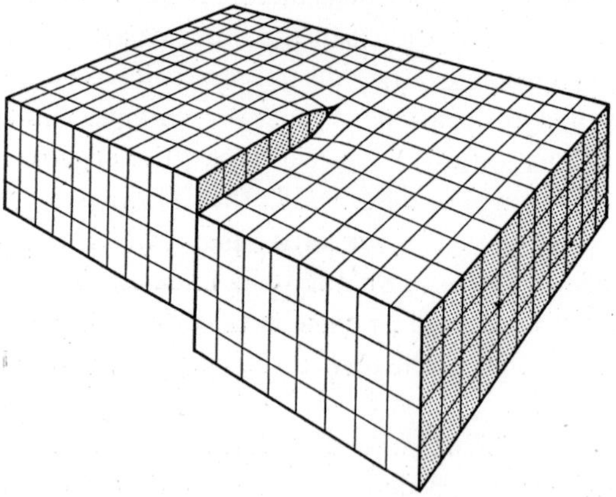

FIG. 59. Screw dislocation (from W. T. Read, Jr., *Dislocations in Crystals*, McGraw-Hill, New York 1953).

FIG. 60. Growth spirals as a consequence of screw dislocations for a SiC crystal [after W. F. Knippenberg, *Philips Res. Rep.* **18,** 161 (1963)].

displays only dislocations and no additional planar defects (see below). In the case of a pure edge dislocation the Burgers vector is perpendicular to the dislocation line, in the case of a screw dislocation it is parallel to this line (cf. Fig. 61). In the general case it makes an angle with the dislocation line of 0 to 90°; the figure shows the steady transition along the dislocation line from edge dislocation to screw dislocation.

Dislocations are generally produced when the crystal is growing, but may also be the result of sufficiently strong external mechanical stress. Dislocations can be made visible by means of chemical etching methods. Specific etching agents destroy the crystal surface predominantly at the end points of dislocation lines where the crystal structure is imperfect. In this way so-called etch pits are formed (cf. Fig. 62). The number of etch pits per unit area is a direct measure of the dislocation density.

Plastic deformation of crystals is a typical effect which cannot be explained by means of the model of a perfect crystal. Although plastic deformations,

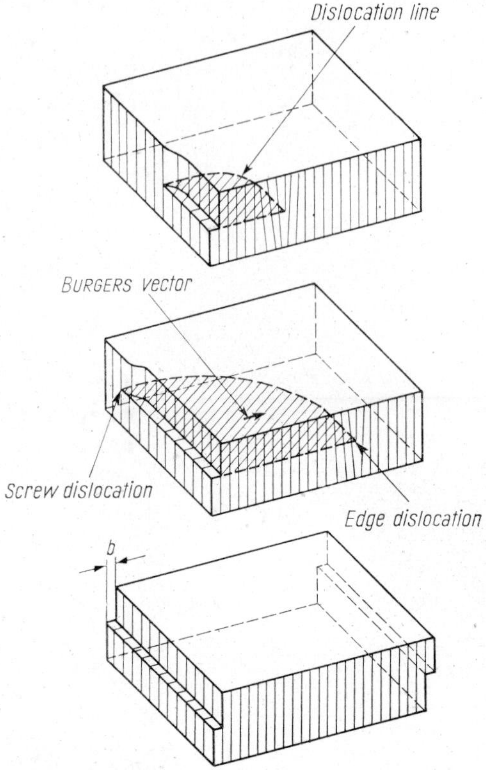

FIG. 61. Edge and screw dislocation; explanation of creep by the m otion of dislocations (after J. J. Gilman).

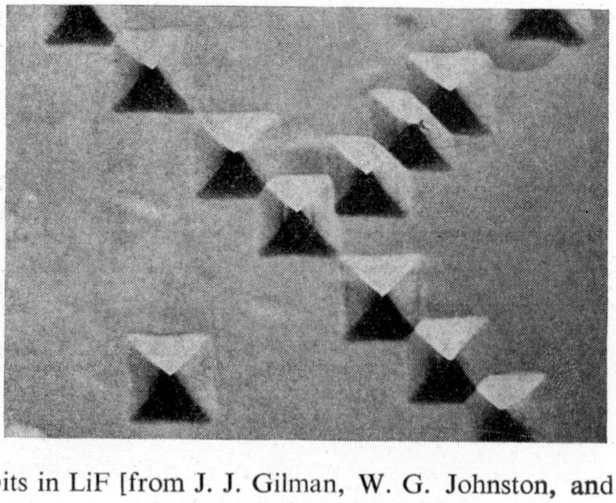

FIG. 62. Etch pits in LiF [from J. J. Gilman, W. G. Johnston, and G. W. Sears, *J. Appl. Phys.* **29**, 747 (1958)].

can be considerable, structure and density of the deformed crystal remain virtually unchanged. In a deformation, parts of the crystal are therefore displaced with respect to one another, i.e., the lattice planes slip. In the case of a perfect crystal the crystal planes should slip as a whole by a lattice constant. The shearing stress necessary for such a process can be estimated; in most cases the corresponding shearing stresses for plastic deformation measured experimentally are smaller by orders of magnitude. This is due to the following: the slipping process does not occur simultaneously in the whole slip plane but stepwise, through the motion of dislocations. This requires much smaller shearing stresses as only the binding forces between the lattice particles in the environment of the dislocation line must be overcome. Figure 61 shows that the motion of dislocations across the crystal is equivalent to the relative displacement of two crystal parts. The effective component of shear causing the slip is parallel to the Burgers vector of the dislocation.

3. PLANAR DEFECTS

Two-dimensional lattice defects (planar defects) are surfaces inside the crystal at which the periodic structure is disturbed. We distinguish essentially two types which may occur also in combinations:

(a) *Grain boundaries.* The orientation of the crystal on one side of the surface is different from that on the other. A grain boundary may be considered

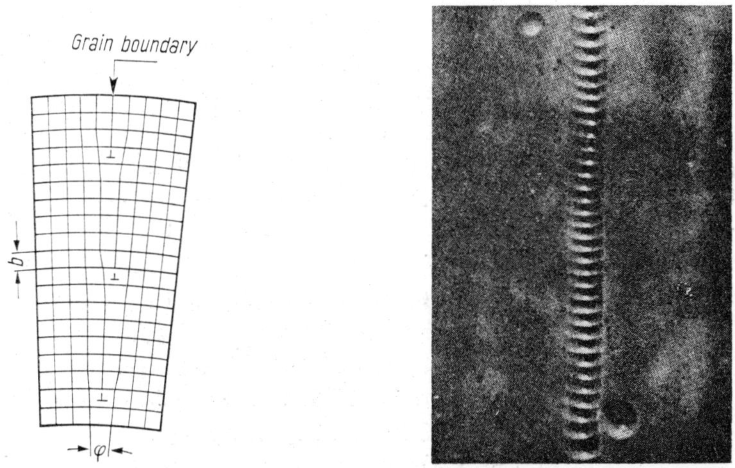

FIG. 63. Grain boundaries in a cubic crystal considered as a series of parallel edge dislocations, (a) schematic picture, (b) made visible by a series of etch pits at the interface between two Ge crystals (from D. Hull, *Introduction to Dislocations*, Pergamon, Oxford 1965)

to be a series of dislocations. A simple case is shown in Fig. 63, where the difference in orientation is small and where the grain boundary can be represented as a series of edge dislocations.

(b) *Stacking faults*. Two lattice planes are mutually shifted by a vector which is *not* a lattice vector. Stacking faults are particularly marked in crystals with close packing. Such crystals are considered as constructed of plane layers of equally large spheres arranged as densely as possible. The spheres of one layer lie in the depressions of the lower layer. There are two possibilities of placing the layers upon each other (cf. Fig. 64):

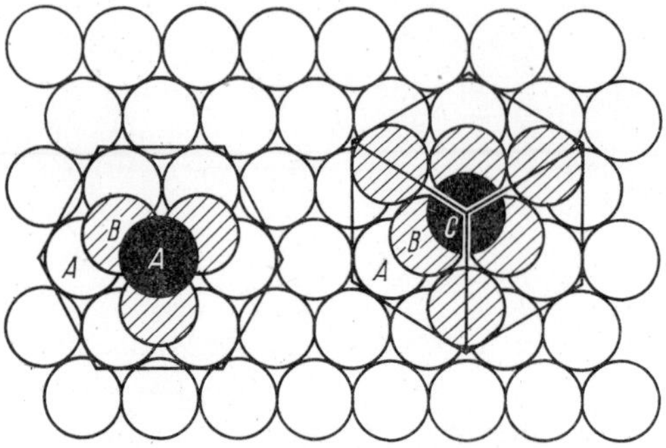

FIG. 64. Hexagonal close packing (layer sequence ABAB ...) and cubic close packing (layer sequence ABCABC...).

1. The spheres of the third layer have the same positions as those of the first layer. The positions being characterized by the letters A, B, C, the layer sequence can be described by ABABAB.... This is the structure of the hexagonal close packing of spheres; the hexagonal axis is normal to the densest layers.

2. The spheres of only the fourth layer have the same positions as those of the first layer. The sequence of layers is thus ABCABCABC... and represents the structure of cubic close packing of spheres (face-centred cubic); the densest layers are normal to the $\langle 1\ 1\ 1 \rangle$ direction.

Stacking faults are the result of irregular sequences of layers, e.g. ABACBAB....

4. THERMODYNAMIC-STATISTICAL THEORY OF ATOMIC IMPERFECTIONS

The density of the line and planar defects depends strongly on the conditions of growth of the crystal and also on external, mechanical stresses. It is not possible to find a quantitative relation between the defect density and the parameters causing the defects. While line and planar defects do not occur in thermodynamic equilibrium, the point defect density above a certain temperature is determined by the thermodynamic equilibrium; we shall discuss this in the following sections.

(a) *ATOMIC IMPERFECTIONS IN MONATOMIC CRYSTALS*

Let us first calculate the density of Schottky defects as a function of temperature under the following assumptions:

1. The crystal consists of N identical atoms; its volume is independent of the temperature.
2. The energy W_S of a Schottky defect is temperature-independent. W_S may be considered to be the difference between the energy necessary to remove a lattice atom to infinity and the energy gained when this atom is brought back to the crystal surface.
3. The number of Schottky defects is n_S; the defects are independent of one another.
4. The eigenfrequencies of the lattice vibrations are not influenced by the defects.

The number n_S of Schottky defects at a temperature T in thermodynamic equilibrium is determined by the condition that the free energy F is a minimum. Since an imperfection increases both the internal energy and the entropy of the crystal, we have instead of eqn. (C. 178):

$$F = U_l + Tg\left(\frac{\Theta}{T}\right) + n_S W_S - kT \ln P_S. \tag{D. 1}$$

The number P_S of possible arrangements of the n_S Schottky vacancies in a crystal of N identical atoms is obtained from combinatorial considerations:

$$P_S = \binom{N}{n_S} = \frac{N!}{(N-n_S)!\,n_S!}. \tag{D. 2}$$

Using Stirling's approximation we can write

$$\ln N! \approx N \ln N - N, \tag{D. 3}$$

which is satisfied for sufficiently large N, and we obtain from (D. 2):

$$\ln P_S \approx N \ln N - (N-n_S) \ln (N-n_S) - n_S \ln n_S. \tag{D. 4}$$

On the basis of the above assumptions a minimization of the free energy in terms of the defect number n_S yields

$$\frac{\partial F}{\partial n_S} = W_S - kT(1 + \ln (N - n_S) - 1 - \ln n_S) = 0. \tag{D. 5}$$

Hence we obtain for the degree γ_S of Schottky imperfection (assuming $n_S \ll N$):

$$\gamma_S \equiv \frac{n_S}{N} = \exp\left(-\frac{W_S}{kT}\right). \tag{D. 6}$$

In thermal equilibrium the number of vacancies is therefore an exponentially growing function of temperature; W_S is the activation energy.

The degree of Frenkel imperfection can be calculated similarly under the same assumptions. The energy of a Frenkel defect, i.e. the energy necessary to bring an atom to an interstitial position, is denoted by W_F. As a Frenkel defect consists of a vacancy and an interstitial atom, the change in entropy consists of two components:

1. n_F vacancies in a crystal of N atoms can be arranged in P_F different ways. The corresponding change in entropy is $k \ln P_F$.

2. The n_F atoms from the vacancies can be distributed in P_I different arrangements over the N_I interstitial positions in the crystal. The corresponding contribution to the entropy is $k \ln P_I$.

The total increase in entropy owing to Frenkel defects will then be given by

$$\Delta S_F = k\left(\ln \frac{N!}{(N-n_F)!\,n_F!} + \ln \frac{N_I!}{(N_I - n_F)!\,n_F!}\right). \tag{D. 7}$$

For the degree γ_F of Frenkel imperfection we now obtain from free energy minimization (under the assumption $n_F \ll N$, $n_F \ll N_I$)

$$\gamma_F \equiv \frac{n_F}{(NN_I)^{1/2}} = \exp\left(-\frac{W_F}{2kT}\right). \tag{D. 8}$$

Unlike eqn. (D. 6), this relation shows that the activation energy for a Frenkel defect is given by *half* the defect energy.

(b) *ATOMIC IMPERFECTION IN IONIC CRYSTALS*

In an ionic crystal both species of ions can display Frenkel defects independently of one another; their concentration is in each case given by eqn. (D. 8). Because of the exponential dependence it will in practice be only the imperfection of the ion species with the smaller defect energy that will manifest itself. We shall now use N to denote the number of molecules.

On the assumption of perfect stoichiometry the Schottky defects in ionic crystals will always appear in pairs (Schottky pairs) so that one vacancy arises for the positive ion and one for the negative ion. This is a necessary condition for electrical neutrality of the crystal surface.

The vacancy left by a positive ion behaves as a negative charge, and vice versa. As a result the vacancies interact electrostatically, and the partners of a Schottky pair can get together and produce a neutral double vacancy. In ionic crystals one thus encounters both dissociated and associated Schottky pairs.

We shall in what follows evaluate the degree of imperfection for dissociated Schottky pairs as in the above cases. In this case also we can treat the vacancies as being independent. The free energy is

$$F = U_l+Tg\left(\frac{\Theta}{T}\right)+n_{SP}W_{SP}-kT\ln P_+ - kT\ln P_-, \tag{D. 9}$$

where

$$P_+ = \binom{N}{n_+} = P_- = \binom{N}{n_-} \tag{D. 10}$$

and

$$n_{SP} = n_+ + n_-; \tag{D. 11}$$

N is the number of molecules, n_{SP} is the number of Schottky pairs, n_+, n_- are the number of vacancies of positive and negative ions, respectively, and W_{SP} is the energy necessary for the production of a dissociated Schottky pair.

From a free energy minimization with respect to the defect number n_{SP} we obtain

$$\gamma_{SP} \equiv \frac{n_{SP}}{N} = \exp\left(-\frac{W_{SP}}{2kT}\right). \tag{D. 12}$$

The degree of imperfection of associated Schottky pairs can be derived similarly. The energy needed to produce an associated Schottky pair is decreased – when compared to W_{SP} – by the energy U_{diss} needed to dissociate the double vacancy, and this energy is thus $W_{SP}-U_{\text{diss}}$. The degree of imperfection of associated Schottky pairs is proportional to $\exp[(U_{\text{diss}}-W_{SP})/kT]$. Comparison with eqn. (D. 12) shows that the degree of association, that is, the ratio of associated to dissociated Schottky pairs, depends in an essential way on the factor $\exp[(U_{\text{diss}}-\frac{1}{2}W_{SP})/kT]$. The dissociation energy U_{diss} is determined by the distance d between different ions of charge q and the dielectric constant ε of the crystal and is approximately given by

$$U_{\text{diss}} \approx \frac{1}{4\pi}\frac{q^2}{\varepsilon_0\varepsilon d}. \tag{D. 13}$$

For alkali-halides $U_{\text{diss}} \approx 1$ eV. It is more complicated to estimate the energy W_{SP} needed to produce a dissociated Schottky pair. For alkali-halides we get as a typical value $W_{SP} \approx 2$ eV. The energies U_{diss} and $\frac{1}{2}W_{SP}$ are often not very different. This indicates that we need, for a reliable evaluation of the degree of association, a more accurate knowledge of these energies than the above estimates give.

So far we have assumed the electric neutrality of ionic crystals to be given by the condition of perfect stoichiometry, that is, for an equal number of positive and negative ions ($N_+ = N_-$). If this condition is violated ($N_+ \neq N_-$), the ionic crystals contain so-called colour centres (cf. pp. 144 ff.). In this case the neutrality of small domains of the crystal is guaranteed by the fact that the vacancies are due to missing *atoms*.

The Schottky and Frenkel defects are special cases of a general atomic imperfection. In practice, because of the exponential dependence of the imperfection degrees, the type of defect with the smaller energy W will predominate. Therefore for a given crystal structure the Frenkel *or* the Schottky defects are typical.

It is possible to estimate theoretically the imperfection energy at absolute zero. Such an estimate yields values of W of the order of 1 eV, that is, 10 to 25% of the lattice energy per building block of the crystal. From a qualitative point of view it is easy to see that the energy for the production of a Schottky defect is proportional to the binding energy of a lattice particle and thus proportional to the lattice energy. The energy for the production of a Frenkel defect will depend additionally on the volume available for interstitials. We therefore have to expect Schottky defects chiefly for close-packed structures.

5. EXTENDED THERMODYNAMIC-STATISTICAL THEORY OF ATOMIC IMPERFECTIONS

We obtain considerable corrections to the degrees of imperfection given above if, instead of the assumptions of p. 113, the changes of the eigenfrequencies near the vacancies and the thermal expansion of the crystal are taken into account.

On the basis of the Einstein model (cf. p. 54) we shall first calculate explicitly the free energy of a perfect crystal. We obtain for the vibrational energy [cf. eqn. (C. 39)]

$$U_v = 3N \frac{\hbar\omega_E}{\exp\left(\frac{\hbar\omega_E}{kT}\right) - 1}. \tag{D. 14}$$

The contribution of the lattice vibrations to the entropy amounts to

$$S = \int_0^{U_v(T)} \frac{dU_v}{T}. \tag{D. 15}$$

Partial integration yields

$$S = \frac{U_v}{T} - 3Nk \ln \left[1 - \exp\left(-\frac{\hbar\omega_E}{kT}\right)\right] \tag{D. 16}$$

and for $T > \hbar\omega_E/k$:

$$S \approx \frac{U}{T} - 3Nk \ln \left(\frac{\hbar\omega_E}{kT}\right). \tag{D. 17}$$

With this we obtain for the free energy of the perfect crystal [cf. eqn. (C. 178)]:

$$F = U_l + 3\, NkT \ln \left(\frac{\hbar\omega_E}{kT}\right). \tag{D. 18}$$

In order to calculate the degree of imperfection we must take into account: (1) that no longer all N particles in the crystal vibrate in the three directions of space with the frequency ω_E. It is assumed that the l nearest neighbours of a defect have a different frequency ω' in the direction of the connection between defect and neighbour. For vacancies $\omega' < \omega_E$, for interstitials the neighbours' frequency ω' as well as the interstitial frequency ω_I is higher: $\omega_I \approx \approx \omega' > \omega_E$. (2) The change in volume of the crystal caused by a variation of temperature results in a temperature dependence of the imperfection energy W.

$$W(T) = W^0 - \beta T. \tag{D. 19}$$

Here W^0 is the imperfection energy at $T = 0$°K and β is a constant. For a monatomic crystal with Schottky imperfections the free energy is given by

$$F = U_l + n_S W_S(T) + kT(3N - n_S l) \ln \left(\frac{\hbar\omega_E}{kT}\right) + n_S lkT \ln\left(\frac{\hbar\omega'}{kT}\right) - kT \ln P_S. \tag{D. 20}$$

From the minimum condition $\partial F/\partial n_S = 0$ we obtain from eqn. (D. 4) for $n_S \ll N$

$$\gamma_S \equiv \frac{n_S}{N} = f \exp\left(-\frac{W_S^0}{kT}\right) = f\gamma_S^0 \tag{D. 21}$$

where

$$f = \left(\frac{\omega_E}{\omega'}\right)^l \exp\left(\frac{\beta}{k}\right). \tag{D. 22}$$

In order to estimate the order of magnitude of the correction factor f we choose the following conditions: a cubic crystal ($l = 6$); Schottky imperfections, $\omega_E/\omega' = 2$, $\beta = 10^{-4}$ eV/deg. We then obtain $f \approx 64 \times 3.2 \approx 200$. In general the correction factor for Schottky imperfection can be of the order of 10^2 to 10^4. In the case of Frenkel imperfections it is generally smaller since $\omega_E/\omega' < 1$. The corrected imperfection degrees γ_{SP} and γ_F are derived similarly as γ_S.

6. EXPERIMENTAL PROOF OF THE EXISTENCE OF ATOMIC IMPERFECTIONS

Density variations. The formation of Schottky vacancies at constant temperature results in an increase of the crystal volume and thus in a density decrease. The difference between measured density and X-ray density is a measure of the Schottky imperfection degree. The X-ray density is given by the atomic mass per unit cell divided by the unit cell volume. Frenkel imperfections leave the density unchanged and cannot be detected by density measurements.

Specific heat anomaly. The generation of lattice defects requires energy and therefore the internal energy of imperfect crystals is higher than that of perfect crystals.

$$U = U_l + U_v + nW, \tag{D. 23}$$

where n is the number of point defects and W is the imperfection energy per defect. The number of defects is given by the degree of imperfection

$$\gamma = \frac{n}{N} = f \exp\left(-\frac{W^0}{kT}\right). \tag{D. 24}$$

Using this result we obtain from eqn. (D. 23) an expression for the specific heat of an imperfect crystal

$$c_V = \left(\frac{\partial U}{\partial T}\right)_V = c_V^{\text{perfect}} + \Delta c_V, \tag{D. 25}$$

where

$$\Delta c_V = Nf \frac{(W^0)^2}{kT^2} \exp\left(-\frac{W^0}{kT}\right). \tag{D. 26}$$

The replacement of W by W^0 in eqn. (D. 23) holds in a good approximation.

The difference between the measured specific heats and the value of c_V^{perfect} (cf. p. 60) increases exponentially as the temperature is increased, and is considerable at higher temperatures (cf. Fig. 66). From eqn. (D. 25) it follows that

$$\ln\left[(c - c_V^{\text{perfect}})T^2\right] = -\frac{W^0}{kT} + \ln\left(Nf\frac{(W^0)^2}{k}\right). \tag{D. 27}$$

When we plot ln $[(c_V - c_V^{\text{perfect}})T^2]$ versus $1/T$ we obtain a straight line whose inclination yields the value of W^0, the imperfection energy at $T = 0°K$, and thus from the ordinate at $1/T = 0$ the correction factor f. In this way the degree γ can be obtained from eqn. (D. 24). Above a certain temperature T_E the imperfection degree satisfies the exponential law derived from thermodynamics. Below this temperature the number of defects remains almost constant, i.e. the imperfections are "frozen in" (cf. Fig. 65). This behaviour is not in con-

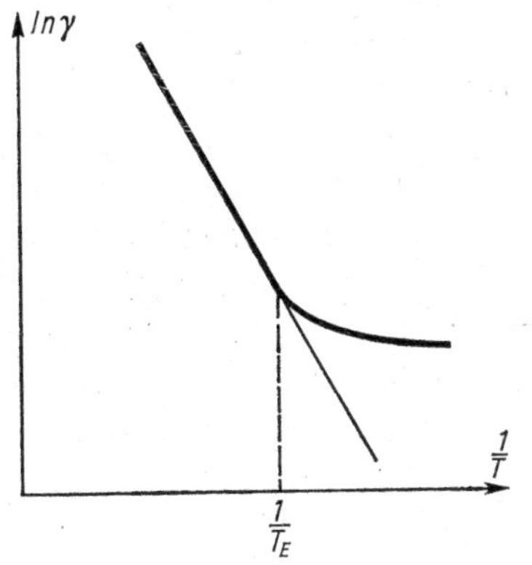

FIG. 65. Degree of imperfection as a function of the inverse temperature. T_E denotes the "freezing-in temperature".

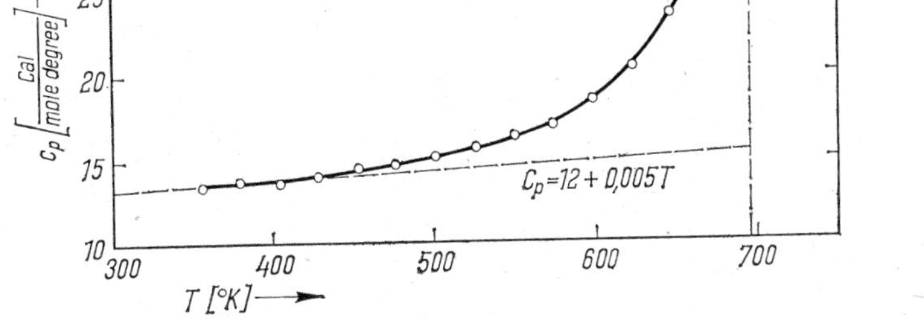

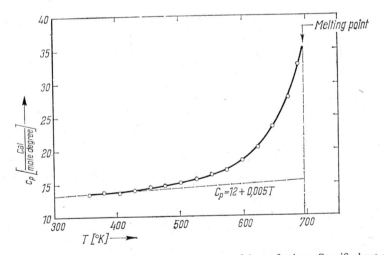

FIG. 66. Specific heat anomaly as a consequence of imperfections. Specific heat at constant pressure of AgBr as a function of temperature [after R. W. Christy and A. W. Lawson, *J. Chem. Phys.* **19**, 517 (1951)].

tradiction with the calculations as they are based on the establishment of thermodynamic equilibrium but do not say anything about the time t the system needs to reach this equilibrium. Below T_E the defect density exceeds the corresponding equilibrium value which is reached as $t \to \infty$.

Measurements of the specific heat anomaly of AgBr (cf. Fig. 66) are in good agreement with the behaviour described above. According to this, AgBr shows Frenkel imperfections with an energy of $W_F \approx 1.3$ eV, the factor $f \approx 10^3$ and the degree of imperfection $\gamma_F \approx 3.7\%$ at the melting point $T_m \approx 700°$K.

7. MASS TRANSPORT IN CRYSTALS

Atomic imperfections are the cause of diffusion as well as ionic conduction in crystals. Point defects possess the important property of a certain mobility under the influence of thermal fluctuations, that is, they may migrate when interacting with phonons. This results in a mass transport in the crystal which is called diffusion. The resulting particle flux is due to either a defect density gradient (true diffusion) or to an applied electric field (ionic conduction).

(a) *DIFFUSION*

Diffusion due to a density gradient can occur in all states of matter. Phenomenological treatments of diffusion problems are based on Fick's laws. The first law

$$\mathbf{s} = -D \operatorname{grad} C \tag{D. 28}$$

together with the continuity equation

$$\operatorname{div} \mathbf{s} + \frac{\partial C}{\partial t} = 0 \tag{D. 29}$$

yields Fick's second law

$$\frac{\partial C}{\partial t} = \operatorname{div} (D \operatorname{grad} C). \tag{D. 30}$$

Here t is the time, $C(x, y, z)$ is the particle density, $\mathbf{s}(x, y, z)$ is the particle flux density, and D is the diffusion coefficient for the particle species considered.

The diffusion coefficient D is generally a second rank tensor. In the following we shall consider D to be a scalar quantity (the diffusion constant) which depends on the temperature and the material but is independent of the density of the migrating particles. We can then rewrite eqn. (D. 30) as

$$\frac{\partial C}{\partial t} = D \nabla^2 C. \tag{D. 31}$$

(b) *ATOMIC DIFFUSION THEORY*

In principle diffusion can take place only in real crystals. The transport of mass is due to a migration of defects; in the case of Schottky and Frenkel defects this form of mass transport is called self-diffusion. In these cases only lattice atoms or ions are involved in the migration process. Impurity diffusion, however, is the result of the motion of impurity atoms or ions.

Essentially we distinguish two mechanisms of atomic diffusion:

1. Vacancy diffusion. An atom or ion moves to a vacancy leaving its original place unoccupied, which results in an opposite migration of vacancies. Thus vacancy diffusion in one direction results in a mass transport in the opposite direction.
2. Interstitial diffusion: Atoms or ions change their interstitial positions.

In the following we shall consider the principles of the theory of atomic diffusion through the example of the vacancy diffusion which is the result of a motion of Schottky defects.

Diffusion of Schottky defects. Figure 67 shows the potential energy curve for a lattice particle in a linear model of a crystal. The energy threshold between the equilibrium positions 1 and 2 is denoted by w. The probability φ_S for a lattice

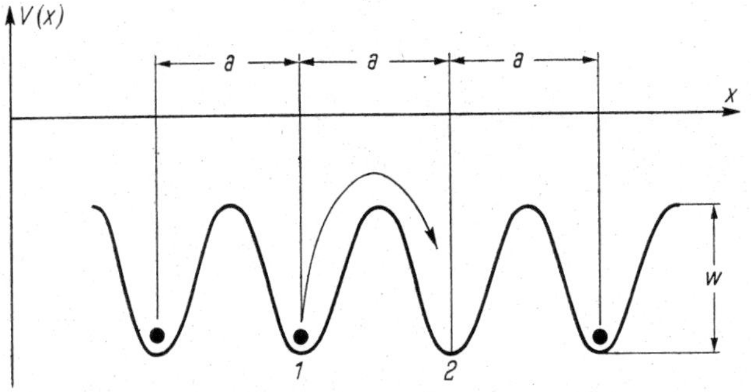

FIG. 67. Potential energy curve of a lattice particle for a linear model of a crystal.

particle to jump from 1 to 2 (the vacancy changing from 2 to 1) is proportional to $\exp(-w/kT)$ and to the jump frequency ν ($\nu \approx 10^{13}\,\text{sec}^{-1}$ is about the highest frequency of lattice vibrations):

$$\varphi_S = \frac{1}{l}\,\nu \exp\left(-\frac{w}{kT}\right), \qquad \text{(D. 32)}$$

where l is the number of nearest neighbours of the defect. This obvious relationship can be explained by quantum-mechanical considerations.†

In order to calculate the particle current density s we use the model described below (Fig. 68); its comparison with the phenomenological equation (D. 28) yields an expression for the diffusion constant. We consider two parallel lattice

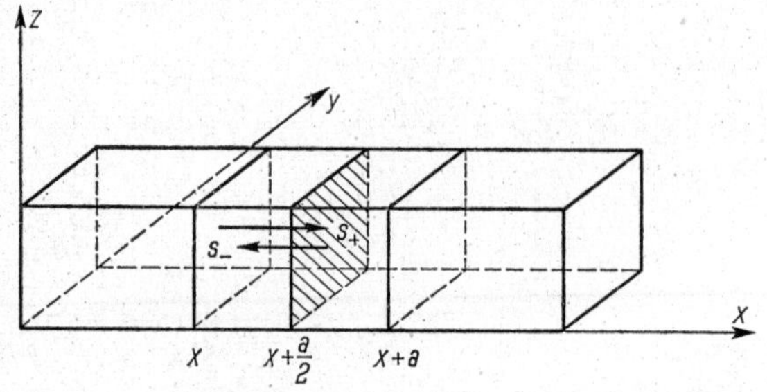

FIG. 68. Model for the calculation of the diffusion current.

planes of separation a. In the x direction (perpendicular to the planes) we assume a density gradient of defects. At the point x there are an_S defects in a volume of size $a \times 1^2$, and at the point $(x+a)$ there are $a(n_S+a\,\partial n_S/\partial x)$ defects. The particle flux s_+ from x to $x+a$ is given by the number of particles $(N-n_S)a$, the jump probability φ_S and the probability γ_S for the particles to find a vacancy in the neighbouring lattice plane:

$$s_+ = (N-n_S)a\varphi_S\gamma_S\,. \tag{D. 33}$$

The particle flux in the opposite direction, from $(x+a)$ to x, is similarly

$$s_- = \left[N-\left(n_S+\frac{\partial n_S}{\partial x}\,a\right)\right]a\varphi_S\gamma_S\,. \tag{D. 34}$$

The net particle flux through the area at $(x+a/2)$ is thus given by

$$s = s_+ - s_- = -a^2\varphi_S\gamma_S\frac{\partial n_S}{\partial x}\,. \tag{D. 35}$$

From a comparison with eqn. (D. 28) and with eqns. (D. 21) and (D. 32) we obtain for the diffusion constant due to vacancy migration

$$D = a^2\varphi_S\gamma_S = \frac{1}{l}\,a^2\nu f_S \exp\left(-\frac{W_S^0+w}{kT}\right) \tag{D. 36}$$

† Cf. J. A. Sussmann, *J. Phys. Chem. Solids* **28**, 1643 (1967).

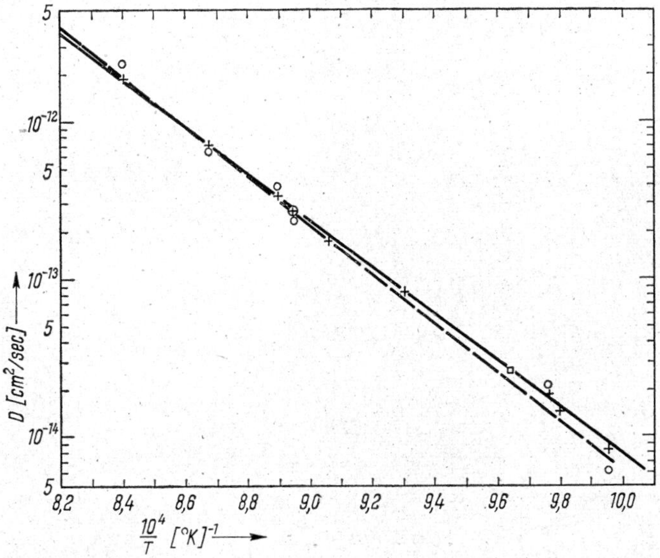

FIG. 69. Self-diffusion coefficient of germanium as a function of the inverse temperature [after H. Widmer and G. R. Gunther-Mohr, *Helv. Phys. Acta* **34,** 635 (1961)].

or

$$D = D_0 \exp\left(-\frac{Q}{kT}\right) \tag{D. 37}$$

with the frequency factor

$$D_0 = \frac{1}{l} a^2 \nu f s \tag{D. 38}$$

and the activation energy of diffusion

$$Q = W_S^0 + w. \tag{D. 39}$$

The most important result is the strong exponential increase of the diffusion constant with the temperature (cf. Figs. 69 and 70). Measurements of the temperature dependence of D yield a value for Q but no information on the imperfection energy W_S^0 or the threshold energy w.

The diffusion constant for interstitial diffusion can be derived in the same way as that for vacancy diffusion. The activation energy Q itself is the sum of a threshold energy (the energy threshold between two neighbouring lattice sites) and *half* the Frenkel defect energy: $Q = w + W_F^0/2$. A self-diffusion of interstitial atoms has so far not been detected with absolute certainty.

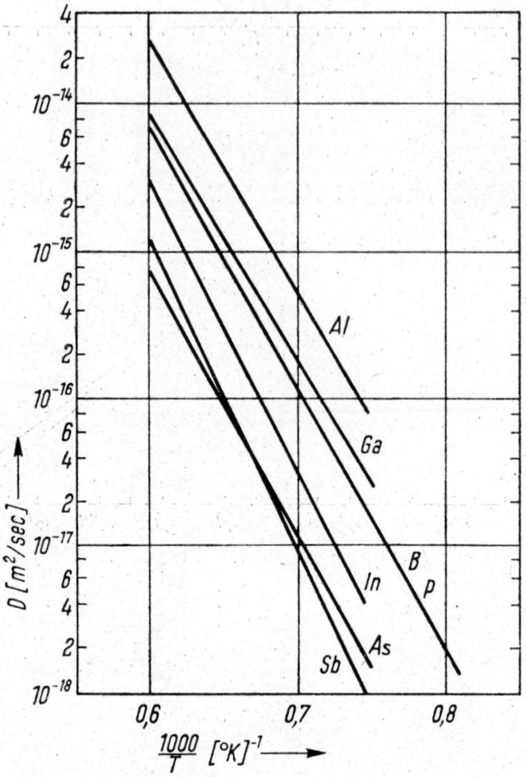

FIG. 70. Diffusion coefficient as a function of the inverse temperature for various elements in silicon (after T. S. Hutchison and D. C. Baird, *The Physics of Engineering Solids*, 2nd ed., Wiley, New York 1968).

The theoretical treatment of impurity diffusion is essentially the same as that of self-diffusion. The jump probability φ_S depends on the nature of the impurities; the threshold energy is a function of the nuclear charge of the foreign atom.

(c) *EXPERIMENTAL DETERMINATION OF THE DIFFUSION CONSTANT*

Especially for practical reasons it is important to know the values of Q and D_0. Low impurity concentrations can decisively determine the electrical properties of crystals, such as semiconductors, or photoconductors. The mechanical properties, too, can be strongly influenced by impurity admixtures which are, for instance, used to improve the properties of metals. The addition of impurities is often performed through diffusion and can be controlled when D or D_0 and Q are known.

The diffusion constant can be immediately determined by means of a tracer method. This is the only method which permits a determination of the self-diffusion constant. The migration is observed by means of "tracers", that is, radioactive isotopes, in terms of time, position, and temperature. The value of D is obtained by solving eqn. (D. 31) which must satisfy the boundary conditions of the experimental set-up.

In such an experiment radioisotopes (for self-diffusion measurements radioisotopes of the host atoms) are evaporated onto the surface of the crystal and the crystal is then heated to the temperature T for a time t in order to accelerate the diffusion of the radioisotopes into the crystal. Subsequently thin layers are cut from the crystal, parallel to the sputtered surface, and their activity is measured.

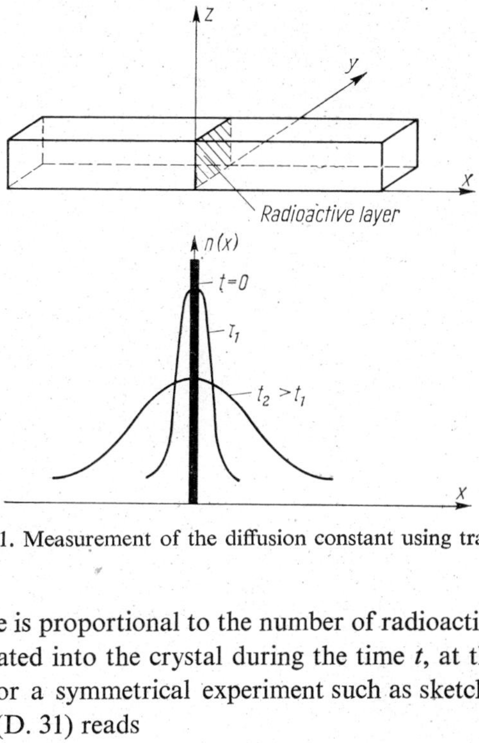

FIG. 71. Measurement of the diffusion constant using tracers.

The counting rate is proportional to the number of radioactive particles $n(x, t)$ which have migrated into the crystal during the time t, at the temperature T, to a depth x. For a symmetrical experiment such as sketched in Fig. 71, the solution of eqn. (D. 31) reads

$$n(x, t) = \frac{n_0}{(\pi D(T)t)^{1/2}} \exp\left(-\frac{x^2}{4D(T)t}\right), \qquad \text{(D. 40)}$$

where n_0 is the radioisotope concentration at $x = 0$, at the time $t = 0$.

Data on the diffusion rate can be obtained from the mean square displacement $\overline{x^2}$ of the migrating particles which, in general, is given by

$$\overline{x^2} = 2Dt. \tag{D. 41}$$

Estimates show that diffusion is a very slow process (cf. diffusion data in Table 14).

TABLE 14. Frequency factors and activation energies for self-diffusion and impurity diffusion. The diffusion coefficient is obtained from eqn. (D. 37) (after B. Chalmers, *Physical Metallurgy* [Wiley, New York 1959], p. 375; N. B. Hannay, *Semiconductors* [Reinhold, New York 1959], p. 244; C. A. Wert and R. M. Thomson, *Physics of Solids* [McGraw-Hill, New York 1964], p. 64)

Material: Crystal	Material: Atoms	D_0 [cm²/sec]	Q [eV]
		Self-diffusion	
Na	Na	0.24	0.45
Cu	Cu	0.20	2.04
Ag	Ag	0.40	1.91
Ge	Ge	10	3.1
U	U	$1.8 \cdot 10^{-3}$	1.20
		Impurity diffusion	
Cu	Zn	0.34	1.98
Ag	Cu	1.2	2.00
	Au	0.26	1.98
	Cd	0.44	1.81
	Pb	0.22	1.65
	Sb	0.17	1.66
Si	Al	8.0	3.47
	Ga	3.6	3.51
	In	16	3.90
	As	0.32	3.56
	Sb	5.6	3.94
	Li	$2.3 \cdot 10^{-3}$	0.66
	Au	$1.1 \cdot 10^{-3}$	1.13
W	Th	1.0	5.4

(d) *IONIC CONDUCTION*

In ionic crystals the diffusion processes are connected with a transport of electric charge. Instead of the density gradient of defects an electric field gives rise to a net ionic current and thus to ionic conduction.

Let us calculate the current of a *single* species of ions (A) in the direction of an internal electric field F. As in this direction the potential energy in the crystal has a gradient (cf. Fig. 72), the threshold energies between one lattice site and neighbouring lattice sites are different; accordingly, the jump probabilities φ_S [cf. eqn. (D. 32)] will be different in different directions. When we use a to

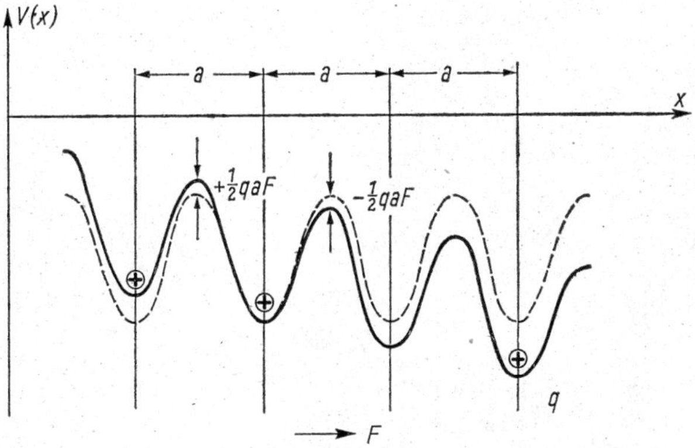

FIG. 72. Potential energy of a positive ion in a one-dimensional crystal in the presence of an electric field.

denote the lattice period parallel to the field direction, the energy threshold for positive ions of charge $+q$ will be lower by $qaF/2$ in the field direction. The jump probability will be

$$\varphi_{SA}^{+} = \frac{1}{l}\,\nu \exp\left(-\frac{w_A - \frac{1}{2}qaF}{kT}\right). \tag{D. 42}$$

Similarly, we shall have for the jump probability in the direction against the field

$$\varphi_{SA}^{-} = \frac{1}{l}\,\nu \exp\left(-\frac{w_A + \frac{1}{2}qaF}{kT}\right). \tag{D. 43}$$

As in eqn. (D. 33) we obtain for the ionic current density in the direction of the field

$$j_A^{+} = q(N - n_A)a\varphi_{SA}^{+}\gamma_{SP} \tag{D. 44}$$

and against the direction of the field

$$j_A^{-} = q(N - n_A)a\varphi_{SA}^{-}\gamma_{SP}\,. \tag{D. 45}$$

Here N is the concentration of molecules, n_A is the vacancy concentration of ions of species A ($n_A = n_{SP}$), and γ_{SP} is the imperfection degree for dissociated Schottky pairs.

Using eqns. (D. 12) and (D. 21) we obtain for the net current of ions of species A, for $n_{SP} \ll N$,

$$j_A = j_A^+ - j_A^- \approx qNf \exp\left(-\frac{W_{SP}^0}{2kT}\right) a(\varphi_{SA}^+ - \varphi_{SA}^-). \tag{D. 46}$$

A current density is generally given by the product of carrier concentration, charge, and drift velocity. In the present case the effective carrier concentration is given by the vacancy concentration

$$n_{SP} = Nf \exp\left(-\frac{W_{SP}^0}{2kT}\right). \tag{D. 47}$$

The drift velocity $\bar{v}$ is obtained from eqns. (D. 42) and (D. 43):

$$\bar{v} = a(\varphi_{SA}^+ - \varphi_{SA}^-) = \frac{2}{l} a\nu \exp\left(-\frac{w_A}{kT}\right) \sinh\left(\frac{qaF}{2kT}\right). \tag{D. 48}$$

In the case of *weak fields* ($qaF/2kT \ll 1$) the function sinh ($qaF/2kT$) can be expanded in a series and the drift velocity becomes proportional to the applied field:

$$\bar{v} \approx \frac{1}{l} a\nu \exp\left(-\frac{w_A}{kT}\right) \frac{qa}{kT} F. \tag{D. 49}$$

In this approximation Ohm's law will hold (the internal field F is assumed to be equal to the applied field):

$$j_A = \sigma_A F \tag{D. 50}$$

with the conductivity

$$\sigma_A = \frac{1}{l} \frac{a^2 q^2 \nu}{kT} Nf \exp\left(-\frac{\frac{1}{2}W_{SP}^0 + w_A}{kT}\right). \tag{D. 51}$$

Thus the ionic conductivity has the same temperature dependence as the diffusion constant [cf. eqn. (D. 37)]:

$$\sigma_A = \sigma_A^0 \exp\left(-\frac{\frac{1}{2}W_{SP}^0 + w_A}{kT}\right) \tag{D. 52}$$

where

$$\sigma_A^0 = \frac{1}{l} \frac{a^2 q^2 \nu}{kT} Nf. \tag{D. 53}$$

In the same way we can describe the conductivity of ions of species B; its difference is mainly due to the different threshold energy w_B. For weak fields the total ionic current in an ionic AB crystal is given by

$$j = (\sigma_A + \sigma_B)F = \sigma F. \tag{D. 54}$$

The weak-field approximation can be applied at $T \approx 300°\text{K}$ for $F \ll 10^6\,\text{V cm}^{-1}$.

In the case of *high field strengths* ($qaF/2kT \gg 1$) the ionic conductivity is a function of the field strength. We obtain from eqns. (D. 46) and (D. 48) for the differential conductivity (only species A is assumed to be mobile):

$$\sigma_A(F) = \frac{\partial j_A}{\partial F} = \sigma_A^0 \exp\left(-\frac{\frac{1}{2}W_{SP}^0 + w_A}{kT}\right) \cosh \frac{qaF}{2kT} \qquad \text{(D. 55)}$$

or

$$\sigma_A(F) \approx \tfrac{1}{2}\sigma_A \exp(\beta F) \qquad \text{(D. 56)}$$

where

$$\beta = \frac{qa}{2kT}. \qquad \text{(D. 57)}$$

In the case of high fields the ionic conductivity increases exponentially with the field strength (Poole's law) until breakdown.

(e) *EXPERIMENTAL RESULTS FOR IONIC CONDUCTION*

The ionic conduction of electric current can be detected through the mass transport. If the electrode consists of the metal which is contained in the crystal in the form of cations (e.g. Ag electrodes on AgCl), the electrode masses are changed by the passage of current: the cathode mass increases, the anode mass decreases. At sufficiently high temperatures the mass changes are equal to the masses deposited according to Faraday's law.

The temperature dependence of ionic conductivity (for $qaF/2kT \ll 1$), represented by a plot of $\ln \sigma$ versus $1/T$, is a curve which consists essentially of two straight lines (cf. Fig. 73). The linear section in the range of high temperatures is characteristic of the substance, while the section at low temperatures displays a "structure-sensitivity", that is, it depends on the impurity concentration and the history of the crystal considered. The clear break in the curves indicates that in the two temperature ranges the ionic conduction is determined by different activation energies.

In the high-temperature range the activation energy corresponds to the sum of half the imperfection energy and the threshold energy [cf. eqn. (D. 52)]. This holds generally also for the case where both ion species contribute to the charge transport through vacancy migration. In this case the threshold energies w_A and w_B are almost equal. For a few typical cases we shall discuss the transport mechanism in the following.

In the range of low temperatures the decisive defect density is no longer determined by the thermal imperfection degree (D. 47) so that the factor $\exp(-W^0/2kT)$ loses its influence on the ionic conductivity. In the simplest

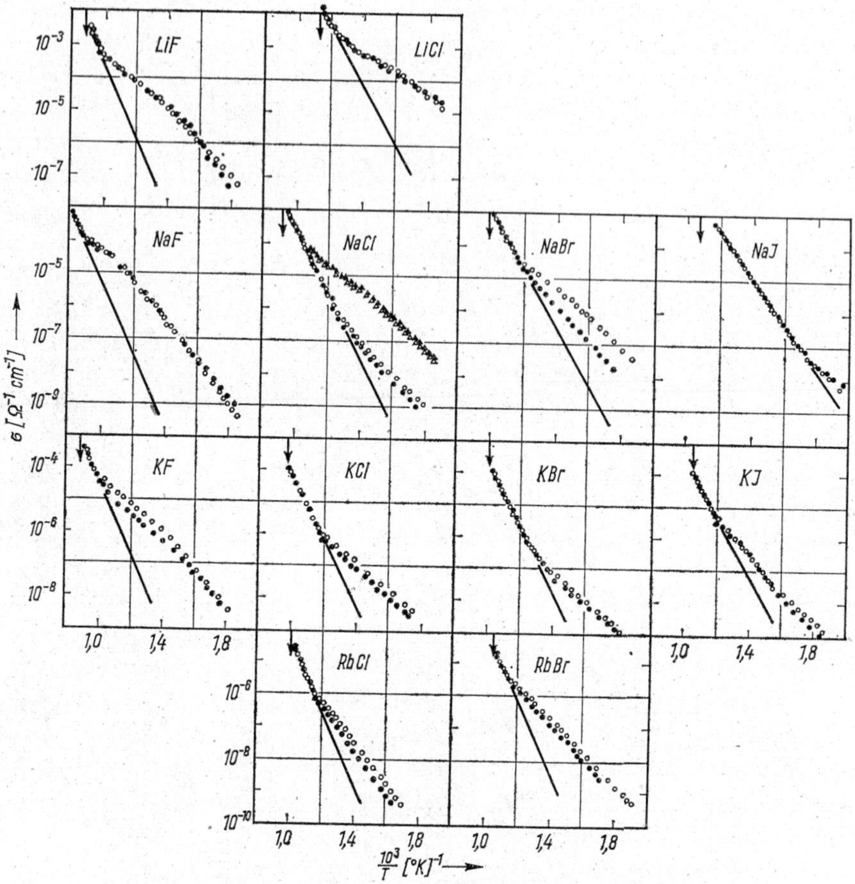

FIG. 73. Ionic conductivity as a function of inverse temperature for various alkali halides. The arrows indicate the melting temperatures. The different points concern different samples [after W. Lehfeldt, *Z. Physik* **85**, 717 (1933)].

case we may assume that the defect concentration below the "freezing-in temperature" T_E (cf. p. 119) remains constant. The activation energy for ionic conduction will then depend only on the threshold energy w. The value of the imperfection energy W^0_{SP} or W^0_F can thus be obtained from the difference of inclination of the straight lines in the ranges of high and low temperatures (cf. Fig. 73).

An incorporation of metal ions of higher valencies raises the freezing-in point T_E and thereby increases the temperature-independent defect concentration. Therefore the ionic conductivity in the low-temperature range increases strongly with the concentration of foreign ions. The influence of bivalent metal ions (Ca^{++}, Sr^{++}, Cd^{++}, Mg^{++}) on the ionic conduction of NaCl, KCl, AgCl, and AgBr has been studied in detail. The conservation of electrical neutrality requires the creation of a vacancy in the cation lattice for each bivalent ion incorporated. The additional vacancy concentration is thus equal to the concentration of the bivalent metal ions and causes the increase in ionic conductivity which, for one species of ions, is given by

$$\Delta\sigma = \frac{1}{l}\,\frac{a^2q^2}{kT}\,\nu\,\Delta N \exp\left(-\frac{w}{kT}\right), \tag{D. 58}$$

where ΔN is the concentration of foreign ions.

The thermal equilibrium concentration of the defects exceeds the concentration of foreign ions only above the temperature T_E. The ionic conductivity is then determined by eqn. (D. 52).

(f) *THE EINSTEIN RELATION*

The ionic conductivity and the diffusion coefficient of a given species of ions are linked by the so-called Einstein relation which, on the assumption of weak fields ($qaF/2kT \ll 1$), can be obtained immediately from eqns. (D. 37) and (D. 52):

$$\frac{\sigma}{D} = \frac{q^2N}{kT}. \tag{D. 59}$$

The Einstein relation is often also given as a relation linking the electrical mobility with the diffusion coefficient of a carrier species. In the case of ionic conduction the mobility b is defined by the relation

$$\sigma = qnb, \tag{D. 60}$$

i.e. the ionic conductivity of species A is expressed in terms of the concentration n and the mobility b of the defects created by the imperfect arrangement of the ions A. From eqns. (D. 59) and (D. 60) we obtain the Einstein relation in its usual form,

$$D_D = \frac{kT}{q}\,b, \tag{D. 61}$$

when we similarly define a diffusion coefficient D_D for defects,

$$D_D = \frac{N}{n}\,D \tag{D. 62}$$

or

$$D_D = a^2\varphi. \tag{D. 63}$$

The diffusion coefficient for defects is obtained in analogy to eqns. (D. 33) to (D. 36) if the defects are treated as migrating particles. Diffusion experiments, however, yield in principle only the diffusion coefficient of the actually migrating particles (cf. pp. 120 ff.).

The Einstein relation has been verified by experiments (cf. Fig. 74). In many cases, however, agreement is found only for the activation energies which have been obtained from measurements of the temperature dependence of diffusion and ionic conduction. Deviations from the Einstein relation are due to, for

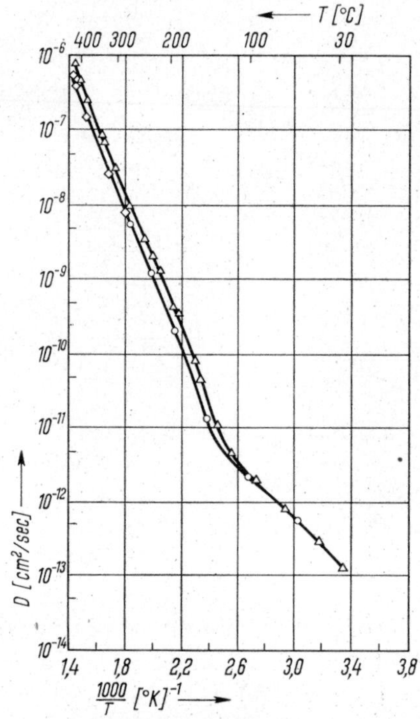

FIG. 74. Temperature dependence of the self-diffusion coefficient of radioactive ^{110}Ag in AgCl; the lower curve shows the diffusion coefficient measured using two probes. The upper curve is calculated from conductivity measurements using the Einstein relation (D. 59) (after W. D. Compton and R. J. Maurer, *J. Phys. Chem. Solids* **1**, 191, 956).

instance, the diffusion mechanisms which do not contribute to charge transport, such as the diffusion of associated Schottky pairs (cf. p. 115). If both ion species are involved in the charge transport, the verification of the Einstein relation requires the knowledge of the so-called transport numbers of the two species of ions. The transport number of ions A is the percentage of σ_A of the total conductivity σ.

Measurements of conductivity and diffusion are not sufficient in order to find out whether the ions migrate via vacancies or via interstitials. This question can be answered by means of additional geometrical considerations (ion radii, structure). According to the present theories the silver halides display Frenkel imperfections. Current transport is due to Ag^+ ions migrating via interstitials or vacancies. The halogen ions do not participate in the charge transport. The alkali halides display Schottky imperfections; here, too, the charge carriers are almost exclusively cations.

8. DIELECTRIC LOSSES IN IONIC CRYSTALS

The atomic imperfection of the crystals entails relaxation effects in their dielectric and elastic properties. We shall restrict ourselves to a treatment of the dielectric losses in ionic crystals which has been related to special atomic imperfections by Breckenridge.

In a dielectric substance electrical energy is consumed if the electric field strength and the displacement density or the polarization are out of phase. One therefore speaks of dielectric losses. They may have the following causes: (1) the electric conductivity of the dielectric, i.e. the phase difference between field and current is smaller than $\pi/2$ (in the perfect dielectric it is equal to $\pi/2$); (2) relaxation effects connected with permanent dipoles, i.e. above a certain frequency of the electric field the orientational polarization is no longer in phase with the field.

The permanent dipoles in pure ionic crystals are associated Schottky pairs (cf. p. 115). In "doped" crystals (e.g. $NaCl+CdCl_2$) additional dipoles exist which are formed by the metal ions of higher valency (Cd^{++}) and the corresponding vacancies. The orientation of the so-called defect dipoles is the result of jumps of ions from neighbouring lattice sites to the vacancies whereby the association with the oppositely charged vacancies or the metal ions of higher valency remains unchanged (cf. Fig. 75). The dielectric relaxation time is a measure of the jumping probability of the ions of the crystal considered.

In the case of constant fields the orientational polarization P_D of the defect dipoles is given by

$$P_D = n_D \alpha_D F, \qquad \text{(D. 64)}$$

where n_D is the concentration of defect dipoles, α_D is the polarizability of the defect dipoles, and F is the internal electric field.

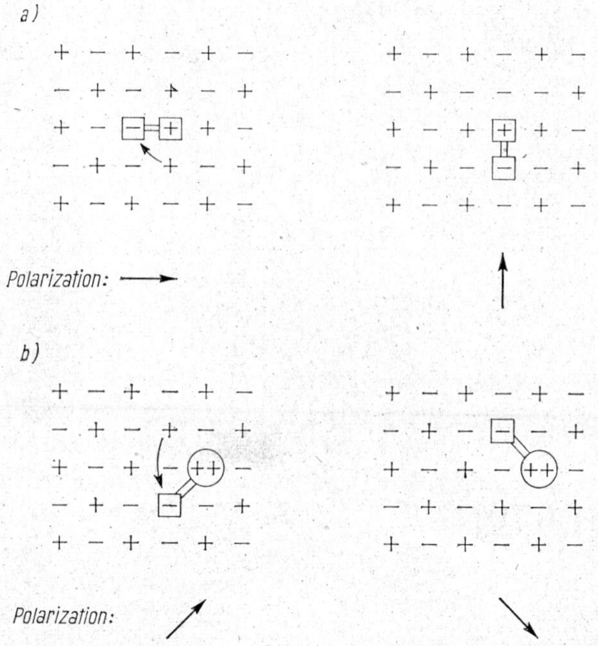

FIG. 75. "Rotation" of dipoles consisting of (a) associated Schottky pairs and (b) foreign ion and associated vacancy.

As in other relaxation effects, the variation with time of P_D is proportional to the difference between the amount of polarization at the time t and some initial or final amount of polarization, P_D^0:

$$\frac{dP_D}{dt} = \frac{1}{\tau}(P_D^0 - P_D(t)); \tag{D. 65}$$

τ is the relaxation time.

In the case of a switch-off process we have

$$t \leqslant 0: \quad \frac{dP_D}{dt} = 0; \qquad P_D^0 = n_D \alpha_D F, \tag{D. 66}$$

$$t > 0: \quad F = 0; \qquad \frac{dP_D}{dt} = -\frac{1}{\tau} P_D(t),$$

$$P_D(t) = P_D^0 \exp\left(-\frac{t}{\tau}\right). \tag{D. 67}$$

In order to obtain a relationship between the relaxation time and the jumping probability of the ions we can use the following model: we denote by n_1 the concentration of the defect dipoles *in* the direction of the field and by n_2 that *opposite* to the direction of the field. The sum of these concentrations must be equal to the total concentration of defect dipoles, n_D:

$$n_D = n_1+n_2. \qquad \text{(D. 68)}$$

The net defect polarization is proportional to the difference between the concentrations n_1 and n_2 and amounts to

$$P_D = qa(n_1-n_2) = qa\,\Delta n. \qquad \text{(D. 69)}$$

Here q is the charge of the ions and a is the separation of the charges of defect dipoles.

The variation with time of P_D is hence obtained as

$$\frac{dP_D}{dt} = qa\left(\frac{dn_1}{dt}-\frac{dn_2}{dt}\right) = qa\frac{d\,\Delta n}{dt}, \qquad \text{(D. 70)}$$

where

$$\frac{dn_1}{dt} = -\varphi_{12}n_1+\varphi_{21}n_2, \qquad \text{(D. 71)}$$

$$\frac{dn_2}{dt} = +\varphi_{12}n_1-\varphi_{21}n_2; \qquad \text{(D. 72)}$$

φ_{12}, φ_{21} are the probabilities for dipole rotation away from and into the direction of the field.

The "rotation" of the dipoles is determined by the jumping probabilities φ_{12} and φ_{21} of the ions in the presence of an electric field (cf. p. 127). According to the jumping direction relative to the field, the dipoles are adjusted in the field direction or opposite to it.

In the case of $F = 0$ [cf. eqn. (D. 42)] we have

$$\varphi_{12} = \varphi_{21} = \varphi = \frac{1}{l'}\nu \exp\left(-\frac{w}{kT}\right), \qquad \text{(D. 73)}$$

where l' is the number of neighbouring ions at equivalent lattice sites which may cause the "rotation" of the defect dipole by a transition to the vacancy.

From this we obtain from eqns. (D. 71) and (D. 72)

$$\frac{dn_1}{dt} = -\frac{dn_2}{dt} \qquad \text{(D. 74)}$$

or from (D. 68)

$$\frac{dn_D}{dt} = 0. \qquad \text{(D. 75)}$$

In analogy to eqns. (D. 66) and (D. 67) we obtain from (D. 69) and (D. 70) for a switch-off process

$$t \leq 0: \quad \frac{dP_D}{dt} = 0; \qquad P_D^0 = qa\,\Delta n_0, \tag{D. 76}$$

$$t > 0: \quad F = 0; \qquad \frac{dP_D}{dt} = qa\frac{d\,\Delta n}{dt}. \tag{D. 77}$$

Using eqns. (D. 69), (D. 71) and (D. 72) we obtain from (D. 77)

$$\frac{dP_D}{dt} = -2\varphi qa\,\Delta n = -2\varphi P_D(t) \tag{D. 78}$$

or

$$P_D(t) = P_D^0 \exp(-2\varphi t). \tag{D. 79}$$

A comparison with eqn. (D. 67) yields the relation between the jumping probability and the relaxation time,

$$\tau = \frac{1}{2\varphi} = \frac{l'}{2\nu}\exp\left(+\frac{w}{kT}\right) = \tau_0 \exp\left(\frac{w}{kT}\right). \tag{D. 80}$$

It is thus possible to determine the threshold energy w independently of the imperfection energy, from the temperature dependence of the relaxation time.

In the experimental determination of τ it is advantageous to use an alternating-field method instead of the switch-off process. In the theory of the relaxation time it is generally accepted that the differential equation (D. 65) can also be applied to an alternating field,

$$F(t) = F_0 \exp(i\omega t). \tag{D. 81}$$

In this case P_D^0 is a function of time and represents the value of P_D in a static field of magnitude $F(t)$, i.e.

$$P_D^0(t) = n_0\alpha_D F_0 \exp(i\omega t). \tag{D. 82}$$

Thus the differential equation (D. 65) takes the form

$$\frac{dP_D}{dt} + \frac{1}{\tau}P_D(t) = \frac{1}{\tau}n_D\alpha_D F_0 \exp(i\omega t). \tag{D. 83}$$

If we put

$$P_D(t) = C\exp(i\omega t) \tag{D. 84}$$

we obtain

$$C = \frac{n_D\alpha_D F_0}{1+i\omega\tau}. \tag{D. 85}$$

The differential equation has the solution

$$P_D(t) = \frac{n_D \alpha_D}{1+i\omega\tau} F_0 \exp(i\omega t). \tag{D. 86}$$

This means that the defect polarization has a periodicity of the same frequency ω as the internal field; the complex factor $(1+i\omega\tau)^{-1}$ indicates a phase shift between field strength and polarization. This phase shift formally yields a complex dielectric constant of the crystal:

$$\varepsilon = \varepsilon' - i\varepsilon''. \tag{D. 87}$$

The conductivity σ of the ionic crystal, which is due to the non-associated defects, has been ignored (cf. p. 140).

The relation between the total polarization P of the crystal and the dielectric constant ε is obtained from a combination of the two general relations

$$P = \varepsilon_0(\varepsilon - 1)E \tag{D. 88}$$

and

$$P = N\alpha F; \tag{D. 89}$$

here E is the external electric field, N is the number of molecules, and α is the total polarizability.

Assuming $E = F$ we can write

$$\varepsilon - 1 = \frac{1}{\varepsilon_0} \frac{P}{F}. \tag{D. 90}$$

The total polarization consists of the lattice polarization P_l and the defect polarization P_D:

$$P = P_l + P_D. \tag{D. 91}$$

The lattice polarization has an ionic and an electronic component: the ionic polarization is due to the mutual displacement of the ions, the electronic polarization is due to the displacement of the electron shell with respect to the nucleus. The response of the lattice polarization to the internal field is virtually inertialess; it is given by

$$P_l = N\alpha_l F, \tag{D. 92}$$

where α_l is the polarizability of the lattice.

A substitution of eqns. (D. 86) and (D. 92) in (D. 90) yields

$$\varepsilon - 1 = \frac{1}{\varepsilon_0} N \left(\alpha_l + \frac{n_D}{N} \frac{\alpha_D}{1+i\omega\tau} \right). \tag{D. 93}$$

The dielectric constant is thus a complex quantity which depends on the frequency of the field. On the assumption of zero electric conductivity, ε will take the real values of ε_S and ε_ω only in the special cases of $\omega = 0$ and $\omega \gg 1/\tau$, respectively. If $\omega = 0$, eqn. (D. 93) takes the form

$$\varepsilon_S - 1 = \frac{1}{\varepsilon_0} N \left(\alpha_I + \frac{n_D}{N} \alpha_D \right). \tag{D. 94}$$

In the case of sufficiently high frequencies $\omega \gg 1/\tau$ the contribution of the orientational polarization can be neglected and eqn. (D. 93) reduces to

$$\varepsilon_\omega - 1 = \frac{1}{\varepsilon_0} N\alpha_I . \tag{D. 95}$$

Both ionic and electronic polarizabilities contribute to the high-frequency dielectric constant ε_ω. It therefore exceeds the optical dielectric constant $\varepsilon_\infty = n^2$ (n is the optical refractive index), which depends only on the electronic polarizability.

Using eqns. (D. 93), (D. 94), and (D. 95) we can express ε in terms of the special values ε_S and ε_ω:

$$\varepsilon = \varepsilon_\omega + \frac{\varepsilon_S - \varepsilon_\omega}{1 + (\omega\tau)^2} - i\omega\tau \frac{\varepsilon_S - \varepsilon_\omega}{1 + (\omega\tau)^2} = \varepsilon' - i\varepsilon'' . \tag{D. 96}$$

A measure of the dielectric loss is the so-called loss angle δ which is defined by

$$\tan \delta = \frac{\text{Im}(\varepsilon)}{\text{Re}(\varepsilon)} . \tag{D. 97}$$

From (D. 96) we hence obtain

$$\tan \delta = \frac{(\varepsilon_S - \varepsilon_\omega)\omega\tau}{\varepsilon_S + \varepsilon_\omega(\omega\tau)^2} . \tag{D. 98}$$

Plotted versus $\omega\tau$, the loss angle has a peak at

$$(\omega\tau)_{\tan \delta_{\max}} = \left(\frac{\varepsilon_S}{\varepsilon_\omega} \right)^{1/2} . \tag{D. 99}$$

As in general ε_S and ε_ω have similar values, we can write to a good approximation

$$(\omega\tau)_{\tan \delta_{\max}} \approx 1. \tag{D. 100}$$

If at different constant temperatures those frequencies $\omega(\tan \delta_{\max})$ are determined for which the loss angles are largest, it is possible to obtain values for the relaxation time $\tau(T)$. A plot $\ln \tau$ vs $1/T$ yields, according to eqn. (D. 80), the threshold energy w as well as the jumping frequency ν (eigenfrequency of the ions). Experimentally it is simpler to determine the position of the loss peak on the temperature scale at different, constant frequencies ω.

The absolute value of tan $\delta_{\max}$ is "structure-sensitive" and proportional to the number n_D of defect dipoles. It follows from eqns. (D. 98) and (D. 99) that

$$\tan \delta_{\max} = \frac{\varepsilon_S - \varepsilon_\omega}{2(\varepsilon_S \varepsilon_\omega)^{1/2}} . \qquad \text{(D. 101)}$$

Subtracting eqn. (D. 95) from (D. 94) we obtain

$$\varepsilon_S - \varepsilon_\omega = \frac{1}{\varepsilon_0} n_D \alpha_D . \qquad \text{(D. 102)}$$

Hence we can express the maximum loss angle in terms of the number of defect dipoles,

$$\tan \delta_{\max} = \frac{1}{2\varepsilon_0(\varepsilon_S \varepsilon_\omega)^{1/2}} n_D \alpha_D . \qquad \text{(D. 103)}$$

A comparison of eqns. (D. 64) and (D. 69) yields the polarizability α_D of the defects:

$$\alpha_D = \frac{qa\,\Delta n}{n_D F} . \qquad \text{(D. 104)}$$

The net concentration Δn of the defect dipoles in the direction of the field depends on the field strength F. Using eqns. (D. 42) and (D. 43) we find for eqns. (D. 71) and (D. 72)

$$\frac{dn_1}{dt} = \frac{1}{l'} \nu \exp\left(-\frac{w}{kT}\right)\left[-n_1 \exp\left(-\frac{qaF}{2kT}\right) + n_2 \exp\left(+\frac{qaF}{2kT}\right)\right], \qquad \text{(D. 105)}$$

$$\frac{dn_2}{dt} = \frac{1}{l'} \nu \exp\left(-\frac{w}{kT}\right)\left[n_1 \exp\left(-\frac{qaF}{2kT}\right) - n_2 \exp\left(+\frac{qaF}{2kT}\right)\right]. \qquad \text{(D. 106)}$$

A substitution in the steady-state condition

$$\frac{d\,\Delta n}{dt} = \frac{dn_1}{dt} - \frac{dn_2}{dt} = 0 \qquad \text{(D. 107)}$$

yields

$$\frac{n_1}{n_2} = \exp\left(\frac{qaF}{kT}\right). \qquad \text{(D. 108)}$$

From eqn. (D. 68) we have

$$\frac{\Delta n}{n_D} = \frac{n_1 - n_2}{n_1 + n_2} = \frac{\dfrac{n_1}{n_2} - 1}{\dfrac{n_1}{n_2} + 1} \qquad \text{(D. 109)}$$

and from this, using (D. 108), we obtain the field dependence of Δn:

$$\Delta n = n_D \tanh\left(\frac{qaF}{2kT}\right). \tag{D. 110}$$

In the approximation of weak electric fields ($qaF/2kT \ll 1$) we can write

$$\Delta n \approx n_D \frac{qaF}{kT}. \tag{D. 111}$$

In this approximation the polarizability of the defect dipoles (D. 104) can be written as

$$\alpha_D \approx \frac{q^2a^2}{kT}. \tag{D. 112}$$

In this context it should be noted that the polarizability α_d of freely rotating, permanent dipoles of magnitude qa in the weak-field approximation can be given by a similar expression (according to the Langevin theory):

$$\alpha_d = \frac{1}{3}\frac{q^2a^2}{kT}. \tag{D. 113}$$

For the maximum loss angle (D. 103) we obtain with (D. 102)

$$\tan\delta_{\max} \approx \frac{1}{2\varepsilon_0(\varepsilon_S\varepsilon_\infty)^{1/2}}\frac{q^2a^2}{kT}n_D. \tag{D. 114}$$

When evaluating the loss angle measurements one has also to take account of the conductivity of the ionic crystals. The electric current passing through a solid is generally the sum of the conduction current and the displacement current:

$$j = \sigma F + \varepsilon_0\varepsilon\frac{dF}{dt}. \tag{D. 115}$$

Here ε is the dielectric constant which is complex because of the relaxation effects [cf. (D. 96)]. For fields which are periodic functions of time

$$F = F_0 \exp(i\omega t) \tag{D. 116}$$

eqn. (D. 115) yields

$$j = \varepsilon_0\varepsilon^*\frac{dF}{dt}, \tag{D. 117}$$

where

$$\varepsilon^* = \varepsilon' - i\left(\varepsilon'' + \frac{\sigma}{\varepsilon_0\omega}\right). \tag{D. 118}$$

Thus the total current is a displacement current in a dielectric substance with dielectric constant ε^*. According to the definition of the loss angle (D. 97) we have now instead of (D. 98)

$$\tan\delta = \frac{\varepsilon''}{\varepsilon'} + \frac{\sigma}{\varepsilon_0\omega\varepsilon'}. \tag{D. 119}$$

Using eqn. (D. 96) we hence obtain for $\varepsilon_S \approx \varepsilon_\omega$

$$\tan\delta \approx \frac{\varepsilon_S - \varepsilon_\omega}{\varepsilon_\omega}\,\frac{\omega\tau}{1+(\omega\tau)^2} + \frac{\sigma}{\varepsilon_0\varepsilon_\omega}\,\frac{1}{\omega}. \tag{D. 120}$$

The loss maxima lie in the range of relatively low frequencies (cf. Fig. 76) so that we can use in a good approximation the d.c. value for the electric conductivity. The frequency dependence of the losses due to the conductivity will therefore be proportional to $1/\omega$, which can be well distinguished

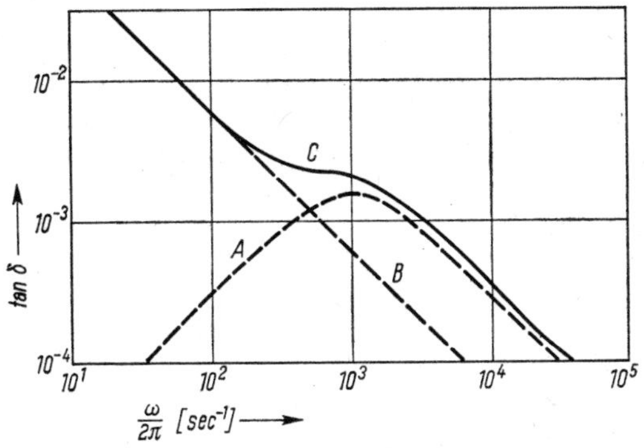

FIG. 76. Frequency dependence of the dielectric losses; A: contribution from the dipoles, B: contribution from the conductivity, C: total losses (after A. B. Lidiard in *Handbuch der Physik*, Vol. 20, Springer, Berlin 1957).

from the frequency dependence $\omega\tau/(1+\omega^2\tau^2)$ of the losses owing to dipole relaxation. The degree of association of defects can be derived from the ratio of the two loss terms at $\omega\tau = 1$.

II. Chemical Imperfections

Chemical imperfections are due to the incorporation of isotopes or impurities which disturb the strict periodicity of the crystal structure. Solid lithium, for example, consists of an isotope mixture of 93% ^{7}Li and 7% ^{6}Li. Every crystal contains foreign substances in a certain percentage which, with the present methods of preparation, cannot be made smaller than 0.1 ppm. Chemical impurities can be detected by neutron or proton activation analyses in concentrations down to about 10^{-3} ppm.

The arrangement of the foreign atoms in a crystal depends essentially on their size relative to that of the "host" atoms. We distinguish two possibilities of incorporation (cf. Fig. 77): (a) substitution, i.e. the foreign atoms are placed at the regular lattice sites of the host atoms, and (b) interstitial incorporation, i.e. the foreign atoms are placed at interstitial positions.

Besides the imperfections due to isotopes and impurities, crystals of chemical compounds may display chemical imperfection owing to deviations from

FIG. 77. Incorporation of foreign atoms X at regular lattice sites and interstitially.

stoichiometry which disturb the periodicity. This type of imperfection has been studied mainly in ionic crystals. It is the cause of the so-called colour centres and the electronic conduction in ionic crystals.

The non-stoichiometric composition of an ionic crystal M^+X^- results in an excess or deficit of positive metal ions M^+ or of negative metalloid ions X^-. Excess or deficit of one sort of ions can be obtained, for example, when an originally stoichiometric crystal is heated in the vapour of one of its components, M or X. The following types of imperfection are due to the non-stoichiometric composition:

1. Metal atoms enter interstitial positions. Every excess atom M behaves like a metal ion M^+ with an electron in its field.
2. Metal ions M^+ occupy available cationic vacancies. The anionic vacancies present in the initially stoichiometric crystal in the same quantity (cf. pp. 114 ff.) behave like positive charges and bind the electrons belonging to the excessive metal ions. Equivalent to this type of imperfection is a deficit of metalloid atoms X, i.e. the lattice of the anions X^- contains vacancies. Neutrality is conserved as every missing X^- ion is replaced by an electron which moves in the field of the "positive" anionic vacancy (missing negative charge).

Similar to these two cases metalloid excess or metal deficit gives rise to the following possibilities of imperfection:

3. Excessive metalloid atoms enter interstitial positions. This type of imperfection, however, is rarely encountered because of the large radii of the metalloid ions (cf. Table 7).
4. In analogy to case 2 the lattice of the M^+ cations contains vacancies. For reasons of neutrality an electron must have been removed from the crystal with every metal ion M^+. This electron may stem from any X^- ion neighbouring the vacancy, which in this way has become a neutral atom X. It is also possible that an M^+ metal ion loses another electron, thus becoming an M^{++} ion. The missing electron will behave like a positively charged "hole" which moves in the field of the "negative" cationic vacancy (missing positive charge).

In the first two cases the imperfection is due to an excess of metal which is chemically due to a reduction. The defect centres created in this way bind electrons. In the other two cases the imperfection is due to a metal deficit which is chemically due to an oxidation. The defect centres will in this case bind "holes".

These are the simplest types of chemical imperfection owing to non-stoichiometric composition of the crystals. More complicated defect centres can be due to different degrees of ionization or certain associations of the simple defect centres mentioned above. Such complex centres are mainly dealt with in the literature on colour centres.

The defect centres listed essentially give rise to two different phenomena: the colour centres (see below) and the electronic conduction in ionic crystals (cf. pp. 149 ff.). This classification is based on the energy necessary to separate the electron or hole bound to the defect centre so that it may move freely in the crystal. This separation energy depends on the electrical polarizability and thus, indirectly, on the type of chemical bond in the crystal. To obtain a rough estimate of this energy on the basis of the hydrogen model (cf. p. 294) the polarizability is taken into account by means of an effective dielectric constant whose value lies between the values of the static and the optical dielectric constants. Such estimates show that for alkali halides the separation energies are of the order of 1 eV, while the ionic crystals of certain oxides, sulphides, selenides, and the like have typical separation energies of the order of 10^{-2} to 10^{-1} eV.

Defect centres with high separation energies appear chiefly as colour centres, while defect centres with low separation energies give rise to the electronic conductivity in ionic crystals.

1. COLOUR CENTRES

Point defects in ionic crystals which, by virtue of their effective charge, bind electrons or holes are called colour centres if an excitation of these electrons or holes gives rise to optical absorption in the visible spectrum. This absorption is particularly clear in the case of alkali halides as they are perfectly transparent from ultraviolet to infrared if they are free from colour centres (i.e. in a stoichiometric state of high purity). The presence of colour centres results in a distinct colouring of the crystal (therefore the term "colour centres") which is due to the presence of absorption bands in an originally transparent range (cf. Fig. 78). Hitherto at least twenty different absorption bands have been found;

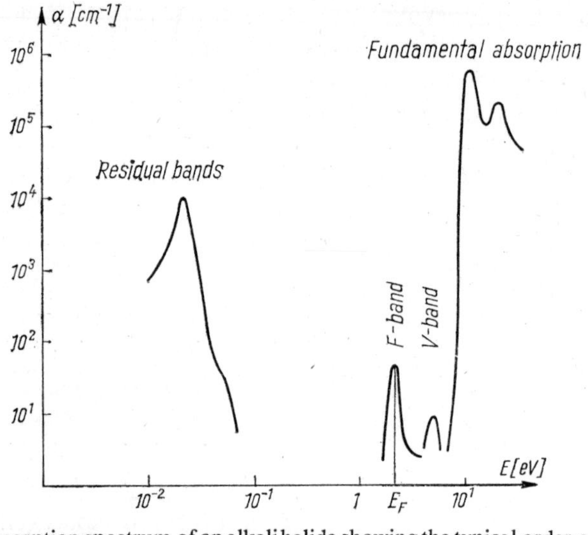

FIG. 78. Absorption spectrum of an alkali halide showing the typical order of magnitude of the absorption coefficient α and the corresponding radiative energy E.

certain configurations of point defects can be attributed to them, which act either as electron traps or as hole traps. Here we shall mention a few of the well-known centres. The following defects bind electrons:

F centre, a vacancy in the anionic lattice with one electron (cf. Fig. 79);

F' centre, a vacancy in the anionic lattice with two electrons;

R_1 centre, two neighbouring vacancies in the anionic lattice with one electron;

R_2 centre, two neighbouring vacancies in the anionic lattice with two electrons;

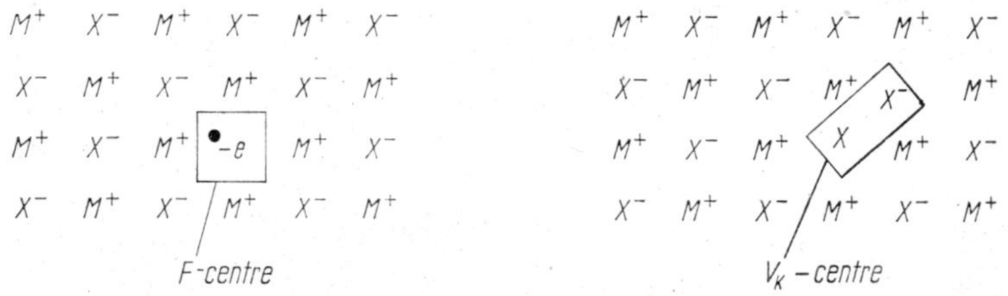

FIG. 79. Colour centres in an ionic M^+X^- crystal.

M centre, two neighbouring vacancies in the anionic lattice and one neighbouring vacancy in the cationic lattice with one electron.

The defects which bind holes comprise the various types of the so-called *V* centres. The configurations originally suggested for their interpretation are "antimorphic" to the electron-binding centres. It was, however, impossible so far to detect and prove these configurations by experimental means (optical polarization, spin resonance). The best-known *V* centre is the V_K centre which was first studied by Känzig. It consists of a molecular ion X_2^- in an otherwise perfect crystal (cf. Fig. 79). The separation of an electron from an anion leads to an association between a neighbouring anion and the metalloid atom formed in this way. The axis of the molecular ion lies in the $\langle 110 \rangle$ direction of the crystal, and the internuclear distance in the molecular ion is smaller than the original distance of the anions in the undisturbed crystal.

The presence of the various types of colour centres depends on the treatment of the crystals and the temperature. The following methods can be used to produce colour centres:

1. *Additive colouring.* Many ionic crystals obtain an excess of one of their components when they are heated in a vapour atmosphere of this component. Atoms from the gas diffuse into the crystal and are incorporated as ions in the lattice; electrons are given off to the crystal or taken up from it. The excess electrons or holes are bound by the vacancies present in the lattice so that colour centres are produced. Cation excess yields electron-binding centres (*F*, *F′*, *R*, . . . centres), anion excess yields hole-binding centres (*V* centres).

2. *Electrolytic colouring.* A suitable arrangement of electrodes makes it possible to "inject" electrons into the crystal or "siphon them off" from it. At elevated temperatures (for alkali halides of the order of 500°C) the ions and thus also the vacancies (cf. pp. 126 ff.) are able to move in the crystal under the influence of an electric field (electrolysis). Injected electrons (high field at the

cathode) are trapped by anion vacancies whereby colour centres (chiefly *F* centres) are produced. For reasons of neutrality anions leave the crystal at the anode so that cation excess occurs. Similarly, a removal of electrons (high field at the anode) results in a deficit of electrons. To conserve neutrality, for every withdrawn electron a cation leaves the crystal which leads to an anion excess or cation deficit and *V* centres are formed.

3. *Treatment with ionizing radiation.* Ionizing radiation is understood to be electromagnetic radiation or corpuscular radiation (such as X-rays, γ-rays, electrons or protons) whose energy is sufficiently high to produce free electrons and holes. Some of these electrons and holes are captured by corresponding vacancies so that colour centres are formed. Unlike the first two methods of additive and electrolytic coloration, the irradiation method normally does not disturb the stoichiometry of the crystal. Moreover, ionizing radiation at low temperatures may produce certain colour centres which are unstable at elevated temperatures and can therefore not be produced by means of the other methods.

It is easy to understand that colour centres produced by irradiation can be destroyed by an exposure to light of suitable wavelength or by heat treatment, that is, the crystal is "bleached". The energy necessary for "bleaching" corresponds to the energy of excitation of a trapped electron or hole which thus becomes able to move in the crystal and can recombine with a hole and electron, respectively.

It is also possible to eliminate colour centres by means of electrolysis with suitable polarity if in this way the stoichiometry of the crystal is recovered.

In the following we shall consider some properties of the simplest and most commonly known colour centre, the *F* centre.

(a) F *CENTRES*

According to a model by J. H. de Boer an *F* centre consists of a vacancy of effectively positive charge in the anion lattice which has bound an electron (cf. Fig. 79). The physical properties of the *F* centre are determined by the interactions of this electron with the ions of the crystal. One of the tasks of the theory is to calculate the energy eigenvalues and eigenfunctions of the electron located in the potential of all lattice particles modified by the vacancy. Here we restrict ourselves to estimate the energy eigenvalues of the electron in a very rough approximation. In a first quantum-mechanical approximation the *F* centre can be described in the same way as the hydrogen atom. Instead of a Coulomb potential we assume a square well potential as the effective positive charge is "smeared out" over the vacancy: all the nearest-neighbour cations of

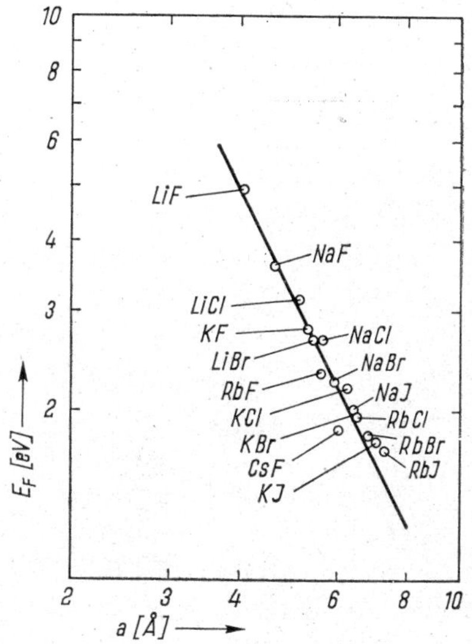

FIG. 80. Position of F-band maxima for alkali halides with NaCl structure as a function of the lattice constant a at 300°K. The straight line shows the equation by Mollwo [(D. 124) with $d = a/2$]. The measured points are taken from Tables 3 and 15.

the anion vacancy contribute to the positive charge of the vacancy. According to this model the trapped electron can move inside a cube of edge d (d is half the lattice constant for crystals with an NaCl structure). Using the Schrödinger equation for a constant potential we obtain from the boundary conditions for standing waves the allowed energy values of the electron [cf. eqn. (E. 11)]:

$$E_n = \frac{\pi^2}{2} \frac{\hbar^2}{md^2} n^2, \qquad \text{(D. 121)}$$

where

$$n^2 = n_1^2 + n_2^2 + n_3^2, \qquad \text{(D. 122)}$$

m is the electron mass, and n is the quantum number (n_1, n_2, n_3 are integers).

The energy difference for a transition from the ground state to the first excited state, according to eqn. (D. 121), is given by

$$\Delta E = \frac{\pi^2}{2} \frac{\hbar^2}{m} \frac{1}{d^2}. \qquad \text{(D. 123)}$$

It corresponds to the energy of electromagnetic radiation in the visible spectrum (cf. Fig. 80 and Table 15).

TABLE 15. Energies E_F of the F-band peaks for alkali halides with NaCl structure at 300°K (after R. K. Dawson and D. Pooley, *Phys. Stat. Sol.* **35**, 95 [1969]).

E_F [eV]	F	Cl	Br	I
Li	4.94	3.16	2.68	
Na	3.60	2.66	2.26	2.01
K	2.79	2.20	1.97	1.78
Rb	2.34	1.97	1.76	1.70
Cs	1.84			

(b) *EXPERIMENTAL PROOFS FOR* F *CENTRES*

de Boer's model of the F centre is supported by the following facts:

The F centres have an absorption peak for electromagnetic radiation of the energy E_F; for the position of this peak Mollwo has found the empirical relation

$$E_F d^2 = \text{const.} \approx 20 \text{ eV Å}^2, \qquad \text{(D. 124)}$$

where d is half the lattice constant for NaCl structures. This relation is well satisfied for alkali halides with NaCl structure (cf. Fig. 80); it is also consistent with the theoretical relation (D. 123). According to it the excitation energies of the F centre depend only on the lattice geometry; in particular, they are independent of the dielectric constant of the crystal considered. The F-band absorption is also independent of the choice of the alkali vapour used to achieve additive colouring (cf. p. 145). Thus, for example, the position of the F band of NaCl is independent of whether the crystal is heated in Na vapour or K vapour. This proves that the electron is not bound to a specific excess cation but to a vacancy.

When irradiated by light crystals with F centres display electronic photoconduction in the absorption band of the F centres. The supply of a sufficient amount of heat can also lead to complete ionization of the F centres. The initially bound electrons are thus rendered capable of moving and, when an electric field is applied to the crystal, electronic conduction occurs in addition to the ionic conduction.

The positive and negative ions in ionic crystals have closed electron shells with paired electrons (with the exception of the ionic crystals which contain transition metals). Therefore ionic crystals are normally diamagnetic (cf. p. 438). The magnetic susceptibility of a crystal with F centres, however, has a paramagnetic contribution which is proportional to the concentration of the F

centres. Every F centre represents an unpaired electron whose spin contributes to paramagnetism (cf. pp. 439 ff.).

Measurements of the paramagnetic electron spin resonance yield more exact information on the electronic structure of the F centre. The values of g obtained for the energy level splitting differ a little from the g factor for free electrons. From this it may be concluded that the wave function of the F centre electron is not of a pure s-type nature. This is attributed to the atomic environment of the F centre.

The electron spin resonance method is generally a valuable aid; in particular, it yields information on the anisotropies of complex centres.

2. ELECTRONIC CONDUCTION IN IONIC CRYSTALS

In a great number of ionic crystals, chiefly crystals of oxides, sulphides, selenides, and tellurides, deviations from the stoichiometric composition are the cause of electronic conduction processes which often appear even below room temperature. In general we must consider these substances to be semi-conductors with preponderantly ionic binding. The excess electrons or holes due to chemical imperfections are set free from the defect centres by the supply of thermal energy and can move in the crystal under the influence of an electric field. An excess of metal ions caused by a reduction (see p. 143) leads to so-called excess conduction; the defect centres responsible for it act as electron sources and are called donors (cf. p. 294). Crystals which display only excess conduction are called excess conductors or reduction semiconductors. Deficit of metal ions due to oxidation leads to so-called deficit conduction; the responsible defect centres act as electron sinks and are called acceptors (cf. p. 296). Crystals which display only deficit conduction are called deficit conductors or oxidation semiconductors. Crystals which may contain both excess metal ions and excess metalloid ions and which can display excess or deficit conduction according to the experimental conditions are called amphoteric semiconductors.

In the following we shall describe the electrical properties of semiconductors with preponderantly ionic binding. These properties, such as conductivity, Hall effect, thermoelectric power, of such compounds depend strongly on the deviations from stoichiometry. For semiconductors with mainly covalent bonds (cf. p. 292), however, the influence of stoichiometry is usually negligible compared with that of the impurities.

(a) *EXCESS CONDUCTORS*

The ZnO crystal is a well-known example for excess conduction. Deviations from stoichiometry in this substance always result in a cation excess. Zn atoms in interstitial positions as well as oxygen vacancies can act as donors.

The electrical properties of ZnO depend essentially on the production conditions, the previous treatment, and the interactions with a gaseous atmosphere (e.g. hydrogen or oxygen). Heating in hydrogen (reduction) increases the conductivity, heating in oxygen (oxidation) reduces it. Thus ZnO is an excess conductor. A decrease in electron concentration (i.e. a reduced conductivity) under the influence of oxygen can be explained by assuming that the excess Zn ions and a corresponding number of electrons migrate from interstitial positions to the crystal surface where they join with the gaseous oxygen to form ZnO. According to whether the Zn ions at the interstitial positions carry one or two charges, the reaction equations are

$$2\,Zn^{+} + 2e + O_2 \rightleftharpoons 2\,ZnO \qquad \text{(D. 125)}$$

or

$$2\,Zn^{++} + 4e + O_2 \rightleftharpoons 2\,ZnO. \qquad \text{(D. 126)}$$

From the law of mass action we obtain from this a relationship between the electron concentration responsible for the conductivity and the oxygen pressure. From eqn. (D. 125) we obtain for it

$$(n^{+}_{Zn})^2 n^2 n_O = \text{const.}\,(n_{ZnO})^2\,. \qquad \text{(D. 127)}$$

The concentration n_{ZnO} of the ZnO molecules exceeds that of the other reaction partners by orders of magnitude so that we can take n_{ZnO} as constant. The concentration n_{O_2} of the O_2 molecules in the gaseous phase is proportional to the oxygen pressure p_{O_2}. Moreover, the electron concentration n is equal to the concentration of the excess Zn^+ ions in interstitial positions. So we can derive from eqn. (D. 127)

$$n = \text{const.}\,p_{O}^{-1/4}\,. \qquad \text{(D. 128)}$$

In the same way (D. 126) yields

$$n = \text{const.}\,p_{O_2}^{-1/6}\,. \qquad \text{(D. 129)}$$

In this case the electron concentration is equal to half the concentration $n_{Zn^{++}}$ of the Zn^{++} ions.

Experiments (cf. Fig. 81) yield a pressure dependence of electric conductivity at constant temperatures above 500°C of the form

$$\sigma \approx \text{const.}\,p_{O_2}^{-1/4.1} \qquad \text{(D. 130)}$$

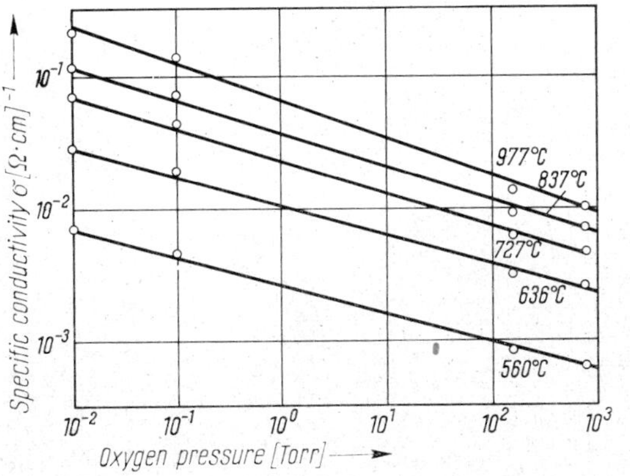

FIG. 81. Oxygen pressure dependence of electric conductivity of ZnO at various temperatures [after C. A. Hogarth, *Z. Phys. Chemie* **198**, 30 (1951)].

and thus give satisfactory agreement with the relation derived from the law of mass action.

The activation energies necessary for the excitation of the electrons are different for different pretreatment of the ZnO. Qualitatively, however, we observe almost always an exponential growth of the electron concentration with temperature; the activation energies may be different in different ranges of temperature. We have the following relation between the carrier concentration n, the activation energy ΔE, and the temperature T:

$$n = \text{const.} \exp\left(-\frac{\Delta E}{kT}\right). \tag{D. 131}$$

As a rule, such relations exist for all semiconductors [cf. eqns. (G. 34) and (G. 92)]. They indicate that carriers are thermally produced by an activation process. Figure 82 shows the temperature dependence of the electron concentration for ZnO crystals which have been heated red hot in hydrogen before the measurements have been performed.

(b) *DEFICIT CONDUCTORS*

Well-known compounds which, in the case of non-stoichiometric composition, have a deficit of cations, are Cu_2O, CuI, Cu_2S, and NiO. In NiO, for example, the metal deficit and thus the deficit of electrons leads to the formation of Ni^{3+} ions but not to the formation of O^- ions. The mechanism of electric conduction is due to electron transitions between Ni^{2+} and Ni^{3+} ions and

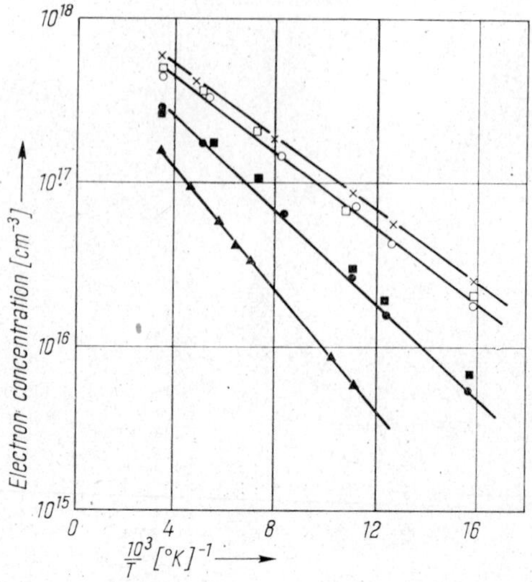

FIG. 82. Temperature dependence of the electron concentration for ZnO crystals after thermal treatment in hydrogen [after H. Rupprecht, *J. Phys. Chem. Solids*, **6**, 144 (1958)]:

▲ heated 20 minutes at 50 atm and 600°C
■● heated 5 minutes at 40 atm and 700°C
○□× heated 5 minutes at 70 atm and 700°C.

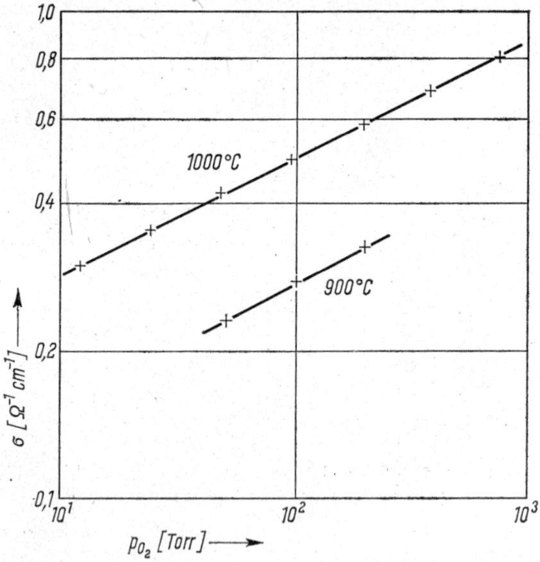

FIG. 83. Electric conductivity of NiO as a function of oxygen pressure [after H. H. von Baumbach and C. Wagner, *Z. phys. Chemie* B **24**, 59 (1934)].

leads to hole conduction (deficit conduction), which manifests itself by the positive sign of the Hall effect and thermoelectric power.

The result of the interaction between NiO and an oxygen atmosphere also indicates deficit conduction. Oxygen is adsorbed at the crystal surface as O^- (chemisorption) as well as incorporated in the crystal as O^{2-}. The latter process leads to the formation of NiO which consists of Ni^{2+} and O^{2-} ions at regular lattice sites. This process is possible only if further Ni^{2+} vacancies are produced. In any case an interaction of NiO with oxygen results in a withdrawal of electrons from the crystal through the production of further Ni^{3+} ions. Thus the hole concentration increases as the oxygen pressure is raised. In this case the oxidation increases the electrical conductivity (cf. Fig. 83), which indicates deficit conduction.

(c) *AMPHOTERIC SEMICONDUCTORS*

Semiconductors which, according to their preparation, display either electron (n-type) or hole (p-type) conduction are generally called amphoteric. Semiconductor crystals with mainly covalent bonds (such as Ge, Si, InSb, GaAs, etc.) usually are amphoteric; according to the nature of the impurities (doping) they may display n-type or p-type conduction (cf. pp. 293 ff.). Here we shall mention only such semiconductors whose amphoteric properties can be explained by deviations from stoichiometry, such as PbS and PbSe. A treatment of PbS in sulphur at low pressure, or *in vacuo*, results in excess conduction which is due to the excess of Pb^{2+} ions. After heating at higher sulphur-vapour pressures PbS dis-

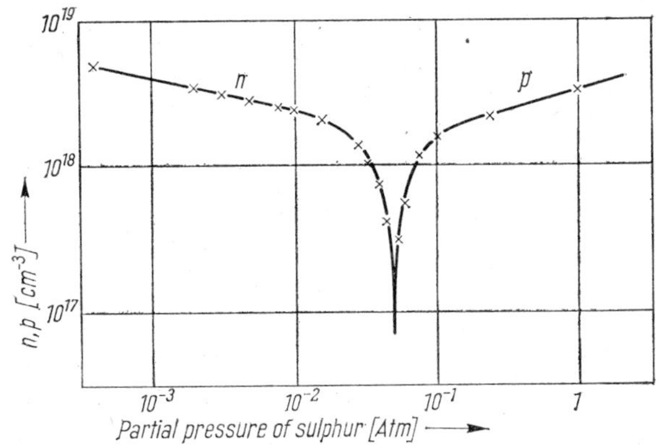

FIG. 84. Carrier concentration in monocrystalline PbS at room temperature, after the crystal has been heated at 1200°K at different partial pressures of sulphur [after J. Bloem, *Philips Res. Rep.* **11,** 273 (1956)].

plays deficit conduction which can be explained by the deficit of Pb^{2+} ions. Here we shall not enter into details of the various possibilities of imperfection in PbS. Figure 84 illustrates the carrier concentration in PbS at room temperature as measured after the crystal has been heated at different partial pressures of sulphur vapour. Note the different pressure dependence (in analogy to Figs. 81 and 83) in the regions of excess and deficit conduction. The minimum carrier concentration corresponds to the stoichiometric composition of the crystal.

References

General

AULEYTNER, J., *X-Ray Methods in Study of Defects in Single Crystals* (Pergamon, Oxford 1967).
CAHN, R. W., *Physical Metallurgy* (North-Holland, Amsterdam 1965).
CHALMERS, B., *Physical Metallurgy* (Wiley, New York 1959).
GOLDMAN, J. E. (Editor), *The Science of Engineering Materials* (Wiley, New York 1957).
GRAY, T. J., *et al.*, *The Defect Solid State* (Interscience, New York 1957).
Handbuch der Physik, edited by FLÜGGE, S. (Springer, Berlin 1955 ff.),
Vol. 6: *Elastizität and Plastizität* (1958),
Vol. 7/1, 2: *Kristallphysik I, II* (1955, 1958),
Vol. 19, 20: *Elektrische Leitungsphänomene I, II* (1956, 1957).
HAUFFE, K., *Reaktionen in und an festen Stoffen*, 2nd ed. (Springer, Berlin 1966).
KRÖGER, F., *The Chemistry of Imperfect Crystals* (North-Holland, Amsterdam 1964).
MCLEAN, D., *Mechanical Properties of Metals* (Wiley, New York 1962).
Progress in Metal Physics, edited by CHALMERS, B. (Butterworth, London 1949 ff.).
SHOCKLEY, W., *et al.* (Editor), *Imperfections in Nearly Perfect Crystals* (Wiley, New York 1952).
VAN BUEREN, H. G., *Imperfections in Crystals* (North-Holland, Amsterdam 1960).

Displacements

AMELINCKX, S., *The Direct Observation of Dislocations*, in *Solid State Physics*, Suppl. 6 (Academic Press, New York 1964).
COTRELL, A. H., *Theory of Crystal Dislocations* (Gordon and Breach, New York 1962).
FRIEDEL, J., *Dislocations* (Pergamon, Oxford 1964).
HULL, D., *Introduction to Dislocations* (Pergamon, Oxford 1965).
READ, W. T., *Dislocations in Crystals* (McGraw-Hill, New York 1953).

Diffusion

BARRER, M., *Diffusion in and through Solids* (Cambridge University Press, London 1951).
JOST, W., *Diffusion und chemische Reaktion in festen Stoffen* (Steinkopff, Leipzig 1937).
SEITH, W., *Diffusion in Metallen* (Springer, Berlin 1955).
SHEWMON, P. G., *Diffusion in Solids* (McGraw-Hill, New York 1963).

Radiation Damage

CORBETT, J. W., *Electron Radiation Damage in Semi-Conductors and Metals*, in *Solid State Physics*, Suppl. 7 (Academic Press, New York 1966).
LEIBFRIED, G., *Bestrahlungseffekte in Festkörpern* (Teubner, Stuttgart 1965)
STRUMANE, R. (Editor), *The Interaction of Radiation with Solids* (North-Holland, Amsterdam 1964).
THOMPSON, M. W., *Defects and Radiation Damage in Metals* (Cambridge University Press, London 1969).

Ionic Crystals

GREENWOOD, N. N., *Ionic Crystals, Lattice Defects and Nonstoichiometry* (Butterworth, London 1968).
LIDIARD, A. B., *Ionic Conductivity*, in *Handbuch der Physik*, Vol. 20 (Springer, Berlin 1957).
MOTT, N. F., and GURNEY, R. W., *Electronic Processes in Ionic Crystals* (Oxford University Press, London 1940).
STASIW, O., *Elektronen- und Ionenprozesse in Ionenkristallen, mit Berücksichtigung photochemischer Prozesse* (Springer, Berlin 1959).
STUMPF, H., *Quantentheorie der Ionenkristalle* (Springer, Berlin 1961).

Colour Centres

FOWLER, W. B. (Editor), *Physics of Color Centers* (Academic Press, New York 1968).
MARKHAM, J. J., *F-Centers in Alkali Halides*, in *Solid State Physics*, Suppl. 8 (Academic Press, New York 1966).
SCHULMAN, J. A., and COMPTON, W. D., *Color Centers in Solids* (Pergamon, Oxford 1963).
SEIDEL, H., and WOLF, H. C., *Physica status solidi* **11**, 3 (1965).
SEITZ, F., *Rev. Mod. Phys.* **18**, 384–408 (1946) and **26**, 7–49 (1954).

E. Foundations of the Electron Theory of Metals

I. Properties of Metals

Of the approximately 100 chemical elements about seventy-five are metals under normal conditions. Besides these monatomic metals alloys and metallic compounds possess the following physical properties which are necessary but singly not sufficient to characterize a metal:

1. A high electric conductivity σ which, above a certain temperature, is inversely proportional to the temperature T.
2. A high thermal conductivity λ which, at sufficiently high temperatures, is independent of T.
3. Validity of the Wiedemann–Franz law, at sufficiently high temperatures, according to which the ratio of thermal and electrical conductivities is equal to a universal constant multiplied by the absolute temperature T:

$$\frac{\lambda}{\sigma} = LT \tag{E. 1}$$

   where L is the Lorenz number,

$$L = \frac{\pi^2}{3}\left(\frac{k}{e}\right)^2. \tag{E. 2}$$

4. A temperature-independent carrier concentration.
5. A high optical absorption which is almost constant in the visible spectrum, and thus a high reflexivity (metallic glance).
6. Ductility, i.e. metals can be rolled and forged.

The metallic properties are determined by the type of chemical bond in the crystal. This type of bond is called a metallic bond and can only be understood with the help of wave mechanics. Qualitatively it can be described as follows. The metal atoms in the crystal are present as ions. The electrons given up (valence electrons) form the so-called electron gas. They are not localized in the lattice as in the ionic crystals but belong to all lattice particles simultaneously. The metal ions are held together by all of the valence electrons. The metallic bond is independent of direction. The number of nearest neighbours (coordination number) is therefore only restricted by the geometrical possibilities of arrangement. For this reason metallic structures have a high coordination number. Most of the monatomic metals crystallize in closest packing of spheres with the coordination number 12 (face-centred cubic or hexagonal lattice) or in the body-centred cubic lattice with the coordination number 8 (cf. Table 3). The directional independence of the metallic bond also makes it possible to produce alloys. An alloy in the strictest sense is a mixture of metals which has metallic properties and one and the same crystal structure over a certain region of composition (cf. pp. 231 ff.).

II. Free-electron Model

At about 1900 the hypothesis of the free-electron gas was first used by Drude to explain both the optical properties and the high electric conductivity and thermal conductivity of metals. He assumed that the valence electrons in the crystal were able to move freely in the crystal and, at the temperature T of the crystal, to behave like the particles of an ideal gas (hence the name "free electron model"). In this model all electron–electron and electron–ion interactions are neglected.

In the following years Lorentz improved the model by assuming a Maxwell–Boltzmann distribution for the electron velocity. The Drude–Lorentz theory was initially supported by the fact that the Wiedemann–Franz law was derived from it which, as to order of magnitude, agrees with the experimental results.

The assumption of a Maxwell–Boltzmann gas, however, proved inconsistent with measurements of specific heats of metals. Such a gas of N electrons should give a contribution to the specific heat of $3Nk/2$. There exists experimental proof (e.g. measurements of the Hall effect, cf. p. 386) that the electron concentration is of the order of the concentration of the atoms. According to this, the electronic contribution to the specific heat of metals should be comparable to the Dulong–Petit value of the lattice heat; the total specific heat of a metal above the Debye temperature Θ should then be $(3+\frac{3}{2})\cdot Nk$. However, metals also show a good agreement of their measured specific heat with the Dulong–Petit value $3Nk$. This indicates that the electron gas cannot be considered to be an ideal Maxwell–Boltzmann gas.

Further contradictions resulting from the Drude–Lorentz theory of the "classical" electron gas will not be discussed here. It can be said, however, that the basic assumptions of Drude were qualitatively correct and were used in later theories. Experiments have shown that the metals are actually pure electronic conductors, without additional ion conduction; electrolytic experiments showed that no mass transport occurred. The presence of mobile electrons in the metals was proved by experiments by Tolman. The presence of alternating currents could be proved in metallic bodies which were rapidly moved to and fro; these currents are due to the inertia of the moving carriers. The specific charge of the carriers determined from these experiments agrees within the limits of experimental accuracy with the e/m values measured for free electrons in cathode rays. We have

$$(e/m \pm 5\%)_{\text{met. elec.}} = (e/m \pm 0.01\%)_{\text{free elec.}}$$

III. Sommerfeld's Theory

According to Sommerfeld the properties of the electron gas are determined by quantum-mechanical principles. Provided one knows the energy and momentum of the electrons, one can theoretically evaluate those properties of metals which are determined by the valence electrons. The calculation of these properties is thus the main task of the theory. In the present case, this problem is reduced to a single-electron problem, and the theory is based on the following assumptions:

1. The valence electrons move independently of one another in the potential of the metal ions and the other electrons.
2. The potential in the interior of the metal is constant, i.e. the electrons are not acted upon by forces provided that no external fields are applied. Therefore the electrons are called "free".
3. A wave function is attributed to every electron which has been obtained as a solution of the time-independent Schrödinger equation.
4. The electron energy distribution is given by the Fermi–Dirac statistics.

Assumption 4 contains the Pauli principle according to which each eigenstate can be occupied by at most two electrons with opposite spins.

In this sense the N electrons are not perfectly independent from one another, even if, according to assumptions 1 and 2, forces acting between the electrons among themselves and the electrons and the metal ions can be neglected.

By virtue of the de Broglie relation a wavelength of $\lambda = h/p$ is attributed to electrons of momentum p. When we introduce the wave number $k = 2\pi/\lambda$, the de Broglie relation takes the form

$$p = \hbar k, \quad \text{or} \quad \mathbf{p} = \hbar \mathbf{k}. \tag{E. 3}$$

The motion of the electrons is considered as the propagation of particle waves which, for the stationary state, is determined by the time-independent Schrödinger equation with suitable boundary conditions (see below):

$$\nabla^2\psi + \frac{2m}{\hbar^2}(E-V)\psi = 0, \tag{E. 4}$$

where ψ is the wave function, E is the total energy of the electron, V is its potential energy, and m is its rest mass. The quantity ψ is also called the probability amplitude. The quantity $\psi\psi^* \, d^3\boldsymbol{r}$ is the probability for an electron to be found in the volume element $d^3\boldsymbol{r}$. The probability for an electron to be anywhere in the system must be unity; this requirement is called the normalization condition and the wave function must satisfy it:

$$\int \psi\psi^* \, d^3\boldsymbol{r} = 1. \tag{E. 5}$$

Here ψ^* is the conjugate complex function to ψ. The integration is carried out over the total volume of the system considered.

As in the present case the potential energy is taken as constant ("free" electrons), and as it is defined to within an additive constant, we can put $V = 0$ in eqn. (E. 4). When we write

$$\psi = C \exp i(\boldsymbol{k}.\boldsymbol{r}) \tag{E. 6}$$

where C is a normalizing constant, $\boldsymbol{k}$ is the wave vector, and $\boldsymbol{r}$ is the position vector, we obtain from eqn. (E. 4)

$$E(\boldsymbol{k}) = \frac{\hbar^2}{2m}k^2 = \frac{\hbar^2}{2m}(k_x^2+k_y^2+k_z^2). \tag{E. 7}$$

This relation $E(\boldsymbol{k})$ represents the dispersion relation for free electrons.

The boundary conditions supply us with a selection of a finite number of $\boldsymbol{k}$ values. Therefore, only certain energy values (eigenvalues) are allowed, and only certain definite functions (eigenfunctions) are solutions to the Schrödinger equation. The boundary conditions themselves are obtained from the homogeneity of the crystal (periodic boundary conditions, cf. p. 56). The spatial periodicity of the wave functions for cells of the size L^3,

$$\psi(x, y, z) = \psi(x+L, y, z) = \psi(x, y+L, z) = \psi(x, y, z+L), \tag{E. 8}$$

yields together with (E. 6) the quantization condition for the $\boldsymbol{k}$ values:

$$k_x = n_x\frac{2\pi}{L}, \quad k_y = n_y\frac{2\pi}{L}, \quad k_z = n_z\frac{2\pi}{L}, \tag{E. 9}$$

or

$$k^2 = \left(\frac{2\pi}{L}\right)^2(n_x^2+n_y^2+n_z^2) = \left(\frac{2\pi}{L}\right)^2 n^2, \tag{E. 10}$$

where n_x, n_y, n_z are integers and n is in general not an integer.

From eqn. (E. 7) we hence obtain the eigenvalues for travelling waves:

$$E_n = 2\pi^2 \frac{\hbar^2}{mL^2} n^2 . \qquad \text{(E. 11)}$$

If we use Z_k^0 to denote the k values from $k = 0$ to $k = k(n)$,

$$Z_k^0 = \frac{L^3}{6\pi^2} k^3 \qquad \text{(E. 12)}$$

[cf. eqn. (C. 51)], we see that we have equally many energy values Z_E^0 from $E = 0$ to $E = E_n$:

$$Z_E^0 = \frac{L^3}{6\pi^2} \frac{(2mE_n)^{3/2}}{\hbar^3} . \qquad \text{(E. 13)}$$

The density of the energy eigenvalues $D(E)$, i.e. the number of energy values per unit energy interval, can be obtained from this by differentiation with respect to E:

$$D(E)\, dE = \frac{dZ_E^0}{dE} dE = \frac{L^3}{4\pi^2} \left(\frac{2m}{\hbar^2}\right)^{3/2} E^{1/2}\, dE. \qquad \text{(E. 14)}$$

These quantization conditions yield the energy eigenvalues allowed for each individual valence electron in the metal.

The application of the Pauli principle and the Fermi–Dirac statistics enables us to find out how many electrons have a certain energy (the so-called energy distribution). According to the Pauli principle only two electrons (with anti-parallel spins) can have the same momentum, the same "state", which is defined by the set of the three quantum numbers n_x, n_y, n_z according to eqn. (E. 10).

The Fermi–Dirac statistics tell us about the occupation probabilities of the states of energy E. The energy distribution, that is, the number $N(E)\, dE$ of electrons with energies between E and $E+dE$, is given by the so-called distribution function

$$N(E)\, dE = 2D(E)\, F(E)\, dE \qquad \text{(E. 15)}$$

with the Fermi–Dirac function

$$F(E) = \frac{1}{\exp\left(\frac{E-\zeta}{kT}\right)+1} , \qquad \text{(E. 16)}$$

where ζ is the chemical potential of the Fermi-gas (see p. 309) or the Fermi energy (see below). From eqns. (E. 14) and (E. 16) we obtain for the distribution function of free electrons (cf. Fig. 85)

$$N(E)\, dE = \frac{L^3}{2\pi^2} \left(\frac{2m}{\hbar^2}\right)^{3/2} \frac{E^{1/2}\, dE}{\exp\left(\frac{E-\zeta}{kT}\right)+1} . \qquad \text{(E. 17)}$$

1. PROPERTIES OF THE FERMI–DIRAC FUNCTION

The function $F(E)$ depends only on an energy difference and is therefore independent of the choice of the origin of the energy scale; $F(E)$ is also independent of the distribution of the quantum states $D(E)$ and is generally valid

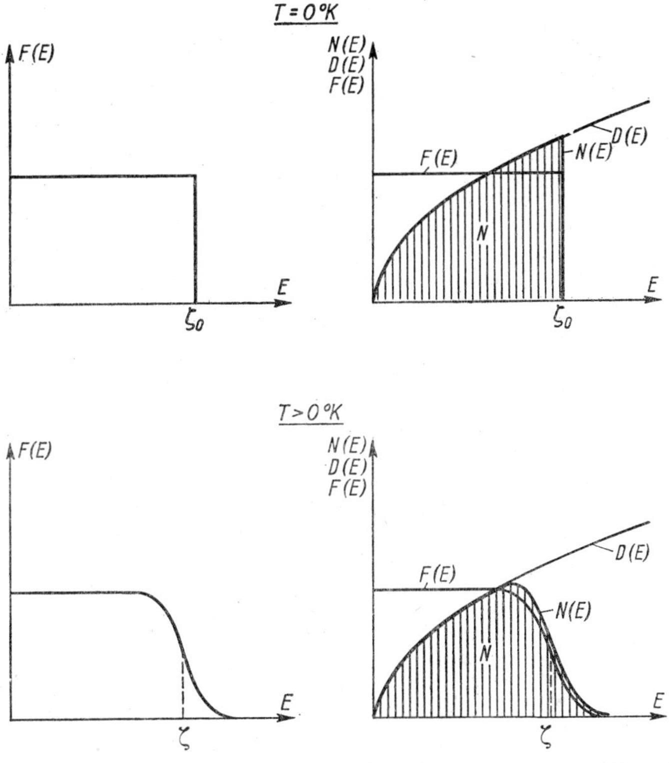

FIG. 85. Fermi–Dirac function $F(E)$, distribution of quantum states $D(E)$ and distribution function $N(E)$ for $T = 0°K$ and $T > 0°K$.

for non-localized electrons. The function $F(E)$ assumes only values between 0 and 1 (cf. Fig. 85):

$$0 \leq F(E) \leq 1. \tag{E. 18}$$

In particular, we have for

$$\begin{array}{lll} T = 0°K & E > \zeta_0; & F(E) = 0, \\ & E \leq \zeta_0: & F(E) = 1, \\ T > 0°K & E \gg \zeta: & F(E) \approx 0, \\ & E \ll \zeta: & F(E) \approx 1, \\ & E = \zeta: & F(E) = \tfrac{1}{2}. \end{array} \tag{E. 19}$$

The energy ζ_0 is the Fermi energy. At $T = 0°K$ all energy states are filled up to ζ_0 with the probability "1"; for higher energies the occupation probability is equal to zero. For $T > 0°K$ the occupation probability for states of the energy $E = \zeta$ is equal to $\frac{1}{2}$. The Fermi energy is a slightly temperature-dependent energy parameter which is determined by the electron density [cf. eqn. (E. 33)].

The function $F(E)$ has the following important property:

$$F(\zeta + \Delta E) = 1 - F(\zeta - \Delta E). \tag{E. 20}$$

This is easily understood when we take into consideration (cf. also Fig. 85) that

$$\frac{1}{\exp\left(\frac{\Delta E}{kT}\right)+1} = 1 - \frac{1}{\exp\left(-\frac{\Delta E}{kT}\right)+1} \tag{E. 21}$$

and

$$\frac{1}{\exp\left(\frac{E-\zeta}{kT}\right)+1} = 1 - \frac{1}{\exp\left(\frac{\zeta-E}{kT}\right)+1}. \tag{E. 22}$$

2. PROPERTIES OF THE ELECTRON GAS AT $T = 0°K$

At absolute zero all states are filled up to the Fermi energy ζ_0, a fact that results from the Fermi–Dirac statistics. The value of this limiting energy depends on the total number N of valence electrons contained in the volume L^3, which occupy the allowed states in agreement with Pauli's exclusion principle. From this we obtain an equation determining ζ_0:

$$N = \int_0^{\zeta_0} N(E)\, dE = 2\int_0^{\zeta_0} D(E)\, F(E)\, dE. \tag{E. 23}$$

Using eqn. (E. 14) and $F(E) = 1$ we obtain

$$\zeta_0 = \frac{\hbar^2}{2m}(3\pi^2 n_e)^{2/3}, \tag{E. 24}$$

where $n_e = N/L^3$ is the electron concentration.

As already mentioned on p. 157, the electron concentration in metals is of the order of the concentration of atoms. For monovalent metals

$$n_e \approx n_{At} \tag{E. 25}$$

is a good approximation. The calculation of ζ_0 from eqns. (E. 24) and (E. 25) yields astonishingly high energy values even at absolute zero (of the order of several eV) for the valence electrons of highest energy (cf. Table 16).

TABLE 16. Fermi energies and degeneracy temperatures for several metals calculated according to the free-electron model [cf. eqns. (E. 24) and (E. 30)]

Metal	ζ_0 [eV]	$T_E = \frac{2}{5}\frac{\zeta_0}{k}$ [°K]
Li	4.72	21 900
Na	3.12	14 480
K	2.14	9 930
Rb	1.82	8 440
Cs	1.53	7 100
Cu	7.04	32 670
Ag	5.51	25 570
Au	5.51	25 570

The total energy of all valence electrons in the volume L^3 at absolute zero amounts to

$$U_0 = \int_0^{\zeta_0} EN(E)\,dE. \tag{E. 26}$$

A substitution of eqns. (E. 17) and (E. 19) yields

$$U_0 = \frac{L^3}{5\pi^2}\left(\frac{2m}{\hbar^2}\right)^{3/2}\zeta_0^{5/2}. \tag{E. 27}$$

From this, using also eqn. (E. 24), we obtain for the mean energy per electron

$$\bar{u}_0 = \frac{U_0}{N} = \frac{3}{5}\zeta_0. \tag{E. 28}$$

In contrast to the classical Maxwell–Boltzmann gas whose particles have no energy at absolute zero, the electron gas has a considerable zero-point energy. It corresponds to a temperature of the order of 10^4°K of a classical gas. This can be easily estimated by means of the following relations:

$$\bar{u}_0 = \tfrac{3}{5}\zeta_0 = \tfrac{3}{2}kT_E, \tag{E. 29}$$

$$T_E = \frac{2}{5}\frac{\zeta_0}{k}, \tag{E. 30}$$

where T_E is the degeneracy temperature (cf. p. 168).

Hence in contrast to the ideal gas, the electron gas at absolute zero has a high pressure P_0 (zero-point pressure). When we substitute the electron concentration n_e and the mean energy $\bar{u}_0$ in the general gas-kinetic relation for the pressure, we obtain the zero-point pressure of the electron gas as

$$P_0 = \tfrac{2}{3} n_e \bar{u}_0 = \tfrac{2}{5} n_e \zeta_0 . \tag{E. 31}$$

This pressure is of the order of 10^5 atm.

3. PROPERTIES OF THE ELECTRON GAS AT $T > 0$°K

The following considerations are based on the assumption that the electron gas is in thermodynamic equilibrium with the metal ions in the lattice. If the temperature is raised, the mean energy of the electron gas is also raised via the interaction with the lattice vibrations (phonons). The excitation energies transferred from the phonons to the electrons amount to about kT. As all states with energies $E < \zeta_0$ are occupied, only the fastest electrons of energies $E \approx \zeta_0$ can be excited to higher levels. Thus the curve of the Fermi–Dirac function is changed only near the Fermi energy, in an energy interval of the order of kT. The value of the Fermi energy is slightly temperature-dependent and is obtained from a normalization condition similar to (E. 23), under the assumption of a temperature-independent electron concentration N/L^3 (cf. Fig. 85):

$$N = \int_0^\infty N(E)\, dE = 2 \int_0^\infty D(E)\, F(E)\, dE. \tag{E. 32}$$

With eqn. (E. 17) we obtain the equation for the determination of ζ:

$$N = \frac{L^3}{2\pi^2} \left(\frac{2m}{\hbar^2} \right)^{3/2} \int_0^\infty \frac{E^{1/2}\, dE}{\exp\left(\dfrac{E-\zeta}{kT} \right) + 1} . \tag{E. 33}$$

With the substitutions $x = E/kT$ and $\alpha = \zeta/kT$ we can rewrite this equation in the form

$$N = \frac{L^3}{2\pi^2} \left(\frac{2mkT}{\hbar^2} \right)^{3/2} \int_0^\infty \frac{x^{1/2}\, dx}{\exp(x-\alpha)+1} \tag{E. 34}$$

or

$$n_e = \frac{2}{\pi^{1/2}} n_0 F_{1/2}(\alpha) \tag{E. 35}$$

with the "effective density of states"

$$n_0 = \frac{1}{4}\left(\frac{2mkT}{\pi\hbar^2}\right)^{3/2} \tag{E. 36}$$

and the Fermi integral

$$F_{1/2}(\alpha) = \int_0^\infty \frac{x^{1/2}\,dx}{\exp(x-\alpha)+1}. \tag{E. 37}$$

The so-called effective density of states means essentially a critical electron concentration (cf. p. 168). In the Fermi statistics we often encounter integrals of the form

$$F(\alpha) = \int_0^\infty \frac{f(x)\,dx}{\exp(x-\alpha)+1}. \tag{E. 38}$$

These Fermi integrals in general cannot be solved analytically; by means of various series expansions they can be given approximately for certain ranges of $\alpha = \zeta/kT$ (cf. also p. 288).

For $\alpha \gg 1$ we can use Sommerfeld's expansion

$$F(\alpha) \approx \int_0^\alpha f(x)\,dx + \frac{\pi^2}{6}f'(\alpha) + \frac{7\pi^4}{360}f'''(\alpha) + \ldots. \tag{E. 39}$$

In particular for $f(x) = x^n$

$$F_n(\alpha) = \int_0^\infty \frac{x^n\,dx}{\exp(x-\alpha)+1} \approx \frac{\alpha^{n+1}}{n+1}\left(1+\frac{\pi^2}{6}(n+1)n\alpha^{-2}+\ldots\right). \tag{E. 40}$$

This series converges rapidly for $\alpha \gg 1$.

In the following we shall discuss eqn. (E. 37) for the two special cases $n_e \gg n_0$ and $n_e \ll n_0$:

1. For metals we have always $n_e \gg n_0$ and thus $F_{1/2}(\alpha) \gg 1$. Accordingly also $\alpha \gg 1$ (cf. Fig. 86). For this range of α values $F_{1/2}(\alpha)$ can be represented by the following series expansion [eqn. (E. 40)]:

$$F_{1/2}(\alpha) \approx \frac{2}{3}\alpha^{3/2}\left(1+\frac{\pi^2}{8\alpha^2}+\ldots\right). \tag{E. 41}$$

Substitution in eqn. (E. 35) yields

$$n_e = \frac{1}{3\pi^2}\left(\frac{2m}{\hbar^2}\right)^{3/2}\zeta^{3/2}\left[1+\frac{\pi^2}{8}\left(\frac{kT}{\zeta}\right)^2+\ldots\right]. \tag{E. 42}$$

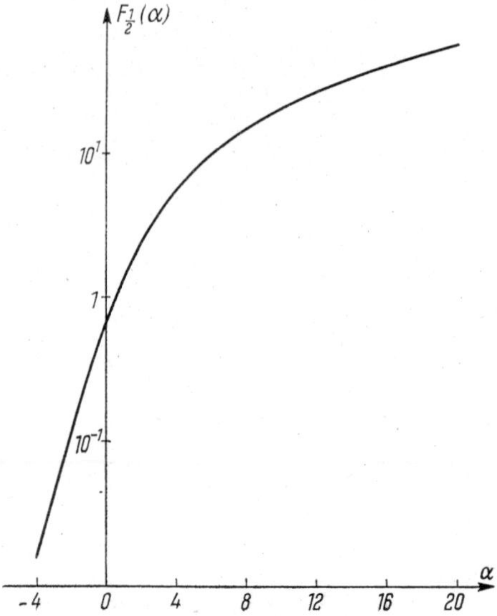

FIG. 86. Fermi integral $F_{1/2}(\alpha)$.

As $\alpha = \zeta/kT \gg 1$ we can replace in the correction term ζ by ζ_0. Repeated application of the binomial expansion and eqn. (E. 24) enable us to obtain

$$\zeta(T) = \zeta_0\left[1-\frac{\pi^2}{12}\left(\frac{kT}{\zeta_0}\right)^2\right] \tag{E. 43}$$

for the temperature dependence of the Fermi energy. Hence we see that it decreases slightly with increasing temperature; the relative drop of ζ between 0 and 300°K amounts to as little as 5×10^{-3} (on the assumption that $\zeta_0 = 5$ eV).

2. Although the case of $n_e \ll n_0$ cannot be realized in metals, it is of great importance for the properties of semiconductors. For the Fermi integral $F_{1/2}(\alpha) \ll 1$ the condition $\alpha \ll 0$ must be satisfied. So we shall formally introduce this range of values; the physical meaning of negative ζ values will become clear later on (see pp. 281 ff.).

When we assume $\zeta \ll 0$, the Fermi–Dirac function goes over into the Maxwell–Boltzmann function

$$F(E) = \frac{1}{\exp\left(\frac{E-\zeta}{kT}\right)+1} \approx \exp\left(-\frac{|\zeta|}{kT}\right)\exp\left(-\frac{E}{kT}\right) \tag{E. 44}$$

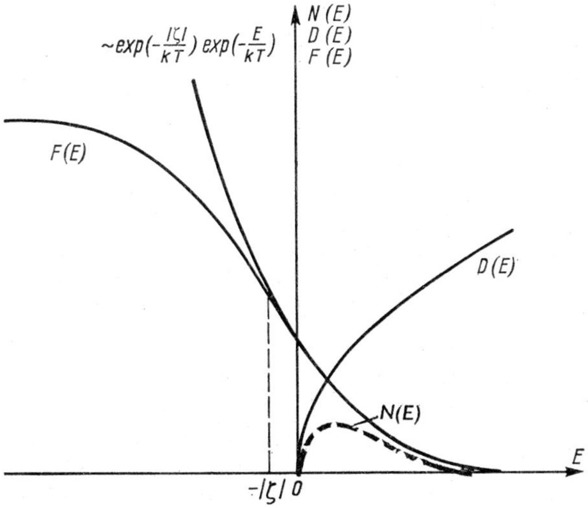

FIG. 87. Maxwell–Boltzmann distribution.

and the energy distribution (E. 17) takes the form of a Maxwell–Boltzmann distribution (cf. Fig. 87):

$$N(E)\,dE = \frac{L^3}{2\pi^2}\left(\frac{2m}{\hbar^2}\right)^{3/2} E^{1/2}\exp\left(-\frac{|\zeta|}{kT}\right)\,\exp\left(-\frac{E}{kT}\right)dE \qquad \text{(E. 45)}$$

or

$$N(E)\,dE \approx \text{const.}\; E^{1/2}\exp\left(-\frac{E}{kT}\right)dE. \qquad \text{(E. 46)}$$

Under the assumption of $\alpha \ll 0$ the Fermi integral $F_{1/2}(\alpha)$ can be solved analytically:

$$F_{1/2}(\alpha) \approx \exp\left(-|\alpha|\right)\int_0^\infty x^{1/2}\exp\left(-x\right)dx = \tfrac{1}{2}\pi^{1/2}\exp\left(-|\alpha|\right). \qquad \text{(E. 47)}$$

With this we obtain from (E. 35) an equation to determine ζ:

$$n_e = n_0 \exp\left(-\frac{|\zeta|}{kT}\right) \qquad \text{(E. 48)}$$

or

$$\frac{|\zeta|}{kT} = \ln\frac{n_0}{n_e} = \ln\frac{1}{4n_e}\left(\frac{2mkT}{\pi\hbar^2}\right)^{3/2}. \qquad \text{(E. 49)}$$

In the case of sufficiently low electron concentrations n_e and high temperatures T the electron gas behaves like a classical gas which is characterized by a Maxwell–Boltzmann distribution. Under these conditions the electron gas is "non-degenerate" (see below). Non-degeneracy is thus observed if

$$n_e \ll n_0 \tag{E. 50}$$

or

$$\frac{\zeta}{kT} > 0 \quad \text{and} \quad \frac{|\zeta|}{kT} \gg 1. \tag{E. 51}$$

4. DEGENERACY OF THE ELECTRON GAS

Deviations of the properties of the electron gas from those of an ideal gas are called degeneracy. The presence of a zero-point energy and thus a zero-point pressure are characteristic features of degeneracy. (The concept of "electron gas degeneracy" is in no way connected with the wave-mechanical concept of "degeneracy" which means the existence of several states of the same energy.)

The criterion for degeneracy is defined by the condition

$$\zeta(T) \approx \zeta_0 \gg kT. \tag{E. 52}$$

The electron gas is therefore degenerate as long as its temperature T is considerably below the degeneracy temperature T_E [cf. eqn. (E. 30)]:

$$T \ll T_E. \tag{E. 53}$$

The criterion (E. 52), together with (E. 24), defines a critical electron concentration n_{crit}: for concentrations given by

$$n_e \gg n_{\text{crit}} = \frac{1}{3\pi^2}\left(\frac{2mk}{\hbar^2}\right)^{3/2} T^{3/2} \tag{E. 54}$$

the electron gas is degenerate. According to this, degeneracy is favoured by low temperatures and small masses (we shall show on p. 218 that the rest mass of the electron must be replaced by the so-called effective mass). Comparison with eqn. (E. 36) shows that n_{crit} is approximately equal to the effective density of states:

$$n_{\text{crit}} = \frac{4}{3\pi^{1/2}} n_0. \tag{E. 55}$$

At $T = 300°\text{K}$ the critical density $n_{\text{crit}} \approx 10^{19}\ \text{cm}^{-3}$; the electron concentration in monovalent metals, however, is $n_e \approx 10^{22}\ \text{cm}^{-3}$. Thus the electron gas of metals is highly degenerate; the degeneracy would disappear only at temperatures considerably above the melting point of the metals [for $T > T_E \approx 10^4\,°\text{K}$ according to eqns. (E. 36) and (E. 50)].

5. SPECIFIC HEAT OF THE ELECTRON GAS

As already mentioned, only the fastest electrons and thus only a small fraction of all electrons can take up thermal energy (because of the Pauli principle). For this reason the specific heat of the degenerate electron gas is much smaller than that of a classical gas (cf. p. 157).

The specific heat c_V is obtained from the temperature dependence of the energy $U(T)$ of the electron gas in the volume L^3,

$$c_V = \left(\frac{\partial U(T)}{\partial T}\right)_{V=L^3}. \tag{E. 56}$$

The total energy $U(T)$ is obtained in analogy to (E. 26) from

$$U(T) = \int_0^\infty E\, N(E)\, dE. \tag{E. 57}$$

Substitution of eqn. (E. 17) yields

$$U(T) = \frac{L^3}{2\pi^2}\left(\frac{2m}{\hbar^2}\right)^{3/2} \int_0^\infty \frac{E^{3/2}\, dE}{\exp\left(\frac{E-\zeta}{kT}\right)+1}. \tag{E. 58}$$

With the substitutions $x = E/kT$ and $\alpha = \zeta/kT$ we have

$$U(T) = \frac{L^3}{2\pi^2}\left(\frac{2m}{\hbar^2}\right)^{3/2} (kT)^{5/2} F_{3/2}(\alpha) \tag{E. 59}$$

where

$$F_{3/2}(\alpha) = \int_0^\infty \frac{x^{3/2}\, dx}{\exp(x-\alpha)+1}. \tag{E. 60}$$

According to eqn. (E. 40) in the case of degeneracy ($\alpha = \zeta/kT \gg 1$) we have

$$F_{3/2}(\alpha) \approx \frac{2}{5}\alpha^{5/2}\left[1+\frac{5}{8}\left(\frac{\pi}{\alpha}\right)^2 + \dots\right]. \tag{E. 61}$$

When we again replace ζ by ζ_0 in the correction term, we obtain with eqns. (E. 59) and (E. 61) for the electron gas energy

$$U(T) = \frac{L^3}{5\pi^2}\left(\frac{2m}{\hbar^2}\right)^{3/2} \zeta^{5/2}\left[1+\frac{5\pi^2}{8}\left(\frac{kT}{\zeta_0}\right)^2 + \dots\right]. \tag{E. 62}$$

Substitution of $\zeta(T)$ from (E. 43) yields

$$U(T) = \frac{L^3}{5\pi^2}\left(\frac{2m}{\hbar^2}\right)^{3/2}\zeta_0^{5/2}\left[1+\frac{5\pi^2}{12}\left(\frac{kT}{\zeta_0}\right)^2+\ \dots\right] \tag{E. 63}$$

and, using (E. 27) and (E. 28), we find

$$U(T) = \frac{3}{5}N\zeta_0\left[1+\frac{5\pi^2}{12}\left(\frac{kT}{\zeta_0}\right)^2+\ \dots\right]. \tag{E. 64}$$

The temperature dependence of the mean energy per electron [cf. eqn. (E. 28)] is given by

$$\bar{u}(T) = \frac{U(T)}{N} = \frac{3}{5}\zeta_0\left[1+\frac{5\pi^2}{12}\left(\frac{kT}{\zeta_0}\right)^2+\ \dots\right]. \tag{E. 65}$$

From (E. 64) we obtain

$$c_V = \frac{\pi^2}{2}\frac{Nk^2}{\zeta_0}T = \gamma T \tag{E. 66}$$

for the specific heat of the electron gas; here

$$\gamma = \frac{\pi^2}{2}\frac{Nk^2}{\zeta_0}. \tag{E. 67}$$

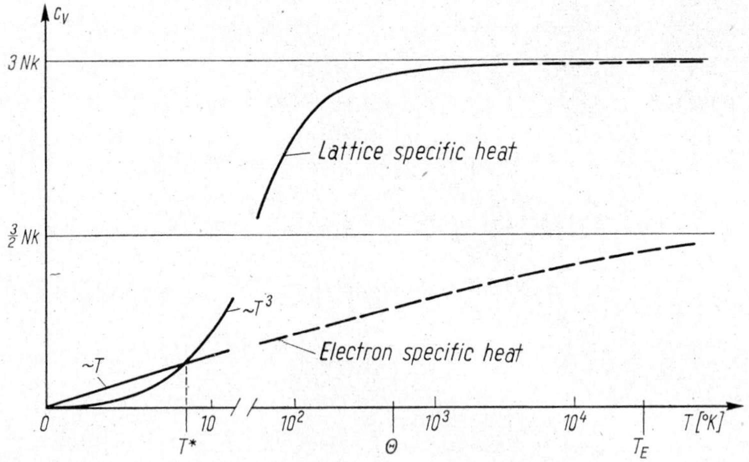

FIG. 88. Lattice specific heat and specific heat of the electron gas as functions of temperature (note the transition from the linear to the logarithmic temperature scale).

Introducing the degeneracy temperature T_E [eqn. (E. 30)], eqn. (E. 66) takes the form

$$c_V = \frac{\pi^2}{5} Nk \frac{T}{T_E}. \tag{E. 68}$$

Unlike the classical gas, the electron gas has a temperature-dependent specific heat, which is directly proportional to the temperature. From (E. 68) we see immediately that the c_V value for temperatures above the Debye temperature Θ (Θ is of the order of 100°K) is only about 1% of the Dulong–Petit value $3Nk$ of the lattice specific heat ($T/T_E \approx 10^{-2}$, cf. p. 163). Only at very low temperatures do the electrons contribute noticeably to the specific heat of metals: as $T \to 0$°K the electron contribution tends to zero as a linear function of T, while the contribution of the lattice vibrations drops as T^3. When we equate these two contributions [(C. 71) and (E. 68)] we obtain the temperature T^* below which the specific heat of the electron gas is predominant (cf. Fig. 88):

$$\frac{12}{5} \pi^4 Nk \left(\frac{T^*}{\Theta}\right)^3 = \frac{\pi^2}{5} Nk \frac{T^*}{T_E} \tag{E. 69}$$

or

$$T^* = \left(\frac{1}{12\pi^2} \frac{\Theta^3}{T_E}\right)^{1/2}. \tag{E. 70}$$

For copper, for instance, with $\Theta = 310$°K and $T_E \approx 3 \times 10^4$°K we find $T^* \approx 3$°K

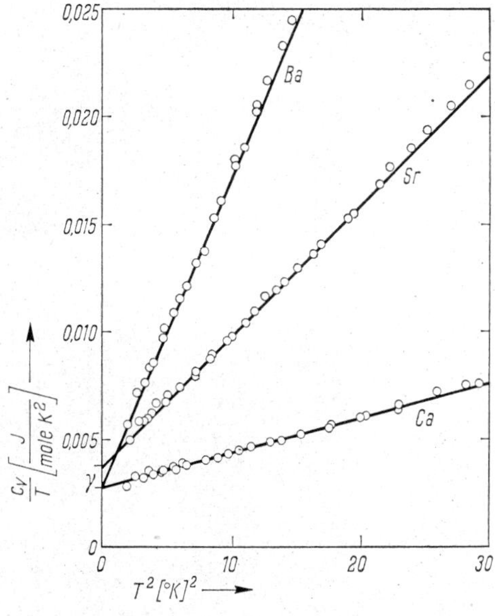

FIG. 89. Total specific heat for alkaline earth metals [after D. H. Parkinson, *Rep. Prog. Phys.* **21,** 226 (1958)].

The total specific heat of metals at $T \ll \Theta$ is therefore

$$c_V^{\text{total}} = \gamma T + \beta T^3 \tag{E. 71}$$

or

$$\frac{c_V^{\text{total}}}{T} = \gamma + \beta T^2 \tag{E. 72}$$

where

$$\beta = \frac{12}{5} \pi^4 \frac{Nk}{\Theta^3}. \tag{E. 73}$$

Results of measurements of the temperature dependence of specific heat plotted as c_V/T vs. T^2 lie on a straight line [cf. eqn. (E. 72)] and thus verify the validity of the Debye T^3 law; extrapolation to $T = 0°\text{K}$ yields the value of γ (cf. Fig. 89). Values of γ obtained from experiments are compiled in Table 17. The values of γ_{free} calculated from eqn. (E. 67) are for many metals except transition metals in good agreement with the experimental values γ_{exp}. The deviations for $\gamma_{\text{exp}} \neq \gamma_{\text{free}}$ can be explained by the fact that the strongly simplifying assumption of a *free* electron gas is often not justified. In particular, the energy eigenvalue density $D(E)$ is changed when one assumes that the electrons move in a periodic rather than a constant potential. The value of γ is, in other words, directly related to the eigenvalue density at the point $E = \zeta_0$ as a combination of eqns. (E. 14), (E. 24), and (E. 67) shows:

$$\gamma = \tfrac{2}{3}\pi^2 k^2 D(\zeta_0). \tag{E. 74}$$

This relation can be obtained quite generally without the restriction of the free electron model. A measurement of the specific heat in the low-temperature range can thus yield information on the eigenvalue density at the Fermi energy ζ_0.

6. ELECTRON EMISSION

Because of the "periodic boundary conditions" (cf. p. 159) the metal properties derived so far apply only to infinite metals. It was therefore not necessary to make any assumptions as to the potential field near the surface of the metal. The energy distribution of the electrons, however, remains unchanged when we consider a finite volume of metal (of size L^3) and use the so-called box model: the electron gas is assumed to be contained in a box with impermeable walls, that is, the potential energy for electrons outside the box is infinitely large compared with that inside the box. We know, however, from experience that the supply of a finite amount of energy is sufficient for electrons

TABLE 17. Specific heats of electrons in metals; γ_{free} has been calculated according to (E. 67) (after C. Kittel, *Introduction to Solid State Physics*, 3rd ed. [Wiley, New York 1966], p. 212; J. M. Ziman, *Electrons and Phonons* [Oxford University Press, London 1960], pp. 114, 117, 127; D. H. Parkinson, *Rep. Progr. Phys.* **21**, 226 [1958], Tables 1, 2, 3)

Metal	γ_{exp} $\left[10^{-3} \frac{\text{W sec}}{\text{mole} \cdot \text{deg}^2}\right]$	$\gamma_{exp}/\gamma_{free}$
Li	1.7	2.3
Na	1.7	1.5
K	2.0	1.1
Cu	0 69	1.37
Ag	0.66	1.02
Au	0.73	1.16
Be	0.22	0.5
Mg	1.35	1.33
Ca	2.73	1.82
Sr	3.64	2.01
Ba	2.7	1.4
Zn	0.65	0.86
Cd	0.71	0.73
Al	1.35	1.6
In	1.69	1.4
Tl	1.47	1.2
Fe	4.98	10.0†
Co	4.73	10.3†
Ni	7.02	15.3†

† Here γ_{free} is calculated for 1 free electron per atom;

to leave the metal (electron emission). It is therefore resonable to modify the box model for a theoretical investigation of electron emission by assuming a finite permeability of the box walls; this means a finite difference of the potential energies for electrons inside and outside the box. The physical cause of this energy difference is the presence of attractive forces between metal ions and electrons. These forces compensate one another in the interior of the crystal but become active as soon as electrons try to leave the crystal (cf. p. 46).

We introduce the following quantities (cf. Fig. 90):

1. The "electron affinity" χ is the difference of the potential energies for electrons inside the metal (E_0) and electrons infinitely distant from the metal surface *in vacuo* (E_∞). This energy difference is strictly defined only for the free electron model, which is based on the assumption of a constant potential in the interior of the metal.

2. The "work function" W is the energy difference between an electron at the Fermi energy $\zeta(T)$ and an electron *in vacuo*, infinitely distant from the metal surface.

According to the way the energy is supplied we distinguish the following emission effects:

1. Thermionic emission due to high temperatures.
2. Field emission, due to high electric field strengths.
3. Photoemission, due to illumination.
4. Emission of secondary electrons due to electron bombardment.
5. Exoemission as a consequence of exothermal reactions at the surfaces or, more generally, mechanical action (such as friction or plastic deformation) on the surfaces or chemical reactions at the surfaces.

The first two effects are limiting cases (electric field $F \to 0$ and temperature $T \to 0$, respectively) of the so-called thermionic field emission (*TF* emission) and can be described satisfactorily within the framework of the Sommerfeld model of free electrons; this will be shown in the following sections. The treatment of the other emission phenomena makes it necessary to take into account the interaction of the electrons with other particles (e.g. photons or phonons, cf. p. 191) and can therefore not be considered within the framework of the free electron model.

(a) *THERMIONIC EMISSION*

Let us calculate the saturation current density $j_x(T, 0)$ for electrons which leave the surface of the metal normal to the x-axis at a temperature T. The saturation current is observed when a sufficiently strong electric field withdraws all electrons, leaving the surface so that the current is not space charge limited. On the other hand, however, the electric field is assumed to be weak enough so that it cannot influence the potential curve (cf. pp. 181 ff.).

Only electrons whose energies are sufficient to surmount the potential barrier of the surface can leave the metal. This necessary condition for thermionic emission refers to the component v_x of the electron velocity which must exceed a critical value v_{x0} (cf. Fig. 90):

$$\frac{m}{2} v_x^2 \geqslant \frac{m}{2} v_{x0}^2 = \zeta + W. \tag{E. 75}$$

The emission current density is given by

$$j_x(T, 0) = e \int_{v_{x0}}^{+\infty} \iint_{-\infty}^{+\infty} \delta(v_x)\, v_x\, n(v_x, v_y, v_z)\, dv_x\, dv_y\, dv_z\,, \tag{E. 76}$$

where $\delta(v_x)$ is the transmission coefficient and $n(v_x, v_y, v_z)$ is the velocity distribution of the electrons; $n(v_x, v_y, v_z)\, dv_x\, dv_y\, dv_z$ is the concentration of electrons whose velocities are in the interval v_x to v_x+dv_x, v_y to v_y+dv_y, v_z to v_z+dv_z. Accordingly, $v_x\, n(v_x, v_y, v_z)\, dv_x\, dv_y\, dv_z$ is the number of electrons coming from inside the metal which, per unit time, hit a unit area of the metal surface. The factor

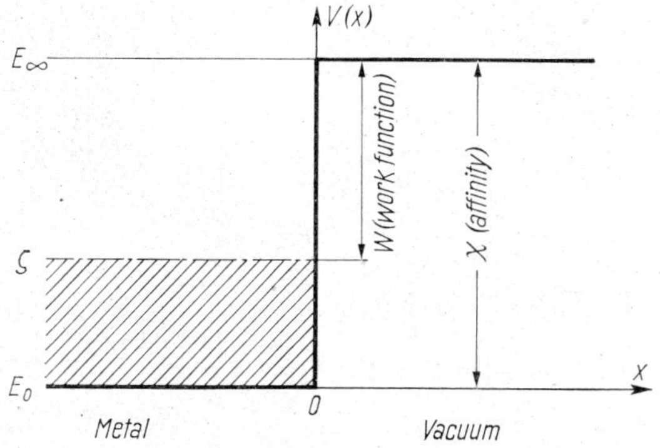

FIG. 90. Potential energy of an electron at a metal surface (model).

$\delta(v_x)$ is the transmissivity of the potential barrier for an electron wave. From the viewpoint of wave mechanics, there exists a certain probability for electrons to be reflected from the boundary surface (cf. pp. 177 ff.) even if their energy is sufficient according to condition (E. 75) for emission.

The velocity distribution can be derived from the energy distribution. By analogy with (E. 15) we have

$$L^3 n(v_x, v_y, v_z)\, dv_x\, dv_y\, dv_z = 2F(v_x, v_y, v_z)\, D(v_x, v_y, v_z)\, dv_x\, dv_y\, dv_z\,. \qquad \text{(E. 77)}$$

From eqns. (E. 3) and (E. 7) it follows that

$$E = \frac{m}{2} v^2 = \frac{m}{2} (v_x^2 + v_y^2 + v_z^2), \qquad \text{(E. 78)}$$

so that we obtain instead of (E. 16)

$$F(v_x, v_y, v_z) = \frac{1}{\exp\left(\dfrac{\frac{m}{2}(v_x^2+v_y^2+v_z^2)-\zeta}{kT}\right)+1} \qquad \text{(E. 79)}$$

and instead of (E. 14)

$$D(v)\, dv = \frac{L^3}{2\pi^2}\left(\frac{m}{\hbar}\right)^3 v^2\, dv = L^3\left(\frac{m}{2\pi\hbar}\right)^3 4\pi v^2\, dv. \qquad \text{(E. 80)}$$

The density of the allowed velocities depends according to the quantization condition (E. 10) only on the magnitude of the velocity. It is proportional to the volume element in the velocity space with the axes v_x, v_y, v_z [$4\pi v^2\,dv$ is the volume of a spherical shell of radius v and thickness dv, cf. eqn. (E. 80)]. We also have

$$D(v_x, v_y, v_z)\,dv_x\,dv_y\,dv_z = L^3\left(\frac{m}{2\pi\hbar}\right)^3 dv_x\,dv_y\,dv_z\,. \tag{E. 81}$$

This relation is a formulation of the well-known law that precisely one quantum state belongs to each volume $h^3 = (2\pi\hbar)^3$ in phase space.

Using eqns. (E. 76), (E. 77), (E. 79), and (E. 81), we obtain for the emission current density

$$j_x(T, 0) = 2e\left(\frac{m}{2\pi\hbar}\right)^3 \int\limits_{v_{x0}}^{+\infty}\int\limits_{-\infty}^{+\infty}\int \frac{\delta(v_x)v_x\,dv_x\,dv_y\,dv_z}{\exp\left(\dfrac{\frac{m}{2}(v_x^2+v_y^2+v_z^2)-\zeta}{kT}\right)+1}. \tag{E. 82}$$

Introducing the polar coordinates

$$\begin{aligned} v_y &= \varrho \sin\varphi, & v_z &= \varrho\cos\varphi, \\ v_y^2+v_z^2 &= \varrho^2, & dv_y\,dv_z &= \varrho\,d\varrho\,d\varphi, \end{aligned} \tag{E. 83}$$

we obtain from eqn. (E. 82)

$$j_x(T, 0) = 2e\left(\frac{m}{2\pi\hbar}\right)^3 \int\limits_{v_{x0}}^{+\infty}\int\limits_{0}^{+\infty}\int\limits_{0}^{2\pi} \frac{\delta(v_x)v_x\,dv_x\varrho\,d\varrho\,d\varphi}{\exp\left(\dfrac{\frac{m}{2}(v_x^2+\varrho^2)-\zeta}{kT}\right)+1}. \tag{E. 84}$$

With the substitutions

$$\frac{m\varrho^2}{2kT} = x, \qquad \varrho\,d\varrho = \frac{kT}{m}\,dx, \tag{E. 85}$$

$$\frac{m}{2}v_x^2-\zeta = \varepsilon, \qquad v_x\,dv_x = \frac{d\varepsilon}{m} \tag{E. 86}$$

eqn. (E. 84) takes the form

$$j_x(T, 0) = \frac{emkT}{2\pi^2\hbar^3}\int\limits_{W}^{\infty}\int\limits_{0}^{\infty}\frac{\delta(\varepsilon)\,dx\,d\varepsilon}{\exp\left(\dfrac{\varepsilon}{kT}+x\right)+1}. \tag{E. 87}$$

The lower limit of integration of ε is according to (E. 75) equal to the work function W. Integration with respect to x yields

$$j_x(T, 0) = \frac{emkT}{2\pi^2\hbar^3} \int_W^\infty \delta(\varepsilon) \ln \left[1+\exp\left(-\frac{\varepsilon}{kT}\right)\right] d\varepsilon. \qquad \text{(E. 88)}$$

This relation applies to every degree of degeneracy and is not restricted to a certain temperature range. Within the framework of the Sommerfeld model it is therefore strictly valid.

For a further evaluation of eqn. (E. 88) we replace the energy-dependent transmission factor $\delta(\varepsilon)$ by a mean value $\bar{\delta}$. Moreover, we know that $W \gg kT$ and thus, in the whole range of integration, $\varepsilon \gg kT$. Therefore, to a good approximation,

$$\ln \left[1+\exp\left(-\frac{\varepsilon}{kT}\right)\right] \approx \exp\left(-\frac{\varepsilon}{kT}\right). \qquad \text{(E. 89)}$$

With these assumptions we obtain from eqn. (E. 88) the famous emission law which was derived independently in 1928 by Sommerfeld and Nordheim:

$$j_x(T, 0) = A_0\, \bar{\delta}\, T^2 \exp\left(-\frac{W}{kT}\right) \qquad \text{(E. 90)}$$

with the universal thermionic constant

$$A_0 = \frac{emk^2}{2\pi^2\hbar^3} = 120 \frac{\text{A}}{\text{cm}^2\ \text{deg}^2}. \qquad \text{(E. 91)}$$

The relation (E. 90) is often called the Richardson equation, as Richardson was the first to treat the problem of thermionic emission and obtained a qualitatively similar relationship. However, instead of using Fermi–Dirac statistics he worked with Maxwell–Boltzmann statistics.

(α) *Transmission or reflection coefficient.* In order to obtain an emission law we have used so far only the height of the potential barrier ($\zeta+W$) between the interior of the metal and the vacuum. The transmission factor δ, however, is unknown as long as we do not know the potential curve for the transition from metal to vacuum.

Let us discuss the concept of the transmission or reflection factor for the simple case of a potential jump at the metal surface ($x = 0$). In the region $x \leq 0$ (metal) the potential energy of the electrons is assumed to be zero, in the region $x > 0$ (vacuum) it is taken to be constant $V = \zeta+W$ (cf. Fig. 90).

As we consider only the motion of the electrons in the x direction, we seek solutions of the one-dimensional Schrödinger equation

$$\frac{d^2\psi}{dx^2}+\frac{2m}{\hbar^2}(E_\perp - V(x))\psi = 0 \tag{E. 92}$$

with

$$E_\perp = E-\frac{\hbar^2}{2m}(k_y^2+k_z^2) = \frac{\hbar^2}{2m}k_x^2 = \frac{m}{2}v_x^2 . \tag{E. 93}$$

With $V(x) = \text{const.}$ the eigenfunctions for the eigenvalue $E_\perp$ can be given, in analogy to eqn. (E. 6). For $x \leqslant 0$ the most general form of the solution is

$$\psi_- = a \exp(ik_x x)+b \exp(-ik_x x). \tag{E. 94}$$

According to this the eigenfunction in the interior of the metal consists of an incident wave of amplitude a and a reflected wave of amplitude b. In the region $x > 0$ an electron wave of amplitude c which leaves the metal is represented by the solution

$$\psi_+ = x \exp(i\varkappa x), \tag{E. 95}$$

where

$$\varkappa = \frac{1}{\hbar}[2m(E_\perp - V)]^{1/2} = \frac{1}{\hbar}[2m(E_\perp - \zeta - W)]^{1/2}. \tag{E. 96}$$

The kinetic energy of the emitted electrons has thus been reduced by $\zeta + W$.

The reflection coefficient R of the boundary surface is defined as the ratio of the squares of the amplitudes

$$R = \frac{b^2}{a^2} . \tag{E. 97}$$

That part of the incident intensity which is not reflected must be emitted. It is called the transmission coefficient

$$\delta = 1-R = 1-\frac{b^2}{a^2} . \tag{E. 98}$$

The amplitudes a, b, c are interrelated by the continuity condition at the boundary surface $x = 0$:

$$\psi_-(0) = \psi_+(0), \tag{E. 99}$$

$$\frac{d\psi_-}{dx}(0) = \frac{d\psi_+}{dx}(0). \tag{E. 100}$$

From these conditions and eqns. (E. 94) and (E. 95) it follows that

$$a+b = c \tag{E. 101}$$

and

$$k_x(a-b) = \varkappa c. \tag{E. 102}$$

When c is eliminated, we obtain from this

$$R = \frac{b^2}{a^2} = \left(\frac{k_x-\varkappa}{k_x+\varkappa}\right)^2 \tag{E. 103}$$

and, from eqn. (E. 98),

$$\delta = 4\frac{k_x\varkappa}{(k_x+\varkappa)^2}. \tag{E. 104}$$

A substitution of (E. 93) and (E. 96) yields the energy dependence of the transmissivity for a potential step of height V:

$$\delta = 4\frac{[E_\perp(E_\perp - V)]^{1/2}}{[E_\perp^{1/2}+(E_\perp - V)^{1/2}]^2}. \tag{E. 105}$$

The transmissivity for a potential step of height $V \approx 10$ eV for electrons of the kinetic energy $E \approx 10.1$ eV, for example, is equal to about 40% (the kinetic energy after the emission into the vacuum is 0.1 eV).

The values obtained from (E. 105) for δ are in principle too low and inconsistent with the experimental results. This is due to the strongly simplified assumption of a potential jump. Taking into account a continuously changing potential curve makes it difficult to calculate δ; however, as shown by numerical evaluations, δ then assumes a value of approximately unity and displays a very weak energy dependence. The approximation $\delta(\varepsilon) \approx \bar{\delta}$ [eqns. (E. 88) to (E. 90)] is therefore justified.

(β) *Comparison with experimental results.* The Richardson formula (E. 90) is in good qualitative agreement with experimental results. The measured values of $j_x(T, 0)$ lie on a straight line when we use the plot $\ln(j_x/T^2)$ vs. $1/T$ (cf. Fig. 91). The ordinate value $A_0\bar{\delta}$ obtained on extrapolating to $1/T = 0$ with a suitable $\bar{\delta}$ value, however, in most cases does not yield a correct value for the universal constant A_0 but a value A which may differ by orders of magnitude.

These deviations are essentially due to the following two reasons:

1. *The temperature-dependence of the work function $W(T)$.* The work function is the difference of two temperature-dependent energies, the electron affinity χ and the Fermi energy ζ. The electron affinity, which corresponds to the bind-

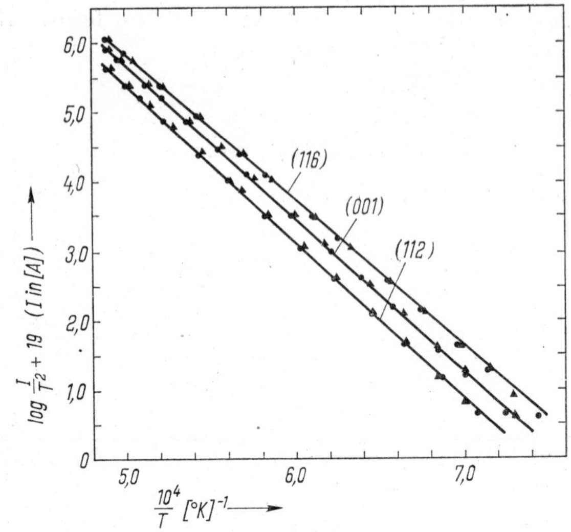

FIG. 91. Sommerfeld–Nordheim law of thermionic emission (Richardson formula) for various orientations of a tungsten crystal. The different data points indicate measurements for different tube geometries [after G. F. Smith, *Phys. Rev.* **94,** 295 (1954)].

ing energy of the electrons in the crystal, decreases with increasing temperature owing to the thermal expansion of the crystal.

The Fermi energy also decreases with increasing temperature [cf. eqn. (E. 43)]. According to the relative variation of the two terms $\chi(T)$ and $\zeta(T)$, the work function can either increase or decrease with increasing temperature. To a first approximation we have

$$W(T) = W_0 \pm aT, \tag{E. 106}$$

where a is a constant of the order of 10^{-4} eV/deg. One must also take into consideration that the Richardson formula does not give an explicit relation between saturation current density and temperature. Substitution of eqn. (E. 106) in (E. 90) yields

$$j_x(T, 0) = A_0 \, \delta \, T^2 \exp\left(\mp \frac{a}{k}\right) \exp\left(-\frac{W_0}{kT}\right). \tag{E. 107}$$

The slope of the straight line ln (j_x/T^2) vs. $1/T$ will thus correspond to the work function W_0 at $T = 0$°K. This value, however, is physically significant only if (E. 106) actually applies down to $T = 0$°K. In many cases the factor $\exp(\pm a/k)$ may explain the deviation of the value A from the constant A_0.

2. *Surface properties.* The electron affinity depends on the atomic array at the surface and thus on the direction of the crystal. The positions of the atoms in the surface of a given crystal orientation can also be changed by impurity adsorption; adsorption will furthermore produce dipole moments and a potential jump at the surface which changes the electron affinity.

The work function $W(T)$ and the experimental values A are not only anisotropic but depend also on the purity of the surface (cf. Table 18). Unambiguous

TABLE 18. Anisotropy of thermionic emission from tungsten (after G. F. Smith, *Phys. Rev.* **94**, 295 [1954], Table 3)

Orientation	W [eV]	A $\left[\frac{\text{A}}{\text{cm}^2\,\text{deg}^2}\right]$
(111)	4.38	52
(112)	4.65	120
(116)	4.29	40
(001)	4.52	105

information on the emission properties requires a measurement of the emission from single-crystal surfaces of definite orientation.

The evaluation of measurements performed with polycrystalline samples is much more difficult. The surface consists of regions of different orientations and thus different work functions (the so-called "patches"). In this case the work function is also a function $W(y, z)$ of the position on the emitting surface. Because of the exponential relation between j_x and W [eqn. (E.90)], the regions with smaller work functions will make the main contribution to the emission current. As the surface is inhomogeneous, the calculation of a current *density* becomes ambiguous. Moreover, because of the dependence $W(y, z)$, electric field components arise parallel to the surface so that the potential near the surface also becomes a function of y and z (the so-called "patch effect"). This phenomenon can become effective also for single-crystal surfaces if, owing to surface imperfections (e.g. roughness, lattice defects), additional regions with different work functions appear.

(b) *THE SCHOTTKY EFFECT*

Sufficiently high electric field strengths at the surface alter the potential outside the metal and thus reduce the work function W. Because of the exponential dependence (E.90) this results in a strong increase of the emission current (the Schottky effect). The reduction of the work function owing to the electric field was first derived by Schottky. The influence of the electric field on an emitted electron can only be estimated if the force acting on the electron is known as a function of the distance x from the surface, i.e. if the potential energy curve is known.

Up to distances of 1 to 2 lattice constants from the geometrical surface ($0 < x < x_0 \approx 10^{-7}$ cm) the emitted electron is under the influence of short-range forces. These forces F_A depend on the atomic environment the electron has left. They determine the specific differences of work functions with respect to crystal orientation, surface conditions, and material. The variation of these forces with distance is known only approximately.

At greater distances ($x > x_0 \approx 10^{-7}$ cm) from the surface long-range forces are acting. The so-called image force is the essential deciding factor; it is of a purely electrostatic character. An electron at a point x *in vacuo* influences a positive charge at the surface of a neutral metal of infinite conductivity; this charge exerts the same force as a real positive charge $+e$ at the mirror point $-x$ of the electron. Thus the electron is attracted by the metal with the image force

$$F_B = \frac{1}{4\pi\varepsilon_0}\frac{e^2}{(2x)^2} = \frac{e^2}{16\pi\varepsilon_0}\frac{1}{x^2}, \tag{E. 108}$$

where ε_0 is the vacuum permittivity.

The work necessary to bring the electron from a point $x > x_0$ to infinity is given by

$$\int_x^\infty F_B\,dx = \frac{e^2}{16\pi\varepsilon_0}\frac{1}{x}. \tag{E. 109}$$

The potential energy as a function of the distance x from the surface for $x > x_0$ (cf. Fig. 92) is given by

$$V(x) = \chi - \frac{e^2}{16\pi\varepsilon_0}\frac{1}{x}. \tag{E. 110}$$

The potential energy inside the metal has been set equal to zero (cf. p. 159); thus its value for an electron at rest, outside the metal, is given by

$$V(\infty) = \chi = \zeta + W. \tag{E. 111}$$

The electron affinity χ is obtained essentially as the sum

$$\chi = \int_0^{x_0} F_A\,dx + \int_{x_0}^\infty F_B\,dx = \int_0^{x_0} F_A\,dx + \frac{e^2}{16\pi\varepsilon_0}\frac{1}{x_0}. \tag{E. 112}$$

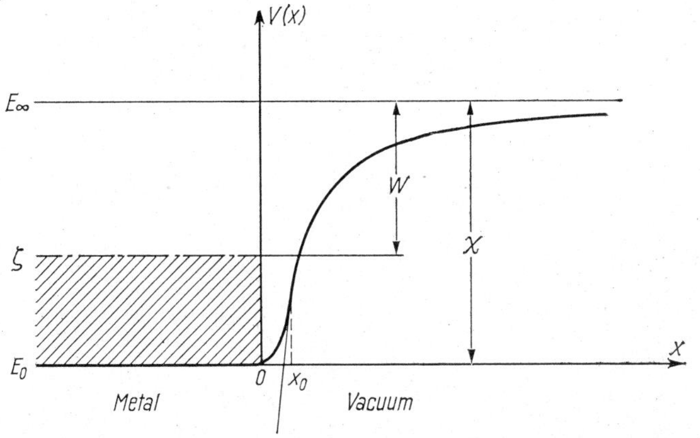

FIG. 92. Potential energy of an electron as a function of the distance from the metallic surface (image potential).

There is thus a contribution to the electron affinity and thus to the work function due to the image force.

The application of an external electric field F creates an additional force F_F opposite to the image force:

$$F_F = -eF. \tag{E. 113}$$

With this we obtain instead of eqn. (E. 110) for the potential energy as a function of the distance $x > x_0$ (cf. Fig. 93)

$$V(x) = \chi - eFx - \frac{e^2}{16\pi\varepsilon_0} \frac{1}{x}. \tag{E. 114}$$

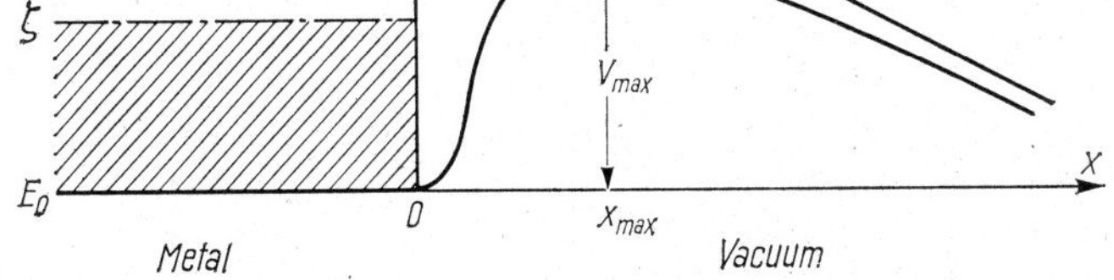

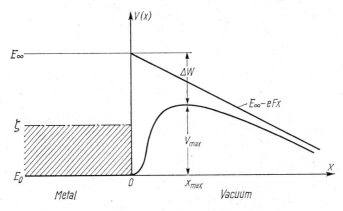

FIG. 93. Potential energy of an electron as a function of the distance from the metallic surface in the presence of an external electric field (decrease of the work function due to the image force).

This function has a maximum at the distance

$$x_{\max} = \frac{1}{4}\left(\frac{e}{\pi\varepsilon_0 F}\right)^{1/2}. \tag{E. 115}$$

At this distance the potential energy is

$$V_{\max} = \chi - \frac{1}{2}\left(\frac{e^3 F}{\pi\varepsilon_0}\right)^{1/2}. \tag{E. 116}$$

In contrast to the case of zero external field, the potential energy does not reach the value χ for $x = \infty$ [cf. eqn. (E. 110)] but only a peak value of $V_{\max}$. The field-dependent work function is defined by $V_{\max}$ [cf. eqn. (E. 111)],

$$V_{\max} = \zeta + W(F). \tag{E. 117}$$

According to this, for $F \neq 0$ the necessary condition for thermionic emission [cf. eqn. (E. 75)] is given by

$$\frac{m}{2} v_x^2(F) \geqslant \zeta + W(F). \tag{E. 118}$$

From eqns. (E. 111), (E. 116), and (E. 117) it follows that

$$W(F) = W - \frac{1}{2}\left(\frac{e^3 F}{\pi\varepsilon_0}\right)^{1/2} = W - \Delta W, \tag{E. 119}$$

i.e. if $F \neq 0$ the work function is reduced by ΔW.

The emission current $j_x(T, F)$ increases with applied electric field; eqn. (E. 90), if we use (E. 119), can be rewritten in the form

$$j_x(T, F) = A_0\,\delta T^2 \exp\left(-\frac{W-\Delta W}{kT}\right) \tag{E. 120}$$

or

$$j_x(T, F) = j_x(T, 0) \exp\left(\frac{1}{2kT}\left(\frac{e^3 F}{\pi\varepsilon_0}\right)^{1/2}\right). \tag{E. 121}$$

The measured values of $j_x(T, F)$ in a plot of $\ln j_x$ vs. $F^{1/2}$ lie on a straight line whose slope, apart from the proportionality to T^{-1}, is universal (cf. Fig. 94). We see that the image force plays an important role. The current extrapolated to $F = 0$ is the saturation value $j_x(T, 0)$ which is given by the Richardson relation (E. 90).

Here we shall not discuss the slight deviations from the Schottky straight line which are periodic functions of the field strength.

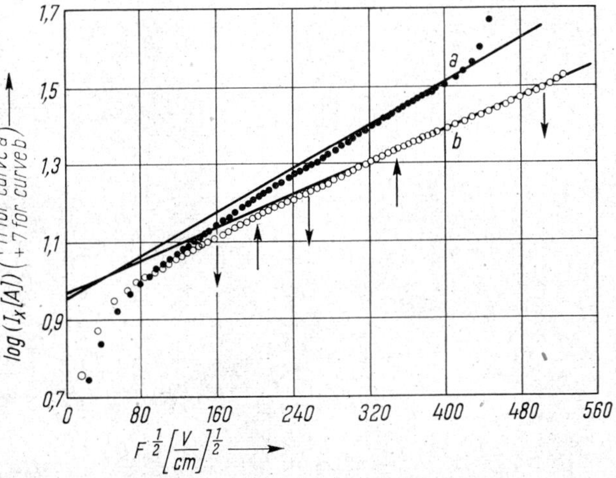

FIG. 94. Thermionic current versus applied electric field (Schottky line) for tungsten (a) at 1373°K, (b) at 1790°K. The solid lines represent the theory according to equation (E. 121). The arrows indicate maxima and minima of the periodic deviations from the Schottky line (after W. B. Nottingham, in *Handbuch der Physik*, Vol. 21, Springer, Berlin 1956).

The relation (E. 121) for Schottky emission does no longer apply in the case of very high fields. Experiments have shown that in fields $F > 10^6$ V cm^{-1} the emission current measured exceeds considerably the current calculated from eqn. (E. 121). This indicates a gradual transition of Schottky emission to field-induced emission.

(c) *FIELD EMISSION*

In the last section we have shown that an external electric field reduces the work function by ΔW. The work function $W(F)$ vanishes at the critical field strength F_{crit} which is given by

$$W(F) = 0 = W - \frac{1}{2}\left(\frac{e^3 F_{\text{crit}}}{\pi\varepsilon_0}\right)^{1/2}. \qquad \text{(E. 122)}$$

Hence we obtain

$$F_{\text{crit}} = \frac{4\pi\varepsilon_0}{e^3} W^2. \qquad \text{(E. 123)}$$

The value of F_{crit} is of the order of 10^8 V cm^{-1}. According to classical physics passing F_{crit} should be accompanied by a sudden onset of the emission current, even at $T = 0$°K, i.e. the electrons would "leak out" of the metal as $W(F) \rightarrow 0$.

Experience has shown that even at fields amounting to only 1% of F_{crit} and at very low temperatures the electron emission is strong and cannot be explained by Schottky emission. Considered from a quantum-mechanical viewpoint, electron emission from the metal is also possible if the electron energy is lower than the potential energy maximum (cf. Fig. 93). The electrons can tunnel through the potential barrier (E. 114) with a certain probability. The condition (E. 118) is no longer valid for sufficiently high field strengths.

The electron current leaving the metal under the influence of an external electric field can be calculated similarly as in the case of thermionic emission. There is, however, an essential difference that now electrons of all velocities $v_x > 0$ can be emitted. Instead of (E. 88) the field emission current is obtained as

$$j_x(T, F) = \frac{emkT}{2\pi^2\hbar^3} \int_{-\zeta}^{\infty} \delta(\varepsilon) \ln \left[1+\exp\left(-\frac{\varepsilon}{kT}\right)\right] d\varepsilon. \quad \text{(E. 124)}$$

The lower limit of integration of the energy ε is obtained from eqn. (E. 86) for $v_x = 0$.

For energies

$$E_\perp = \frac{m}{2} v_x^2 < \zeta + W(F) \quad \text{(E. 125)}$$

or for

$$\varepsilon < W(F) \quad \text{(E. 126)}$$

the transmission coefficient $\delta(\varepsilon)$ depends strongly on the energy. Therefore, it is not allowed to average over the energy as in the case of thermionic emission [cf. eqns. (E. 88) to (E. 90)].

(α) *Transmission coefficient.* The transmissivity of a potential barrier is obtained from the ratio of the particle flux densities in front of and behind the barrier. The particle flux densities in the x direction are determined by the solutions $\psi(x)$ of the one-dimensional Schrödinger equation (E. 92). Solutions $\psi(x)$ can be given in analytical form only for a few particular potentials $V(x)$. By means of the WKB (Wentzel, Kramers, and Brillouin) method an approximate eigenfunction is obtained under the assumption

$$\frac{1}{2\pi} \left| \frac{d\lambda(x)}{dx} \right| \ll 1, \quad \text{(E. 127)}$$

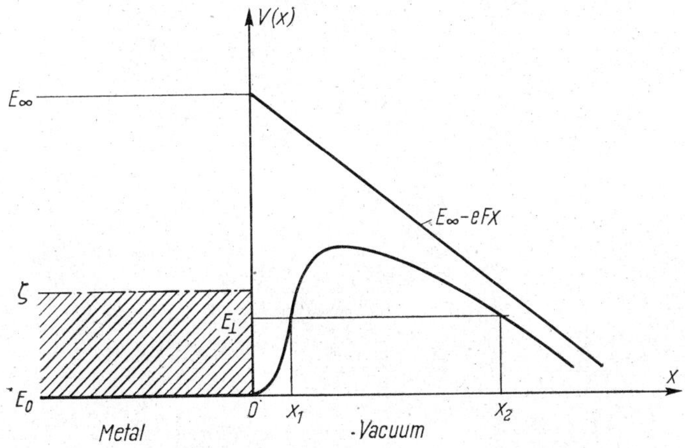

FIG. 95. Tunnelling probability through a potential barrier $V(x)$.

that is, the de Broglie wavelength is assumed to vary only insignificantly over a distance of the wavelength. With

$$\frac{2\pi\hbar}{\lambda} = |p_x| = [2m(|E_\perp - V(x)|)]^{1/2} \tag{E. 128}$$

we obtain from condition (E. 127)

$$\frac{\hbar m}{|p_x|^3}\left|\frac{dV(x)}{dx}\right| \ll 1. \tag{E. 129}$$

The WKB approximate solutions $\psi(E_\perp, x)$ hold only for regions of the coordinate x in which $V(x)$ varies but little and where $|p_x(x)|$ is sufficiently large.

Using the WKB approximation we obtain for the transmission coefficient

$$\delta(E_\perp) = \exp\left\{-\frac{2}{\hbar}\int_{x_1}^{x_2}[2m(V(x)-E_\perp)]^{1/2}\,dx\right\}. \tag{E. 130}$$

The coordinates x_1 and x_2 (cf. Fig. 95) are determined by the relation

$$V(x) - E_\perp = 0, \tag{E. 131}$$

i.e. an electron of energy $E_\perp$ hits the potential barrier at a distance $x_1(E_\perp)$ from the metal surface and leaves it at a distance $x_2(E_\perp)$, with the probability $\delta(E_\perp)$. Equation (E. 130) applies only to energies $E_\perp < V_{max}$ and only if eqn. (E. 129) is satisfied. The transmissivity for electrons with $E \approx V_{max}$ can therefore no longer be determined by means of (E. 130).

A substitution of the potential $V(x)$ characteristic for field emission, (E. 114), in eqns. (E. 130) and (E. 131) yields

$$\delta(E_\perp) = \exp\left\{-\frac{2}{\hbar}\int_{x_1}^{x_2}\left[2m\left(\chi - E_\perp - eFx - \frac{e^2}{16\pi\varepsilon_0}\frac{1}{x}\right)\right]^{1/2} dx\right\} \quad \text{(E. 132)}$$

where

$$x_{1,2} = \frac{\chi - E_\perp}{2eF}\left[1 \mp \left(1 - \frac{4e^3F}{16\pi\varepsilon_0(\chi - E_\perp)^2}\right)^{1/2}\right]. \quad \text{(E. 133)}$$

The integral in eqn. (E. 132) is a complete elliptical integral which was calculated for the first time by Nordheim. The result of an evaluation of (E. 132) is

$$\delta(E_\perp) = \exp\left(-\frac{4}{3}\frac{(2m)^{1/2}}{e\hbar F}(\chi - E_\perp)^{3/2}\,\phi\right) \quad \text{(E. 134)}$$

where

$$\phi = \phi\left(\frac{(e^3F)^{1/2}}{\chi - E_\perp}\right). \quad \text{(E. 135)}$$

The function ϕ assumes values between 0 and 1; for $F \neq 0$ we have always $\phi < 1$. Using eqns. (E. 86) and (E. 111) we find

$$\delta(\varepsilon) = \exp\left(-\frac{4}{3}\frac{(2m)^{1/2}}{e\hbar F}(W - \varepsilon)^{3/2}\,\phi\right). \quad \text{(E. 136)}$$

When we use instead of (E. 114) the strongly simplified potential dependence

$$V(x) = \chi - eFx \quad \text{(E. 137)}$$

for $V(x)$, which ignores the image force, the integral in (E. 132) can be evaluated elementary. We then obtain immediately eqn. (E. 134) or (E. 136) for the transmissivity with $\phi = 1$. The influence of the image force, i.e. of $\phi < 1$, is strong as ϕ appears in the exponent. Thus the image force increases the transmissivity by lowering the potential barrier.

For further calculations of the field emission current we assume $T \approx 0$°K. The upper limit of integration in (E. 124) is then $\varepsilon = 0$ since, by virtue of (E. 19), no states of energies $\varepsilon > 0$ ($E_\perp > \zeta$) are occupied at $T \approx 0$°K. The lower limit of integration may be replaced by $\varepsilon = -\infty$ since $\delta(\varepsilon) \to 0$ exponentially as ε decreases. As we have taken $T \approx 0$°K we always have $\varepsilon < 0$ and

$$\ln\left[1 + \exp\left(-\frac{\varepsilon}{kT}\right)\right] \approx -\frac{\varepsilon}{kT}. \quad \text{(E. 138)}$$

With this eqn. (E. 124) takes the form

$$j_x(0, F) = \frac{em}{2\pi^2\hbar^3}\int_{-\infty}^{0}\delta(\varepsilon)\,\varepsilon\,d\varepsilon. \quad \text{(E. 139)}$$

In this approximation the field emission current does not depend on temperature and it can be shown that, also generally, the electron emission under the influence of strong fields is largely independent of the temperature.

The main contribution to the field emission current is due to electrons of energy $E_{\perp} \approx \zeta$ (i.e. $\varepsilon \approx 0$) for which the tunnelling probability is highest. It is therefore possible to expand the exponent of the transmissivity at $\varepsilon = 0$; eqn. (E. 136) takes the form

$$\delta(\varepsilon) \approx \exp\left[-\frac{4}{3}\frac{(2m)^{1/2}}{e\hbar F}\left(W^{3/2}-\frac{3}{2}W^{1/2}\varepsilon+\ \ldots\right)\phi_0\right] \qquad \text{(E. 140)}$$

where

$$\phi_0 = \phi\left(\frac{(e^3F)^{1/2}}{W}\right). \qquad \text{(E. 141)}$$

With this we obtain from eqn. (E. 139)

$$j_x(0, F) \approx -\frac{em}{2\pi^2\hbar^3}\exp\left(-\frac{4}{3}\frac{(2m)^{1/2}}{e\hbar F}W^{3/2}\phi_0\right)\int\limits_{-\infty}^{0}\varepsilon\exp\left(\frac{2(2m)^{1/2}}{e\hbar F}W^{1/2}\varepsilon\right)d\varepsilon \qquad \text{(E. 142)}$$

and hence

$$j_x(0, F) \approx \alpha\frac{F^2}{W}\exp\left(-\beta\phi_0\frac{W^{3/2}}{F}\right) \qquad \text{(E. 143)}$$

with the universal constants

$$\alpha = \frac{e^3}{16\pi^2\hbar}, \qquad \text{(E. 144)}$$

$$\beta = \frac{4}{3}\frac{(2m)^{1/2}}{e\hbar}. \qquad \text{(E. 145)}$$

The relation between the field emission current and the applied field strength (E. 143) is called the Fowler–Nordheim equation. It resembles the Richardson relation (E. 90): the temperature dependence is replaced by an analogous field dependence.

(β) *Experimental results.* The high field strengths necessary to prove the existence of field emission can be reached with a needle-shaped emitter of a radius of curvature of the order of 10^{-4} cm. The quantitative verification of the Fowler–Nordheim law assumes an accurate knowledge of the form of the emitter to permit a change of the directly measurable relation between

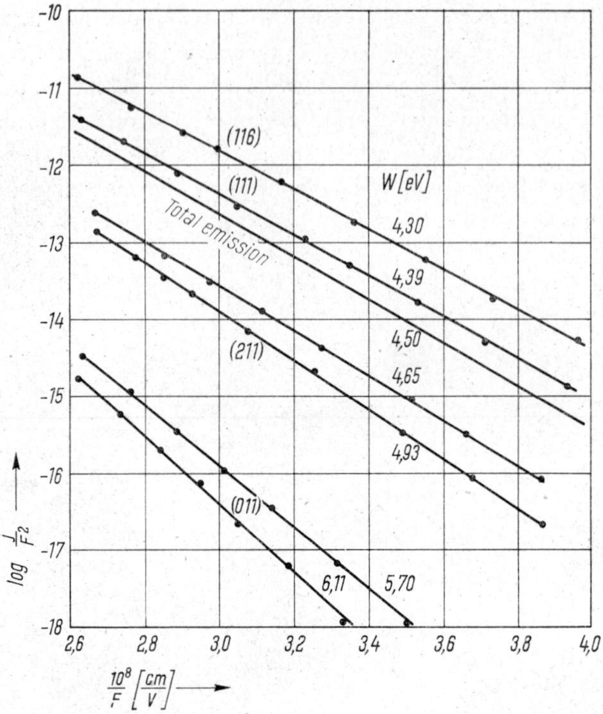

FIG. 96. Field emission current as a function of the applied field (Fowler–Nordheim lines) for various crystallographic directions of a pure tungsten single crystal. For the (011)- and (211)-planes the result depends sensitively on the thermal treatment of the emission point (after E. W. Müller).

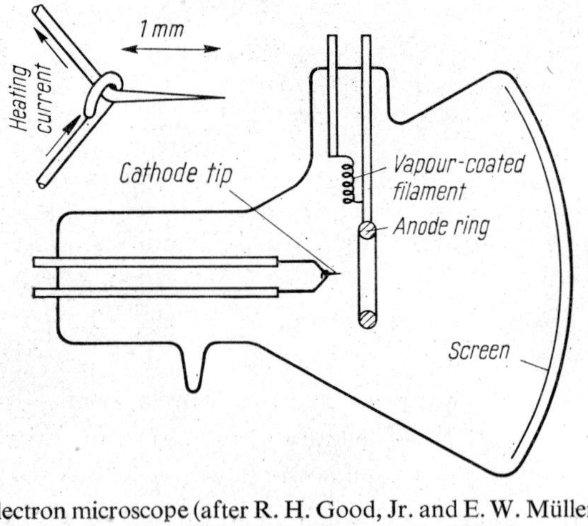

FIG. 97. Field electron microscope (after R. H. Good, Jr. and E. W. Müller, in *Handbuch der Physik*, Vol. 21, Springer, Berlin 1956).

current and voltage to the relation between current density and field strength (cf. Fig. 96).

Because of the dependence of the work function on the crystallographic direction (cf. p. 181) the current density distribution over the needle emitter is not homogeneous. This effect can be immediately observed in a field electron microscope. Such a microscope is obtained when the emitter needle is surrounded by a fluorescent screen kept at the anode potential (cf. Fig. 97). The electrons leave the emitter in an almost radial direction and produce on the screen a strongly enlarged "picture" of the emitter surface. The enlargement is essentially given by the ratio of the distance between screen and needle point and the radius of curvature of the latter; it is of the order of 10^5 to 10^6. Bright and dark areas in the emission picture correspond to different crystallographic directions with low and high work functions. The "pictures" of pure single-crystal emitters show immediately the corresponding crystallographic symmetries. The field electron microscope is suited for investigations of the adsorption and desorption properties of gases in dependence on the crystallographic direction, and of the surface migration of atoms in the range of elevated temperatures.

(d) *PHOTO-EMISSION*

Compared with the thermionic and field-induced emissions the theoretical treatment of photo-emission and secondary-electron emission is basically more difficult. In this case the emission processes are possible only after a primary process. While for thermionic and field-induced emissions the electrons have an equilibrium distribution, the photo- and secondary electron emission is based on a non-equilibrium distribution which is due to the interaction between electrons and photons or primary electrons. The collision processes which supply the energy necessary for electron emission must be known if the emission currents are to be calculated. The basis of all collision processes is the simultaneous conservation of energy and momentum.

Using the photo-effect as an example we shall show in the following that the simultaneous validity of energy and momentum conservation is practically impossible if only interactions between photons and *free* electrons are taken into account.

Let E_0 and E be the energies of the free electron before and after the absorption of a photon of energy $\hbar\omega$. The energy conservation law requires that

$$E = E_0 + \hbar\omega. \qquad \text{(E. 146)}$$

The momenta must simultaneously satisfy the momentum conservation law

$$(2mE)^{1/2} = (2mE_0)^{1/2} + \frac{\hbar\omega}{c}. \tag{E. 147}$$

When we square (E. 147) and substitute (E. 146) we obtain

$$1 = \left(\frac{2E_0}{mc^2}\right)^{1/2} + \frac{\hbar\omega}{2mc^2}. \tag{E. 148}$$

For metal electrons we have always $E_0 \ll mc^2$ and also $\hbar\omega \ll mc^2$; the condition of eqn. (E. 148) for photon absorption is therefore never satisfied for free electrons. It is thus impossible to explain the photo-effect by means of Sommerfeld's free-electron model. The laws of energy and momentum conservation can be satisfied simultaneously only if further interactions with the phonons are taken into account. The conditions are similar in the case of the emission of secondary electrons.

In contrast to the theory of thermionic and field-induced emission, the theory of the other emission processes has not yet reached an entirely satisfactory state although more complex models than the Sommerfeld one have been used. The interested reader may be referred to the specialist literature.

7. LIMITS OF SOMMERFELD'S FREE-ELECTRON MODEL

The free-electron model and the application of Fermi statistics have proved very successful in improving our physical comprehension especially of the monovalent metals. It is an astonishing fact that, in spite of the neglect of the strong electrostatic forces between the positive ions and the electrons, Sommerfeld's model may explain satisfactorily the following properties of metals:

1. the specific heat of the electron gas (cf. pp. 169 ff.),
2. thermionic emission (cf. pp. 174 ff.),
3. field emission (cf. pp. 185 ff.),
4. the Wiedemann–Franz law (cf. pp. 365 ff.),
5. the dia- and paramagnetism of the electron gas (cf. pp. 449 ff.).

The theoretical calculation of galvanomagnetic effects, such as the magneto-resistance and the Hall effect, may deviate considerably from the experimental results. Thus, for example, the magneto-resistance measured for tungsten is higher by a factor of 10^{12} than that calculated on the basis of the free electron model. Similarly, the positive Hall coefficients of some metals (e.g. Zn, Cd) cannot be explained by this model. The difficulties arising in connection with the emission processes have been indicated already in the last section. These examples show that the free-electron model idealizes the real conditions in too strong a way. This can also be understood when we take into consideration that this theory uses only two quantities to characterize the metal: the electron concentration and the work function.

In the next chapter we base our discussion of solid-state properties on a more general model. This automatically answers the question why certain substances are metals or insulators, semiconductors or semimetals, and it is no longer necessary to use specific models which apply, for instance, only to metals, or only to ionic crystals.

References

General

JENKINS, R. O., and TRODDEN, W. G., *Electron and Ion Emission from Solids* (Dover, New York 1965).

Thermionic Emission

EISENSTEIN, A. S., *Oxide Coated Cathodes*, in *Advances in Electronics and Electron Physics*, Vol. 1 (Academic Press, New York 1948).
FOMENKO, V. S., and SAMSONOV, G. W., *Handbook of Thermionic Properties* (Plenum, New York 1966).
HOUSTON, J. M., and WEBSTER, H. F., *Thermionic Energy Conversion*, in *Advances in Electronics and Electron Physics*, Vol. 17 (Academic Press, New York 1962).
MURPHY, E. L., and GOOD, R. H., *Thermionic Emission, Field Emission and the Transition Region*, *Phys. Rev.* **102**, 1464 (1956).
NERGAARD, L. S., *Electron and Ion Motion in Oxide Cathodes*, in *Halbleiter-Probleme*, Vol. 3 (Vieweg, Braunschweig 1956).
NICHOLS, M. H., and HERRING, C., *Thermionic Emission*, *Rev. Mod. Phys.* **21**, 185 (1949).
NOTTINGHAM, W. B., *Thermionic Emission*, in *Handbuch der Physik*, Vol. 21 (Springer, Berlin 1956).

Field Emission

DRECHSLER, M., and MÜLLER, E. W., *Feldemissionsmikroskopie* (Springer, Berlin 1963).
DYKE, W. P., and DOLAN, W. W., *Field Emission*, in *Advances in Electronics and Electron Physics*, Vol. 8 (Academic Press, New York 1956).
GOMER, R., *Field Emission and Field Ionisation* (Harvard University Press, Cambridge, Mass. 1961).
GOOD, R. H., and MÜLLER, E. W., *Field Emission*, in *Handbuch der Physik*, Vol. 21 (Springer, Berlin 1956).
MURPHY, E. L., and GOOD, R. H., *Thermionic Emission, Field Emission and the Transition Region*, *Phys. Rev.* **102**, 1464 (1956).

Photo-emission

GÖRLICH, P., *Recent Advances in Photoemission*, in *Advances in Electronics and Electron Physics*, Vol. 11 (Academic Press, New York 1959).

SOMMER, A. H., *Photoemissive Materials* (Wiley, New York 1968).

WEISSLER, G. L., *Photoionisation in Gases and Photoelectric Emission from Solids*, in *Handbuch der Physik*, Vol. 21 (Springer, Berlin 1956).

Secondary Electron Emission

DEKKER, A. J., *Secondary Electron Emission*, in *Solid State Physics*, Vol. 6 (Academic Press, New York 1958).

HACHENBERG, O., and BRAUN, W., *Secondary Electron Emission from Solids*, in *Advances in Electronics and Electron Physics*, Vol. 11 (Academic Press, New York 1959).

KOLLATH, R., *Sekundärelektronenemission fester Körper bei Bestrahlung mit Elektronen*, in *Handbuch der Physik*, Vol. 21 (Springer, Berlin 1956).

MCKAY, K. G., *Secondary Emission*, in *Advances in Electronics and Electron Physics*, Vol. 1 (Academic Press, New York 1948).

Data on the free electron model may be found in various references on pp. 4, 277 and 404.

F. Electrons in a Periodic Potential

I. Assumptions of the Single-electron Approximation

The motion of electrons in crystalline solids is described by a much more refined theory, where one does no longer assume—as in the previous chapter—that the electrons are free, but takes into account the interaction between the electrons and the atomic nuclei. We have then instead of a constant potential in the interior of the solid a periodic potential with the same period and symmetry as the crystal considered. The starting-point for a more refined theory is therefore the knowledge of the crystal structure. This can only be determined experimentally as it has so far not been possible to derive theoretically the crystal structure for a given element or for elements in a given ratio of abundances. For instance, one cannot predict the crystal structure iodine will assume when it changes from a vapour into a solid.

Here we shall restrict ourselves to the following problem which, at any rate, can only be solved on simplifying assumptions: let us consider an element of nuclear charge Z whose crystal structure is given. We have to find the stationary spatial charge distribution of the ZN electrons and the energy states of the perfect crystal of N atoms. This problem means that we have to solve an $(N+ZN)$-body problem, which is generally impossible. With the following assumptions we can reduce it to a one-body problem (the so-called single-electron approximation):

1. The nuclei at their lattice sites are perfectly at rest. Interactions between electrons and phonons are thus neglected.
2. The interactions between the *individual* electrons (Coulomb forces, exchange forces) are ignored. Every electron is in the potential field which is due to the positive nuclei and an averaged charge distribution of the other electrons. The strong fields of the nuclei are thus more or less screened by the electrons. For many general results the amount of screening is unimportant.

The consequence of these two assumptions, however, is decisive: each particular electron is in the same strictly periodic potential which has the periodicity of the structure considered. We can therefore describe each electron by the same

Schrödinger equation

$$\nabla^2\psi + \frac{2m}{\hbar^2}(E - V(\boldsymbol{r}))\psi = 0 \tag{F. 1}$$

where

$$V(\boldsymbol{r}) = V(\boldsymbol{r} + n_1\boldsymbol{a}_1 + n_2\boldsymbol{a}_2 + n_3\boldsymbol{a}_3) \tag{F. 2}$$

where $\boldsymbol{r}$ is the position vector, $\boldsymbol{a}_1$, $\boldsymbol{a}_2$, $\boldsymbol{a}_3$ are translation vectors, and n_1, n_2, n_3 are integers. Now we have to find the eigenvalues and eigenfunctions for one electron in the periodic potential $-V(\boldsymbol{r})/e$.

II. Bloch Waves

Bloch has proved that the solutions of the Schrödinger equation with a periodic potential are of the form (Bloch theorem):

$$\psi = u(\boldsymbol{r}) \exp i(\boldsymbol{K} \cdot \boldsymbol{r}) \tag{F. 3}$$

where

$$u(\boldsymbol{r}) = u(\boldsymbol{r} + n_1\boldsymbol{a}_1 + n_2\boldsymbol{a}_2 + n_3\boldsymbol{a}_3), \tag{F. 4}$$

and $\boldsymbol{K}$ is a "free" wave vector. The solutions (the Bloch functions) are travelling plane waves modulated by the function $u(\boldsymbol{r})$ with the periodicity of the potential, that is, with the period of the lattice. Such waves are also called Bloch waves (cf. Fig. 98). The electrons in the periodic potential are thus non-localized. The probability that they are in the volume element $d^3\boldsymbol{r}$ is, according to (F. 3),

$$\psi\psi^* \, d^3\boldsymbol{r} = uu^* \, d^3\boldsymbol{r}. \tag{F. 5}$$

In the Bloch functions (F. 3) the wave vector $\boldsymbol{K}$ can assume values between $-\infty$ and $+\infty$. However, $\boldsymbol{K}$ is not given unambiguously by eqns. (F. 3) and (F. 4)

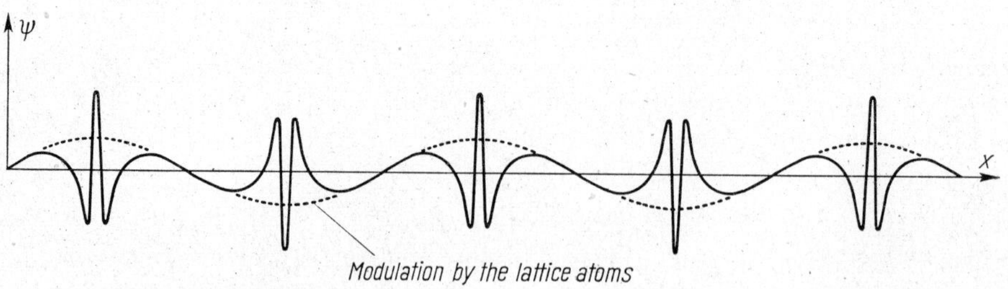

FIG. 98. Bloch wave (the figure shows a Bloch wave with wavelength $\lambda = 2d$; d is the distance between two adjacent atoms in the direction of the wave vector).

since (in the one-dimensional case)

$$\begin{aligned}\psi(K) &= u(K, x) \exp(iKx)\\ &= u(K, x) \exp\left(-i2\pi m \frac{x}{a}\right) \exp\left[ix\left(K+\frac{2\pi m}{a}\right)\right], \end{aligned} \tag{F. 6}$$

where K is the x component of the wave vector, a is the translation period in the x direction, and m is an integer. The function

$$u(K_m, x) = u(K, x) \exp\left(-i2\pi m \frac{x}{a}\right) \tag{F. 7}$$

has also the periodicity a of the lattice [cf. eqn. (F. 4)]. With $K_m = K+2\pi m/a$ we obtain from (F. 6)

$$\psi(K) = \psi\left(K+\frac{2\pi m}{a}\right), \tag{F. 8}$$

that is, the same wave function is represented by different values of K. We obtain all wave functions in terms of K when we restrict the K values to an interval of the length $2\pi/a$. The wave number limited in this way is called the "reduced" wave number k, in contrast to the "free" wave number K [cf. eqn. (C. 151) where a denoted half the translation period].

We choose either

$$0 \leqslant k \leqslant \frac{2\pi}{a} \tag{F. 9}$$

as the range of k or a symmetrical range about $k = 0$,

$$-\frac{\pi}{a} \leqslant k \leqslant +\frac{\pi}{a}. \tag{F. 10}$$

The interval defined by (F. 10) is identical with the first Brillouin zone in the reciprocal lattice (cf. p. 91). In the three-dimensional case all reduced wave vectors are contained in a polyhedron in $\boldsymbol{K}$ space which, in general, also has the form of the first Brillouin zone. The reduction of the $\boldsymbol{K}$ values to a limited range of values is a consequence of the periodicity of the potential.

Moreover, just as in the case of the free electrons, the distribution of the $\boldsymbol{k}$ values is discontinuous (cf. p. 159). Owing to the boundary conditions only discrete $\boldsymbol{k}$ values are allowed, i.e. there exist only a finite number of reduced wave vectors. We shall calculate this number in the following for the one-dimensional case. The periodic boundary conditions (cf. p. 159) have, in principle, nothing to do with the periodicity of the potential. Physically, they express the homogeneity of the crystal. We subdivide the one-dimensional crystal into basic regions of length L containing N atoms each. Thus we have

$$L = Na. \tag{F. 11}$$

From the boundary condition

$$\psi(x) = \psi(x+L) \tag{F. 12}$$

we obtain with (F. 3) and (F. 4)

$$u(x) \exp(ikx) = u(x+L) \exp[ik(x+L)] = u(x) \exp[ik(x+L)] \tag{F. 13}$$

and from this

$$k = \frac{2\pi}{L} n_x = \frac{2\pi}{Na} n_x, \tag{F. 14}$$

where n_x is an integer. With eqns. (F. 9) and (F. 10) we obtain for the range of values of n_x

$$0 \leqslant n_x \leqslant N$$

or (F. 15)

$$-N/2 \leqslant n_x \leqslant +N/2.$$

Altogether we have N reduced k values. In the three-dimensional case N is the number of unit cells in a basic region of volume L^3. Only the simple structures of monatomic crystals contain one atom per unit cell; in this case N is the number of atoms in the basic region. A generalization of eqns. (F. 10) and (F. 15) to the three-dimensional case means that the number of states in the first Brillouin zone is identical with the number of unit cells in a basic region.

III. Eigenvalues and Energy Bands

As in the free electron case we obtain information on the eigenvalue spectrum $E_n(\boldsymbol{k})$, if we substitute the solutions $\psi(\boldsymbol{k})$ of (F. 3) in the Schrödinger equation (F. 1). Now we obtain no explicit relation $E(\boldsymbol{k})$ but an equation determining $u(\boldsymbol{r})$:

$$\nabla^2 u + 2i(\boldsymbol{k}\cdot\text{grad}_r)u + \frac{2m}{\hbar^2}\left(E - \frac{\hbar^2}{2m}k^2 - V(\boldsymbol{r})\right)u = 0. \tag{F. 16}$$

This is an eigenvalue equation like the Schrödinger equation. The value of the wave vector $\boldsymbol{k}$ enters it as a parameter so that, according to eqns. (F. 14) and (F. 15), we have N different eigenvalue equations of the form of (F. 16). Each of these equations has solutions for a discrete sequence of eigenvalues $E_n(\boldsymbol{k})$. The eigenvalues which belong to a certain value of $\boldsymbol{k}$ are enumerated by the quantum number n, the so-called band index n. Thus the two quantities n and $\boldsymbol{k}$ characterize the eigenvalues and thereby the eigenfunctions. Instead of $E_n(\boldsymbol{k})$ one often uses the symbol $E_{n\boldsymbol{k}}$ so that the solution of the Schrödinger equation reads

$$\psi_{n,\,\boldsymbol{k}} = u_{n,\,\boldsymbol{k}}(\boldsymbol{r}) \exp i(\boldsymbol{k} \cdot \boldsymbol{r}). \tag{F. 17}$$

An arbitrary eigenvalue $E_n(\boldsymbol{k})$ will vary only little if $\boldsymbol{k}$ varies little. The N discrete values of $\boldsymbol{k}$ lie very close to one another, i.e. they are quasi-continuous, if N is taken to be sufficiently large [cf. eqns. (F. 14) and (F. 15)]. To each value of n pertain N energy values $E_n(\boldsymbol{k})$ which, for reasons of continuity, appear in a quasi-continuous sequence forming a so-called energy band. Thus the energy spectrum consists of bands with different band indices n. Each energy band contains N eigenvalues and to each energy band there belong N eigenfunctions as long as we neglect the electron spin. The individual energy bands are arranged on the energy scale in such a way that they are either separated from one another by energy gaps, or they overlap one another partially (see Fig. 99). The distribution of energy bands leads to fundamental conclusions as to the solid state properties (cf. pp. 225 ff.).

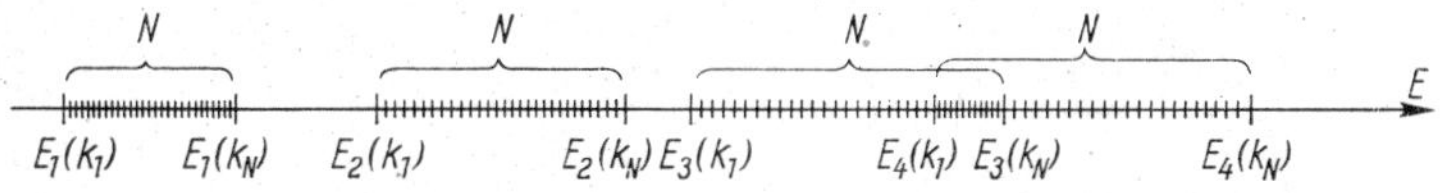

FIG. 99. Allowed energy levels on the energy scale.

From eqn. (F. 16) we can derive further information on the eigenvalue spectrum $E_n(\boldsymbol{k})$ within the band n. The conjugate complex equation to (F. 16) reads

$$\nabla^2 u^*_{n,\,\boldsymbol{k}} - 2i(\boldsymbol{k} \cdot \operatorname{grad}_{\boldsymbol{r}})\, u^*_{n,\,\boldsymbol{k}} + \frac{2m}{\hbar^2}\left(E - \frac{\hbar^2}{2m} k^2 - V(\boldsymbol{r})\right) u^*_{n,\,\boldsymbol{k}} = 0. \tag{F. 18}$$

The same differential equation is obtained for $u_{n,\,-\boldsymbol{k}}$ if $\boldsymbol{k}$ is replaced by $-\boldsymbol{k}$ in (F. 16). This yields

$$u^*_{n,\,\boldsymbol{k}} = u_{n,\,-\boldsymbol{k}}\,. \tag{F. 19}$$

As $u_{n,\,\boldsymbol{k}}$ and $u^*_{n,\,\boldsymbol{k}}$ have the same eigenvalues $E_n(\boldsymbol{k})$ for each parameter $\boldsymbol{k}$, this yields the relation

$$E_n(\boldsymbol{k}) = E_n(-\boldsymbol{k}). \tag{F. 20}$$

The $E_n(\boldsymbol{k})$ are thus even functions with respect to $\boldsymbol{k} = 0$. Accordingly the solutions $\psi_{n,\,\boldsymbol{k}}$ and $\psi_{n,\,-\boldsymbol{k}}$ pertain to the same eigenvalue $E_n(\boldsymbol{k})$, i.e. the eigenvalues $E_n(\boldsymbol{k})$ are doubly degenerate.

In the one-dimensional case it is easy to show that the $E_n(k)$ curve has an extremum at either edge of the band, provided that $E_n(k)$ can be differentiated at these points. Let us assume the band edges to be defined by the wave numbers $k = 0$ and $k = \pm\pi/a$. Because of the periodicity of the wave number [cf. eqn. (F. 8)] we have

$$E_n(k) = E_n\left(k+\frac{2\pi m}{a}\right). \tag{F. 21}$$

Under the assumption that $E_n(k)$ can be differentiated we obtain from this

$$\frac{dE_n(k)}{dk} = \frac{dE_n\left(k+\frac{2\pi m}{a}\right)}{dk}. \tag{F. 22}$$

Moreover, owing to the symmetry of the $E_n(k)$ relation (F. 20) we have

$$\frac{dE_n(k)}{dk} = -\frac{dE_n(-k)}{dk}. \tag{F. 23}$$

For $k = 0$ we hence obtain immediately

$$\frac{dE_n(0)}{dk} = -\frac{dE_n(0)}{dk} = 0. \tag{F. 24}$$

For $k = \pm\pi/a$ eqn. (F. 22) yields ($m = \mp 1$)

$$\frac{dE_n\left(-\frac{\pi}{a}\right)}{dk} = \frac{dE_n\left(+\frac{\pi}{a}\right)}{dk} \tag{F. 25}$$

and (F. 23)

$$\frac{dE_n\left(-\frac{\pi}{a}\right)}{dk} = -\frac{dE_n\left(+\frac{\pi}{a}\right)}{dk}. \tag{F. 26}$$

We therefore have

$$\frac{dE_n\left(\pm\frac{\pi}{a}\right)}{dk} = 0. \tag{F. 27}$$

Usually $dE_n(k)/dk$ has no further zeros inside a band. Figure 100 shows qualitatively the form of the $E_n(k)$ curve. Without explicit data on the periodic potential $V(r)$ it cannot be decided whether there appear for two subsequent bands at the same band edge two maxima, two minima, or only one maximum and one minimum.

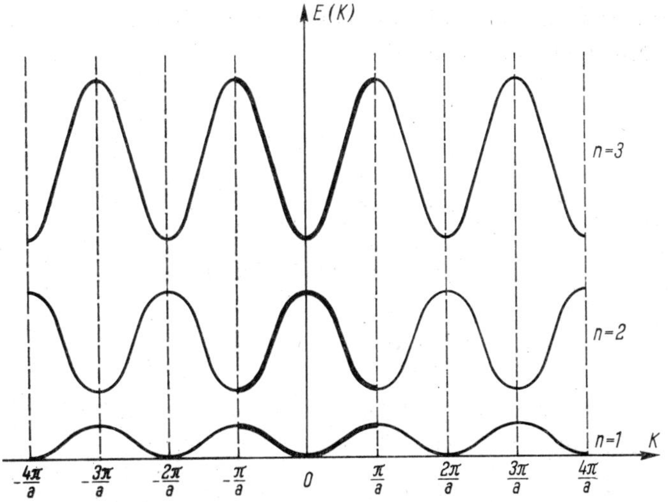

FIG. 100. Periodicity of the functions $E_n(k)$.

IV. Special Cases of Potentials

1. MATHIEU'S DIFFERENTIAL EQUATION; FLOQUET'S SOLUTION

For the one-dimensional case we can give a qualitative description of the influence the potential exerts on the distribution of the allowed and forbidden energy ranges. When we describe the potential by a Fourier series

$$V(x) = \sum_{n=-\infty}^{+\infty} V_n \exp\left(2\pi i n \frac{x}{a}\right) \tag{F. 28}$$

and choose

$$V_n = 0 \quad \text{for} \quad n \geqslant 2 \tag{F. 29}$$

and

$$V_1 = V_{-1}, \tag{F. 30}$$

we obtain the special periodic potential

$$V(x) = V_0 + 2V_1 \cos\left(2\pi \frac{x}{a}\right). \tag{F. 31}$$

In this case the Schrödinger equation takes the form

$$\frac{d^2\psi}{dx^2} + \frac{2m}{\hbar^2}\left[E - V_0 - 2V_1 \cos\left(2\pi \frac{x}{a}\right)\right]\psi = 0. \tag{F. 32}$$

With the potential chosen in this way the Schrödinger equation corresponds to the Mathieu differential equation

$$\frac{d^2\varphi}{d\xi^2}+(\eta+\gamma\cos 2\xi)\varphi = 0, \tag{F. 33}$$

which, according to Floquet, has solutions of the form

$$\varphi(\xi) = A(\xi)\exp(\mu\xi)+A(-\xi)\exp(-\mu\xi), \tag{F. 34}$$

where

$$A(\xi) = A(\xi+l\pi), \tag{F. 35}$$

with l an integer. According to the value of the parameters η and γ we obtain two types of solution:

1. undamped waves modulated by $A(\xi)$, i.e. μ is purely imaginary: $\mu = i\beta$;
2. damped or aperiodically attenuated waves, i.e. μ is complex or real: $\mu = \alpha+i\beta$.

The regions of existence of these two types of solution are shown in Fig. 101. Only the case of the undamped modulated waves is physically interesting (shaded regions). A comparison of eqns. (F. 32) and (F. 33) yields the following attribution to the physical quantities:

$$\varphi = \psi, \tag{F. 36}$$

$$\xi = \pi\frac{x}{a}, \tag{F. 37}$$

$$\eta = \frac{2ma^2}{\pi^2\hbar^2}(E-V_0), \tag{F. 38}$$

$$\gamma = -\frac{4ma^2}{\pi^2\hbar^2}V_1. \tag{F. 39}$$

For $\gamma = 0$ (constant potential) undamped waves exist for all energy values $\eta > 0$ $(E > V_0)$. This case corresponds to the energy spectrum of free electrons. For $\gamma > 0$ (periodic potential) only finite regions of η are possible for undamped waves. This means the existence of energy bands. As γ increases, that is as the amplitude of the periodic potential grows, the energy bands become narrower and narrower until with discrete energy values the case of bound electrons is reached. We show in Fig. 101 for a given potential (γ = constant) a horizontal dash–dot line along which one can read off the widths of the allowed and for-

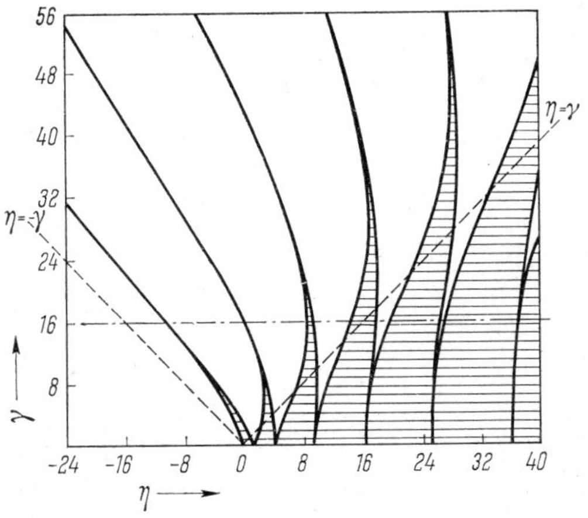

FIG. 101. Floquet solutions of the Mathieu differential equation (from L. Brillouin, *Wave Propagation in Periodic Structures*, McGraw-Hill, New York 1946).

bidden energy bands. The behaviour described here is qualitatively true for general periodic potentials and can be transferred to three-dimensional problems.

2. BRILLOUIN'S APPROXIMATION FOR WEAKLY BOUND ELECTRONS

A calculation of the "band structure" (that is, the $E_n(\boldsymbol{k})$ curve) requires special assumptions about the periodic potential $V(\boldsymbol{r})$. We assume the potential energy of the electrons to be small compared with their kinetic energy. This means that the periodic contribution to the energy can be taken as a small perturbation. An electron in a position-independent potential is free, one in a potential with a small position-dependent amplitude is "weakly bound".

In the following we restrict ourselves to a one-dimensional problem and calculate $E(K)$ for a linear monatomic crystal with lattice constant a. The potential energy is again expanded in a Fourier series [cf. eqn. (F. 28)] which is resolved into a constant zero-order term and a perturbation term (periodic fluctuations):

$$V(x) = \sum_{n=-\infty}^{+\infty} V_n \exp\left(2\pi i n \frac{x}{a}\right) = V_0 + \sum_{\substack{n=-\infty \\ n\neq 0}}^{+\infty} V_n \exp\left(2\pi i n \frac{x}{a}\right). \quad \text{(F. 40)}$$

The origin of the energy scale is chosen in such a way that the mean value of the potential energy is equal to zero, i.e.

$$\bar{V} = \frac{1}{a}\int_0^a V(x)\,dx = 0 \quad \text{and} \quad V_0 = 0. \tag{F. 41}$$

As $V(x)$ is a real quantity, the Fourier coefficients V_n and V_{-n} are each other's conjugate complex. For the sake of simplicity we assume that all coefficients V_n are also real. Hence

$$V_n^* = V_{-n} \quad \text{and} \quad V_n^* = V_n. \tag{F.42}$$

The solutions of the Schrödinger equation are Bloch functions according to (F. 3):

$$\psi(K, x) = u(K, x)\exp(iKx). \tag{F. 43}$$

For the following considerations it is suitable to use the free wave number K. The periodic modulation factor $u(K, x)$ can be expanded in a Fourier series (just as the potential energy):

$$\begin{aligned} u(K, x) &= \sum_{n=-\infty}^{+\infty} c_n(K)\exp\left(2\pi in\frac{x}{a}\right) \\ &= c_0(K) + \sum_{\substack{n=-\infty \\ n\neq 0}}^{+\infty} c_n(K)\exp\left(2\pi in\frac{x}{a}\right). \end{aligned} \tag{F. 44}$$

Hence we obtain from eqn. (F. 43)

$$\psi(K, x) = c_0(K)\exp(iKx) + \sum_{\substack{n=-\infty \\ n\neq 0}}^{+\infty} c_n(K)\exp\left[ix\left(K+\frac{2\pi n}{a}\right)\right]. \tag{F. 45}$$

The first term of this equation corresponds to the solution for free electrons [cf. eqn. (E. 6)], the second term accounts for the position-dependent amplitude of the potential energy and can here be taken as a perturbation term.

Information on the $E(K)$ relation is again obtained on substituting the solutions (F. 45) in the Schrödinger equation:

$$\begin{aligned} &\sum_{n=-\infty}^{+\infty} -\left(K+\frac{2\pi n}{a}\right)^2 c_n \exp\left(2\pi in\frac{x}{a}\right) \\ &\quad + \frac{2m}{\hbar^2}\left[E - \sum_{\substack{n=-\infty \\ n\neq 0}}^{+\infty} V_n \exp\left(2\pi in\frac{x}{a}\right)\right] \sum_{n=-\infty}^{+\infty} c_n \exp\left(2\pi in\frac{x}{a}\right) = 0. \end{aligned} \tag{F. 46}$$

Comparison of the coefficients yields

$$-\left(K+\frac{2\pi n}{a}\right)^2 c_n+\frac{2m}{\hbar^2}E(K)c_n-\frac{2m}{\hbar^2}\sum_{p=-\infty}^{+\infty} V_p c_{n-p} = 0. \quad \text{(F. 47)}$$

Thus we obtain for the coefficients c_n of the Bloch function

$$c_n = \frac{\sum\limits_{p=-\infty}^{\infty} V_p c_{n-p}}{E(K)-\frac{\hbar^2}{2m}\left(K+\frac{2\pi n}{a}\right)^2}. \quad \text{(F. 48)}$$

On the basis of the behaviour of the free electrons we make a perturbation-theoretical substitution in the denominator,

$$E(K) \approx E_0(K) = \frac{\hbar^2}{2m}K^2, \quad \text{(F. 49)}$$

and discuss the behaviour of c_n. According to the value of K we must distinguish the following two cases:

(1) $$E_0(K) \neq \frac{\hbar^2}{2m}\left(K+\frac{2\pi n}{a}\right)^2. \quad \text{(F. 50)}$$

The denominator is non-zero. In the weak-potential approximation the coefficients V_n and V_p tend to zero and thus also the coefficients c_n of the Bloch function, with the exception of c_0. For K values which satisfy eqn. (F. 50) we have in a good approximation

$$\psi(K) \approx c_0(K)\exp(iKx) \quad \text{(F. 51)}$$

with the eigenvalues [cf. eqn. (E. 7)]:

$$E(K) \approx \frac{\hbar^2}{2m}K^2. \quad \text{(F. 52)}$$

We see that in this case the electrons behave like free electrons.

(2) $$E_0(K) = \frac{\hbar^2}{2m}\left(K+\frac{2\pi n}{a}\right)^2. \quad \text{(F. 53)}$$

From (F. 49) we hence obtain the critical K values

$$K_{\text{crit}} = -\frac{\pi}{a}n, \quad \text{(F. 54)}$$

for which the corresponding coefficients c_n assume very high values. From the behaviour of the coefficients c_0 and c_n in the neighbourhood of the critical values K_{crit} we can draw conclusions as to the $E(K)$ curve near K_{crit}. According to (F. 48) we have near $K = -\pi n/a$

$$c_n = \frac{V_n c_0 + \ldots}{E(K) - \frac{\hbar^2}{2m}\left(K + \frac{2\pi n}{a}\right)^2} \tag{F. 55}$$

and

$$c_0 = \frac{V_{-n} c_n + \ldots}{E(K) - \frac{\hbar^2}{2m} K^2}. \tag{F. 56}$$

The sum in the numerator can essentially be replaced by a single term. Independently of the value of K, the coefficient c_0 is in principle very large as for it the denominator in (F. 48) always tends to zero because of (F. 49). Particularly for $K = -\pi n/a$ the coefficient c_n reaches a high value. All other coefficients can be neglected for $K = -\pi n/a$.

According to perturbation theory we put

$$K = -\frac{\pi}{a} n + \varkappa, \tag{F. 57}$$

and

$$E(K) = \frac{\hbar^2}{2m}\left(\frac{\pi}{a} n\right)^2 + \varepsilon. \tag{F. 58}$$

With this eqns. (F. 55) and (F. 56) can be written in the form

$$c_n\left[\frac{\hbar^2}{2m}\left(\frac{\pi}{a} n\right)^2 + \varepsilon - \frac{\hbar^2}{2m}\left(\varkappa + \frac{\pi}{a} n\right)^2\right] = c_0 V_n \tag{F. 59}$$

and

$$c_0\left[\frac{\hbar^2}{2m}\left(\frac{\pi}{a} n\right)^2 + \varepsilon - \frac{\hbar^2}{2m}\left(\varkappa - \frac{\pi}{a} n\right)^2\right] = c_n V_{-n}. \tag{F. 60}$$

These homogeneous and linear equations for c_0 and c_n have a non-trivial solution if the determinant of the coefficients vanishes. Taking into account that $V_n = V_{-n}$ (F. 42) we obtain from this condition a quadratic equation for ε with the solutions

$$\varepsilon_{1,2} = \frac{\hbar^2}{2m}\varkappa^2 \pm \left[V_n^2 + \left(\frac{\hbar^2}{m}\frac{\pi}{a} n\varkappa\right)^2\right]^{1/2}. \tag{F. 61}$$

According to this the energy E near $K = -\pi n/a$ or $\varkappa = 0$ can be described by the function

$$E(\varkappa) = \frac{\hbar^2}{2m}\left(\frac{\pi}{a} n\right)^2 + \frac{\hbar^2}{2m}\varkappa^2 \pm \left[V_n^2 + \left(\frac{\hbar^2}{m}\frac{\pi}{a} n\varkappa\right)^2\right]^{1/2}. \tag{F. 62}$$

For small values of $\varkappa$ the root can be expanded:

$$E(\varkappa) = \frac{\hbar^2}{2m}\left(\frac{\pi}{a}n\right)^2 + \frac{\hbar^2}{2m}\varkappa^2 \pm V_n\left[1 + \frac{1}{2V_n^2}\left(\frac{\hbar^2}{m}\frac{\pi}{a}n\varkappa\right)^2\right] \tag{F. 63}$$

or

$$E(\varkappa) = E_\pm + \left[1 \pm \frac{\hbar^2}{m}\left(\frac{\pi}{a}n\right)^2 \frac{1}{V_n}\right]\frac{\hbar^2}{2m}\varkappa^2 \tag{F. 64}$$

where

$$E_\pm = \frac{\hbar^2}{2m}\left(\frac{\pi}{a}n\right)^2 \pm V_n. \tag{F. 65}$$

We see that above and below the energy gap the energy curve $E(\varkappa)$ or $E(K)$ has a parabolic form.

At the point $K = -\pi n/a$, or $\varkappa = 0$, the energy has two different values, E_+ and E_-. At this point we have an energy jump ΔE, twice as large as the

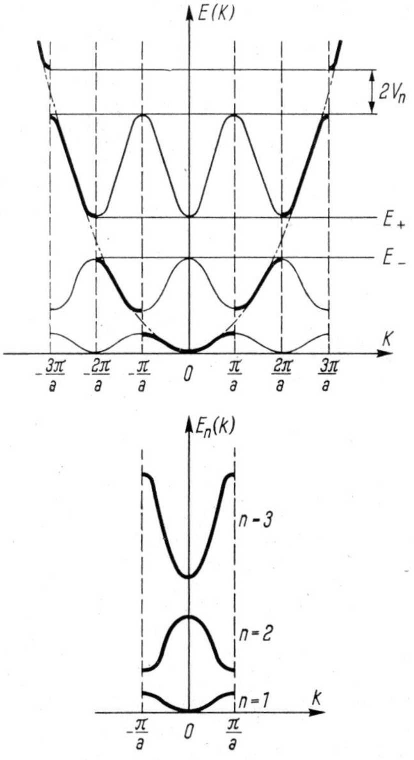

FIG. 102. The functions $E(K)$ and $E_n(K)$ in the weak-potential approximation.

Fourier coefficient V_n of the potential energy (cf. Fig. 102):

$$\Delta E\left(-\frac{\pi}{a}n\right) = 2\,|V_n|. \tag{F. 66}$$

As $E(K)$ is an even function [cf. eqn. (F. 20)], the same energy jump will occur at the point $K = +\pi n/a$. The derivative of (F. 64) with respect to $\varkappa$ vanishes at $\varkappa = 0$ or $K = \pm\pi n/a$:

$$\left.\frac{dE}{d\varkappa}\right|_{\varkappa=0} = \left.\frac{dE}{dK}\right|_{K=\pm\frac{\pi}{a}n} = 0. \tag{F. 67}$$

With this we have proved (F. 24) and (F. 27).

These derivations apply to every $|n| \geqslant 1$. Energy jumps, however, need not appear for every n; if the nth Fourier coefficient $V_n = 0$, the $E(K)$ curve will become continuous at the points $K = \pm\pi n/a$.

From eqn. (F. 66) we see that the appearance of energy gaps ΔE, or of energy bands, is the consequence of the periodic potential. The energy jumps increase as the amplitude of the periodic potential grows (cf. p. 202). Quantitatively, however, this result (F. 66) is valid only within the framework of the assumptions of perturbation theory.

For the solution of the Schrödinger equation at $K = -\pi n/a$ we obtain from (F. 45)

$$\psi\left(-\frac{\pi}{a}n,\, x\right) = c_0\left(-\frac{\pi}{a}n\right)\exp\left(-i\frac{\pi}{a}nx\right) + c_n\left(-\frac{\pi}{a}n\right)\exp\left(+i\frac{\pi}{a}nx\right). \tag{F. 68}$$

With this particular choice of K all other coefficients of the sum in (F. 45) can be neglected. Using (F. 42), (F. 59), and (F. 60), eqn. (F. 68) can be written in the form

$$\psi_\pm\left(-\frac{\pi}{a}n,\, x\right) = c_0\left(-\frac{\pi}{a}n\right)\left[\exp\left(-i\frac{\pi}{a}nx\right) \pm \exp\left(+i\frac{\pi}{a}nx\right)\right]. \tag{F. 69}$$

For the wave function a standing wave has been obtained. This result is generally valid if at $\boldsymbol{K}_{\text{crit}}$ the energy curve has a discontinuity (cf. p. 217).

(a) *THE E(K) AND E_n(k) CURVES*

(α) *The one-dimensional case.* The $E(K)$ function obtained from eqns. (F. 1) and (F. 51) is single-valued and parabolic, as in the case of free electrons, except for the critical points $K = \pm\pi n/a$ where the $E(K)$ curve has discontinuities and the energy is split into two values [eqn. (F. 64), cf. also Fig. 102, thick solid line]. In this representation the nth energy band is attributed to the nth Brillouin zone. In the one-dimensional case the nth Brillouin zone has the range of values

$$\pm\frac{\pi}{a}n \gtrless K \gtrless \pm\frac{\pi}{a}(n-1). \tag{F. 70}$$

Because of the periodicity of the wave number (F. 9), the $E(K)$ curve displays a periodicity of $2\pi/a$. When we restrict ourselves to the reduced wave number k, we attribute all energy bands to the first Brillouin zone (the reduced zone). This makes the $E(k)$ relation multiple-valued and the energy bands must be marked by the corresponding quantum number n: $E_n(k)$ (cf. Fig. 102).

(β) *The two- and three-dimensional cases.* In the two- and three-dimensional cases the energy curve depends on the propagation vector $\boldsymbol{K}$, i.e. $E(\boldsymbol{K})$ depends on the direction. For each direction of $\boldsymbol{K}$ the energy has a qualitative behaviour as shown in Fig. 102. The allowed and forbidden energy regions, however, are different for different directions of $\boldsymbol{K}$ (cf. Fig. 103), which may result in an over-

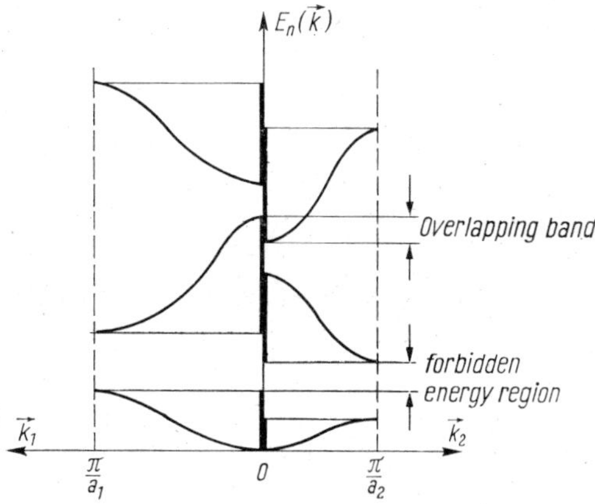

FIG. 103. The function $E_n(\boldsymbol{k})$ for various directions of the wave vector.

lapping of bands. This means that the same energy values in subsequent energy bands belong to different wave vectors. In this case an electron can go over from one energy band to the next higher band without energy supply only through a suitable change in direction.

In order to discuss the $E(\boldsymbol{K})$ relations in two- and three-dimensional problems it is expedient to use the curves or surfaces of constant energy in $\boldsymbol{K}$-space.

This $\boldsymbol{K}$-space is subdivided into Brillouin zones which depend only on the geometry of the lattice (cf. pp. 39 ff.). By analogy with the one-dimensional case [cf. eqns. (F. 9) and (F. 10)] it can be shown that the propagation vector $\boldsymbol{K}$ is defined to within a lattice vector $\boldsymbol{h}$ of the reciprocal lattice

$$\boldsymbol{K} = \boldsymbol{k}+n\boldsymbol{h}, \tag{F. 71}$$

where n is an integer. It is therefore possible to limit the values of the wave vector to a minimum reduced region of $\boldsymbol{K}$-space, i.e. to the first Brillouin zone.

The curves and surfaces of constant energy in the interior of the first Brillouin zone are concentric circles and spheres, respectively, as there, for the corresponding $\boldsymbol{K}$-values, the free-electron approximation is satisfied:

$$E(\boldsymbol{K}) \approx \frac{\hbar^2}{2m} K^2 = \frac{\hbar^2}{2m} (K_x^2+K_y^2+K_z^2). \tag{F. 72}$$

This is not valid near the boundaries of the Brillouin zone (cf. Fig. 104).

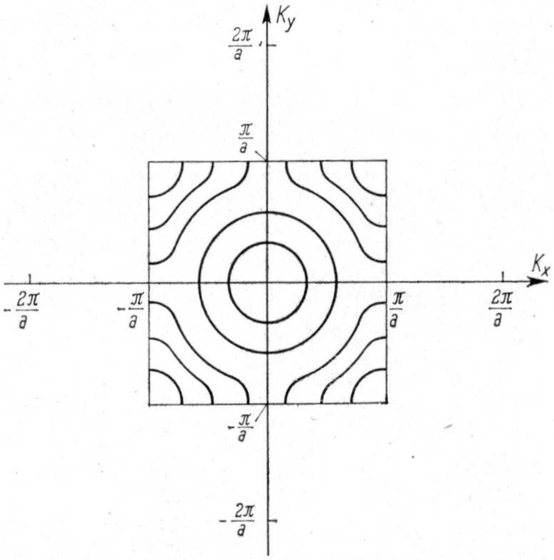

FIG. 104. Curves of constant energy in the $K_x K_y$ plane.

As in the one-dimensional case, the energy may have discontinuities at critical values $\boldsymbol{K}_{\text{crit}}$ of the propagation vector. By analogy with (F. 53) we have the condition for $\boldsymbol{K}_{\text{crit}}$

$$E(\boldsymbol{K}) = \frac{\hbar^2}{2m} K_{\text{crit}}^2 = \frac{\hbar^2}{2m} (\boldsymbol{K}_{\text{crit}}+\boldsymbol{h})^2 . \tag{F. 73}$$

Hence it follows that

$$2(\boldsymbol{K}_{\text{crit}}.\boldsymbol{h}) = h^2. \qquad \text{(F. 74)}$$

This relation is identical with eqn. (B. 51) and defines the boundary surfaces of Brillouin zones. Accordingly the necessary condition for Bragg reflection of X-rays is identical with the necessary condition for the appearance of energy gaps. The curves or surfaces of constant energy are perpendicular to those zone boundaries which, in $\boldsymbol{K}$-space, indicate the positions of energy jumps; there the derivative of the energy in the direction perpendicular to the zone boundary must vanish [cf. eqn. (F. 67) in the one-dimensional case].

Energy discontinuities, however, appear only if the corresponding Fourier coefficients of the potential are non-zero. The Fourier expansion of the potential energy in the case of three-dimensional crystals with a simple primitive lattice is

$$V(\boldsymbol{r}) = \sum_n V_n \exp[i(\boldsymbol{h}_n.\boldsymbol{r})]. \qquad \text{(F. 75)}$$

For a monatomic crystal having a Bravais lattice with basis, the potential energy is assumed to be the sum of the potential energies of the individual Bravais lattices which penetrate one another. Instead of (F. 75) we obtain

$$V_s(\boldsymbol{r}) = V(\boldsymbol{r}) + \sum_{j=1}^{l} V(\boldsymbol{r}+\boldsymbol{q}_j), \qquad \text{(F. 76)}$$

where the $\boldsymbol{q}_j$ are the basis vectors [cf. (B. 44)]. From (F. 75) we have

$$V_s(\boldsymbol{r}) = \sum_n V_n \exp[i(\boldsymbol{h}_n.\boldsymbol{r})]\left\{1+\sum_{j=1}^{l}\exp[i(\boldsymbol{h}_n.\boldsymbol{q}_j)]\right\} \qquad \text{(F. 77)}$$

or

$$V_s(\boldsymbol{r}) = \sum_n V_n S_n \exp[i(\boldsymbol{h}_n.\boldsymbol{r})] \qquad \text{(F. 78)}$$

with

$$S_n = 1+\sum_{j=1}^{l}\exp[i(\boldsymbol{h}_n.\boldsymbol{q}_j)]. \qquad \text{(F. 79)}$$

Except for the atomic scattering factor, the expression for S_n is equal to the structure factor [cf. eqn. (B. 46)] which, for certain reciprocal lattice vectors $\boldsymbol{h}_n$, can become identically equal to zero. Then the corresponding Fourier coefficients $V_n S_n$ of the periodic potential $V_s(\boldsymbol{r})$ are also vanishing. Thus we have the following correlation between the appearance of X-ray reflections and energy discontinuities: in the case of no reflections from the lattice planes $(h_1h_2h_3)$ there is no energy jump at the boundary surface of the Brillouin zone which, from eqn. (F. 74), is defined by the reciprocal lattice vector $\boldsymbol{h} = h_1\boldsymbol{b}_1+h_2\boldsymbol{b}_2+h_3\boldsymbol{b}_3$.

If, for example, the body-centred cubic lattice is considered to be a simple primitive lattice with a basis, the first Brillouin zone is a cube whose faces are the mid-normal planes to the reciprocal lattice vectors with $h_1, h_2, h_3 = \pm 1, 0, 0; 0, \pm 1, 0; 0, 0, \pm 1$. The structure factor for the body-centred cubic lattice is according to (B. 48)

$$S_{bcc} = 1 + \exp [i\pi(h_1 + h_2 + h_3)], \tag{F. 80}$$

i.e. for $h_1 + h_2 + h_3 = 1$ we have $S_{bcc} = 0$. Therefore the energy at the boundary of the first Brillouin zone defined in this way is continuous. The first surfaces for which the energy has a jump are the boundaries of the second Brillouin zone of the simple cubic lattice. They are identical with those of the first Brillouin zone derived from the simple primitive unit cell of the body-centred cubic lattice.

3. BLOCH'S APPROXIMATION FOR STRONGLY BOUND ELECTRONS

The electrons of the inner shells of the lattice atoms have a potential energy which is much larger than their kinetic energy. Therefore they are predominantly near the nuclei, deep in the "potential well" of the atom. The eigenvalues of such electrons are calculated by an approximation method due to Bloch. This approximation method is based on the bound electrons of free atoms and takes into account the atom–atom interaction as a small perturbation.

The electron in the neighbourhood of a certain lattice particle with the lattice vector $\boldsymbol{r}_j$ is assumed to be insignificantly affected by the presence of the neighbouring atoms. Its wave function near this lattice particle can therefore be replaced in a good approximation by the atomic eigenfunction $\psi_{at}(r - \boldsymbol{r}_j)$ which applies strictly to a bound electron of a free atom. This eigenfunction depends on the relative distance between nucleus and electron and decreases rapidly to zero as the distance from the lattice particle increases. The wave function of the electron in the crystal is taken as a linear combination of the atomic eigenfunctions of all lattice particles (LCAO method: *L*inear *C*ombination of *A*tomic *O*rbitals):

$$\psi(\boldsymbol{r}) = \sum_j c_j \psi_{at}(\boldsymbol{r} - \boldsymbol{r}_j). \tag{F. 81}$$

The coefficients c_j must be chosen in such a way that the Bloch function properties are satisfied. This is achieved for

$$c_j = c \exp (i\boldsymbol{K}.\boldsymbol{r}_j). \tag{F. 82}$$

The coefficients are all of the same magnitude; physically this means that all atoms in the lattice are equivalent. Using eqn. (F. 82) the ansatz for the solution of the Schrödinger equation has the form

$$\psi(\boldsymbol{K}, \boldsymbol{r}) = c \sum_j \exp(i\boldsymbol{K}.\boldsymbol{r}_j)\, \psi_{at}(\boldsymbol{r}-\boldsymbol{r}_j). \qquad \text{(F. 83)}$$

The energy eigenvalues are again obtained by means of perturbation calculations, the results of which are the following. The interatomic separation in the crystal is so small that the eigenfunctions of neighbouring atoms overlap. Because of this overlapping each discrete eigenvalue for the electron of the free atom splits into a band of eigenvalues for the electron in the crystal. Each energy band contains again N states (cf. p. 199). The splitting of the eigenvalues into energy bands depends on the magnitude of the matrix elements

$$B = \int \psi_{at}^*(\boldsymbol{r}-\boldsymbol{r}_j)\, [V(\boldsymbol{r})-V_{at}(\boldsymbol{r})]\, \psi_{at}(\boldsymbol{r})\, d^3\boldsymbol{r} \qquad \text{(F. 84)}$$

where $V(\boldsymbol{r})-V_{at}(\boldsymbol{r})$ is the difference between the real potential and the free atom potential. One needs an overlap between $\psi_{at}(\boldsymbol{r}-\boldsymbol{r}_j)$ and $\psi_{at}(\boldsymbol{r})$, in order that B is non-zero. To a first approximation one thus considers only matrix elements between electrons of the nearest-neighbour atoms.

V. Summary

Many important qualitative properties of the eigenfunctions and the energy spectrum are a consequence of the periodicity of the potential. The solutions of the Schrödinger equation are Bloch functions. The energy eigenvalues are periodic functions in $\boldsymbol{K}$-space with the period of an arbitrary reciprocal lattice vector.

The energy scale consists of allowed and forbidden energy regions, the energy bands and the energy gaps. Each band contains the same number of states.

The binding of the electrons to the nuclei is the stronger, the higher the amplitude of the periodic potential. Weak binding manifests itself in broad allowed energy bands and narrow gaps; as the binding becomes stronger, the forbidden bands become broader and the allowed bands become narrower.

Energy bands of many substances have in recent years been calculated, using various sophisticated methods and employing modern computers. We refer to the specialist literature for such methods.

VI. Motion of an Electron in the Periodic Potential

The motion of the electron in the crystal is determined by the time-dependent Schrödinger equation:

$$-\frac{\hbar}{i}\frac{\partial \Psi}{\partial t} + \frac{\hbar^2}{2m}\nabla^2\Psi - V(\boldsymbol{r})\Psi = 0. \qquad \text{(F. 85)}$$

Using the substitution

$$\Psi = \psi(\boldsymbol{r}) \exp\left(-i\frac{E}{\hbar}t\right) \tag{F. 86}$$

we obtain from it the time-independent Schrödinger equation (F. 1) whose solutions have been discussed in the previous section. The solutions of (F. 85) are then

$$\Psi_{n,\boldsymbol{k}}(\boldsymbol{r}, t) = u_{n,\boldsymbol{k}}(\boldsymbol{r}) \exp\left[i\left((\boldsymbol{k}.\boldsymbol{r}) - \frac{E_n(\boldsymbol{k})}{\hbar}t\right)\right]. \tag{F. 87}$$

According to the Heisenberg relations

$$\Delta p_x\, \Delta x \gtrsim \hbar, \quad \Delta p_y\, \Delta y \gtrsim \hbar, \quad \Delta p_z\, \Delta z \gtrsim \hbar \tag{F. 88}$$

the position of an electron whose momentum is known exactly ($\Delta p \rightarrow 0$) is absolutely uncertain. The motion of an electron, however, can be followed only if its position is a definite function of time. At the expense of exact momentum the position of an electron can be defined if, instead of a monochromatic Bloch wave, a wave packet is attributed to it. A wave packet consists of a sum of monochromatic waves which differ only slightly in their wave numbers. The electron is then represented by a Bloch function averaged over a certain range of k. When Bloch waves of the same energy band with wave numbers between $k_0 - \Delta k$ and $k_0 + \Delta k$ are superimposed, the following function for an electron with wave number k_0 is obtained (one-dimensional):

$$\Psi_{n,\bar{k}}(x, t) = \frac{1}{2\,\Delta k} \int_{k_0-\Delta k}^{k_0+\Delta k} u_{n,k}(x) \exp\left[i\left(kx - \frac{E_n(k)}{\hbar}t\right)\right] dk. \tag{F. 89}$$

In this equation it is justified to replace the summation by an integration because the allowed k-values have a quasicontinuous distribution (cf. p. 199). The modulation factor $u_{n,k}(x)$ varies only little with k and can therefore be replaced by $\bar{u}_{n,k_0}$ and taken in front of the integral sign. Putting

$$k = k_0 + \delta k \tag{F. 90}$$

and expanding $E(k)$ near k_0,

$$E_n(k) = E_n(k_0) + \left(\frac{\partial E_n}{\partial k}\right)_{k_0} \delta k + \ldots, \tag{F. 91}$$

we can rewrite eqn. (F. 89) in the form

$$\Psi_{n,\bar{k}}(x, t)$$

$$\approx \frac{u_{n,\bar{k}}(x)}{2\,\Delta k} \exp\left[i\left(k_0 x - \frac{E(k_0)}{\hbar}t\right)\right] \int_{-\Delta k}^{+\Delta k} \exp\left\{i\,\delta k\left[x - \left(\frac{\partial E_n}{\partial k}\right)_{k_0} \frac{t}{\hbar}\right]\right\} d(\delta k). \tag{F. 92}$$

Hence it follows that

$$\Psi_{n,\bar{k}}(x,t) \approx A(x,t)\,\bar{u}_{n,k_0}(x) \exp\left[i\left(k_0 x - \frac{E_n(k_0)}{\hbar}t\right)\right], \tag{F. 93}$$

where

$$A = \frac{\sin(\xi\,\Delta k)}{\xi\,\Delta k} \tag{F. 94}$$

and

$$\xi = x - \frac{1}{\hbar}\left(\frac{\partial E}{\partial k}\right)_{k_0} t. \tag{F. 95}$$

The amplitude of a wave packet of Bloch functions depends not only on the periodic lattice factor $\bar{u}_{n,k_0}(x)$ but also, and essentially, on the additional factor $A(x, t)$ which is independent of the periodicity of the lattice. The value of A is at most equal to unity for $\Delta k = 0$ (that is, monochromatic Bloch waves), or for $\xi = 0$. The amplitude of the wave packet is high only if

$$x = \frac{1}{\hbar}\left(\frac{\partial E}{\partial k}\right)_{k_0} t. \tag{F. 96}$$

For all $|\xi| \gg 0$ the amplitude of the wave packet goes to zero. This indicates that the wave packet in the crystal is localized in a region whose position is a function of time. The centre of mass of the wave packet ($\xi = 0 \rightarrow A = 1$) is identified with the position of the electron.

1. MEAN PARTICLE VELOCITY

The wave packet, or the electron with the wave number k_0 and the energy $E_n(k_0)$, moves with the group velocity v which is given by (F. 96) as

$$v = \frac{dx}{dt} = \frac{1}{\hbar}\left(\frac{\partial E}{\partial k}\right)_{k_0}. \tag{F. 97}$$

Thus we have shown that the group velocity of Bloch waves is equal to that of plane waves. If we put

$$E = \hbar\omega, \tag{F. 98}$$

we can derive the group velocity immediately from the well-known relation

$$v = \frac{\partial \omega}{\partial k} = \frac{1}{\hbar}\left(\frac{\partial E}{\partial k}\right). \tag{F. 99}$$

For the three-dimensional case we have instead of (F. 97)

$$\boldsymbol{v}(n, \boldsymbol{k}) = \frac{1}{\hbar}(\mathrm{grad}_{\boldsymbol{k}}\, E_n(\boldsymbol{k}))_{\boldsymbol{k}}. \tag{F. 100}$$

The relations (F. 97) and (F. 100) can also be derived via quantum-mechanical averaging. Strictly speaking, the electron is accelerated and decelerated in the periodic potential so that it has a spatially periodic momentary velocity. The averaging of the momentary velocity over the crystal or over a lattice period yields the macroscopic observable mean velocity of the electron which is equal to the group velocity.

According to eqn. (F. 100) the mean electron velocity depends only on the energy $E_n(\boldsymbol{k})$ and the wave vector, but is constant in space and time. This means that the average motion of the electron is that of a free one, and electrons are not scattered by atoms *at rest*, i.e. a strictly periodic crystal has no resistivity.

The magnitude of the electron velocity depends on the $E_n(\boldsymbol{k})$ dependence. On the basis of the free electron model we obtain for the velocity from (E. 7) and (F. 100) the classical result

$$v = \frac{\hbar \boldsymbol{k}}{m}. \tag{F. 101}$$

Also for electrons in the periodic potential the velocity increases with the free wave vector if the latter does not satisfy Bragg's reflection condition (F. 74). In other words, for a certain reduced wave vector within the first Brillouin zone the velocities normally increase from band to band with increasing quantum number n.

At the band edges, however, apart from the exceptions described above, the velocity component normal to the edge of the Brillouin zone is equal to zero as there the normal component of grad $E_n(\boldsymbol{k})$ vanishes (cf. p. 211). This case corresponds to the Bragg reflection of electron waves with the critical propagation vectors $\boldsymbol{K}_{\text{crit}}$. These waves undergo total reflection at the individual lattice planes $(h_1h_2h_3)$ which are determined by the reciprocal lattice vector $\boldsymbol{h}$ by eqn. (F. 74). In this way a standing electron wave arises, as was shown already by eqn. (F. 69) for the one-dimensional case. Electrons with the propagation vectors $\boldsymbol{K}_{\text{crit}}$ cannot propagate through a strictly periodic crystal.

Figure 105 shows the electron velocity as a function of the wave number for the one-dimensional case; the velocity v_x is obtained by differentiating the function $E(K_x)$ with respect to K_x [cf. eqn. (F. 99)].

2. THE CRYSTAL ELECTRON UNDER THE INFLUENCE OF AN EXTERNAL FORCE

The influence of an external force results in a variation with time of the electron energy. In order to calculate the law of motion we must start from the time-dependent Schrödinger equation and derive the time-dependent solutions

$\Psi(\boldsymbol{r}, t)$. One often assumes—in the quasi-classical sense—that also in the case of an external force acting upon the electron the electron can be represented by a wave function $\Psi(\boldsymbol{r}, \boldsymbol{k}(t))$ of the stationary Schrödinger equation, where, however, the wave vector is a function of time. The time dependence $\boldsymbol{k}(t)$ can be easily obtained from the following consideration: for an electron with the velocity $\boldsymbol{v}$ the force $\boldsymbol{F}$ changes during dt the energy by dE:

$$dE = (\boldsymbol{F}.\boldsymbol{v})\, dt. \tag{F. 102}$$

As E is a function of $\boldsymbol{k}$, the energy change (within an energy band) is given by

$$dE = (\text{grad}_{\boldsymbol{k}}\, E.d\boldsymbol{k}), \tag{F. 103}$$

where $d\boldsymbol{k}$ is the change of the wave vector during dt. When we equate (F. 102) and (F. 103) and use (F. 100) we obtain

$$\boldsymbol{F} = \hbar \frac{d\boldsymbol{k}}{dt} = \hbar \dot{\boldsymbol{k}}. \tag{F. 104}$$

This differential equation yields the variation with time of the wave vector and, together with the $E(\boldsymbol{k})$ relation, that of the energy. Since these two quantities determine the behaviour of the electron, eqn. (F. 104) can be considered the equation of motion of the crystal electron. It is, though, not identical with Newton's equation of motion according to which the force is equal to the momentum variation with time (see below).

3. MEAN ELECTRON MOMENTUM

The mean momentum of an electron is connected to its mean velocity by the relation

$$\boldsymbol{v} = \frac{1}{m} \boldsymbol{p}, \tag{F. 105}$$

where m is the rest mass of the electron.

This is a general relation since the velocity operator $(\hbar/im)$ $\text{grad}_{\boldsymbol{k}}$ and the momentum operator $(\hbar/i)$ $\text{grad}_{\boldsymbol{k}}$ differ only by the factor $1/m$. Using eqn. (F. 100) we obtain for the mean momentum

$$\boldsymbol{p} = \frac{m}{\hbar} \text{grad}_{\boldsymbol{k}}\, E(\boldsymbol{k}). \tag{F. 106}$$

For *free* electrons only we obtain then from eqn. (E. 7)

$$\boldsymbol{p} = \hbar \boldsymbol{k} \tag{F. 107}$$

and from (F. 104) Newton's law

$$\boldsymbol{F} = \hbar \dot{\boldsymbol{k}} = \dot{\boldsymbol{p}}. \tag{F. 108}$$

4. CRYSTAL MOMENTUM

We obtain a formal analogy to classical mechanics by introducing a crystal momentum $\boldsymbol{P}$ defined as

$$\boldsymbol{P} = \hbar \boldsymbol{k}. \tag{F. 109}$$

With this eqn. (F. 104) can be rewritten in the form

$$\boldsymbol{F} = \dot{\boldsymbol{P}}. \tag{F. 110}$$

Like the wave vector, the crystal momentum is quantized and differs from the former only by the constant factor $\hbar$. Instead of $\boldsymbol{k}$-space a $\boldsymbol{P}$-space is introduced, which is called the crystal momentum space.

The mean momentum $\boldsymbol{p}$ of the electron and the crystal momentum $\boldsymbol{P}$ are different: $\boldsymbol{p}$ is a quantum-mechanical mean value while $\boldsymbol{P}$ is to be considered a quantum number. Only in the case of free electrons are these two quantities identical.

5. MEAN ACCELERATION; EFFECTIVE MASS

From eqn. (F. 100) we obtain for the mean acceleration

$$\frac{d\boldsymbol{v}}{dt} = \left(\frac{d\boldsymbol{k}}{dt} . \operatorname{grad}_{\boldsymbol{k}}\right) \boldsymbol{v} \tag{F. 111}$$

and with eqns. (F. 100) and (F. 104)

$$\frac{d\boldsymbol{v}}{dt} = \left(\boldsymbol{F} . \frac{1}{\hbar^2} \operatorname{grad}_{\boldsymbol{k}}\right) \operatorname{grad}_{\boldsymbol{k}} E(\boldsymbol{k}). \tag{F. 112}$$

On the basis of the analogy with Newton's law an effective mass m^* is defined by the relation

$$\frac{1}{m^*} = \frac{1}{\hbar^2} \operatorname{grad}_k \operatorname{grad}_k E(\boldsymbol{k}). \tag{F. 113}$$

The effective mass defined in this way is not a scalar quantity in the usual sense which has a single fixed value for the crystal electron, but a second rank tensor. It is the inverse of the tensor of the reciprocal effective mass derived from (F. 113):

$$\frac{1}{m^*} = \frac{1}{\hbar^2} \begin{pmatrix} \dfrac{\partial^2 E}{\partial k_1 \, \partial k_1} & \dfrac{\partial^2 E}{\partial k_1 \, \partial k_2} & \dfrac{\partial^2 E}{\partial k_1 \, \partial k_3} \\ \dfrac{\partial^2 E}{\partial k_2 \, \partial k_1} & \dfrac{\partial^2 E}{\partial k_2 \, \partial k_2} & \dfrac{\partial^2 E}{\partial k_2 \, \partial k_3} \\ \dfrac{\partial^2 E}{\partial k_3 \, \partial k_1} & \dfrac{\partial^2 E}{\partial k_3 \, \partial k_2} & \dfrac{\partial^2 E}{\partial k_3 \, \partial k_3} \end{pmatrix}. \tag{F. 114}$$

Since the differentiations are interchangeable, this tensor is symmetrical and can be transformed to principal axes. It then contains only the diagonal elements which define the so-called principal masses:

$$\frac{1}{m_{xx}} = \frac{1}{\hbar^2}\frac{\partial^2 E}{\partial k_x^2} = \frac{\partial^2 E}{\partial P_x^2},$$
$$\frac{1}{m_{yy}} = \frac{1}{\hbar^2}\frac{\partial^2 E}{\partial k_y^2} = \frac{\partial^2 E}{\partial P_y^2}, \qquad \text{(F. 115)}$$
$$\frac{1}{m_{zz}} = \frac{1}{\hbar^2}\frac{\partial^2 E}{\partial k_z^2} = \frac{\partial^2 E}{\partial P_z^2}.$$

In principle the effective mass is thus completely determined by the three principal masses. If the energy $E(\boldsymbol{k})$ depends only on the square of the wave vector, i.e. if the energy surfaces display spherical symmetry, all principal masses are identical. In this case the effective mass is a scalar quantity. In particular for free electrons we obtain with eqn. (E. 7)

$$m_{\text{free}}^* = m. \qquad \text{(F. 116)}$$

Essential properties of the effective masses can be derived from the one-dimensional crystal model. In this case

$$\frac{1}{m^*} = \frac{1}{\hbar^2}\frac{d^2E}{dk^2}. \qquad \text{(F. 117)}$$

A double differentiation of the $E(k)$ function yields the reciprocal effective mass as a function of the wave number (cf. Fig. 105). According to this, the effective mass is positive in the lower part of the energy band and negative in the upper part. An electron is thus accelerated by an external force if its energy is in the lower part of the band and decelerated when it is in the upper part. Its mean velocity in the case of deceleration reaches zero at precisely the moment when its energy has reached the upper edge of the band (cf. p. 216).

The value of the effective mass is higher in narrow energy bands than in broad ones, since in narrow bands the second derivative d^2E/dk^2 is always smaller than in broad bands. According to p. 213 the allowed energy bands are the narrower, the stronger the electrons are bound. "Strongly bound" electrons are thus "heavier" than "weakly bound" electrons and are therefore less influenced by the external force.

In the three-dimensional case the tensor character of the effective mass also plays a role. If the force $\boldsymbol{F}$ acts in an arbitrary direction, the acceleration will in general be in a direction different from that of $\boldsymbol{F}$. Acceleration and force

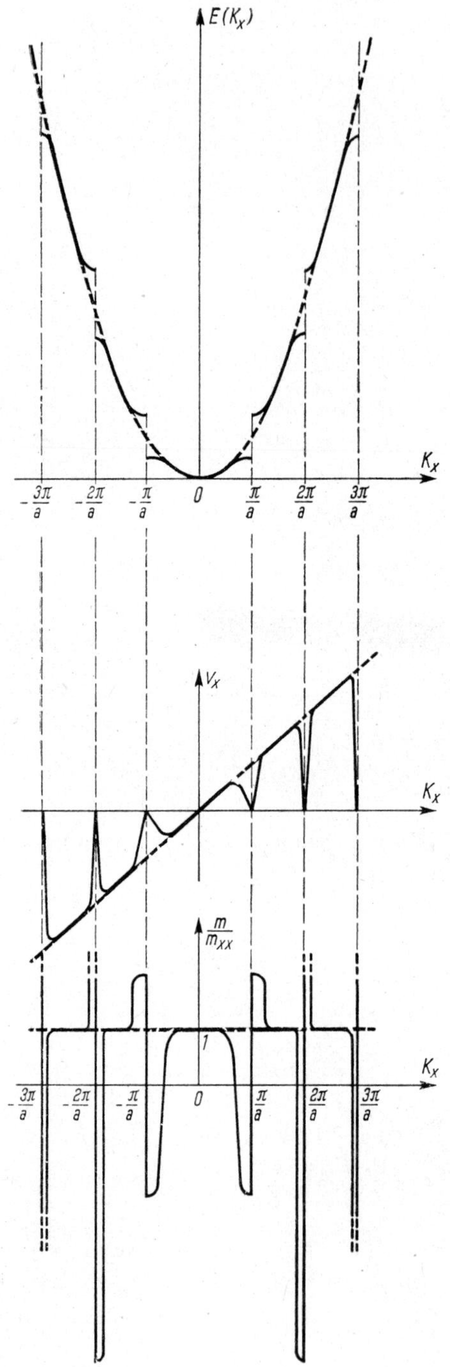

FIG. 105. Energy, velocity and inverse effective mass as functions of the wave number. The dashed curves are those for free electrons (after H. Fröhlich, *Elektronentheorie der Metalle*, Springer, Berlin 1936).

will be parallel only if the force acts in the direction of one of the three principal axes of the effective masses, m_{xx}, m_{yy}, m_{zz}, or if the effective mass is a scalar.

The properties of the effective mass derived from the definition equation (F. 113) are based on the fact that the electron in the crystal is subjected not only to the external force $\boldsymbol{F}$ but also to the lattice force $\boldsymbol{F}_l$. In practical cases, however, $\boldsymbol{F}_l$ is not known. Therefore the equation of motion of the electron

$$\frac{dv}{dt} = \frac{1}{m}(\boldsymbol{F}+\boldsymbol{F}_l) \qquad \text{(F. 118)}$$

is replaced by the relation

$$\frac{dv}{dt} = \frac{1}{m^*}\boldsymbol{F}. \qquad \text{(F. 119)}$$

In this way the influence of the unknown lattice force is taken into account by replacing the true mass of the electron by an effective mass. Comparison of eqns. (F. 118) and (F. 119) yields

$$\left(\frac{m}{m^*}-1\right)\boldsymbol{F} = \boldsymbol{F}_l. \qquad \text{(F. 120)}$$

This relation shows explicitly that the difference between m and m^* is due to the lattice forces. The effective mass describes the behaviour of a crystal electron in the field of an external force $\boldsymbol{F}$. The term "crystal electron" expresses the fact that the laws of motion of the electron depend also on the lattice forces.

The effective mass can be determined experimentally (cf. pp. 236 ff.). The measurements permit conclusions as to the structure of the energy spectrum $E_n(\boldsymbol{k})$. Unfortunately, even with detailed data about the effective mass, it is not possible to draw precise conclusions as to the exact form of the periodic potential which is actually unknown. Theoretically it is only possible to derive the energy curve and hence the effective mass from the periodic potential. In practice this way of going about is also impossible as the periodic potential cannot be determined experimentally.

6. EIGENVALUE DENSITY AND EFFECTIVE MASS

The possible values for the wave vectors have a quasicontinuous distribution in $\boldsymbol{k}$-space (cf. p. 197). Therefore only a finite number of states exists for which the wave vectors are between $\boldsymbol{k}$ and $\boldsymbol{k}+\boldsymbol{dk}$. The $E(\boldsymbol{k})$ relation defines thus a finite number of eigenvalues in the energy interval between E and $E+dE$.

We have shown on p. 198 that the number of states in each Brillouin zone is equal to the number N of elementary cells, or of atoms in the volume L^3. Each Brillouin zone corresponds to an energy band which contains N eigenvalues. Narrow bands have a high eigenvalue density [cf. eqn. (E. 14)]; for higher energies the band width increases and the eigenvalue density decreases.

When we generalize (F. 12) to the three-dimensional case we obtain for cubic symmetry instead of (F. 14)

$$k_x = \frac{2\pi}{L} n_x, \quad k_y = \frac{2\pi}{L} n_y, \quad k_z = \frac{2\pi}{L} n_z. \tag{F. 121}$$

A single state in the $\boldsymbol{k}$-space occupies the volume $(2\pi/L)^3$. The volume $dk_x dk_y dk_z$ contains dZ_k states:

$$dZ_k = \left(\frac{L}{2\pi}\right)^3 dk_x\, dk_y\, dk_z. \tag{F. 122}$$

The eigenvalue density $D(E) = dZ_E/dE$ can be obtained from the number of states contained in $\boldsymbol{k}$-space between the surfaces of constant energy E = const.

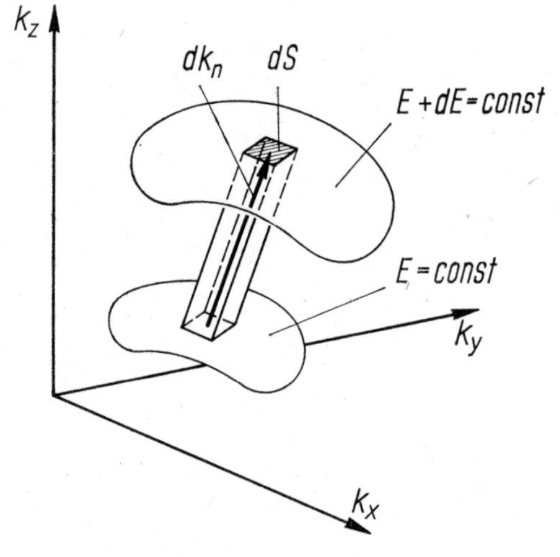

FIG. 106. Computation of the eigenvalue density $D(E)$.

and $E+dE$ = const. (cf. Fig. 106). Under the assumption that $E(\boldsymbol{k})$ is an arbitrary but single-valued and differentiable function,

$$D(E)\, dE = \int_E^{E+dE} dZ_k = \left(\frac{L}{2\pi}\right)^3 \int_E^{E+dE} dk_x\, dk_y\, dk_z. \tag{F. 123}$$

The integration is performed over the volume dV_k in $\boldsymbol{k}$ space between the two surfaces of constant energy. We have

$$dV_k = \int_E^{E+dE} dk_x\, dk_y\, dk_z\,. \tag{F. 124}$$

Instead of the coordinates k_x, k_y, k_z we choose orthogonal coordinates on the surface $E(\boldsymbol{k})$. So we obtain for the volume element

$$dk_x\, dk_y\, dk_z = dS\, dk_n \tag{F. 125}$$

where

$$dk_n = \frac{dE}{|\text{grad}_k\, E|}\,; \tag{F. 126}$$

here dS is an element of area on the surface $E(\boldsymbol{k})=\text{const}$. With eqns. (F. 125) and (F. 126) we obtain for the eigenvalue density the general relation

$$D(E)\, dE = \left(\frac{L}{2\pi}\right)^3 dE \int\limits_{E=\text{const.}} \frac{dS}{|\text{grad}_k\, E|}\,. \tag{F. 127}$$

An analytical treatment is possible only for a few special $E(\boldsymbol{k})$ relations.

For free electrons we obtain from this, using (E. 7),

$$D(E)\, dE = \left(\frac{L}{2\pi}\right)^3 dE\, \frac{m}{\hbar^2} \int\limits_{E=\text{const.}} \frac{dS}{k}\,. \tag{F. 128}$$

In this special case the surfaces of constant energy are spheres of radius k. i.e. the integration with $k = \text{const.}$ yields the spherical surface. From eqn, (F. 128) we obtain the eigenvalue density for free electrons which has been derived already within the framework of Sommerfeld's theory, cf. eqn. (E. 14):

$$D(E)\, dE = \frac{L^3}{4\pi^2}\left(\frac{2m}{\hbar^2}\right)^{3/2} E^{1/2}\, dE. \tag{F. 129}$$

Let us now calculate the eigenvalue density of crystal electrons for which the following $E(\boldsymbol{k})$ relation is assumed to hold:

$$E(\boldsymbol{k}) = E_R + \frac{\hbar^2}{2}\left(\frac{k_x^2}{m_{xx}} + \frac{k_y^2}{m_{yy}} + \frac{k_z^2}{m_{zz}}\right). \tag{F. 130}$$

Here E_R is a constant energy, for example, the energy at a band edge, and m_{xx}, m_{yy}, m_{zz} are the principal masses of the effective mass m^*.

According to this equation the surfaces of constant energy in $\boldsymbol{k}$-space are ellipsoids with the major axes b_x, b_y, b_z which are given by

$$b_x^2 = \frac{2}{\hbar^2} m_{xx}(E-E_R),$$
$$b_y^2 = \frac{2}{\hbar^2} m_{yy}(E-E_R), \qquad \text{(F. 131)}$$
$$b_z^2 = \frac{2}{\hbar^2} m_{zz}(E-E_R).$$

According to (F. 123) the eigenvalue density is determined by the volume dV between the two ellipsoids E = const. and $E+dE$ = const. The volume of the ellipsoid E = const. amounts to

$$V_k = \frac{4\pi}{3} b_x b_y b_z = \frac{4\pi}{3}\left(\frac{2}{\hbar^2}\right)^{3/2} (E-E_R)^{3/2}(m_{xx}m_{yy}m_{zz})^{1/2}. \qquad \text{(F. 132)}$$

Hence follows

$$dV_k = \frac{dV_k}{dE}\,dE = 2\pi\left(\frac{2}{\hbar^2}\right)^{3/2} (E-E_R)^{1/2}(m_{xx}m_{yy}m_{zz})^{1/2}\,dE. \qquad \text{(F. 133)}$$

Using eqns. (F. 123) and (F. 124) we obtain for the eigenvalue density

$$D(E)\,dE = \frac{L^3}{4\pi^2}\left(\frac{2}{\hbar^2}\right)^{3/2} (m_{xx}m_{yy}m_{zz})^{1/2}(E-E_R)^{1/2}\,dE. \qquad \text{(F. 134)}$$

We see that the effective mass has an essential influence on the eigenvalue density. High values of m^* mean high eigenvalue densities and thus narrow energy bands since the number of eigenvalues per band is always equal to N. This is in agreement with the relation between the intensity of the periodic potential, the width of the energy bands, and the value of the effective mass (cf. p. 219).

In particular for $m_{xx} = m_{yy} = m_{zz} = m^*$ the surfaces of constant energy are spheres and eqn. (F. 134) takes the form

$$D(E)\,dE = \frac{L^3}{4\pi^2}\left(\frac{2m^*}{\hbar^2}\right)^{3/2} (E-E_R)^{1/2}\,dE. \qquad \text{(F. 135)}$$

In this case the effective mass is a scalar and the $E(\boldsymbol{k})$ relation reads

$$E(\boldsymbol{k}) = E_R + \frac{\hbar^2}{2m^*} k^2. \qquad \text{(F. 136)}$$

The crystal electrons will then behave like free electrons of mass m^* and one speaks of "quasi-free" electrons.

Figure 107 shows the $D(E)$ curves for different m^* values.

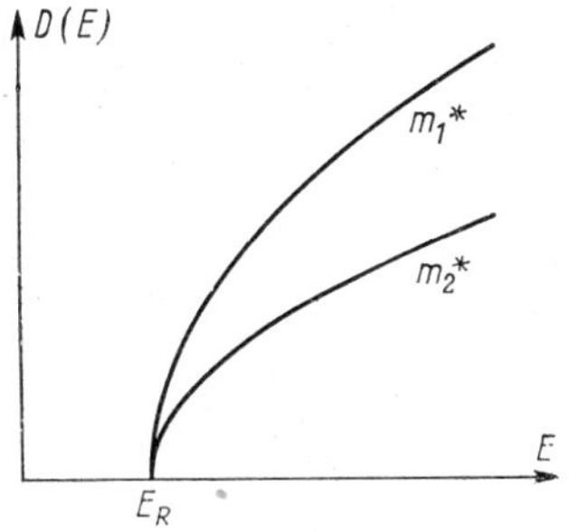

FIG. 107. Eigenvalue densities as a function of energy for different effective masses ($m_1^* > m_2^*$).

VII. Ensemble of Electrons

Hitherto we calculated the possible states and eigenvalues of the crystal electrons on the basis of the assumptions of the one-electron approximation without asking which of the states (n, $\boldsymbol{k}$) actually are occupied. From the occupation of the band structure important properties of the crystals can be derived. This makes it necessary to pass over from a consideration of the behaviour of a single electron to that of the system of all the electrons contained in the crystal. This transition is possible within the framework of the one-electron approximation: the energy levels obtained in this approximation are assumed to be occupied according to the Pauli principle and the Fermi–Dirac statistics (cf. p. 160). This means that the system of electrons is treated on the basis of the following postulates:

1. The nuclei are at rest at the lattice sites; therefore electron–phonon interactions do not exist.
2. All electrons are in the same potential which is due to the nuclei and an averaged charge distribution of the electrons. Electron–electron interactions are neglected.
3. The potential has the strict periodicity of the crystal structure.
4. The electron ensemble is subject to Pauli's exclusion principle: Each state (n, $\boldsymbol{k}$) can be occupied by at most two electrons with antiparallel spins.
5. The electron distribution over the possible energy levels is in agreement with the Fermi–Dirac statistics.

At absolute zero all energy states are occupied by two electrons each, up to some maximum energy. Its value – as in the free electron model – depends on the total number of electrons in the crystal. A crystal with N unit cells, each cell containing one atom of nuclear charge Z, contains NZ electrons. As each energy band contains N eigenstates, it may be filled by $2N$ electrons. At $T = 0°K$ the NZ electrons occupy all energy levels of the lowest energy bands. The number of the necessary energy bands is $Z/2$. According to whether Z is even or odd, all energy bands are completely filled or one-half of the energy levels in the highest occupied band remains empty.

In the case of more complex structures or overlapping bands the nuclear charge Z is not the only parameter which is decisive for the filling of the highest band. In all cases, however, the degree of filling of the highest energy band at $T = 0°K$ indicates whether the crystal is an insulator or a metal.

1. DISTRIBUTION FUNCTION

The electron distribution function [cf. eqn. (E. 15)] is given by

$$N(E)\,dE = 2D(E)\,F(E)\,dE, \tag{F. 137}$$

where in the general case $D(E)$ is given by (F. 127) and $F(E)$ is the Fermi–Dirac function

$$F(E) = \frac{1}{\exp\left(\frac{E-\zeta}{kT}\right)+1}. \tag{F. 138}$$

A calculation of (F. 137) is only possible if we know exactly the $E_n(\boldsymbol{k})$ relation which enters the eigenvalue density function. The calculations must be carried out separately for each particular case and in general require more complex approximation methods than those treated on pp. 203 and 212. In principle the eigenvalue density has a minimum number of singularities, the so-called van Hove singularities at which the derivatives of the eigenvalue density with respect to the energy have discontinuities. It can be shown in general that such singularities are due to the periodicity of the lattice. This applies not only to the eigenvalue density of the crystal electrons but also to the frequency spectrum $\varrho(\omega)$ of the phonons which also possesses van Hove singularities.

Independent of the exact form of the distribution function for a certain substance, several types of distribution functions can be given which are characteristic of metals, insulators, semiconductors, and semimetals (cf. Fig. 108).

Since for many problems the completely filled lower energy bands are insignificant, it is possible, without any loss of generality, to restrict oneself to the upper energy bands pertaining to the valence electrons.

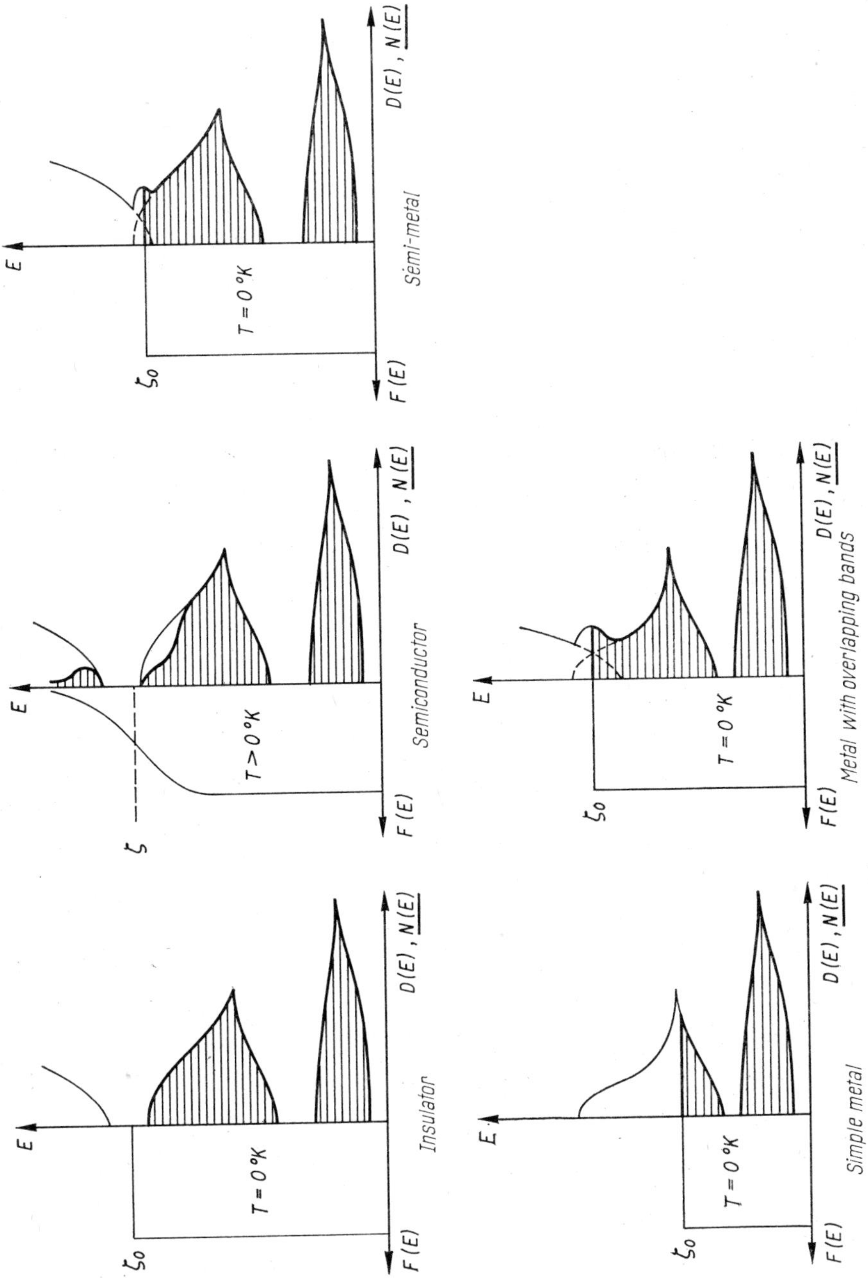

FIG. 108. Distribution functions for insulators, metals, semiconductors and semimetals.

2. INSULATORS AND METALS

In the case of an insulator the highest occupied energy band is completely filled; the maximum energy of the electrons coincides exactly with the upper band edge. The next higher energy band is completely empty. The energy gap between it and the filled band is so broad that the thermal energy is insufficient to excite electrons up to the empty band.

The additional momentum gained by all electrons of a completely occupied energy band under the influence of an external force (e.g. the influence of an external electric field) is equal to zero, i.e. the electrons of a completely filled band do not contribute to the electric current. This is easy to prove. The momentum variation of an electron in the state $(n, \boldsymbol{k})$ is given by

$$dp_x = d(mv_x) = m\frac{dv_x}{dt}\,dt. \tag{F. 139}$$

Using (F. 112) and (F. 113) we can write

$$dp_x = \frac{m}{m_{xx}}F_x\,dt. \tag{F. 140}$$

The momentum variation of all electrons of a band, averaged over the x component, amounts to

$$\overline{dp_x} = \frac{1}{\frac{\pi}{a}-\left(-\frac{\pi}{a}\right)}\int\limits_{-(\pi/a)}^{+(\pi/a)} dp_x\,dk_x. \tag{F. 141}$$

Using eqns. (F. 115) and (F. 140) we obtain

$$\overline{dp_x} = \frac{a}{2\pi}\frac{m}{\hbar^2}F_x\,dt\int\limits_{-(\pi/a)}^{+(\pi/a)}\frac{d^2E}{dk_x^2}\,dk_x = \frac{a}{2\pi}\frac{m}{\hbar^2}F_x\,dt\left.\frac{dE}{dk_x}\right|_{-(\pi/a)}^{+(\pi/a)}. \tag{F. 142}$$

Because of (F. 67)

$$\overline{dp_x} = 0. \tag{F. 143}$$

Analogous calculations hold for the y and z directions. The result of this calculation is a consequence of the Pauli principle: if all states of an energy band (of a Brillouin zone) are occupied, the electrons can only exchange their states but cannot occupy new states.

In the case of metals the highest occupied band is only partly filled. If energy is supplied from an external field the electrons of the highest band can pass over to free states and thus obtain an additional momentum, i.e. an electric current will flow.

The alkali metals which all have odd numbers of electrons (Z) are typical examples of metals. Crystals of elements with even Z can – but need not – be insulators. Typical examples of elements with insulator properties are the solid noble gases whose electron shells are closed and whose electron numbers are even. On the other hand, elements of group IIA of the periodic system (the alkaline earth metals) which also possess even electron numbers have metal properties. The decisive factor is in any case not the number Z but whether bands overlap. In the case of overlapping bands, as, for example, for the alkaline earth metals, the highest band is only partially occupied in spite of even electron numbers, since in the overlapping energy range the number of available states is twice as large.

In the case of insulators and semiconductors (see below) it is usual to call the highest occupied bands which are completely filled at $T = 0°K$ the valence bands. The next highest bands, which at $T = 0°K$ are empty, are called the conduction bands, as they can for $T \neq 0°K$ contain electrons which lead to electron conduction. Each of these bands can contain a maximum of $2N$ electrons (p. 226). The maximum of the highest valence band and the minimum of the lowest conduction band are separated by an energy gap. As valence and also conduction bands partly overlap on the energy scale, we shall in what follows often consider a so-called two-band model, in which valence and conduction bands are combined in a single valence and a single conduction band. The bands below the valence bands are in general completely filled and hence do not affect the electronic properties of solids.

In the case of metals the highest, partly filled bands might be called conduction bands. In more complicated cases it is, however, often more useful to characterize these bands directly by the nature of their charge carriers (electrons or holes).

3. SEMICONDUCTORS AND SEMIMETALS

The strict distinction between insulator and metal is valid only at absolute zero. At temperatures $T > 0°K$ the electrons of the valence band of insulators can be thermally excited over the energy gap ΔE into the conduction band with a probability of $\exp(-\Delta E/kT)$. With suitable values of ΔE and T also insulators may have a detectable conductivity. It is due to the incompletely occupied valence band as well as to the electrons excited into the conduction band. "Insulators" with electronic conductivity are called electronic semiconductors which are of fundamental importance for technical electronics. There is no qualitative but only a quantitative difference between insulators and semiconductors, in that the energy gap is relatively smaller for semiconductors ($\Delta E \lesssim 3$ eV). The properties of semiconductors (cf. pp. 279 ff.) are in principle different from those of the metals.

A crystal is said to be a semi-metal if its valence and conduction bands overlap in a very narrow energy range. The electrical conductivity of semimetals (such as arsenic, antimony, and bismuth) is smaller than that of normal metals by one to two orders of magnitude.

4. THE FERMI SURFACE

The Fermi energy ζ_0 or $\zeta(T)$ of metals is determined by means of equations of the form of (E. 23) or (E. 32). The effective electron concentration will correspond to the number of electrons in the incompletely filled shells.

In the case of insulators and semiconductors the use of such equations is insufficient. At $T = 0$°K the value of the integral remains unchanged if the upper limit of integration (ζ_0) is chosen somewhere in the energy gap between valence and conduction bands. There the eigenvalue density is equal to zero so that the Fermi energy would have an uncertainty of the width ΔE of the energy gap. We shall show on pp. 281 ff. that the Fermi energy can be determined by means of the neutrality condition and has a precisely defined position between the valence and conduction bands.

The special surface of constant energy in $\boldsymbol{k}$ space

$$E(\boldsymbol{k}) = \zeta = \text{const.} \tag{F. 144}$$

is called the Fermi surface. According to the definition of the Fermi energy the Fermi surface is a boundary surface which, at $T = 0$°K, separates in $\boldsymbol{k}$-space all occupied states from the unoccupied ones.

For the properties of the metals the topology of the Fermi surface is very important. The electrical conductivity, the thermal conductivity, and many other properties are essentially determined by only the relatively few electrons whose energies are in the range $\zeta \pm kT$. The external forces are in general so small that, because of the Pauli principle, only those electrons can change their states. For the treatment of transport phenomena it is therefore important to know the band structure particularly near $E = \zeta$. From the shape of the Fermi surface and its position relative to the first Brillouin zone important conclusions can be drawn as to the properties of a metal. The properties of the electrons in different directions of $\boldsymbol{k}(\zeta)$ depend on the position of the electron on the Fermi surface. For directions in which the Fermi surface is far away from the edge of the Brillouin zone, the electrons behave more "freely" than for directions in which the Fermi surface is near the zone edge. We shall show on pp. 236 ff. that the topology of the Fermi surface can be determined by means of various experiments.

For free and quasi-free electrons the Fermi surface is a sphere inside the first Brillouin zone. The radius $k(\zeta)$ of the Fermi sphere increases with the concentration n_e of the valence electrons. From eqns. (E. 23), (F. 135), and (F. 136) we find

$$E - E_R = \zeta = \frac{\hbar^2}{2m^*} k^2(\zeta) = \frac{\hbar^2}{2m^*} (3\pi^2 n_e)^{2/3} \tag{F. 145}$$

or

$$k(\zeta) = (3\pi^2 n_e)^{1/3}. \tag{F. 146}$$

The condition that the Fermi sphere touches the Brillouin zone determines the stability limits of the various crystal structures in an alloy series. We shall discuss this in the following.

For insulators and non-degenerate semiconductors (cf. p. 283) the Fermi energy has a value inside the energy gap within which strictly speaking there are no allowed states for the electrons. The concept of a Fermi surface thus loses its meaning. Instead of Fermi surfaces we are dealing in insulators and non-degenerate semiconductors with constant energy surfaces just above the lowest conduction band edge or just below the upper valence band edge.

5. METALS AND ALLOYS (HUME-ROTHERY RULES)

An alloy is a metallic mixed crystal in which two or more sorts of atoms are statistically distributed over the lattice sites. It is assumed that the lattice periodicity is not disturbed by the incorporation of "foreign" atoms. The wave functions and eigenvalues of the electrons are assumed to remain essentially unchanged in a certain miscibility range; only the electron concentration n_e and thus also the Fermi energy can be altered by additive alloying. A higher electron concentration means that, according to (F. 146), the Fermi sphere is larger. It has been shown that a structure is stable until the Fermi sphere, growing with increasing electron concentration, touches the boundaries of the first Brillouin zone. Although for the electrons of further added atoms there still are free states in the first Brillouin zone, it is often energetically advantageous if, instead of an occupation of the free states of the original zone, a rearrangement takes place which results in a structure with a larger Brillouin zone.

This new crystal structure remains stable with increasing electron density until the Fermi surface again touches the boundaries of the Brillouin zone. The crystal structure belonging to a certain miscibility range is called a phase.

The phase transitions can be explained from the eigenvalue density curve $D(E)$ which is different for different structures (cf. Fig. 109). As long as the eigenvalue density increases with increasing electron density, i.e. with increas-

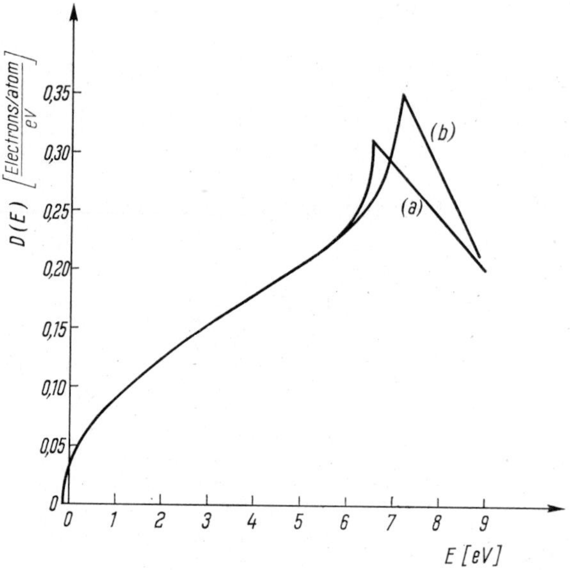

FIG. 109. Eigenvalue density for (a) face-centred cubic structure, (b) body-centred cubic structure [after H. Jones, *Proc. Phys. Soc.* **49**, 250 (1937)].

ing energy because of the Pauli principle, the additional occupation of states consumes little energy. If, however, towards the upper band edge, the eigenvalue density *drops* as the energy increases, the energy necessary to fit in further electrons would increase strongly. Compared with this, a rearrangement to a new phase requires less energy. The maximum eigenvalue density corresponds to the value ζ of the Fermi energy for which the Fermi sphere touches the Brillouin zone.

These considerations are a convincing theoretical basis for the Hume-Rothery rules. According to these rules an alloy series of several metals will always then enter a new phase (change its structure), when the mean number of valence electrons per atom has reached a certain value. This value is characteristic of the transition to the phase considered and independent of the alloy elements. From the condition that the Fermi sphere touches the Brillouin zone boundary the critical electron density is obtained. Dividing this density by the total number of atoms we find the corresponding characteristic number of valence electrons per atom given by Hume-Rothery.

THE COPPER–ZINC ALLOY SYSTEM

Let us consider the system of the copper–zinc alloys as an example. Pure copper has a face-centred cubic structure (lattice constant a). The pertinent reciprocal lattice is body-centred cubic (lattice constant $2\pi/a$). The construc-

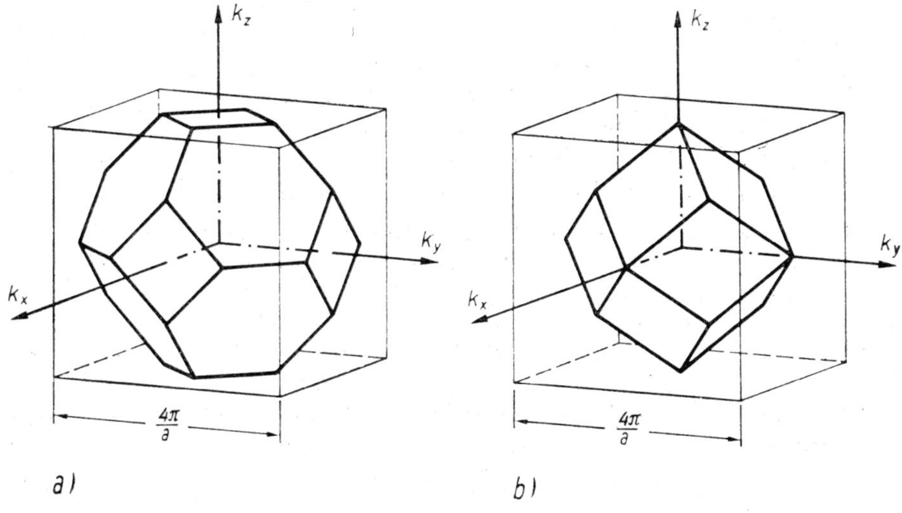

FIG. 110. Brillouin zones for (a) the face-centred cubic, (b) the body-centred cubic lattice [after A. B. Pippard, *Rep. Prog. Phys.* **23**, 176 (1960)].

tion of the first Brillouin zone yields an octahedron with truncated corners (cf. Fig. 110 a).

In the first Brillouin zone exactly N states are available for a crystal of volume L^3 with N atoms (cf. p. 198). For atoms of valence z the density of the valence electrons is given by

$$n_e = \frac{zN}{L^3}. \qquad \text{(F. 147)}$$

The multiple primitive unit cell of volume a^3 of the face-centred cubic lattice contains four atoms so that

$$n_e = \frac{4z}{a^3}. \qquad \text{(F. 148)}$$

For copper $z = 1$.

From eqn. (F. 146) we obtain for the volume of the Fermi sphere

$$V_{F.\,s.} = \frac{4\pi}{3} k^3(\zeta) = 4\pi^3 n_e \qquad \text{(F. 149)}$$

or

$$V_{F.\,s.} = \left(\frac{2\pi}{L}\right)^3 \frac{zN}{2} = 16\left(\frac{\pi}{a}\right)^3 z. \qquad \text{(F. 150)}$$

Since according to (F. 122) a single state in $\boldsymbol{k}$ space needs a volume $(2\pi/L)^3$, the Fermi sphere contains $Z_{F.\,s.}$ states, where

$$Z_{F.\,s.} = \frac{zN}{2}. \qquad \text{(F. 151)}$$

Hence it follows (cf. p. 226) that for monovalent metals ($z = 1$) half the available states in the first Brillouin zone are occupied.

When Zn ($z = 2$) is added to form the alloy, the mean number of valence electrons per atom, i.e. the mean valence $\bar{z}$ of all atoms, increases. The concentration of the valence electrons grows and so does the radius of the Fermi sphere. The radius $k_{\max}(\zeta)$ for which the Fermi sphere touches the Brillouin zone boundaries is given by

$$k_{\max}(\zeta) = 3^{1/2}\frac{\pi}{a}. \qquad \text{(F. 152)}$$

Using eqns. (F. 149) and (F. 150) we obtain the critical mean number of valence electrons per atom:

$$\bar{z}_\alpha = \frac{\pi}{4} 3^{1/2} = 1.362. \qquad \text{(F. 153)}$$

For $1 < \bar{z} \leqslant z_\alpha$ copper–zinc alloys have a face-centred cubic lattice (α-phase).

When $\bar{z} > \bar{z}_\alpha$, the face-centred cubic phase becomes unstable. It gradually passes over to the β-phase as the electron concentration increases; the β-phase has a body-centred cubic lattice. The stability limit of this phase can be derived in a similar way as in the first case. The reciprocal lattice of the body-centred cubic lattice is face-centred cubic. The first Brillouin zone is a dodecahedron (cf. Fig. 110 b). The radius $k_{\max}(\zeta)$ of the Fermi sphere in it is

$$k_{\max}(\zeta) = 2^{1/2}\frac{\pi}{a}. \qquad \text{(F. 154)}$$

The multiple primitive unit cell of volume a^3 of the body-centred cubic lattice contains two atoms so that we have instead of (F. 148)

$$n_e = \frac{2\bar{z}}{a^3}. \qquad \text{(F. 155)}$$

Using eqns. (F. 154) and (F. 155) we obtain from (F. 149)

$$\bar{z}_\beta = \frac{\pi}{3} 2^{1/2} = 1.480. \qquad \text{(F. 156)}$$

According to it, the β-phase (bcc lattice) is stable for $\bar{z} \leqslant \bar{z}_\beta$.

Further phase transitions follow as the zinc concentration is raised. We get then more complicated structures than for the α- and β-phase. The γ-phase has a cubic structure with fifty-two atoms per unit cell. The Fermi sphere touches the Brillouin zone at a mean number of valence electrons per atom of $\bar{z}_\gamma = 1.54$. The ε-phase and the η-phase are hexagonal close packings with different ratios c/a of the axes. The ε-phase is stable up to $\bar{z}_\varepsilon = 1.868$. The η-phase finally reaches up to pure zinc ($z = 2$).

The transitions to the new phases are gradual. Beyond the critical values of $\bar{z}_{\alpha, \beta, \gamma, \varepsilon,}$ in a certain concentration range two phases coexist: $\alpha+\beta$, $\beta+\gamma$, etc.

From the $\bar{z}$ values the concentration of the z-valent component can be easily determined for alloys with monovalent metals. From the relations

$$N_1 + zN_z = \bar{z}N \qquad \text{(F. 157)}$$

and

$$N_1 + N_z = N \qquad \text{(F. 158)}$$

we obtain for the concentration

$$\frac{N_z}{N} = \frac{\bar{z}-1}{z-1}. \qquad \text{(F. 159)}$$

TABLE 19. Stability limits for the phases of diatomic alloys

Alloy	Stability limit of the phases							
	α		β		γ		ε	
	$\bar{z}_\alpha$	% B	$\bar{z}_\beta$	% B	$\bar{z}_\gamma$	% B	$\bar{z}_\varepsilon$	% B
$A^{(1)}$–$B^{(2)}$	1.36	36	1.48	48	1.54	54	1.87	87
Cu–Zn	1.38	38	1.48	48	1.58–1.66	58–66	1.78–1.87	78–87
Ag–Zn	1.38	38			1.58–1.63	58–63	1.67–1.90	67–90
Ag–Cd	1.42	42	1.50	50	1.59–1.63	59–63	1.65–1.82	65–82
$A^{(1)}$–$B^{(3)}$	1.36	18	1.48	24	1.54	27	1.87	43
Cu–Al	1.41	20	1.48	24	1.63–1.77	31–38		
Cu–Ga	1.41	20						
Ag–Al	1.41	20					1.55–1.80	27–40
$A^{(1)}$–$B^{(4)}$	1.36	12	1.48	16	1.54	18	1.87	29
Cu–Si	1.42	14	1.49	16				
Cu–Ge	1.36	12						
Cu–Sn	1.27	9	1.49	16	1.60–1.63	20–21	1.73–1.75	25

Here N_1 is the number of monovalent atoms in the volume L^3, N_z is the number of z-valent atoms in L^3, and N is the number of all atoms in L^3.

Substituting the special values of $\bar{z}_{\alpha, \beta, \gamma, \varepsilon}$ we obtain the critical concentrations up to which the phases are stable. The concentrations determined in this way agree satisfactorily with experiments (cf. Table 19). This means that the assumption of spherical Fermi surfaces is justified for these metals and alloys.

VIII. Principles and Methods of Band Structure Analysis

There exist numerous experiments which prove the occurrence of energy bands in crystals. Only a combination of theory and experiment permits quantitative conclusions as to the band structure. When the band structure is known the electrical, magnetic, optical, and elastic properties of the crystals can be understood. This explains the extraordinarily great number of theoretical and experimental papers dealing with the band structures of various substances. Here we shall not enter into details of the special theoretical methods; we only want to mention some experimental methods which yield information as to the band structure:

Specific heat of electrons (cf. p. 172).
Emission and absorption of soft X-rays (cf. p. 239).
Optical or magneto-optical absorption (cf. p. 244 or 274).
Anomalous skin effect (cf. p. 248).
Cyclotron resonance and Azbel–Kaner resonance (cf. p. 264 or 269).
de Haas–van Alphen effect (cf. p. 271).
Galvanomagnetic effects (cf. p. 396).

Most of these methods are based on radiative transitions in crystals, on the one hand, and on the influence of an external magnetic field on the energy spectrum of crystal electrons on the other hand. We refer to the specialist literature for other effects which also give information about the band structure, such as electron–phonon interaction effects, anisotropy of the electrical conductivity in strong electric fields ("hot electrons"), or positron annihilation.

1. RADIATIVE TRANSITIONS IN CRYSTALS

Transitions of the crystal electrons or of phonons to other energy states are called radiative transitions if phonons are absorbed or emitted in these processes. "Reststrahlen" (cf. p. 93), for example, are the result of radiative transitions of phonons and can be explained as photon–phonon interactions in which the crystal electrons do not participate. Information on the energy distribution of the crystal electrons can be obtained from investigations of electron radiative transitions in which, additionally, electron–phonon interactions may be involved.

The so-called selection rules tell us whether a radiative transition is "allowed" or "forbidden", i.e. whether its probability is high or vanishing in a first approximation. The calculation of the transition probability is as a rule based on a quantum-mechanical perturbation analysis. The behaviour of an electron in the periodic potential of the crystal is studied under the additional influence of the electromagnetic field which is considered to be a time-periodic perturbation. The transition probability depends on the value of the additional potential which is periodic in time and on the unperturbed wave functions corresponding to the initial and final states of the electron. The selection rules obtained (except for the parity conservation which shall not be discussed here) are the classical conservation laws of energy and momentum. As a result of the Pauli principle there is the additional selection rule, that a transition is only possible if the final state is not yet occupied.

For an electron transition between the states characterized by the band indices n and n' and the wave vectors $\boldsymbol{k}$ and $\boldsymbol{k}'$ the energy conservation law reads

$$E_{n'k'} - E_{nk} = \hbar\omega, \tag{F. 160}$$

where ω is the angular frequency of the electromagnetic wave.

The momentum law refers to the conservation of the crystal momentum (cf. p. 218). For direct transitions which are based only on a photon–electron interaction, we have

$$\boldsymbol{k}' - \boldsymbol{k} = \boldsymbol{s} \tag{F. 161}$$

where

$$|\boldsymbol{s}| = s = \frac{2\pi}{\lambda_{cr}} = \frac{2\pi}{\lambda n_{cr}} = \frac{\omega}{c_{cr}} = \frac{\omega}{c n_{cr}}, \tag{F. 162}$$

$\boldsymbol{s}$ is the wave vector of the electromagnetic wave, λ_{cr} is the wavelength in the crystal, λ is the wavelength *in vacuo*, c_{cr} is the phase velocity in the crystal, c is the velocity of light, and n_{cr} is the refractive index of the crystal.

In the case of wavelengths in the X-ray range ($\lambda < 10^{-6}$ cm) the transition spectrum is not influenced by the selection rule (F. 161). The corresponding radiation energies $\hbar\omega$ are so large that transitions take place only between energy bands which are far away from one another. The lower energy band is generally so low that it may be considered to be a discrete energy level (cf. p. 213). In this "band" the energy E_n is almost independent of the wave vector $\boldsymbol{k}$.

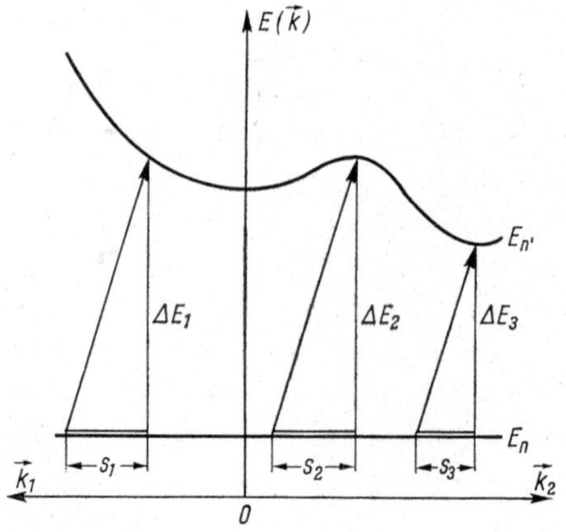

FIG. 111. Radiative transitions in the X-ray range ($\Delta E = E_{n'} - E_n = \hbar\omega = \hbar c_{cr} s$).

In this case transitions between the level E_n and the band with the quantum number n' are allowed for all values $\hbar\omega$ of the energy difference $E_{n'} - E_n$ (cf. Fig. 111).

In the case of direct transitions in the optical range, however, momentum conservation gives a stricter energy conservation condition. For $\lambda > 10^{-6}$ cm the photon momentum $\hbar s = 2\pi\hbar/\lambda$ becomes negligible with respect to the crystal momentum $\hbar k = 2\pi\hbar/a$ which is determined by the lattice constant $a \approx 10^{-8}$ cm. Hence we obtain from (F. 161)

$$\boldsymbol{k} = \boldsymbol{k}' \qquad \text{(F. 163)}$$

and instead of (F. 160) the stricter condition

$$E_{n'k} - E_{nk} = \hbar\omega \qquad \text{(F. 164)}$$

where $n \neq n'$.

Direct optical transitions are always "vertical" in the $E(\boldsymbol{k})$ plot and can take place only between different energy bands n and n' since in one and the same energy band only one energy value pertains to one $\boldsymbol{k}$ value. According to the relative behaviour of the $E_n(\boldsymbol{k})$ and $E_{n'}(\boldsymbol{k})$ curves certain values $\hbar\omega$ of the energy differences $E_{n'} - E_n$ do not occur in direct transitions (cf. Fig. 112). But also with these values of $\hbar\omega$ optical transitions are observed; they are realized by the additional participation of phonons.

For a one-phonon process (cf. p. 71) the momentum law (F. 161) is replaced by the relation

$$\boldsymbol{k}' - \boldsymbol{k} = \boldsymbol{s} \pm \boldsymbol{q}, \qquad \text{(F. 165)}$$

where $\boldsymbol{q}$ is the wave vector of the phonon. The energy law takes the form

$$E_{n'k} - E_{nk} = \hbar\omega \pm u_P, \qquad \text{(F. 166)}$$

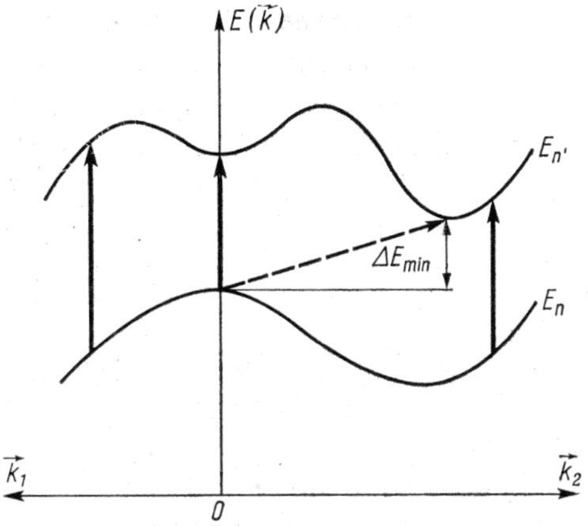

FIG. 112. Direct and indirect radiative transitions in the optical range.

where u_p is the phonon energy; the negative (positive) sign refers to emission (absorption) of a phonon.

Theoretically it is most probable that one phonon participates in the radiative transition. Many-phonon interactions have essentially smaller probabilities for a radiative transition. Radiative transitions with additional electron–phonon interactions are called indirect transitions; they are "oblique" in the $E(\boldsymbol{k})$ diagram (cf. Fig. 112). The indirect transition is assumed to occur via a virtual state with a very short lifetime: the electron reaches the virtual state in a direct transition, from where it passes over to the final state, emitting or absorbing a phonon with a wave vector $\pm \boldsymbol{q} = \boldsymbol{k}' - \boldsymbol{k}$. By virtue of the Heisenberg relations the energy of the virtual state is uncertain because of its short lifetime. The energy conservation applies only to the complete process. For transitions between different bands ($n \neq n'$) the phonon energy u_p is generally negligible compared with the energy differences $E_{n'} - E_n$.

(a) *X-RAY EMISSION AND ABSORPTION*

The spectra of X-ray transitions in crystals are an immediate and impressive proof for the existence of energy bands. Emission spectra yield information on the distribution function $N(E)$ of the electrons in the highest occupied band (occupied states) and absorption spectra on the eigenvalue density $D(E)$ above the Fermi energy ζ (available states).

1. *Emission.* When a crystal is bombarded by electrons of suitable energy (of the order of $10^2 - 10^3$ eV), electrons of a low energy band are excited. This band is very narrow compared with the valence and conduction bands so that

it may be considered to be a discrete energy level. Electrons from the conduction or valence band fall down to the vacant inner level, causing X-ray emission. The intensity distribution $I(E)$ of the emitted X-rays as a function of the energy E is directly proportional to the number of electrons of energy E and to the probability $\varphi(E)$ of a transition from a state of energy E to the vacant inner level (the zero of the energy scale is assumed to be the lower edge of the valence or conduction bands)

$$I(E) = C\varphi(E)N(E). \tag{F. 167}$$

Here C is a constant. The transition probability is a continuous slowly varying function of E; the distribution functions for metals and insulators differ characteristically from each other (cf. Fig. 108). The spectral distribution of the X-ray emission therefore essentially reflects the occupation of the valence and conduectin bands.

Theimo ssion spectra of metals are characterized by an abrupt intensity drop at the high-energy side; this emission edge corresponds to the position of the Fermi energy ζ; for $E > \zeta$ the function $F(E)$ and thus also $N(E)$ and $I(E)$ drop steeply to zero. On the low-energy side of the spectrum the intensity drop corresponds qualitatively to the gradual decrease of the eigenvalue density $D(E)$ to zero. The exact course of $I(E)$ is modified also by $\varphi(E)$ and additional effects which shall not be discussed. The width of the emission band is comparable with the value of the Fermi energy. Under the assumption that all valence electrons are quasi-free, we can equate the measured band width and the Fermi energy ζ [eqn. (F. 145)] and so obtain a value for the effective mass m^*. Table 20 shows that the ratios m/m^* obtained in this way are almost equal to unity, that is, the observed band widths of metals are in good agreement with the Fermi energy obtained from Sommerfeld's relation (E. 24).

TABLE 20. Fermi energies of various metals calculated on the basis of the free-electron model (E. 24) and determined experimentally from X-ray emission spectra (after N. F. Mott and H. Jones, *The Theory of the Properties of Metals and Alloys* [Oxford University Press, London 1936], p. 127)

Element	ζ [eV] theoretical	ζ [eV] experimental	m/m^*
Li	4.7	4.2±0.6	0.89
Na	3.2	3.5±1	1.09
Be	13.8	13.5±2.5	0.98
Mg	7.2	4.0±1.5	0.55
Al	12.0	16.0±2	1.33

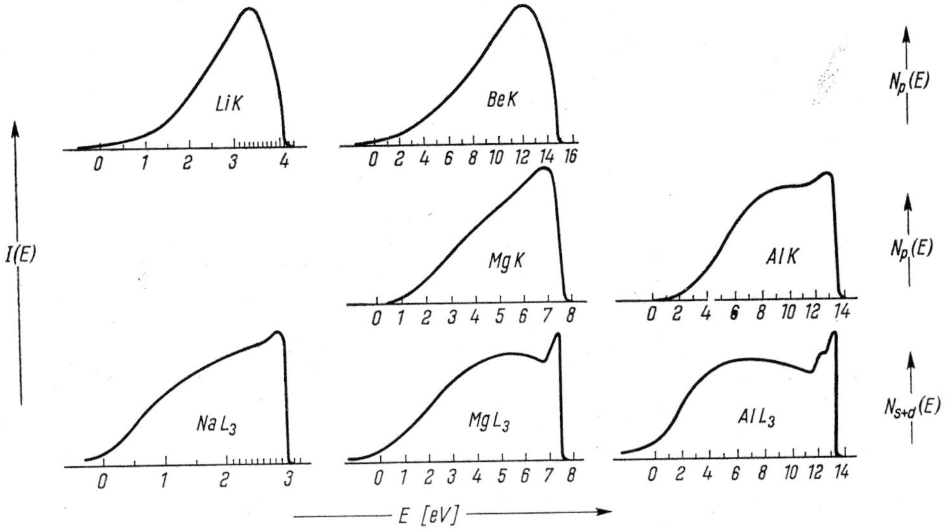

FIG. 113. X-ray emission spectra of several metals. K and L_3 designate the energy terms, into which electrons from the conductivity band are jumping under the emission of X-rays [after H. W. B. Skinner, *Rep. Prog. Phys.* **5,** 257 (1938)].

Figure 113 shows X-ray emission spectra of several metals. In all cases we observe the typical strong drop of emission intensity at high energies. The spectra of Mg and Al indicate the overlapping of two energy bands.

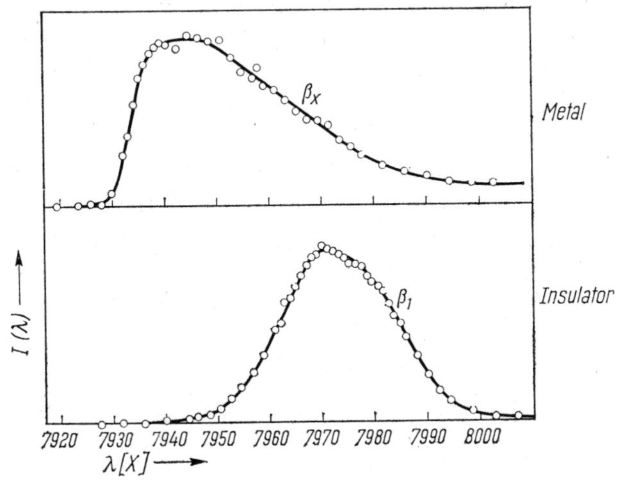

FIG. 114. X-ray emission spectrum for a metal and for an insulator [after B. Nordfors, *Ark. Fysik* **10,** 279 (1956)]; X is the Siegbahn unit (1 X $= 1.00202 \times 10^{-3}$ Å).

In the case of insulators and semiconductors the form of the emission band corresponds to complete occupation of the valence band. The intensity drops gradually on either side since the eigenvalue density $D(E)$ and thus also $N(E)$ and $I(E)$ drop to zero at the band edges. The width of the emission band is equal to the width of the valence band.

According to the occupation of the highest energy band the difference between a metal and an insulator is clearly seen in the form of the X-ray emission spectrum (cf. Fig. 114).

2. *Absorption.* The absorption of X-rays is due to the excitation of electrons from a low energy level to unoccupied states. The absorption coefficient for a certain energy E of X-rays is proportional to a transition probability and to the number of unoccupied states of energy E, that is, the eigenvalue density above a maximum energy. Therefore the energy distribution of the unoccupied states manifests itself in the absorption spectra. Since with increasing electron energy the energy bands overlap more and more, the spectra do not display marked absorption bands.

The absorption coefficient of metals displays a sudden and strong increase above a certain excitation energy (cf. Fig. 115): the spectrum has an absorption edge. Its energetic position coincides with that of the emission edge and also indicates the position of the Fermi energy ζ. The high-energy side of the absorption edge displays a fine structure of the absorption spectrum: the absorption coefficient has a series of maxima and minima in an energy range of the order of 100 eV (in Fig. 115 denoted by A, B, C, α, β). For an investigation of the unoccupied energy levels in the valence or conduction bands the energy range of the order of 10 eV above the absorption edge is particularly interesting. Because of experimental difficulties (radiation source; thin, homogeneous, high-purity samples; energy resolution), this range of the absorption spectrum of soft X-rays (Kossel structure) has been investigated only in a few cases. The fine structure up to several 100 eV (the so-called Kronig structure) shows that there exists no upper limit for discontinuities in the electron-energy spectrum. The maxima and minima in the fine structure are attributed to Bragg reflections of the highly excited electrons in certain directions of the crystal. An unambiguous attribution of the peaks to energy bands is not possible. But it can be shown that the energy position of the extrema depends only on the type of the lattice and the lattice constant, not, however, on the type of the atoms. This proves that the band structure in the range of high energies, that is, for "weakly bound" electrons, depends only on the lattice period (cf. pp. 203 ff.).

The absorption edge of insulators and semiconductors lies at higher energies than the emission edge. This shift is due to the energy gap between valence

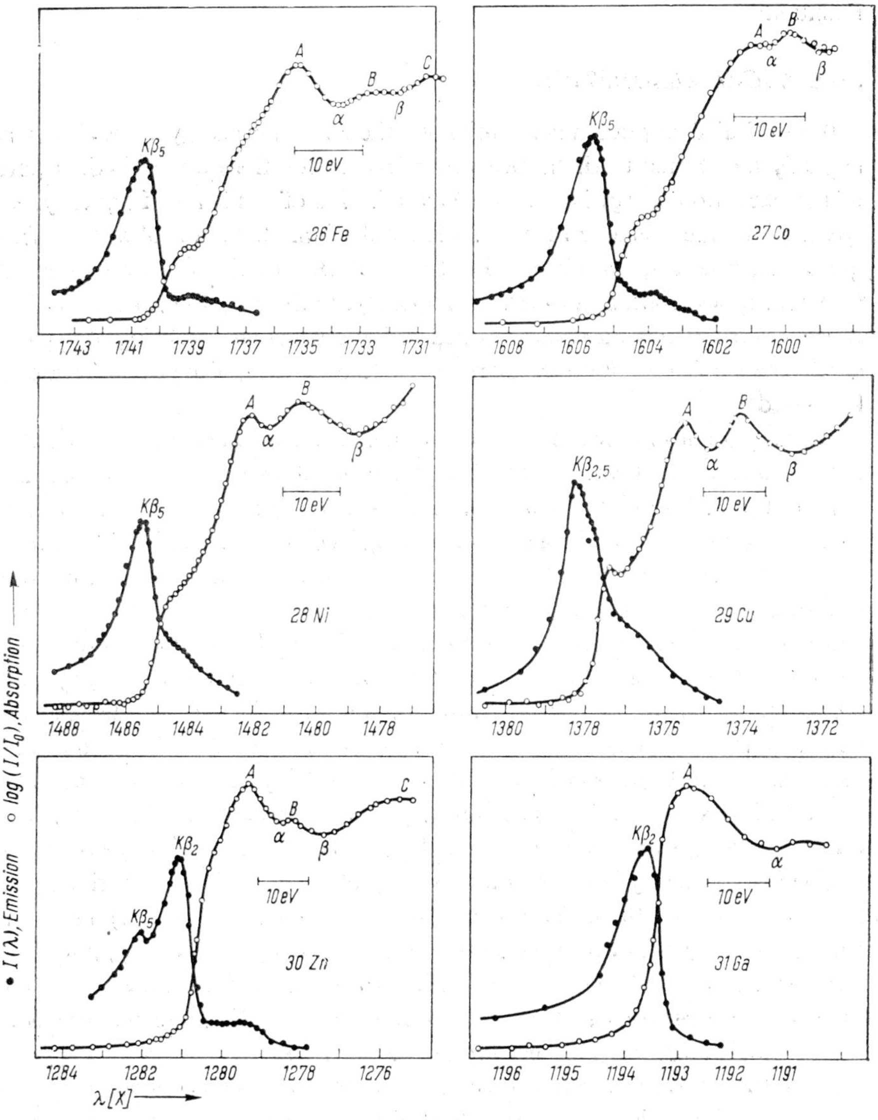

FIG. 115. X-ray emission and absorption spectra of different metals [after W. W. Beeman and H. Friedman, *Phys. Rev.* **56,** 392 (1939)]; $K\beta_2$, $K\beta_5$, $K\beta_{2,5}$ are the designations for the characteristic X-ray emission.

and conduction band. Experimental results are available for silicon and germanium.†

(b) *OPTICAL ABSORPTION*

For optical absorption it is essential that the excitation energy of the electrons is partly transferred to the lattice where lattice vibrations are excited. If part of the excitation energy is re-emitted as radiation of a different frequency, we speak of luminescence or recombination radiation. Radiation of a given frequency is, however, not absorbed if the total excitation energy is re-emitted in the form of radiation of the same frequency. In this case the interaction between electrons and photons is called either incoherent scattering, if the excited state is of short lifetime, or resonant fluorescence, if the excited state is long-lived.

The energy of photons of the optical range is only sufficient for the excitation of transitions between occupied and unoccupied states of one and the same band (intraband transitions) or between neighbouring bands (interband transitions between valence and conduction bands). In this spectral range 10^{-5} cm $<$ $< \lambda < 10^{-3}$ cm) the absorption behaviour of metals is entirely different from that of insulators or semiconductors.

Metals absorb electromagnetic radiation in the whole spectrum up to wavelengths of the middle ultraviolet (i.e. for $\lambda > 10^{-5}$ cm) where they become "transparent". Electrons with the Fermi energy ζ can be excited with a minimum energy supply since the states above ζ are almost unoccupied. These intraband transitions can in principle only be understood as indirect transitions with additional electron–phonon interactions. The absorption decreases strongly as soon as the excited electrons in one period of the electromagnetic wave hardly interact with phonons, that is, as soon as the period $2\pi/\omega$ is small compared with the electron–lattice relaxation time τ (cf. p. 397). Now the excitation energy is not transferred to the lattice (i.e. to the phonons) but gives rise to the emission of radiation of the incident wavelength. Here we shall not discuss further properties of absorption in metals since they are not immediately related to the energy spectrum of the electrons.

From the optical absorption spectra of insulators or semiconductors, however, we can immediately derive a value for the energy gap between valence and conduction bands. Electrons of the valence band can be excited into the conduction band only if the radiation energy is at least as high as the minimum energy gap. Therefore strong absorption in insulators or semiconductors sets

† Cf. D. H. Tomboulian and D. E. Bedo, *Phys. Rev.* **104**, 590 (1956).

in above a certain minimum energy $\hbar\omega_{min}$, which is essentially given by the width ΔE of the energy gap. We have

$$\Delta E \approx \hbar\omega_{min}. \tag{F. 168}$$

The strong increase of absorption at the energy $\hbar\omega_{min}$ is called the long-wave-length limit or the absorption edge of fundamental absorption. The behaviour of the absorption coefficient as a function of the photon energy near $\hbar\omega_{min}$ depends on the form of the $E(\boldsymbol{k})$ curve in the valence and conduction bands. According to the positions of the extrema with respect to the wave vector $\boldsymbol{k}$ indirect transitions also play a role besides the direct transitions.

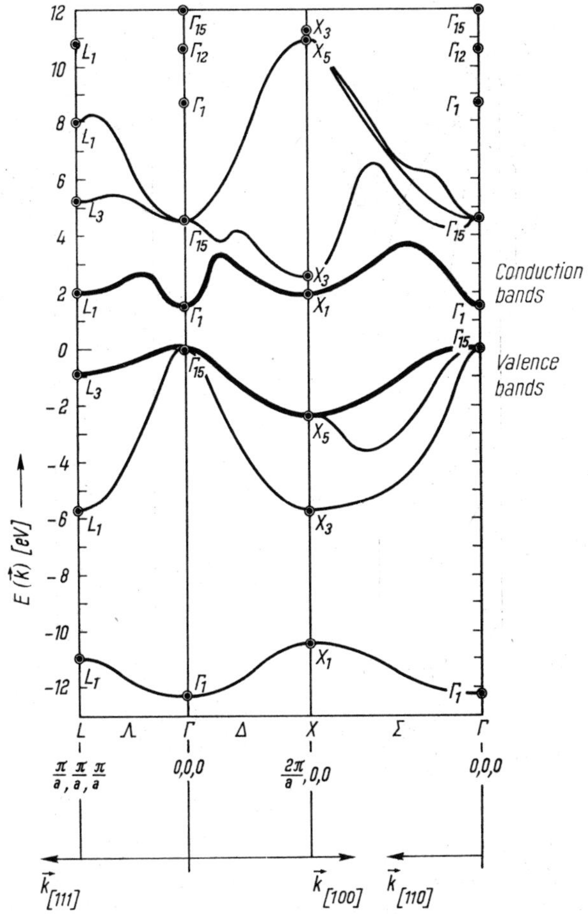

FIG. 116. Band structure of GaAs [after F. Herman and W. E. Spicer, *Phys. Rev.* **174,** 906 (1968)]. The letters L, Δ, Γ are symbols of group theory and are referring to symmetry lines and symmetry points in the Brillouin zone.

In the simplest case the maximum of the highest valence band and the minimum of the lowest conduction band lie at the same wave vector $\boldsymbol{k}$ (e.g. for the intermetallic compound GaAs at $\boldsymbol{k} = 0$; cf. Fig. 116). In this case the absorption is determined mainly by direct transitions. The absorption edge of the fundamental absorption is very steep; its position corresponds to the minimum energy difference between valence and energy band.

In many cases, however, the minimum energy difference is not the smallest vertical distance between valence and conduction bands (e.g. for Ge, Si, cf. Fig. 117). The onset of fundamental absorption will then also correspond to the minimum energy difference between the bands but it is then due to indi-

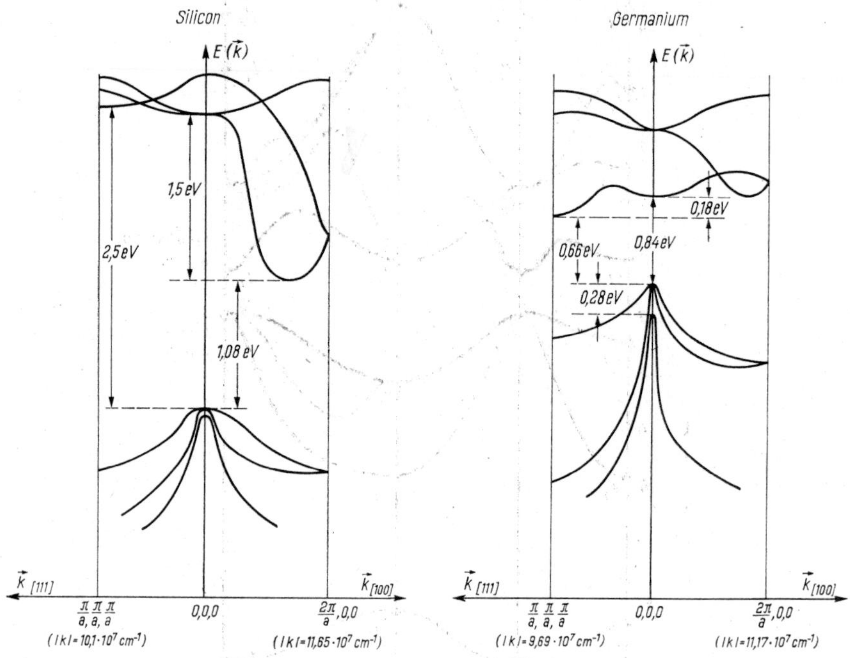

FIG. 117. Band structure (valence and conduction bands) of Si and Ge [after F. J. Blatt, *Solid State Physics* **4**, 199 (1957)].

rect transitions. They have a smaller transition probability. The absorption curve, therefore, is less steep with increasing energy up to a value that corresponds to the minimum *vertical* energy separation. At this energy the absorption increases strongly since then the direct transitions set in (Fig. 118).

The absorption curve near $\hbar\omega_{\min}$ displays a fine structure which makes it difficult to determine the energy gap exactly. The fine structure is caused by

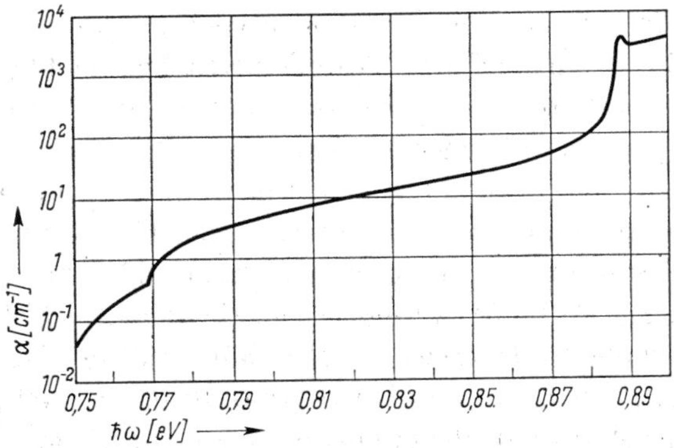

FIG. 118. Absorption coefficient of Ge as a function of quantum energy at 20°K (after T. P. McLean, *Progress in Semiconductors*, Vol. 5, Heywood, London 1962).

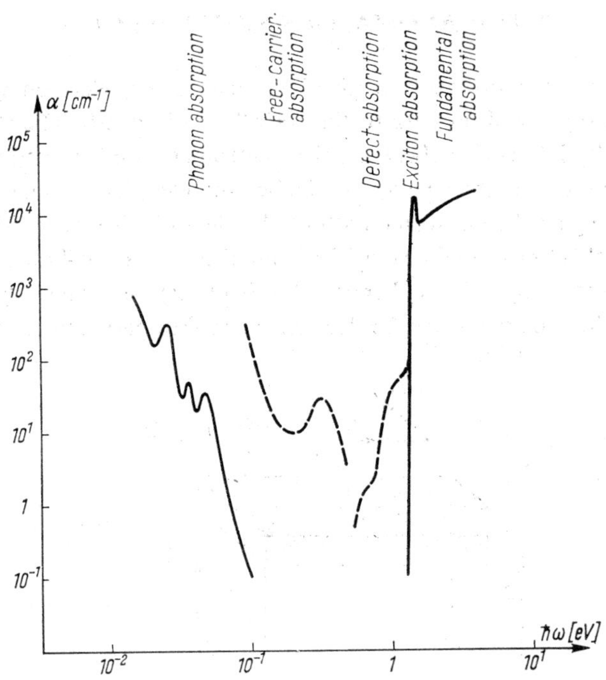

FIG. 119. Absorption spectrum of a semiconductor; the figure is schematic and shows the typical order of magnitude of the absorption constants as a function of the radiation energy for various absorption mechanisms. Fundamental absorption, exciton absorption and phonon absorption (solid lines) are characteristic for each semiconductor; carrier absorption and defect absorption (broken lines) are dependent on the doping of the semiconductor (values for GaAs from R. K. Willardson and A. C. Beer (Eds.), *Semiconductors and Semimetals*, Vol. 3, *Optical Properties of III-V Compounds*, Academic Press, New York 1967).

exciton† excitation, by transitions to additional energy states (the so-called defect levels, cf. p. 279) which are due to lattice imperfections, and by various possibilities of electron–phonon interactions in indirect transitions (emission or absorption of acoustic or optical phonons).

Apart from the fundamental absorption insulators or semiconductors have additional absorption ranges at lower energies. The pertinent absorption coefficients are smaller than those of fundamental absorption by orders of magnitude (cf. Fig. 119). The various absorption bands, which mostly lie in the infrared, are due to the excitation of residual radiation (for crystals with ionic binding, cf. p. 93), to transitions from or to defect levels, and to intraband transitions. As regards the evaluation and interpretation of these effects we refer the reader to the specialist literature.

2. THE ANOMALOUS SKIN EFFECT

The skin effect is observed in metals; it means that electromagnetic fields penetrate into a metal only up to the so-called skin depth. The skin depth is determined by interactions between the electromagnetic field and the conduction electrons and is the smaller the higher the electrical conductivity of the metal. As long as the mean free path of the electrons is small with respect to the skin depth, electrons with velocities in all directions contribute to the screening of the electromagnetic field (normal skin effect). In sufficiently pure samples and at low temperatures the electron mean free path can exceed the skin

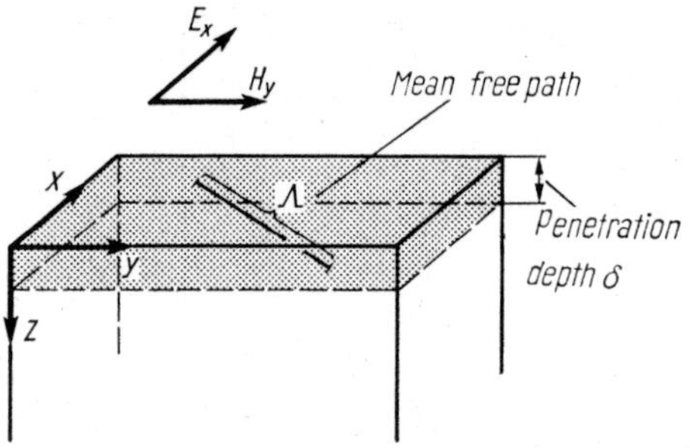

FIG. 120. The anomalous skin effect.

† An exciton is a bound electron-hole pair, in which the electron and hole are mutually bound due to their Coulomb attraction. Excitons correspond to excited crystal atoms with hydrogen-like energy states just below the conduction band. The energy needed for the excitation of an exciton is somewhat less than that needed for the excitation of a separate electron-hole pair (fundamental absorption).

depth, and we shall show that in this case only electrons whose velocity vectors are almost parallel to the metal surface may contribute to field screening (anomalous skin effect). These electrons correspond to a narrow region of the Fermi surface. This enables us to study the effect of electrons with special wave vectors and to draw conclusions about the Fermi surface.

In order to treat the anomalous skin effect we start from the relations for the normal skin effect. It follows from classical electrodynamics that the amplitude of an external electromagnetic field (components E_x, 0, 0; 0, H_y, 0; cf. Fig. 120) in a material of d.c. conductivity σ decreases exponentially with increasing distance z from the surface:

$$E_x(z) = E_x(0) \exp\left(-\frac{1+i}{\delta} z\right) \exp(i\omega t) \qquad \text{(F. 169)}$$

where

$$\delta = \left(\frac{2}{\mu_0 \mu \omega \sigma}\right)^{1/2}. \qquad \text{(F. 170)}$$

Here δ is the classical skin depth, μ_0 is the vacuum permeability, μ is the relative permeability (for non-ferromagnetic substances $\mu \approx 1$), ω is the angular frequency of the electromagnetic field, and σ is the d.c. conductivity.

Experimentally the skin depth is determined from the real part of the surface impedance which is defined by

$$Z = \frac{E_x(0)}{\int_0^\infty j_x(z)\, dz}. \qquad \text{(F. 171)}$$

Using the Maxwell equations

$$\text{curl}\, \boldsymbol{E} = -\mu_0 \mu \dot{\boldsymbol{H}} \qquad \text{(F. 172)}$$

and

$$\text{curl}\, \boldsymbol{H} = \sigma \boldsymbol{E} \qquad \text{(F. 173)}$$

and Ohm's law

$$\boldsymbol{j} = \sigma \boldsymbol{E} \qquad \text{(F. 174)}$$

we can rewrite eqn. (F. 171):

$$Z = -i\omega\mu_0\mu \left(\frac{E_x}{\frac{\partial E_x}{\partial z}}\right)_{z=0}. \qquad \text{(F. 175)}$$

Using (F. 169) we obtain for the real part of the surface impedance, that is, for the surface resistance,

$$R = \text{Re}\,(Z) = \tfrac{1}{2}\mu_0\mu\omega\delta \qquad \text{(F. 176)}$$

or, using eqn. (F. 170),

$$R = \left(\frac{\mu_0 \mu \omega}{2\sigma}\right)^{1/2}. \tag{F. 177}$$

The surface resistance is determined, for example, from the increase in temperature of a sample under the influence of an electomagnetic field or from the Q-factor of a resonance cavity of the material considered.

According to eqn. (F. 177) the plot $1/R$ vs. $\sigma^{1/2}$ yields a straight line. Above a certain d.c. conductivity, however, $1/R$ begins to become independent of σ (cf. Fig. 121). This deviation from the "normal" behaviour, which is called

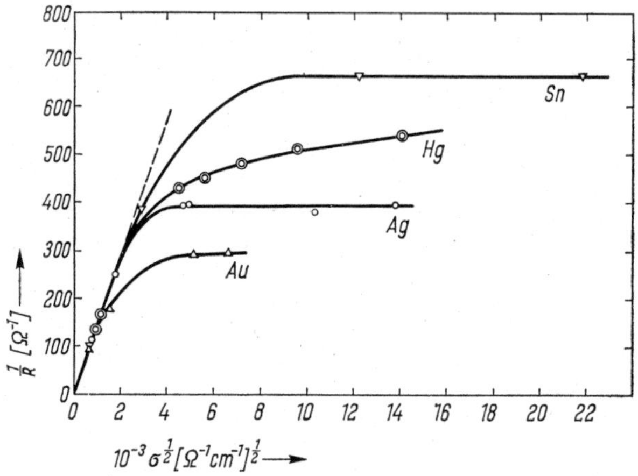

FIG. 121. Surface conductance of gold, silver, mercury, and tin at 1200 MHz [after A. B. Pippard, *Proc. Roy. Soc.* A, **191,** 385 (1947)].

the anomalous skin effect, is observed if the mean free path Λ of the electrons becomes larger than the skin depth δ. The mean free path Λ of the metal electrons is linked with the electrical conductivity σ by the relation

$$\sigma = \frac{ne^2}{m^* v_F} \Lambda = \frac{ne^2}{m^*} \tau, \tag{F. 178}$$

where

$$\Lambda = \tau v_F ; \tag{F. 179}$$

n is the electron density, τ the relaxation time (cf. pp. 396 ff.) and v_F the velocity of electrons of the Fermi energy ζ (cf. p. 215).

The mean free path increases as the interaction between electrons and phonons decreases; it is therefore larger at low temperatures. At sufficiently low temperatures and sufficiently high frequencies the condition

$$\delta_{\text{anom}} \lessdot \Lambda \qquad \text{(F. 180)}$$

for the anomalous skin effect is satisfied.

The anomalous skin effect is due to the fact that the spatial variation of the electric field strength $E_x(z)$ over a distance of one mean free path is no longer negligible. The variation with time of the electrical field strength is sufficiently small under the conditions of the anomalous skin effect so that no relaxation effects occur. This has been assumed already in eqn. (F. 173), in which the displacement current is neglected, which entails a complex conductivity; accordingly in eqn. (F. 174) only the d.c. conductivity is decisive. The neglect of relaxation effects is normally satisfied if the period $2\pi/\omega$ of the applied field is large compared with the relaxation time τ of the electrons. Particularly in the case of the anomalous skin effect it can be shown that $2\pi/\omega$ must be large only in comparison with the time δ_{anom}/v_F the electrons need to penetrate the anomalous skin depth perpendicular to the metal surface. From eqns. (F. 179) and (F. 180) it follows that the time δ_{anom}/v_F is smaller than τ. Substituting typical values for the anomalous skin depth ($\delta_{\text{anom}} \approx 10^{-5}$ cm) and the Fermi velocity ($v_F \approx 10^8$ cm/sec) we obtain a value of $\omega \approx 10^{14}$ sec^{-1} for the critical angular frequency of the electromagnetic field, that is, relaxation effects do not play an important role for wavelengths $\lambda > 10^{-2}$ cm. As a rule microwaves ($\lambda \approx 1$ cm) are used in investigations of the anomalous skin effect.

According to the above condition (F. 180) for the anomalous skin effect only such electrons can absorb electromagnetic energy whose free paths lie entirely within the skin layer, that is, electrons which move more or less parallel to the crystal surface (cf. Fig. 120). The d.c. conductivity, which determines the screening of the electromagnetic field, is therefore reduced. According to Pippard the "effective" d.c. conductivity σ_{eff} is determined by the fraction $\delta_{\text{anom}}/\Lambda$ of the electron concentration,

$$\sigma_{\text{eff}} = \frac{3}{2}\beta\frac{\delta_{\text{anom}}}{\Lambda}\sigma, \qquad \text{(F. 181)}$$

where β is a correction factor of the order of 1. Instead of eqn. (F. 170) for the classical skin depth δ we obtain from (F. 181) for the anomalous skin depth

$$\delta_{\text{anom}} = \left(\frac{2}{\mu_0\mu\omega\sigma_{\text{eff}}}\right)^{1/2}. \qquad \text{(F. 182)}$$

Substituting (F. 181) and solving for δ_{anom} we obtain

$$\delta_{\text{anom}} = \left(\frac{4}{3\beta\mu_0\mu\omega} \frac{\Lambda}{\sigma} \right)^{1/3}. \qquad \text{(F. 183)}$$

We see that the anomalous skin depth depends on the ratio Λ/σ.

For free or quasi-free electrons this ratio is a material constant which depends only on the electron concentration [cf. eqn. (F. 178); according to (E. 24) the electron concentration determines the Fermi energy ζ and thus also the Fermi velocity v_F]. In this case the surface resistance is completely independent of the d.c. conductivity (cf. Fig. 121) as well as of the orientation of the crystal.

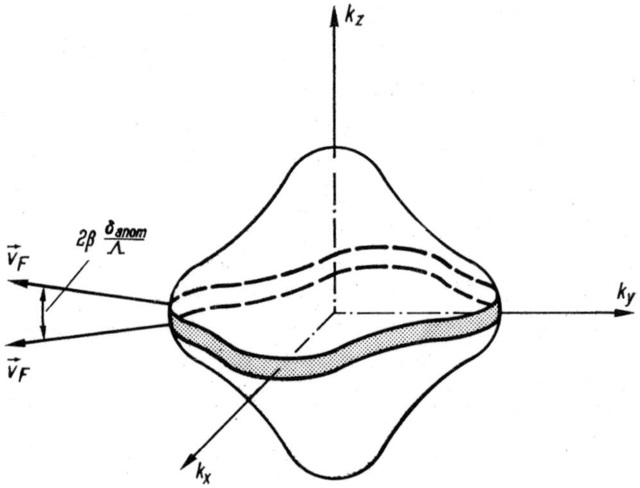

FIG. 122. Belt on the Fermi surface for electrons contributing to the anomalous skin effect.

In the general case of non-spherical Fermi surfaces the anomalous skin depth depends on the crystallographic direction of the surface. Only electrons in states of a narrow region of the Fermi surface contribute to the effective conductivity σ_{eff}; the size of this region depends on the orientation of the crystal.

The electric conductivity for metals is given by [cf. eqn. (J. 61)]

$$\sigma = \frac{e^2}{4\pi^3\hbar} \int\limits_{E=\zeta} \frac{\tau v_x^2}{v_F} dS = \frac{e^2}{4\pi^3\hbar} \int\limits_{E=\zeta} \tau v_F \cos^2 \varphi \, dS, \qquad \text{(F. 184)}$$

where $\tau(\zeta)$ is the relaxation time for $E = \zeta$ (cf. pp. 396 ff.), v_x is the component of the Fermi velocity in the direction of the electric field E_x, dS is a surface element on the Fermi surface, and φ is the angle between the velocities v_x and $\boldsymbol{v}_F$.

The velocity vectors $\boldsymbol{v}_F$ are always perpendicular to the Fermi surface [cf. eqn. (F. 100)]. Only electrons whose velocity vectors make angles up to $\pm\beta\delta_{\text{anom}}/\Lambda$ with the k_xk_y plane (cf. Fig. 122) contribute to the effective conductivity σ_{eff} which is essential for the anomalous skin effect. The integration with respect to dS will therefore cover only a "belt" on the Fermi surface, the position of which is determined by the condition $v_z = 0$ and which has a height $2\beta r\delta_{\text{anom}}/\Lambda$ (r is the radius of curvature of the Fermi surface in the plane defined by $\boldsymbol{v}_F$ and the k_z axis). The surface element dS is therefore given by

$$dS = 2\,|r|\,\beta\,\frac{\delta_{\text{anom}}}{\Lambda}\,ds, \qquad \text{(F. 185)}$$

where ds is a line element along the "belt" on the Fermi surface for $v_z = 0$.

Equation (F. 184) then takes the form

$$\sigma_{\text{eff}} = \frac{e^2}{2\pi^3\hbar}\,\beta\delta_{\text{anom}} \oint\limits_{E=\zeta} |r|\cos^2\varphi\, ds \qquad \text{(F. 186)}$$

or

$$\sigma_{\text{eff}} = \frac{e^2}{2\pi^3\hbar}\,\beta\delta_{\text{anom}} \oint\limits_{E=\zeta} |r_y|\, dk_y, \qquad \text{(F. 187)}$$

where

$$r_y = r\cos\varphi, \qquad \text{(F. 188)}$$

$$ds = dk_y\cos\varphi; \qquad \text{(F. 189)}$$

r_y is the projection of the radius of curvature on the k_xk_z plane.

Equation (F. 187) shows that the effective conductivity and thus also the anomalous skin effect depends only on the geometry of the Fermi surface. The integral is particularly large if the integration over the "belt" contains relatively flat regions of the Fermi surface.

A combination of eqns. (F. 177), (F. 182), and (F. 187) yields the surface resistance

$$R = \left(\frac{2\pi^3\hbar}{e^2\beta}\right)^{1/3} \left(\frac{\mu_0\mu\omega}{2}\right)^{2/3} \left(\oint\limits_{E=\zeta} |r_y|\, dk_y\right)^{-1/3}. \qquad \text{(F. 190)}$$

It is obvious that in the case of a non-spherical Fermi surface R has different values for different orientations of the crystal surface. It is, however, difficult to obtain information about the shape of the Fermi surface from the orienta-

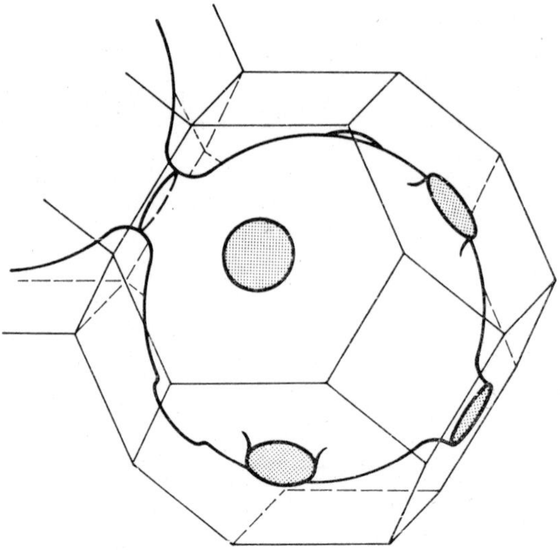

FIG. 123. Fermi surface of copper (after A. B. Pippard).

tional dependence of the surface resistance or of the integral. From measurements of the anomalous skin effect Pippard succeeded in determining the Fermi surface of copper (cf. Fig. 123), which in contrast to the expected behaviour (cf. p. 231) is non-spherical and in contact with the Brillouin zone.

3. ENERGY SPECTRUM OF CRYSTAL ELECTRONS IN AN EXTERNAL MAGNETIC FIELD

The methods used to determine the shape of constant energy surfaces are based on the application of forces in certain directions relative to the crystal orientation. This makes it possible to "separate" those electrons whose wave vectors lie in definite directions and to follow the motion of these particular electrons. Such experiments must therefore be performed with high-quality single crystals and, generally, at low temperatures to avoid electron–phonon interactions which would influence the motion of the electrons. Measurements of the same effect in different crystallographic directions will then yield information on the shape of the Fermi surface (for metals) or of the characteristic constant energy surfaces (for semiconductors). Apart from the method of the anomalous skin effect, most of the other methods require the application of magnetic fields.

An external magnetic field affects the crystal electrons not only by the Lorentz force but also by a quantization of the cyclotron orbits (see below).

From eqn. (F. 104) we have if there is no external electric field

$$F = \hbar \dot{k} = e[v \wedge B], \tag{F. 191}$$

where v is the electron velocity and B is the magnetic induction.

Since the velocity vectors v of the electrons are always perpendicular to the surfaces of constant energy in k space [cf. eqn. (F. 100)], the Lorentz force is tangential to these surfaces. The energy of the electrons is therefore not influenced by the Lorentz force; the electrons move along surfaces of constant energy. The end-point of the wave vector k of an electron with energy E draws a curve which is the curve of intersection of the surface E = const. and the plane $k_{\parallel}$ = const. perpendicular to the magnetic field ($k_{\parallel}$ is the component of k parallel to the magnetic induction B; cf. Fig. 124).

The curves defined by $k_{\parallel}$ depend on the topology of the constant energy surfaces; they are either closed curves within the first Brillouin zone, closed

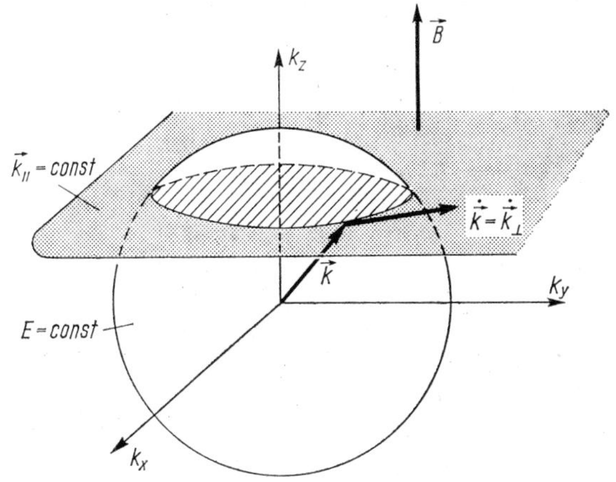

FIG. 124. The motion of an electron in a magnetic field.

curves which are extended over several Brillouin zones (extended orbits), or even open curves which are not closed in k-space (open orbits).

For closed orbits the time of revolution $2\pi/\omega_c$ of the electrons can be given, if the electrons are not scattered. From eqn. (F. 191) we obtain the time of revolution by integrating over the closed orbit:

$$\frac{2\pi}{\omega_c} = \oint \frac{dk}{\dot{k}_\perp} = \frac{\hbar}{eB} \oint \frac{dk}{v_\perp(k)}, \tag{F. 192}$$

where ω_c is the cyclotron frequency, $k_\perp$ is the component of the wave vector perpendicular to the magnetic induction $\boldsymbol{B}$, and $v_\perp(\boldsymbol{k})$ is the velocity component perpendicular to $\boldsymbol{B}$. For free or quasi-free electrons the closed orbits are circles on the spherical energy surfaces. The integration in (F. 192) yields immediately

$$\omega_c = \frac{e}{m^*}B, \tag{F. 193}$$

where m^* is the scalar effective mass.

The relaxation time τ is the mean time during which the electrons are not scattered, i.e. the time during which they do not interact with phonons or defects. According to the relation between the relaxation time τ and the revolution time $2\pi/\omega_c$, we can distinguish the following two cases:

1. $\omega_c\tau < 1$, *the magnetic fields are weak:* the electrons are scattered several times within the time $2\pi/\omega_c$ so that their orbits are incomplete. In this case the quantization of the cyclotron orbits (see below) is in most cases negligible. In this approximation the galvanomagnetic effects (cf. pp. 380 ff.) can be treated satisfactorily.
2. $\omega_c\tau > 1$, *the magnetic fields are strong:* the electrons have closed orbits within periods of the order of the relaxation time. This means that the motion is periodic in planes perpendicular to the direction of the magnetic induction. According to quantum theory a periodic motion is only allowed for a certain selection of the classically possible energies. The energy component $E_\perp$ which determines the electron motion in directions perpendicular to the magnetic field must therefore be quantized. It follows that only certain electron orbits are allowed (quantization of the cyclotron orbits).

Landau was the first to calculate the energy spectrum of free electrons in a magnetic field within the framework of a theory of diamagnetism. The solution of the Schrödinger equation in the presence of a magnetic field yields the differential equation for the harmonic oscillator whose allowed energy values are equidistant on the energy scale. This means that the quasi-continuous energy spectrum of the electrons becomes discrete in the presence of a magnetic field; in this discrete spectrum the so-called Landau levels also display an equidistant distribution; the kinetic energy of the electrons remains unchanged only in the direction parallel to the magnetic induction. For quasi-free electrons of the effective mass m^* we have instead of

$$E = E_\pm \pm \frac{\hbar^2}{2m^*}(k_x^2+k_y^2+k_z^2) = E_\pm \pm \frac{\hbar^2}{2m^*}k^2 \tag{F. 194}$$

for $\boldsymbol{B}(0, 0, B_z) \neq 0$:

$$E_m = E_{\pm} \pm \left[\left(l + \frac{1}{2} \right) \hbar\omega_c + \frac{\hbar^2}{2m^*} k_z^2 \right], \tag{F. 195}$$

where $E_{\pm}$ is the energy of a band edge [cf. eqn. (F. 64)] and $l = 0, 1, 2, \ldots$ is the Landau magnetic quantum number. In this way each quasi-continuous energy band is split into a series of one-dimensional partial bands (cf. Fig. 125). In the direction of k_z parallel to the magnetic field the energy dependence $E(k_z, l)$ is in each partial band the same as $E(k_z)$ without external magnetic field. For directions of $\boldsymbol{k}_\perp$ perpendicular to the magnetic field only discrete values of $|\boldsymbol{k}_\perp|$ are allowed [see eqns. (F. 196) and (F. 200) as well as Fig. 127]. The states of the crystal electrons in the magnetic field are characterized no longer by the band index n and the wave vector $\boldsymbol{k}(k_x, k_y, k_z)$ but by the band index n, the Landau quantum number l, and the component k_z of the wave vector in the direction of the magnetic field.

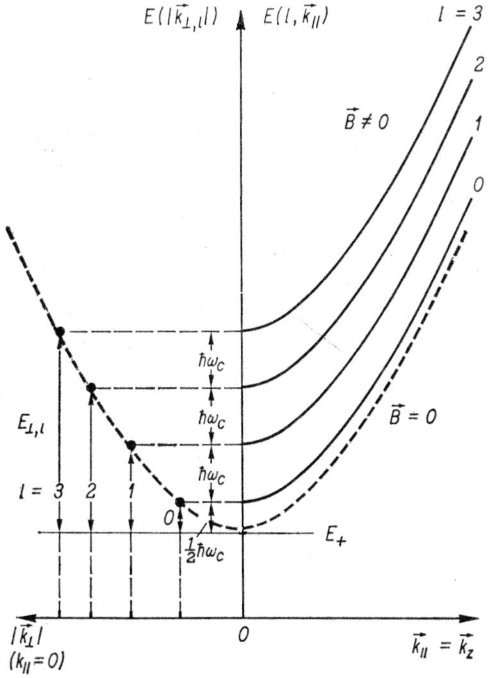

FIG. 125. Splitting of an energy band in a magnetic field. The function $E(\boldsymbol{k})$ for a wave vector vertical and parallel respectively to the magnetic field $\boldsymbol{B}(0, 0, B_z)$. Compare with the $\boldsymbol{E}(\boldsymbol{k})$ curve for $\boldsymbol{B} = 0$ (dashed curve).

According to the radical change of the $E(\boldsymbol{k})$ dependence in the presence of a magnetic field, the surfaces of constant energy in $\boldsymbol{k}$-space are also completely changed. It is advantageous to consider the surfaces of constant energy $E_\perp$. We have

$$E_\perp = (l+\tfrac{1}{2})\hbar\omega_c. \tag{F. 196}$$

The surfaces $E_\perp$ = const. pertaining to the total energy E = const. are obtained from the following construction (cf. Fig. 126). We determine the curves of intersection of the energy surfaces E = const. for $\boldsymbol{B} = 0$ and the planes

$$k_z = \left[\frac{2m^*}{\hbar^2}(E-E_\perp)\right]^{1/2} = \text{const.} \tag{F. 197}$$

for $\boldsymbol{B}(0, 0, B_z) \neq 0$. These curves of intersection enclose the cross-sections A_l which have discrete values according to the quantum number l. The surfaces $E_\perp$ = const. are cylinders whose axes are parallel to $\boldsymbol{B}$ and whose cross-sections A_l are constant.

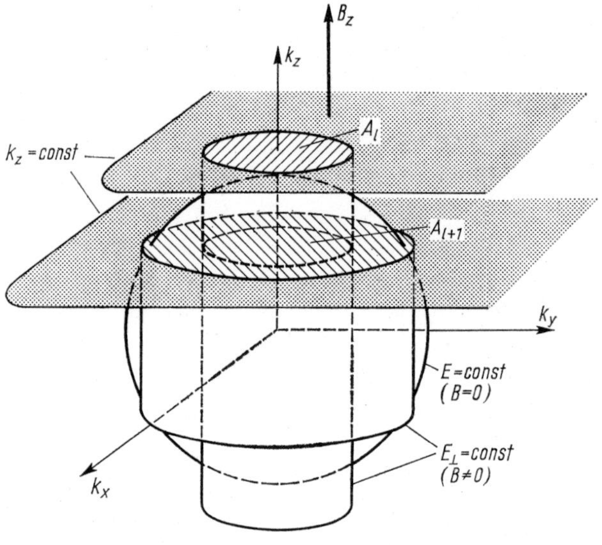

FIG. 126. Energy surfaces for $B = 0$ and $B \neq 0$.

The distribution of the states over the individual cylinders depends on the density of states in the magnetic field. We shall compare in the following the regions in $\boldsymbol{k}$-space which contain the same number $Z(E_0) = Z(E_m)$ of states without and with magnetic field. The energies E_0 and E_m at which the same number of states has been reached without and with magnetic field, respectively, are in general different from one another.

When $\boldsymbol{B} = 0$ the number $D(E)\,dE$ of states with energies between E and $E+dE$ is given by the volume in $\boldsymbol{k}$-space between the constant energy surfaces $E =$ const. and $E+dE =$ const. (cf. p. 222). Assuming isotropic, scalar effective masses, we can use (F. 135) and have

$$D(E)\,dE = \frac{L^3}{(2\pi)^2}\left(\frac{2m^*}{\hbar^2}\right)^{3/2} E^{1/2}\,dE. \tag{F. 198}$$

The zero of the energy scale is assumed to be a band edge. The integration of eqn. (F. 198) yields immediately the number of states up to a maximum energy E_0:

$$Z(E_0) = \frac{L^3}{6\pi^2}\left(\frac{2m^*}{\hbar^2}\right)^{3/2} E_0^{3/2}. \tag{F. 199}$$

If $\boldsymbol{B} \neq 0$ the distribution of states parallel to the k_z axis is the only one that remains unchanged; in the planes perpendicular to k_z it is basically changed. In the $k_x k_y$ planes only states with transverse energies $E_\perp = (l+\frac{1}{2})\hbar\omega_c$ are allowed. The quasi-continuously distributed states for $\boldsymbol{B} = 0$ are "condensed" on the orbits defined by $E_\perp =$ const. Each energy curve $E_\perp = (l+\frac{1}{2})\hbar\omega_c =$ const.

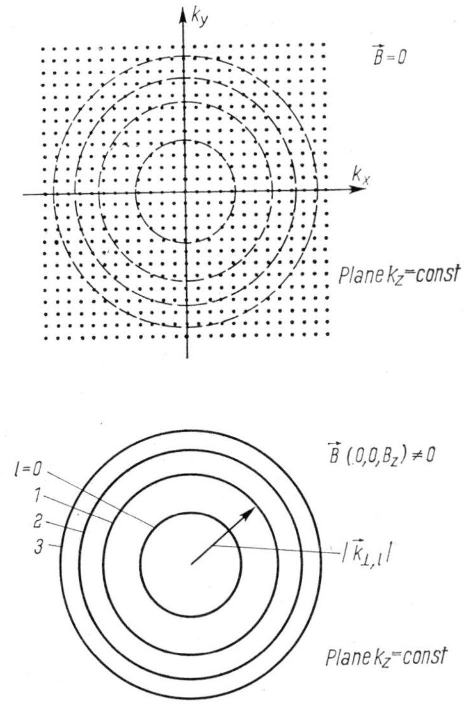

FIG. 127. Distribution of states in the plane $k_z =$ const. for $B = 0$ and $B \neq 0$.

contains the same number $D(l, k_z)\,dk_z$ of states, which for $\boldsymbol{B} = 0$ lie between the curves $E_\perp = l\hbar\omega_c = \text{const.}$ and $E_\perp = (l+1)\hbar\omega_c = \text{const.}$ (cf. Fig. 127). In order to calculate $D(l, k_z)\,dk_z$ we determine the number of states $D(E_\perp, k_z)\,dE_\perp dk_z$ for $\boldsymbol{B} = 0$ between two curves of constant energy $E_\perp = \text{const.}$ and $E + dE_\perp = \text{const.}$ in a plane $k_z = \text{const.}$ The energy $E_\perp$, which determines the motion of the electrons in the $k_x k_y$ plane, is given by

$$E_\perp = \frac{\hbar^2}{2m^*}(k_x^2 + k_y^2) = \frac{\hbar^2}{2m^*}k_\perp^2 \tag{F. 200}$$

under the assumption of quasi-free electrons. As in (F. 123) we obtain for the density of states

$$D(E_\perp, k_z)\,dE_\perp\,dk_z = \left(\frac{L}{2\pi}\right)^3 2\pi k_\perp\,dk_\perp\,dk_z \tag{F. 201}$$

or, using eqn. (F. 200),

$$D(E_\perp, k_z)\,dE_\perp\,dk_z = \frac{L^3}{(2\pi)^2}\,\frac{m^*}{\hbar^2}\,dE_\perp\,dk_z\,. \tag{F. 202}$$

For $\boldsymbol{B} \neq 0$ according to (F. 196) only certain transverse energies $E_\perp$ are allowed. The number of states in the smallest possible energy interval $\Delta E_\perp = \hbar\omega_c$, that is along an orbit with the quantum number l, is given by

$$D(l, k_z)\,dk_z = \frac{L^3}{(2\pi)^2}\,\frac{m^*}{\hbar}\,\omega_c\,dk_z\,. \tag{F. 203}$$

According to this the number of states is the same along all orbits of different quantum number l, and it is also independent of k_z.

From (F. 203) we obtain immediately the number of states with total energy E in a partial band with quantum number l if, using eqn. (F. 195), we introduce the energy E instead of the variable k_z:

$$D(l, E)\,dE = \frac{L^3}{(2\pi)^2}\left(\frac{2m^*}{\hbar^2}\right)^{3/2}\frac{\hbar\omega_c}{2}\left[E - \left(l + \frac{1}{2}\right)\hbar\omega_c\right]^{-1/2} dE. \tag{F. 204}$$

The relation for the eigenvalue density in the magnetic field which is analogous to (F. 198) is obtained by a summation over all partial bands whose edges are below the energy E, that is, by summation over all l, for which the root in (F. 204) remains positive:

$$D(E)\,dE = \sum_{l=0}^{l'} D(l, E)\,dE. \tag{F. 205}$$

The eigenvalue density is infinite at each edge of a partial band and decreases at higher energies proportional to $E^{-1/2}$ (cf. Fig. 128). Moreover, the edges of the original energy bands are shifted by $\hbar\omega_c/2$ under the influence of the

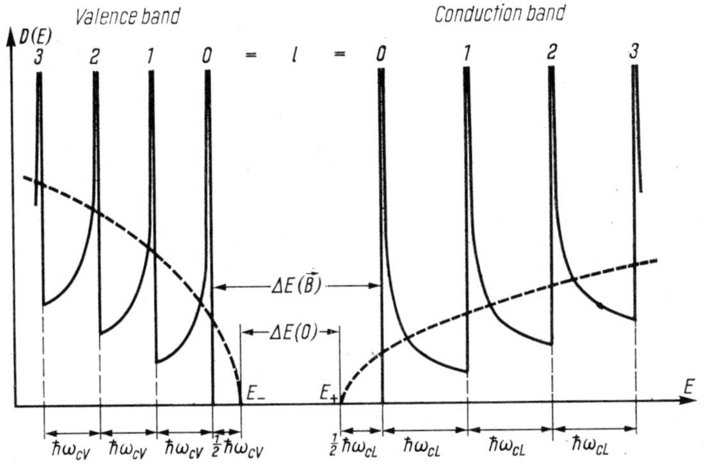

FIG. 128. Eigenvalue density in the valence band and conduction band for $B = 0$ (dashed curves) and $B \neq 0$ (solid curves).

magnetic field: the lower band edges to higher energies and the upper edges to lower energies. This model ignores the splitting of the energy levels owing to the electron spin (cf. pp. 453 ff.).

The region $+k_{zl}$ to $-k_{zl}$ on each cylinder of constant quantum number l which contains states of energies $E \leqslant E_m$ is given by the relation

$$E_m = \left(l+\frac{1}{2}\right)\hbar\omega_c + \frac{\hbar^2}{2m^*}k_{zl}^2. \tag{F. 206}$$

When $E_m/\hbar\omega_c$ is given, the quantum number l' of the maximum transverse energy is fixed by the relation

$$E_m - \left(l'+\frac{1}{2}\right)\hbar\omega_c = \frac{\hbar^2}{2m^*}k_{zl'}^2 > 0 \tag{F. 207}$$

or

$$l' < \frac{E_m}{\hbar\omega_c} - \frac{1}{2}. \tag{F. 208}$$

The number of states with energies $E \leqslant E_m$ is obtained by means of a summation over the states on the individual cylinders with the quantum numbers from 0 to l':

$$Z(E_m) = \sum_{l=0}^{l'} \int_{-k_{zl}}^{+k_{zl}} D(l, k_z)\, dk_z. \tag{F. 209}$$

From the equation

$$Z(E_0) = Z(E_m) \tag{F. 210}$$

a relation is obtained which links the energies E_0 and E_m. A substitution of eqns. (F. 199) and (F. 209) and application of (F. 194) and (F. 203) yields

$$\frac{E_0}{\hbar\omega_c} = \frac{3}{2}\,\frac{\sum\limits_0^{l'} k_{zl}}{k_0}. \tag{F. 211}$$

Using (F. 206) we obtain

$$\left(\frac{E_0}{\hbar\omega_c}\right)^{3/2} = \frac{3}{2}\sum_{l=0}^{l'}\left[\frac{E_m}{\hbar\omega_c} - \left(l+\frac{1}{2}\right)\right]^{1/2}. \tag{F. 212}$$

Knowing $E_m/\hbar\omega_c$ we can calculate $E_0/\hbar\omega_c$ and also E_m/E_0. With these values we can determine the regions $+k_{zl}$ to $-k_{zl}$ of available states on each cylinder with the quantum number l. Assuming that $E\ \ = 0$, and combining eqns. (F. 206) and (F. 194) we obtain for $l < l'$

$$\left(\frac{k_{zl}}{k_0}\right)^2 = \frac{E_m}{E_0} - \left(l+\frac{1}{2}\right)\frac{\hbar\omega_c}{E_0}. \tag{F. 213}$$

Figure 129 shows the regions in $\boldsymbol{k}$-space which contain the same number of states of quasi-free electrons with and without magnetic field, for various values of $E_m/\hbar\omega_c$. Let us assume, for example, that the surface $E = E_0$ is the Fermi sphere. In weak magnetic fields, that is, if $E_m \gg \hbar\omega_c$, the Fermi sphere is only slightly changed; in high magnetic fields ($E_m \approx \hbar\omega_c$), however, the distribution of states is fundamentally different.

We see that the magnetic field gives rise to an essential change in the energy band structure and thus in the shape of the energy surfaces. None the less measurements in magnetic fields yield valuable information about the shape of the energy surfaces for $\boldsymbol{B} = 0$. This is possible since the shape of the energy surface cross-sections perpendicular to the direction of the magnetic field remains essentially unchanged; the quantization of the electron orbits only distorts the cross-sections so that the geometrical similarity is maintained. In the case of non-spherical energy surfaces the shape of the preferred energy surface cross-sections depends on the direction of the magnetic field with respect to the orientation of the crystal. In this way a measurement of the direction dependence of a certain effect in a magnetic field enables us to determine all energy surface cross-sections which exist for $\boldsymbol{B} = 0$.

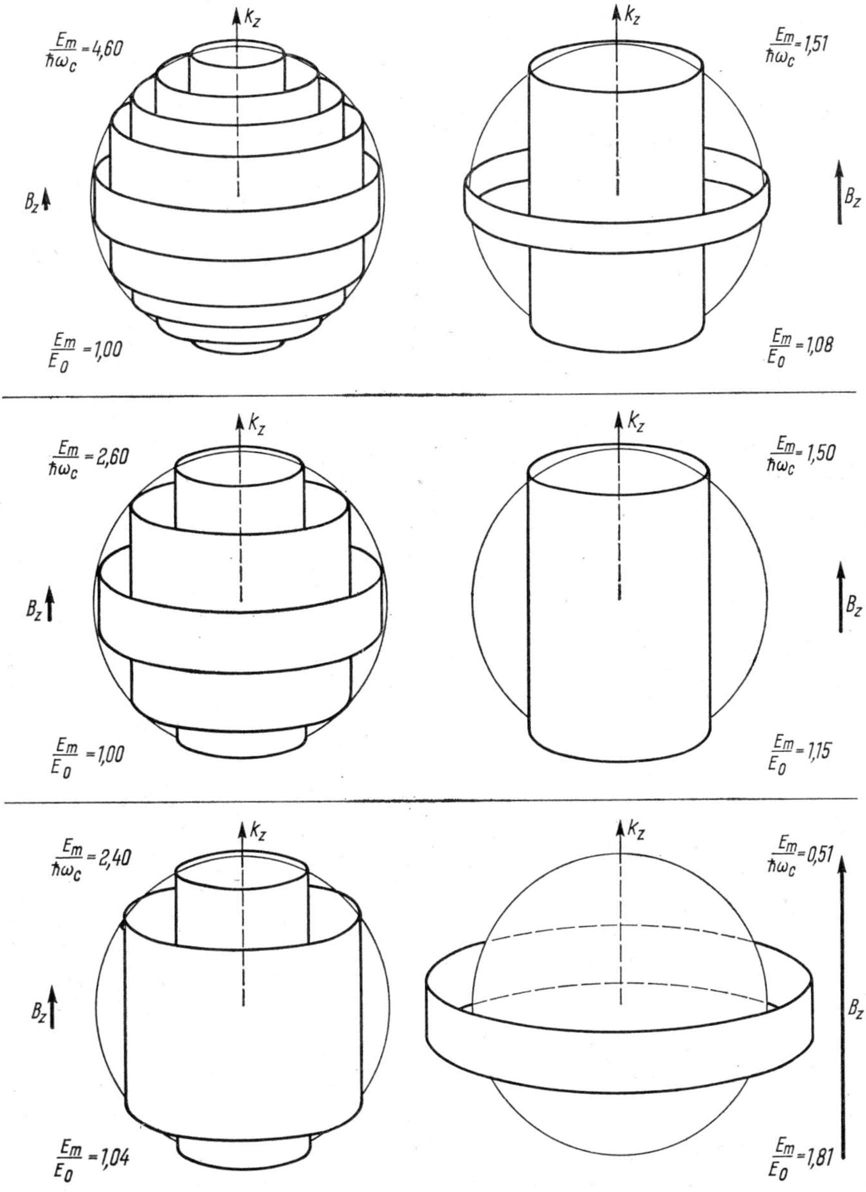

FIG. 129. Density of states of quasi-free electrons for different magnetic fields. The cylinders contain the same number $Z(E_m)$ of states as the energy sphere $E = E_0$.

(a) *CYCLOTRON RESONANCE*

The method of cyclotron resonance makes it possible to determine the components of the effective mass tensor and hence the curvature of the corresponding energy surface. In this way information is obtained about the shape of the Fermi surface or the constant energy surface which corresponds to the upper valence band edge or the lower conduction band edge.

In the following we shall describe the principle of cyclotron resonance for the simple case of quasi-free electrons. Under the action of a static magnetic field (given by the magnetic induction B_z) the electrons move in $\boldsymbol{k}$-space along circular orbits in the planes $k_z = \text{const}$. The angular frequency is given by the so-called cyclotron resonance frequency ω_c (F. 193). When an additional a.c. electric field of angular frequency ω is applied in a direction perpendicular to the magnetic field, the electrons are excited to oscillate in this direction (that is, in the planes $k_z = \text{const.}$). They take up energy from the electric field in passing over to unoccupied states of higher energies. By virtue of the quantization of the electron orbits in the planes perpendicular to the magnetic field, in which the a.c. electric field should be the only field that is active, the nearest available states have an energy which is higher by $\hbar\omega_c$. The electromagnetic wave is therefore strongly absorbed at the "resonance frequency" ω_R:

$$\omega_R = \omega_c = \frac{e}{m^*} B. \tag{F. 214}$$

Cyclotron resonance means radiative transitions between the partial bands of one and the same energy band. Such transitions are also called direct intraband transitions.

Knowing the magnetic induction and having determined the frequency of the absorption maximum we obtain immediately a value for the effective mass m^*. In the experiment an electric microwave field is used and the absorption is measured as a function of the magnetic field.

For cyclotron resonance measurements it is necessary that $\omega_c\tau > 1$ (cf. p. 256). The quantization of the translational energy manifests itself only in this case in a marked absorption peak. This condition is satisfied at sufficiently low temperatures and for pure crystals, that is, in the case of insignificant scattering of electrons by phonons and defects.

Another condition with respect to the electron concentration must be fulfilled. On the one hand, the sensitivity of the apparatus used requires a minimum number of electrons (or holes, cf. p. 279) to achieve a clear signal. On the other hand, the number of carriers must in principle be so small that the electric field can penetrate into the sample, that is the skin depth must be sufficiently large. Estimates show that this applies to semiconductors but not to metals (cf. pp. 269 ff.).

In the general case for electrons in an arbitrary lattice potential the resonance frequency $\omega_R = \omega_c$ depends on the direction of the magnetic field with respect to the orientation of the crystal. By analogy with eqn. (F. 193) we define the cyclotron mass m_c as

$$\omega_c = \frac{e}{m_c} B. \tag{F. 215}$$

Substitution into (F. 192) yields

$$m_c = \frac{\hbar}{2\pi} \oint \frac{dk}{v_\perp(\boldsymbol{k})} \tag{F. 216}$$

or, using (F. 100),

$$m_c = \frac{\hbar^2}{2\pi} \oint \frac{dk_\perp}{dE} dk, \tag{F. 217}$$

where $dk_\perp$ is the distance in **k**-space between two orbits of constant energy E and $E+dE$ in a plane perpendicular to the magnetic field. From Fig. 130 we see immediately that the contour integral is equal to the difference dA of the

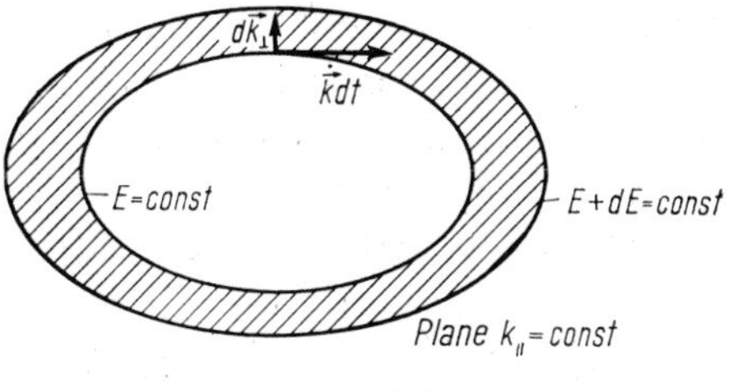

FIG. 130. Computation of the cyclotron mass.

cross-sectional areas which are enclosed by the orbits E = const. and $E+dE$ = = const. in the plane $k_{||}$ = const. Using this eqn. (F. 217) can be rewritten in the form

$$m_c = \frac{\hbar^2}{2\pi} \left(\frac{dA}{dE}\right)_{k_{||}=\text{const}}. \tag{F. 218}$$

The cyclotron mass is thus connected with the energy dependence of the cross-sections A of closed orbits on constant energy surfaces. It is in general *not* identical with the effective mass m^* defined by (F. 113), which belongs to a certain state $(n, \boldsymbol{k})$.

Only in the case of quasi-free electrons is $m^* = m_c$. Combining the relations

$$E = \frac{\hbar^2}{2m^*}(k_\perp^2 + k_{||}^2) \tag{F. 219}$$

and

$$A(k_{||}) = \pi k_\perp^2 \tag{F. 220}$$

we find

$$\left(\frac{dA}{dE}\right)_{k_{||}=\text{const}} = \frac{2\pi m^*}{\hbar^2}. \tag{F. 221}$$

Substitution in (F. 218) shows that for quasi-free electrons with a scalar effective mass m^*

$$m_c = m^*. \tag{F. 222}$$

Many semiconductors, such as, for instance, germanium and silicon, display a directional dependence of the cyclotron resonance, that is, an anisotropy of the cyclotron masses. Accordingly the corresponding energy surfaces are not spherical. According to the Fermi–Dirac statistics the charge carriers (conduction electrons and holes) occupy in semiconductors mainly the states very close to the energy gap at low temperatures. Only charge carriers in these states contribute to the cyclotron resonances. The determining energy surfaces for conduction electrons correspond to the absolute minima of the conduction band. In many cases, several maxima and minima can occur within one band. In cases in which several minima or several so-called valleys occur one speaks of a "many-valley" structure of the energy band. As soon as the minima of the conduction band are absolute minima, the $E(\boldsymbol{k})$ behaviour near these minima $\boldsymbol{k}_0$ determines the semiconductor properties. Near a minimum $E(\boldsymbol{k})$ is a quadratic function of $\boldsymbol{k}$. We have (the energy is reckoned from the minimum)

$$E(\boldsymbol{k}) = \frac{\hbar^2}{2}\left(\frac{\Delta k_1^2}{m_1} + \frac{\Delta k_2^2}{m_2} + \frac{\Delta k_3^2}{m_3}\right), \tag{F. 223}$$

where $\Delta k_{1,2,3}$ are the coordinates in $\boldsymbol{k}$-space reckoned from the point $\boldsymbol{k} = \boldsymbol{k}_0$ of the minimum. The local axes 1, 2, 3 are chosen in such a way that the effective mass $m^*(\boldsymbol{k})$ is given by the principal masses m_1, m_2, m_3. According to eqn. (F. 223) the surface $E = \text{const.}$ is an ellipsoid with centre $\boldsymbol{k} = \boldsymbol{k}_0$ and principal axes

$$a = \frac{1}{\hbar}(2m_1E)^{1/2}, \quad b = \frac{1}{\hbar}(2m_2E)^{1/2}, \quad c = \frac{1}{\hbar}(2m_3E)^{1/2}. \tag{F. 224}$$

The number of the additional minima depends on the symmetry of the crystal considered. The conduction band of silicon, for example, has six absolute minima at $\boldsymbol{k}_0 = \pm(k_{0x}, 0, 0), \pm(0, k_{0y}, 0), \pm(0, 0, k_{0z})$. The essential energy surfaces are six ellipsoids of revolution inside the Brillouin zone, whose major axes coincide with the ⟨100⟩ directions (cf. Fig 131). Germanium has a similar structure of the energy surfaces. The conduction band has eight absolute minima in the ⟨111⟩ directions. The corresponding energy surfaces are eight revolution ellipsoids, at the zone boundary with major axes along the ⟨111⟩ directions (cf. Fig. 131).

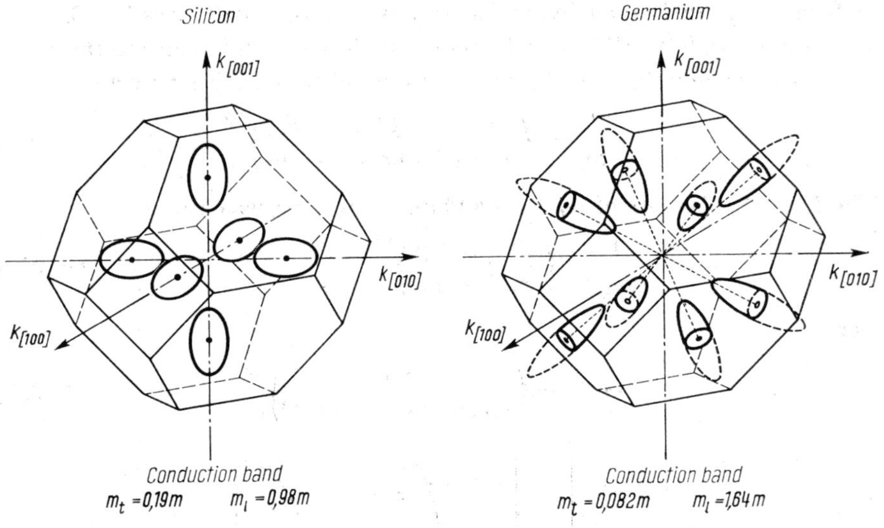

FIG. 131. Energy surfaces for conduction electrons in Si and Ge.

In both cases there are only two different principal masses, namely

$$m_l = m_1 \tag{F. 225}$$

and

$$m_t = m_2 = m_3 . \tag{F. 226}$$

The application of a magnetic field defines preferred orbits in planes perpendicular to $\boldsymbol{B}$. When we consider a single ellipsoid, the times of revolution of the electrons are the same along all orbits in parallel planes. For an energy surface which consists of a single ellipsoid, a single direction-dependent resonance frequency belongs to one direction of $\boldsymbol{B}$. If, however, the energy surface consists of several ellipsoids (as, for example, in the case of germanium and silicon), the preferred orbits on the various ellipsoids are not all equivalent. Therefore several resonance frequencies are obtained for almost all directions of $\boldsymbol{B}$. From this and the special directions for which a single resonance frequency appears, information is obtained about the connection and position of the pertinent energy surfaces.

The cross-section A of an orbit on an ellipsoid E = const. is given by

$$A = \frac{2\pi}{\hbar^2} \frac{E}{\left(\dfrac{\alpha^2}{m_2 m_3} + \dfrac{\beta^2}{m_1 m_3} + \dfrac{\gamma^2}{m_1 m_2}\right)^{1/2}}, \tag{F. 227}$$

where α, β, γ are the direction cosines of $\boldsymbol{B}$ with respect to the axes 1, 2, 3.

Together with (F. 218) this relation yields the connection between the cyclotron mass m_c and the components m_1, m_2, m_3 of the effective mass m^*:

$$\frac{1}{m_c} = \left(\frac{\alpha^2}{m_2 m_3} + \frac{\beta^2}{m_1 m_3} + \frac{\gamma^2}{m_1 m_3}\right)^{1/2}. \tag{F. 228}$$

In the particular case of $m_l = m_1$ and $m_t = m_2 = m_3$ we have

$$\frac{1}{m_c} = \left(\frac{\alpha^2}{m_t^2} + \frac{\beta^2 + \gamma^2}{m_t m_l}\right)^{1/2}, \tag{F. 229}$$

or

$$m_c = m_t \left(\frac{m_l}{m_l \cos^2 \Theta + m_t \sin^2 \Theta}\right)^{1/2} \tag{F. 230}$$

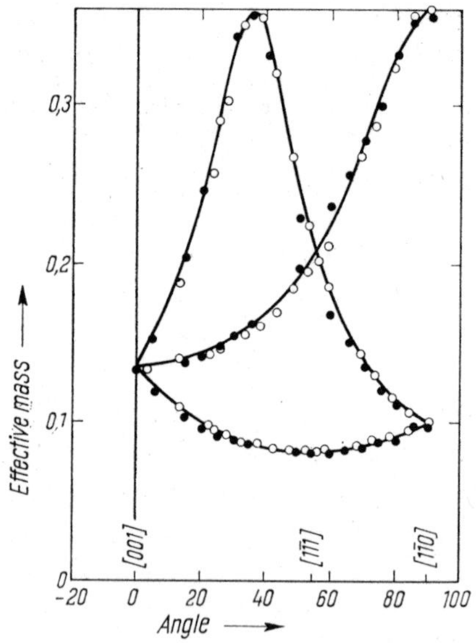

FIG. 132. Relative effective electron masses m_c/m in Ge at T = 4°K for magnetic field directions in the (110)-plane [after G. Dresselhaus, A. F. Kip and C. Kittel, *Phys. Rev.* **98**, 368 (1955)].

where

$$\alpha^2 = \cos^2 \Theta \quad \text{and} \quad \alpha^2+\beta^2+\gamma^2 = 1, \tag{F.231}$$

Θ being the angle between $\boldsymbol{B}$ and the major axis of the ellipsoid considered. Figure 132 shows the dependence of the cyclotron mass on the direction of the magnetic field for germanium. The agreement between theory (F. 230) and experiment is very good.

(b) *AZBEL–KANER RESONANCE*

Cyclotron resonance in metals can be measured with another arrangement of the external fields. The magnetic field is applied parallel to the metal surface; the vector of the a.c. electric field is also parallel to the metal surface and parallel or perpendicular to the magnetic field (cf. Fig. 133). The resonance effect proposed by Azbel and Kaner is based on the anomalous skin effect (cf. pp. 248 ff.). Assuming that

$$\Lambda > r > \delta_{\text{anom}}, \tag{F.232}$$

where Λ is the mean free path, r the radius of the cyclotron orbit, and δ_{anom} the anomalous skin depth, electrons which orbit about the magnetic field return several times into the skin layer. The electric field can only there exert

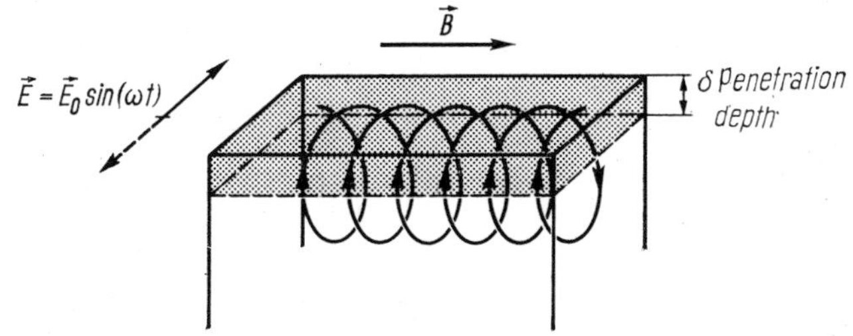

FIG. 133. The Azbel–Kaner resonance.

an additional force on them. If after each revolution the electrons have the same phase relation with the alternating electric field, some of them will be accelerated after each revolution in the skin layer. In this way the electrons absorb energy from the field. The suitable phase relation and thus also the condition for resonance absorption reads

$$\omega = n\omega_c, \tag{F.233}$$

where $n = 1, 2, 3, \ldots$. The a.c. field frequency can thus also be equal to higher harmonics of the revolution frequency ω_c. This means that several maxima appear in the absorption spectrum (cf. Fig. 134).

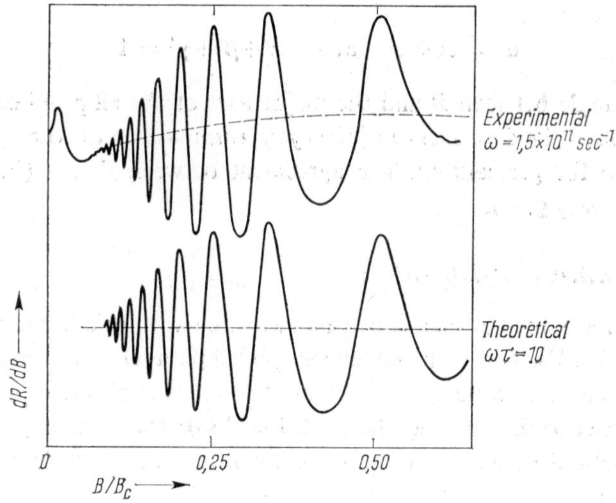

FIG. 134. Azbel–Kaner resonances in copper. Field dependence of the derivative of the surface resistance R with respect to the magnetic induction B. $B_c = m_c \omega_c / e$ designates the magnetic induction for the resonance $n = 1$ [from A. F. Kip, D. N. Langenberg and T. W. Moore, *Phys. Rev.* **124,** 359 (1961)].

Measuring the absorption of microwaves of constant frequency ω as a function of the magnetic field, one obtains Azbel–Kaner resonances for values B_R of the magnetic induction which are proportional to $1/n$. From (F. 215) and (F. 233) we obtain

$$B_R = \frac{m_c \omega}{e} \frac{1}{n} . \qquad \text{(F. 234)}$$

By virtue of (F. 216) this relation is direction-dependent.

An evaluation of the measurements is in principle more difficult than in the case of ordinary cyclotron resonances since the $E(\boldsymbol{k})$ relation near the Fermi energy is generally not a quadratic function of $\boldsymbol{k}$. The relevant energy surfaces (Fermi surfaces) are therefore not ellipsoids. Those electrons whose orbits enclose maximum or minimum cross-sections of the Fermi surface give the main contribution to the resonance. The electrons on the other orbits are generally out of phase and do not contribute to the resonance.

(c) *DE HAAS–VAN ALPHEN EFFECT*

The de Haas–van Alphen effect, like cyclotron resonance, yields information on the shape of the Fermi surface and the effective masses of metals. de Haas and van Alphen were the first to observe in the case of bismuth that, at sufficiently low temperatures, the magnetic susceptibility (cf. p. 408) is a periodic function of the reciprocal magnetic field strength. This oscillatory variation of the magnetic susceptibility, or of the magnetization, with the magnetic field strength, the so-called de Haas–van Alphen effect, could be detected for a great number of metals.

The cause of this effect is the quantization of the cyclotron orbits (cf. p. 256) and thus the oscillatory variation of the total electron energy as a function of the magnetic field strength. We shall show in the following that the oscillations are equidistant on the scale of the reciprocal magnetic field strength and that the corresponding constant period is inversely proportional to the extreme values of the sectional areas of planes perpendicular to the magnetic field and the Fermi surface. A measurement of the periods of the de Haas–van Alphen effect for different crystal orientations with respect to the magnetic field direction yields information about the shape of the Fermi surface.

In general the magnetization of a system is determined by the thermodynamic relation

$$M = -\frac{1}{V}\left(\frac{\partial F}{\partial B}\right)_{T,V}, \qquad \text{(F. 235)}$$

where

$$F = U - TS; \qquad \text{(F. 236)}$$

F is the free energy, U is the internal energy, S is the entropy, V is the volume considered, and B is the magnetic induction.

At absolute zero the magnetization is given by the variation of the internal energy with the magnetic induction. The decisive contribution of the internal energy (cf. p. 45) is the total energy of the electrons in a metal of volume V. The contribution due to the electron spins (cf. p. 454) is not taken into account in the following considerations.

So far we have shown that in the presence of a magnetic field the quasi-continuous energy spectrum of the electrons changes to a series of equidistant Landau levels which are strongly degenerate (cf. pp. 256 ff.). The principle of the de Haas–van Alphen effect becomes obvious when we consider a two-dimensional model perpendicular to the magnetic field. The total electron energy as a function of the magnetic induction is obtained from a summation over the energies of all occupied Landau levels [cf. eqn. (F. 196)]:

$$U(B) = \sum_{l=0}^{\lambda} 2D(B)\,\hbar\omega_c\left(l+\frac{1}{2}\right)+[N-(\lambda+1)\,2D(B)]\hbar\omega_c\left(\lambda+\frac{3}{2}\right) \quad \text{(F.237)}$$

with the number of states per Landau level (degeneracy), cf. eqns. (F. 203) and (F. 193):

$$D(B) = \left(\frac{L}{2\pi}\right)^2 \frac{2\pi e}{\hbar} B. \quad \text{(F.238)}$$

Here $\lambda(B)$ is the highest Landau quantum number for which the corresponding Landau level is still completely filled and N is the number of electrons in the two-dimensional model. The first term of eqn. (F. 237) denotes the energy of the $\lambda+1$ completely filled Landau levels, the second term the energy of the next higher incompletely filled level. From eqn. (F. 237) it follows immediately that all N electrons are in completely occupied Landau levels if

$$\frac{1}{B} = 2\left(\frac{L}{2\pi}\right)^2 \frac{2\pi e}{\hbar} \frac{\lambda+1}{N}. \quad \text{(F. 239)}$$

Then the number of completely filled levels decreases as the magnetic induction increases. Above the critical induction B_0, which is given by

$$2D(B_0) = N \quad \text{or} \quad \frac{1}{B_0} = 2\left(\frac{L}{2\pi}\right)^2 \frac{2\pi e}{\hbar} \frac{1}{N}, \quad \text{(F. 240)}$$

all electrons are in a single Landau level. In this case the total energy increases proportional to $B(B > B_0)$, and the magnetization has then the constant saturation value $-N\mu_B$ (μ_B is the value of the Bohr magneton, cf. p. 425).

For $B < B_0$ the total energy of the electrons (F. 237) is a periodic function of $1/B$ as shown by the following considerations. Each Landau level contains as many states as are contained in the zero-field case between the hypothetical energy surfaces $E_\perp = \hbar\omega_c l =$ const. and $E_\perp = \hbar\omega_c(l+1) =$ const. (cf. p. 259). We consider the position of the highest completely filled Landau level (quantum number λ) with respect to the Fermi energy ζ_0. For a certain value of the magnetic induction ζ_0 is assumed to lie exactly in the middle between the levels λ and $\lambda+1$. Then the number of occupied states below ζ_0 as well as the total energy of the electron gas are equal to the corresponding values for the zero-field case. With increasing field strength the energy of the Landau levels increases, and thus also the total energy of the electron gas, until the highest level λ surpasses the value ζ_0. Because of the Fermi distribution and the increasing degeneracy more electrons pass over to lower levels and the total energy drops until ζ_0 again comes to lie in the middle of two Landau levels. It is thus easy to see that the total-energy variation increases with the magnetic field strength, that is, when the number of Landau levels below ζ_0 decreases (cf. Fig. 135).

The state of minimum total energy is given by the condition (F. 239) and has a constant period of repetition

$$\Delta\left(\frac{1}{B}\right) = 2\left(\frac{L}{2\pi}\right)^2 \frac{2\pi e}{\hbar} \frac{1}{N}. \tag{F. 241}$$

Because of the periodic variation of the total energy the magnetization of the electron gas also displays oscillations (cf. Fig. 135) according to (F. 235).

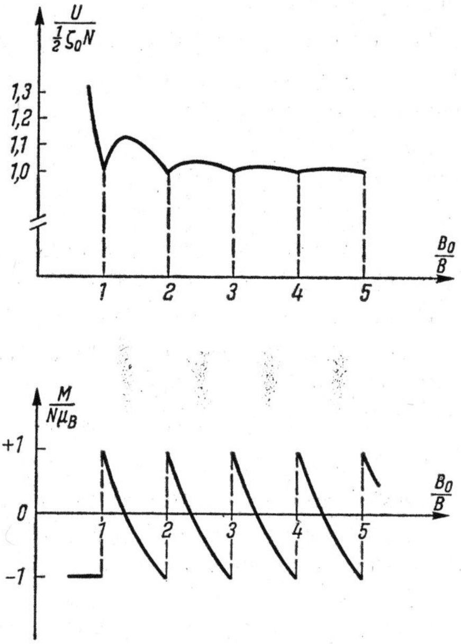

FIG. 135. Energy and magnetization of the electron gas as a function of the inverse magnetic induction at $T = 0$°K (two-dimensional model) (after D. Shoenberg, in *Progress in Low Temperature Physics*, Vol. 2, North-Holland, Amsterdam 1957).

The period of these oscillations depends on the cross-section of the Fermi surface perpendicular to the magnetic field. In the two-dimensional model (xy plane perpendicular to B) the number of electrons at $T = 0$°K is equal to twice the number of states in the $k_x k_y$ plane inside the energy "surface" $E = \zeta$, that is [cf. eqns. (E. 23) and (F. 123)],

$$N = 2\left(\frac{L}{2\pi}\right)^2 \int\limits_{E=\zeta} dk_x\, dk_y = 2\left(\frac{L}{2\pi}\right)^2 A_\zeta, \tag{F. 242}$$

where A_ζ is the cross-section of the Fermi surface perpendicular to the magnetic field. Substitution in eqn. (F. 241) yields

$$\frac{1}{B} = \frac{2\pi e}{\hbar} \frac{1}{A_\zeta}. \tag{F. 243}$$

Because of the electron velocity components parallel to the magnetic field it is much more difficult to describe the de Haas–van Alphen effect for a three-dimensional system. The discontinuities in the magnetization, which are characteristic of the two-dimensional model, have vanished; the oscillatory dependence on the magnetic induction, however, is observed also in the three-dimensional model. The oscillation period is inversely proportional to the maximum or minimum cross-sections $A_{\max}$ or $A_{\min}$ of the Fermi surface perpendicular to the magnetic induction B:

$$\Delta\left(\frac{1}{B}\right) = \frac{2\pi e}{\hbar} \frac{1}{A_{\max}} \quad \text{or} \quad \Delta\left(\frac{1}{B}\right) = \frac{2\pi e}{\hbar} \frac{1}{A_{\min}}. \tag{F. 244}$$

In principle the electrons in all cross-sections of the Fermi surface perpendicular to the magnetic field give rise to de Haas–van Alphen oscillations with the appropriate periods. However, one observes essentially only contributions of electrons in extreme cross-sections, since in their vicinity there exist more parallel sections with approximately the same cross-sectional areas than in the vicinity of arbitrary cross-sections. In certain field directions even several oscillation periods may be observed simultaneously, which correspond to maximum and minimum cross-sections of the Fermi surface. On the $1/B$ scale short periods correspond to maximum cross-sections and long periods to minimum cross-sections of the Fermi surface (cf. Fig. 136).

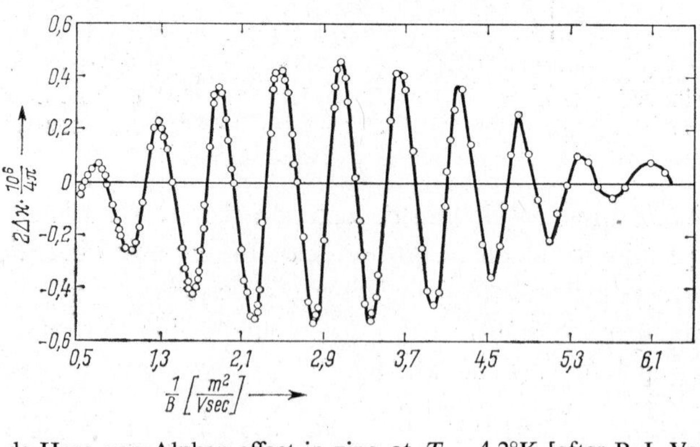

FIG 136. de Haas–van Alphen effect in zinc at $T = 4.2°K$ [after B. I. Verkin and I. M. Dmitrenko, *Bull. Acad. Sci. USSR. Phys. Ser.* **19**, 365 (1955)].

(d) *MAGNETO-OPTICAL ABSORPTION*

Besides electron transitions between the partial bands of the same main band (cyclotron resonance) there exist also transitions between the Landau levels of different main bands. Such interband transitions cause the so-called magneto-optical absorption.

The transitions from the partial bands of the valence bands to those of the conduction band have been studied already for many semiconductors. The necessary photon energies $\hbar\omega$ are given by

$$\hbar\omega \geqslant \Delta E(0)+(l+\tfrac{1}{2})\hbar\omega_c' \qquad \text{(F. 245)}$$

where

$$\omega_c' = \omega_{cv}+\omega_{cc}, \qquad \text{(F. 246)}$$

ΔE being the width of the energy gap between valence and conduction bands when there is no magnetic field and ω_{cv}, ω_{cc} the electron cyclotron frequencies for the valence and the conduction bands. By virtue of (F. 215) and (F. 246) we have

$$\omega_c' = \frac{e}{m'}B, \qquad \text{(F. 247)}$$

where

$$m' = \frac{m_{cv}m_{cc}}{m_{cv}+m_{cc}}; \qquad \text{(F. 248)}$$

m' is the reduced cyclotron mass and m_{cv}, m_{cc} are the cyclotron masses in the valence and conduction bands.

The possible transitions are determined by the selection rules. The theory of magneto-absorption is analogous to the theory of optical absorption without magnetic field. We have again to distinguish between direct and indirect transitions; the nature of the transition determines the absorption coefficient as a function of the photon energy.

On the basis of the conservation of energy (F. 245) the following conclusions can be drawn about the spectrum of magneto-optical absorption:

The application of a magnetic field shifts the long-wavelength absorption edge from the energy ΔE to higher energies by $\hbar\omega_c'/2$ (we ignore the energy term splitting due to electron spin). According to eqns. (F. 245) and (F. 247), the displacement of the absorption edge is proportional to the magnetic induction B (cf. Fig. 137). When we first neglect the motion of electrons in the direction of $\boldsymbol{k}_{\parallel}$ parallel to $\boldsymbol{B}$, the equality sign must be taken in (F. 245). Corresponding

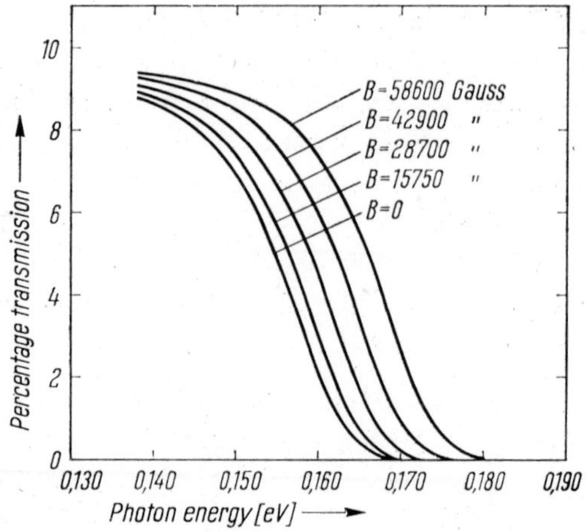

FIG. 137. Displacement of the absorption edge of InSb in a magnetic field [after E. Burstein, G. S. Picus, R. F. Wallis and F. J. Blatt, *Phys. Rev.* **103,** 826 (1956)].

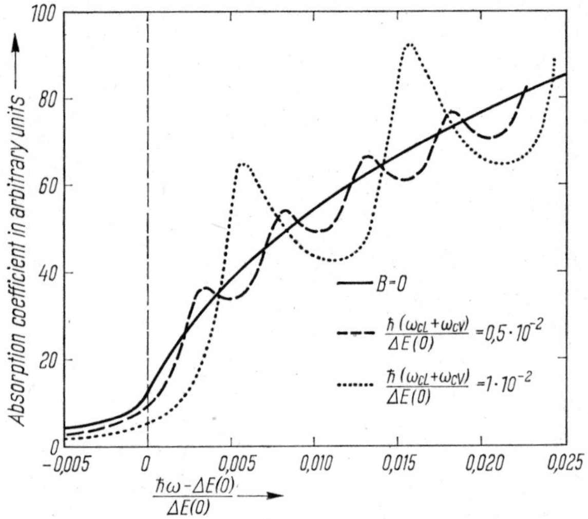

FIG. 138. Magneto-optic absorption of a semiconductor [after E. Burstein, G. S. Picus, R. F. Wallis and F. J. Blatt, *Phys. Rev.* **113,** 15 (1959)].

to the integral values of l only photons of certain energies would be absorbed and the absorption spectrum would be a line spectrum. The quasi-continuous energy distribution of the states in the direction of $\boldsymbol{k}_{\parallel}$ as well as indirect transitions cause the appearance of marked peaks on the high energy side of fundamental absorption in the spectrum instead of lines (cf. Fig. 138). The peaks have an equidistant distribution along the energy scale; their spacing is proportional to the magnetic induction and inversely proportional to the reduced cyclotron mass. This is an immediate proof of the existence of the Landau levels. Plotting the energy of the absorption edge and the absorption peaks versus the magnetic induction, we obtain a set of straight lines which intersect at $\boldsymbol{B} = 0$. The extrapolation to $\boldsymbol{B} = 0$ yields with good accuracy the width of the energy gap; from the inclination of the straight lines (with the magnetic quantum number l as a parameter) we obtain the value of the reduced cyclotron mass.

The determination of the effective masses from magneto-optical measurements is less direct than that from cyclotron measurements. One obtains in principle only the value of the reduced mass (F. 247) since in interband transitions the effective masses of different bands act simultaneously. The masses m_{cc} and m_{cv} and thus also the reduced m_c' depend on the direction of the magnetic field if the energy surfaces are not spherical for $\boldsymbol{B} = 0$. An anisotropic band structure for $\boldsymbol{B} = 0$ thus leads to an anisotropic Landau splitting at $\boldsymbol{B} \neq 0$. Therefore magneto-absorption generally depends on the direction of the magnetic field. A measurement of the directional dependence yields information about the anisotropy of the band structure.

The effect of magneto-absorption makes it possible to study also transitions to such states which, because of lacking electron population, cannot be investigated by means of cyclotron resonance measurements. In this way one can obtain information about energy surfaces which are normally unoccupied.

References

BLOUNT, E. I., *Formalisms of Band Theory*, in *Solid State Physics*, Vol. 13 (Academic Press, New York 1962).

BRAUER, W., *Einführung in die Elektronentheorie der Metalle* (Vieweg, Braunschweig 1966).

BRILLOUIN, L., *Wave Propagation in Periodic Structures* (McGraw-Hill, New York 1946).

BUSCH, G., *Experimentelle Methoden zur Bestimmung effektiver Massen in Metallen und Halbleitern*, in *Halbleiterprobleme*, Vol. 6 (Vieweg, Braunschweig 1961).

CALLAWAY, J., *Energy Band Theory* (Academic Press, New York 1964).

FLETCHER, G. C., *The Electron Band Theory of Solids* (North-Holland, Amsterdam 1971).

FRÖHLICH, H., *Elektronentheorie der Metalle* (Springer, Berlin 1936).

HARRISON, W. A., *Pseudopotentials in the Theory of Metals* (Benjamin, New York 1966).

HARRISON, W. A., and WEBB, M. B. (Editor), *The Fermi Surface* (Wiley, New York 1960).

HERMAN, F., *Theoretical Investigations of the Electronic Band Structure of Solids*, *Rev. Mod. Phys.* **30,** 102–121 (1958).

JONES, H., *The Theory of Brillouin Zones and Electronic States in Crystals* (North-Holland, Amsterdam 1960).

LAX, B., *Experimental Investigations of the Electronic Band Structure of Solids*, *Rev. Mod. Phys.* **30,** 122–154 (1958).

LOUCKS, T., *Augmented Plane Wave Method* (Benjamin, New York 1967).

MERCOUROFF, W., *La surface de Fermi des métaux* (Masson, Paris 1967).

MOTT, N. F., and JONES, H., *Theory of the Properties of Metals and Alloys* (Oxford University Press, London 1936).

PINCHERLE, L., *Band Structure Calculations in Solids*, in *Reports on Progress in Physics*, Vol. 23 (Physical Society, London 1960).

PIPPARD, A. B., *Experimental Analysis of the Electronic Structure of Metals*, in *Reports on Progress in Physics*, Vol. 23 (Physical Society, London 1960).

PIPPARD, A. B., *The Dynamics of Conduction Electrons* (Gordon and Breach, New York 1965).

RAIMES, S., *The Wave Mechanics of Electrons in Metals* (North-Holland, Amsterdam 1961).

SLATER, J. C., *The Electronic Structure of Solids*, in *Handbuch der Physik*, Vol. 19 (Springer, Berlin 1956).

SLATER, J. C., *Quantum Theory of Molecules and Solids*, Vol. 2: *Symmetry and Energy Bands in Crystals* (McGraw-Hill, New York 1965).

SOMMERFELD, A., and BETHE, H., *Elektrontheorie der Metalle*, in *Handbuch der Physik*, Editors GEIGER, H. und SCHEEL, K., Vol. 24/2 (Springer, Berlin 1933); Reprint in Heidelberger Taschenbücher, Vol. 19 (Springer, Berlin 1967).

WILSON, A. H., *The Theory of Metals* (Cambridge University Press, London 1953).

ZIMAN, J. M., *Electrons in Metals* (Taylor and Francis, London 1963).

G. Semiconductors

THE principal difference between metals and insulators has been defined according to the occupation of the highest energy bands (cf. pp. 226 ff.). At sufficiently high temperatures the electrons in insulators can be excited from the valence band over the energy gap to the conduction band. In this way insulators become ideal or intrinsic semiconductors. The incompletely occupied energy bands (*both* the conduction and the valence bands) give rise to electric conduction, the so-called intrinsic conduction (cf. pp. 280 ff.). The number of electrons lacking in the valence band is precisely equal to the number of electrons in the conduction band.

Inevitable imperfections in a crystal, that is, impurities and lattice defects, disturb the strict periodicity of the crystal potential. This disturbance gives rise to additional allowed energy levels (defect levels) in the energy gap between valence and conduction bands. An excitation of electrons from the valence band to unoccupied defect levels, or from occupied defect levels to the conduction band, also results in the appearance of incompletely occupied energy bands (in the simplest case conduction band *or* valence band) which make conduction possible; this type of conduction is called impurity conduction. Semiconductors displaying this type of conductivity are called extrinsic or impurity semiconductors.

At low temperatures each semiconductor shows impurity conduction (cf. pp. 300 ff.), which can usually be neglected compared with the increasing intrinsic conduction as the temperature rises.

I. Holes

The unoccupied states of the valence band are called holes. It is reasonable to introduce such quasi-particles since the contribution of the valence band to the electric conductivity depends only on the number of vacant states. These states behave like particles with positive charge and positive effective mass. The effective mass of the hole is, as to its magnitude, equal to the corresponding effective mass of the initially present electron, but has the opposite sign (on p. 219 we have shown that electrons with energies near an upper band edge have always negative effective masses).

The equivalence between the few holes and the many electrons of the valence band is always valid in the one-electron approximation (cf. p. 195), that is, when electron–electron interactions are negligible.

In this way external forces act not only on the electrons in the conduction band but also on the holes in the valence band. The energy distribution of the holes, like that of the electrons, is governed by the Fermi–Dirac statistics. The probability for a state of energy E in thermodynamic equilibrium to be unoccupied, i.e. to be occupied by a hole, is given by

$$1-F(E) = 1-\frac{1}{\exp\left(\frac{E-\zeta}{kT}\right)+1} = \frac{1}{\exp\left(\frac{\zeta-E}{kT}\right)+1}. \tag{G. 1}$$

The Fermi–Dirac function of the holes is thus a mirror-image of that of the electrons about the line $E = \zeta$. High electron energies correspond to low hole energies.

In the following we shall calculate the carrier concentrations of intrinsic and of impurity semiconductors in thermodynamic equilibrium, in the absence of external forces, and at temperatures $T > 0°K$.

II. Intrinsic Semiconductors

According to eqn. (E. 32) the concentration n of the conduction electrons is given by

$$n = 2\int_{E_c}^{\infty} D_n(E)F(E)\,dE, \tag{G. 2}$$

where $D_n(E)$ is the eigenvalue density in the conduction band per unit volume, and E_c is the energy of the bottom of the conduction band.

Strictly speaking, the integration is only over the conduction band. Because of the rapid drop of the Fermi–Dirac function, however, the main contribution to the integral is due to states near the lower band edge so that in a good approximation it is possible to shift the upper limit of integration to infinity. Near the lower band edge the $E(\boldsymbol{k})$ relation is given by (F. 130) or (F. 223):

$$E(\boldsymbol{k}) = E_c+\frac{\hbar^2}{2}\left(\frac{k_x^2}{m_{xx}}+\frac{k_y^2}{m_{yy}}+\frac{k_z^2}{m_{zz}}\right). \tag{G. 3}$$

The eigenvalue density of the conduction electrons per energy valley (cf. p. 266) is then given by [cf. eqn. (F. 134)]

$$D_n(E) = \frac{1}{4\pi^2}\left(\frac{2}{\hbar^2}\right)^{3/2}(m_{xx}m_{yy}m_{zz})^{1/2}(E-E_c)^{1/2}. \tag{G. 4}$$

In the case of an isotropic effective mass $m_n = m_{xx} = m_{yy} = m_{zz}$ we have

$$D_n(E) = \frac{1}{4\pi^2}\left(\frac{2m_n}{\hbar^2}\right)^{3/2}(E-E_c)^{1/2}. \tag{G. 5}$$

Similarly we obtain from (G. 1) the hole concentration p,

$$p = 2\int_{E_V}^{-\infty} D_p(E)\,[1-F(E)]\,dE, \tag{G. 6}$$

where

$$D_p(E) = \frac{1}{4\pi^2}\left(\frac{2m_p}{\hbar^2}\right)^{3/2}(E_V-E)^{1/2}. \tag{G. 7}$$

$D_p(E)$ is the eigenvalue density in the valence band per unit volume, E_V is the energy at the top of the valence band, and m_p is the isotropic effective mass of the holes.

Also for the case of anisotropic effective masses the eigenvalue density is expressed by eqns. (G. 5) and (G. 7). The scalar masses m_n and m_p are then mean values averaged over the tensor components of the effective masses; these mean values are then called density-of-states effective masses. For that of the electrons, m_n, for example, we have

$$m_n^3 = m_{xx}m_{yy}m_{zz} \tag{G. 8}$$

in the case of ellipsoidal energy surfaces (F. 130).

In semiconductors in thermodynamic equilibrium and in sufficiently weak electric fields normally only states at the bottom of the conduction band are occupied and the states at the top of the valence band are vacant. In these cases the effective masses are independent of the energy since the $E(\boldsymbol{k})$ relations near the band edges are as a rule parabolic [cf. eqns. (F. 115) and (F. 223)].

In eqns. (G. 2) and (G. 6) for the concentrations n and p the Fermi energy ζ is so far contained as an unknown parameter. It can only be determined by means of the neutrality condition of the semiconductor. An ideal semiconductor is neutral, i.e. without charge excess, if the number of negative electrons is equal to the number of positive holes:

$$n = p = n_i, \tag{G. 9}$$

where n_i is the intrinsic density (inversion-density). From eqns. (G. 2) and (G. 4) or (G. 6) and (G. 7) it follows that

$$n = \frac{1}{2\pi^2}\left(\frac{2m_n}{\hbar^2}\right)^{3/2}\int_{E_c}^{\infty}\frac{(E-E_c)^{1/2}\,dE}{\exp\left(\frac{E-\zeta}{kT}\right)+1}, \tag{G. 10}$$

or

$$p = \frac{1}{2\pi^2}\left(\frac{2m_p}{\hbar^2}\right)^{3/2}\int_{E_V}^{-\infty}\frac{(E_V-E)^{1/2}\,dE}{\exp\left(\frac{\zeta-E}{kT}\right)+1}. \tag{G. 11}$$

The equation (G. 9) for the determination of ζ can generally not be solved analytically. Let us introduce the substitutions

$$\frac{E-E_c}{kT} = x, \qquad \frac{E_V-E}{kT} = y; \tag{G. 12}$$

$$\frac{\zeta-E_c}{kT} = \alpha, \qquad \frac{E_V-\zeta}{kT} = \beta. \tag{G. 13}$$

Hence it follows that

$$\frac{E-\zeta}{kT} = x-\alpha, \qquad \frac{\zeta-E}{kT} = y-\beta; \tag{G. 14}$$

$$dE = kT\,dx, \quad dE = -kT\,dy \tag{G. 15}$$

and

$$a+\beta = -\frac{E_c-E_V}{kT} = -\frac{\Delta E_i}{kT}, \tag{G. 16}$$

where ΔE_i is the width of the energy gap between valence and conduction bands. Substituting (G. 12) and (G. 13) in (G. 10) and (G. 11) we obtain

$$n = \frac{1}{2\pi^2}\left(\frac{2m_n kT}{\hbar^2}\right)^{3/2}\int_0^{\infty}\frac{x^{1/2}\,dx}{\exp(x-\alpha)+1} = \frac{2}{\pi^{1/2}}\,n_0 F_{1/2}(\alpha) \tag{G. 17}$$

and

$$p = \frac{1}{2\pi^2}\left(\frac{2m_p kT}{\hbar^2}\right)^{3/2}\int_0^{\infty}\frac{y^{1/2}\,dy}{\exp(y-\beta)+1} = \frac{2}{\pi^{1/2}}\,p_0 F_{1/2}(\beta) \tag{G. 18}$$

with the effective densities of state [cf. eqn. (E. 36)]

$$n_0 = \frac{1}{4}\left(\frac{2m_n kT}{\pi\hbar^2}\right)^{3/2}, \tag{G. 19}$$

$$p_0 = \frac{1}{4}\left(\frac{2m_p kT}{\pi\hbar^2}\right)^{3/2} \tag{G. 20}$$

and the Fermi integrals [cf. eqn. (E. 37)]

$$F_{1/2}(\alpha) = \int_0^\infty \frac{x^{1/2}\,dx}{\exp(x-\alpha)+1}, \tag{G. 21}$$

$$F_{1/2}(\beta) = \int_0^\infty \frac{y^{1/2}\,dy}{\exp(y-\beta)+1}. \tag{G. 22}$$

Let us assume the normal carrier concentration in semiconductors to be so small that

$$n \ll n_0, \quad p \ll p_0. \tag{G. 23}$$

This means that the electron and hole concentrations are non-degenerate (cf. p. 168). In this case, according to (E. 51), we have

$$\alpha \ll -1, \quad \beta \ll -1 \tag{G. 24}$$

or, according to (G. 13),

$$E_c - \zeta \gg kT, \quad \zeta - E_V \gg kT, \tag{G. 25}$$

that is, the Fermi energy ζ lies above the valence band and below the conduction band, at distances many times kT from the band edges. Under this assumption the Fermi integrals can be solved analytically (E. 47) and one obtains for the carrier concentrations instead of (G. 17) and (G. 18)

$$n = n_0 \exp(\alpha), \tag{G. 26}$$

$$p = p_0 \exp(\beta). \tag{G. 27}$$

Then the neutrality condition (G. 9) yields the position of ζ:

$$\exp(\alpha-\beta) = \frac{p_0}{n_0} \tag{G. 28}$$

or, from (G. 13),

$$\zeta - E_c - E_V + \zeta = kT \ln\left(\frac{p_0}{n_0}\right), \tag{G. 29}$$

or

$$\zeta = \frac{E_c+E_V}{2} + \frac{1}{2} kT \ln\left(\frac{p_0}{n_0}\right). \tag{G. 30}$$

Substitution of eqn. (G. 19) and (G. 20) yields

$$\zeta = \frac{E_c+E_V}{2} + \frac{3}{4} kT \ln\left(\frac{m_p}{m_n}\right). \tag{G. 31}$$

This relation applies to an ideal semiconductor under the assumption of non-degeneracy. At $T = 0°K$ the Fermi energy ζ lies exactly in the middle of the energy gap between valence and conduction bands. At $T > 0°K$ the Fermi energy ζ increases or drops as a linear function of the temperature according to whether $m_n < m_p$ or $m_n > m_p$; if $m_n = m_p$ the Fermi energy ζ lies in the middle of the energy gap (cf. Fig. 139), independent of T.

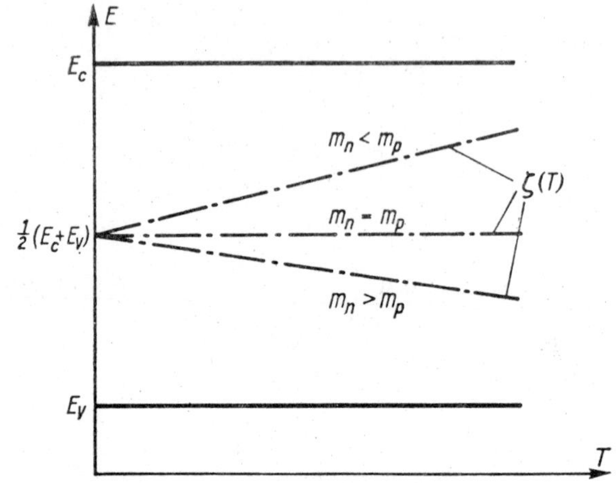

FIG. 139. Temperature dependence of the Fermi energy for a non-degenerate ideal semiconductor. [The temperature dependence of the energy gap (see eqn. G. 57) has not been taken into account.]

From the combination of eqns. (G. 26) and (G. 27) we obtain immediately the temperature dependence of the carrier concentration of an ideal semiconductor, when we take the neutrality condition (G. 9) into account. We obtain

$$np = n_0 p_0 \exp(\alpha + \beta). \tag{G. 32}$$

Substitution of eqns. (G. 16), (G. 19), and (G. 20) yields

$$np = \frac{1}{16}\left(\frac{2kT}{\pi\hbar^2}\right)^3 (m_n m_p)^{3/2} \exp\left(-\frac{\Delta E_i}{kT}\right). \tag{G. 33}$$

From this and (G. 9) we obtain

$$n_i = \frac{1}{4}\left(\frac{2kT}{\pi\hbar^2}\right)^{3/2} (m_n m_p)^{3/4} \exp\left(-\frac{\Delta E_i}{2kT}\right). \tag{G. 34}$$

This relation, which was first derived by Wilson, is again valid only in the case of non-degeneracy, i.e. if

$$n_i \ll n_0 \quad \text{and} \quad n_i \ll p_0. \tag{G. 35}$$

In this approximation the carrier concentrations increase strongly with the temperature; they grow exponentially as $1/T$ decreases. A plot of $\ln (n_i/T^{3/2})$ versus $1/T$ yields a straight line whose slope is proportional to the width ΔE_i of the energy gap. At $T = 0°K$ the carrier concentration is exactly equal to zero.

1. DEGENERACY OF THE CARRIER CONCENTRATION

The conditions (G. 35) for non-degeneracy are no longer satisfied, if the carrier concentration becomes comparable to the effective density of states or exceeds it. According to (G. 17) we have $\alpha \approx 0$ or $\beta \approx 0$ in the case of $n \approx n_0$ or $p \approx p_0$, that is, the Fermi energy almost coincides with the bottom of the conduction band or the top of the valence band. We shall show that for $\alpha = 0$ the electron concentration is almost equal to the critical concentration which is defined by (E. 54). Using $F_{1/2}(0) = 0.678 \approx \frac{2}{3}$ (cf. Table 21) we obtain from eqn. (G. 17)

$$n \approx \frac{4}{3\pi^{1/2}} n_0 . \tag{G. 36}$$

Comparison with (E. 55) and substitution of (G. 19) yields

$$n \approx n_{\text{crit}} = \frac{1}{3\pi^2}\left(\frac{2m_n kT}{\hbar^2}\right)^{3/2} . \tag{G. 37}$$

We see that degeneracy is favoured by small effective masses.

For an intrinsic semiconductor with $\Delta E_i \gg kT$ the condition of neutrality shows that a carrier concentration will be degenerate only if the masses m_n and m_p are essentially different. For $\alpha = 0$ we obtain from eqn. (G. 16)

$$\beta = -\frac{\Delta E_i}{kT} \ll -1. \tag{G. 38}$$

The hole concentration is then given by

$$p = p_0 \exp (\beta) \tag{G. 39}$$

according to (G. 27). Using the neutrality condition we obtain from (G. 36) and (G. 39)

$$n_0 \approx p_0 \exp (\beta), \tag{G. 40}$$

or, because of (G. 38),

$$n_0 \ll p_0 . \tag{G. 41}$$

Substitution of eqns. (G. 19) and (G. 20) yields

$$m_n \ll m_p . \tag{G. 42}$$

TABLE 21. Fermi integrals $F_n(\alpha) = \int_0^\infty \frac{x^n\,dx}{\exp(x-\alpha)+1}$ for the range $-4 < \alpha < 20$ [after O. Madelung, *Halbleiter*, in *Handbuch der Physik*, Vol. 20 (Springer, Berlin 1957, p. 58)]

α	$\frac{1}{1+e^{-\alpha}}$	$F_{-1/2}$	$\ln(1+e^{\alpha})$	$F_{1/2}$	F_1	$F_{3/2}$	F_2	$F_{5/2}$
−4	0.1799 −1	0.3204 −1	0.1815 −1	0.16128 −1	0.0182 0	2.42685 −2	0.0366 0	6.07731 −2
−3	0.4743 −1	0.8526 −1	0.4859 −1	0.43366 −1	0.0492 0	6.56115 −2	0.0990 0	1.64742 −1
−2	1.1920 −1	2.1918 −1	1.2693 −1	1.14588 −1	0.1310 0	1.75800 −1	0.2662 0	4.44554 −1
−1	2.6894 −1	5.2114 −1	3.1326 −1	2.90501 −1	0.3387 0	4.60848 −1	0.7050 0	1.18597 0
0	5.0000 −1	1.0722 0	6.9315 −1	6.78094 −1	0.8225 0	1.15280 0	1.8030 0	3.08259 0
1	7.3106 −1	1.8204 0	1.3137 0	1.39638 0	1.8062 0	2.66168 0	4.3120 0	7.62653 0
2	8.8079 −1	2.5954 0	2.1270 0	2.50246 0	3.5135 0	5.53725 0	9.4450 0	1.75294 1
3	9.5257 −1	3.2852 0	3.0486 0	3.97699 0	6.0957 0	1.03537 1	1.8870 1	3.69321 1
4	9.8201 −1	3.8743 0	4.0182 0	5.77073 0	9.6267 0	1.76277 1	3.4592 1	7.13480 1
5	9.9331 −1	4.3832 0	5.0067 0	7.83797 0	1.4138 1	2.78024 1	5.8120 1	1.27489 2
6	9.9753 −1	4.8338 0	6.0025 0	1.01443 1	1.9642 1	4.12610 1	9.1744 1	2.13098 2
7	9.9909 −1	5.2416 0	7.0009 0	1.26646 1	2.6144 1	5.83422 1	1.3736 2	3.36814 2
8	9.9967 −1	5.6170 0	8.0003 0	1.53805 1	3.3645 1	7.93526 1	1.9699 2	5.08084 2
9	9.9988 −1	5.9674 0	9.0001 0	1.82776 1	4.2145 1	1.04574 2	2.7261 2	7.37087 2
10	9.9995 −1	6.2972 0	1.0000 1	2.13445 1	5.1645 1	1.34270 2	3.6623 2	1.03468 3
11	9.9998 −1	6.6096 0	1.1000 1	2.45718 1	6.2145 1	1·68688 2	4.7986 2	1.41237 3
12	9.9999 −1	6.9076 0	1.2000 1	2.79518 1	7.3645 1	2.08062 2	6.1548 2	1.88225 3
13	1.0000 0	7.1930 0	1.3000 1	3.14775 1	8.6145 1	2.52616 2	7.7510 2	2.45700 3
14	1.0000 0	7.4672 0	1.4000 1	3.51430 1	9.9645 1	3.02564 2	9.6072 2	3.14983 3
15	1.0000 0	7.7314 0	1.5000 1	3.89430 1	1.1415 2	3.58112 2	1.1743 3	3.97448 3
16	1.0000 0	7.9868 0	1.6000 1	4.28730 1	1.2965 2	4.19458 2	1.4180 3	4.94522 3
17	1.0000 0	8.2342 0	1.7000 1	4.69286 1	1.4615 2	4.86794 2	1.6936 3	6.07677 3
18	1.0000 0	8.4744 0	1.8000 1	5.11061 1	1.6365 2	5.60305 2	2.0032 3	7.38433 3
19	1.0000 0	8.7076 0	1.9000 1	5.54019 1	1.8215 2	6.40171 2	2.3488 3	8.88359 3
20	1.0000 0	8.9350 0	2.0000 1	5.98128 1	2.0165 2	7.26568 2	2.7325 3	1.05906 4

The last (separated) figure of each line gives the power of ten by which the corresponding value must be multiplied.

While for $\alpha = 0$ the electron concentration becomes degenerate, the hole concentration is far below the concentration of degeneracy, p_{crit}:

$$p_{\text{crit}} = \frac{4}{3\pi^{1/2}} p_0 \gg p_0 \exp(\beta) = p. \tag{G. 43}$$

The value of α is to be considered a measure of the electron gas degeneracy:

$\alpha \ll 0$: non-degeneracy, that is, ζ lies below the bottom of the conduction band;

$\alpha \approx 0$: beginning degeneracy, that is, ζ lies near the bottom of the conduction band;

$\alpha \gg 0$: strong degeneracy, that is, ζ lies in the conduction band.

In the same way the parameter β determines the degeneracy of the hole gas.

The electron concentration plotted as a function of the position of the Fermi energy [eqn. (G. 17)] is shown in Fig. 140. It shows also the two limiting cases

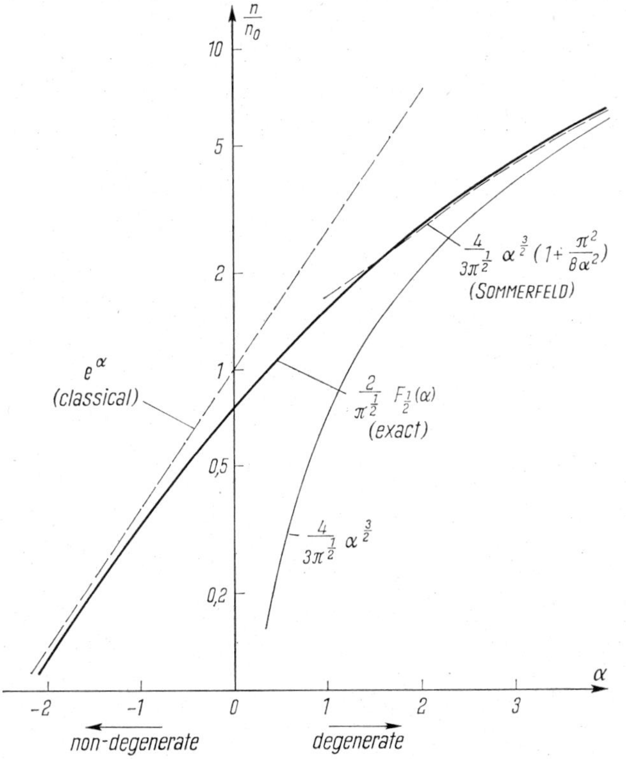

FIG. 140. Electron concentration as a function of the Fermi energy, measured in units of kT.

of non-degeneracy (G. 26) and strong degeneracy (E. 42). This graph applies also to the hole concentration p/p_0 as a function of β. In the case of semiconductors generally $\alpha < 0$ and $\beta < 0$; in the case of metals $\alpha \approx 10^3$.

The Fermi integrals are tabulated for $\alpha \gtrless 0$ (cf. Table 21). For various ranges of α-values the following series expansions are applicable:

1. $\alpha \leqslant 0$ (after McDougall and Stoner):

$$F_{1/2}(\alpha) \approx \frac{\pi^{1/2}}{2} \exp(\alpha) \sum_{n=0}^{\infty} (-1)^n \frac{\exp(n\alpha)}{(n+1)^{3/2}}. \qquad \text{(G. 44)}$$

2. $\alpha \leqslant 2$ (after Busch and Labhart):

$$F_{1/2}(\alpha) \approx \frac{\pi^{1/2}}{2} \exp(\alpha) [1+a \exp(\alpha)-b \exp(2\alpha)]^{-1} \qquad \text{(G. 45)}$$

where $a = 0.3694$; $b = 0.02796$.

3. $\alpha \gg 1$ [after Sommerfeld, cf. eqn. (E. 41)]:

$$F_{1/2}(\alpha) \approx \frac{2}{3} \alpha^{3/2} \left(1+\frac{\pi^2}{8} \frac{1}{\alpha^2}+ \ldots\right). \qquad \text{(G. 46)}$$

Whether and at which temperatures the carrier gas of a semiconductor becomes degenerate depends on the effective mass of the carriers and on the temperature dependence of the carrier concentration. For both intrinsic and impurity semiconductors (cf. pp. 292 ff.) at low temperatures the carrier concentration in general decreases exponentially with decreasing temperature [cf. eqn. (G. 34) or (G. 92)]. This decrease is stronger than the decrease of the critical concentration which follows a $T^{3/2}$ law (G. 37). The degeneracy of a semiconductor will therefore vanish at sufficiently low temperatures (cf. Fig. 141). It can also vanish

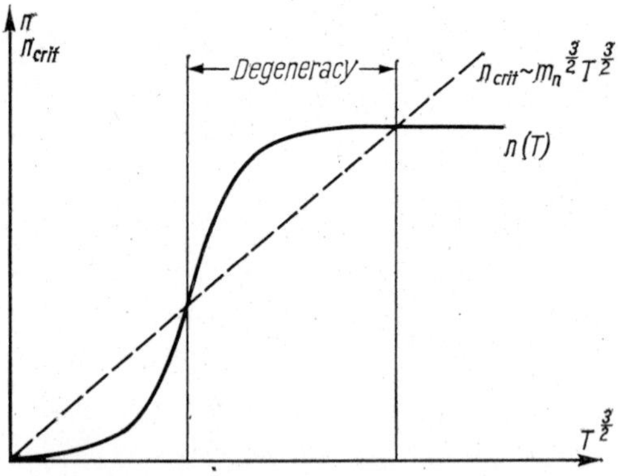

FIG. 141. Temperature dependence of the electron concentration and of the critical concentration. Degeneracy of the electron concentration for $n > n_{crit}$.

at sufficiently high temperatures when the carrier concentration tends towards a constant limit. For an intrinsic semiconductor this limit is given by the atomic density and would be reached only above the melting point. For an impurity semiconductor, however, the carrier concentration may reach saturation as a consequence of defect exhaustion (see p. 302) so that an existing degeneracy can vanish again at higher temperatures. In the case of semiconductors degeneracy may therefore be expected neither at very high temperatures nor at very low ones, but rather at intermediate temperatures. Figure 141 shows that the temperature range of degeneracy depends on both the effective mass of the carrier concentration (→ slope of the straight line $n_{crit} \propto T^{3/2}$) and the temperature dependence of the carrier concentration $n(T)$. For a given semiconductor the carrier gas is degenerate only if the corresponding curves $n(T)$ and $n_{crit}(T)$ intersect each other.

2. EXPERIMENTAL PROOF FOR THE TEMPERATURE DEPENDENCE OF THE CARRIER CONCENTRATION (ELECTRICAL CONDUCTIVITY OF AN INTRINSIC SEMICONDUCTOR)

One of the characteristic differences between semiconductors and metals manifests itself in the temperature dependence of the carrier concentration [cf. eqns. (E. 42) and (G. 34)] and the corresponding electrical conductivity. The current density j in a semiconductor is the sum of the electron current density j_n and the hole current density j_p. These current densities are given by the charge, concentration, and velocity of the carriers. We have

$$j = j_n + j_p = |e\bar{v}_{Dn}|\, n + |e\bar{v}_{Dp}|\, p_n, \tag{G. 47}$$

where $\bar{v}_{Dn}$ and $\bar{v}_{Dp}$ are the mean drift velocities of electrons and holes, respectively. For isotropic crystals and sufficiently small electric field strengths F, we have

$$\bar{v}_{Dn} = b_n F \quad \text{and} \quad \bar{v}_{Dp} = b_p F, \tag{G. 48}$$

where b_n and b_p are the mobilities of electrons and holes, respectively. The drift velocity is a mean value averaged over time and over all electrons or holes. It has, like the mobility (see p. 369), a different sign for electrons and holes. A combination of eqns. (G. 47) and (G. 48) yields the electrical conductivity:

$$\sigma = |eb_n|\, n + |eb_p|\, p. \tag{G. 49}$$

Using this we obtain for an intrinsic semiconductor ($n = p = n_i$)

$$\sigma_i = (|eb_n| + |eb_p|) n_i. \tag{G. 50}$$

A measurement of $\sigma_i(T)$ gives us $n_i(T)$, if the temperature dependence of the mobilities b_n and b_p are known. It can be estimated as follows. The mobility is a measure for the "friction" of the carriers moving through the crystal. This "friction" is due to interactions with defects, phonons, or other disturbances of the strict periodicity of the crystal (cf. pp. 396 ff.). The mobility is proportional to the mean free time τ of a carrier, that is, the mean time between two collisions. This time is inversely proportional to the thermal velocity of the carrier and the collision probability

$$b_{n,\,p} \propto \tau \propto \frac{1}{v}\,\frac{1}{W_{\text{coll}}}\,. \qquad \text{(G. 51)}$$

When we consider only the interactions between a non-degenerate carrier gas and phonons, we obtain the following temperature dependences of the thermal velocity and the collision probability: for Maxwell–Boltzmann statistics we have

$$\frac{m}{2}\,v^2 = \frac{3}{2}\,kT \qquad \text{(G. 52)}$$

or

$$v \propto T^{1/2}\,. \qquad \text{(G. 53)}$$

The collision probability is proportional to the square of the amplitude of the lattice vibrations. This value is proportional to the vibrational energy U_v and therefore, for $T > \Theta_{\text{Debye}}$, proportional to the temperature [cf. eqn. (C. 67)]. Thus

$$W_{\text{coll}} \propto T. \qquad \text{(G. 54)}$$

With the proportionalities (G. 53) and (G. 54) we obtain from (G. 51)

$$b_{n,\,p} \propto T^{-3/2}\,. \qquad \text{(G. 55)}$$

A combination of eqns. (G. 34), (G. 50), and (G. 55) yields the temperature dependence of the electrical conductivity of a non-degenerate intrinsic semiconductor:

$$\sigma_i = \text{const.}\exp\left(-\frac{\Delta E_i}{2kT}\right). \qquad \text{(G. 56)}$$

When we also take into account the temperature dependence of the energy gap

$$\Delta E_i = \Delta E_{i0} - \gamma T, \qquad \text{(G. 57)}$$

the constant in (G. 56) must be multiplied by a factor $\exp(-\gamma/2k)$ and we obtain

$$\sigma_i = \text{const.}\exp\left(-\frac{\Delta E_{i0}}{2kT}\right). \qquad \text{(G. 58)}$$

This relation has been verified by experiments at sufficiently high temperatures; in a graph of ln σ_i versus $1/T$ the measured values $\sigma_i(T)$ lie on a straight line whose slope yields the width of the energy gap extrapolated to $T = 0°K$ (cf. Fig. 142). This straight line is characteristic of the substance considered. According to purity and pretreatment of the substance, the experimental results

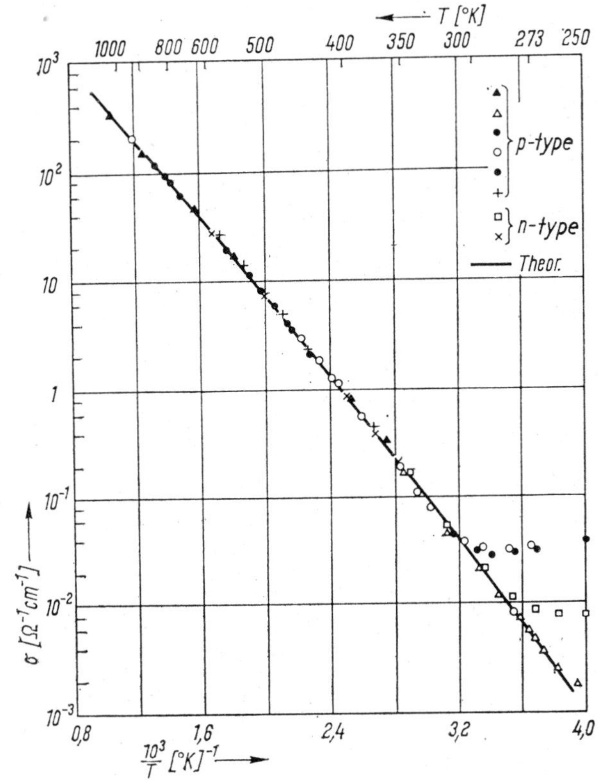

FIG. 142. Temperature dependence of the electric conductivity of Ge [after F. J. Morin and J. P. Maita, *Phys. Rev.* **94,** 1525 (1954)].

below certain temperatures deviate far from the straight line, that is, the electrical conductivity exceeds the values obtained from (G. 58) by many orders of magnitude. The temperature range within which eqn. (G. 58) is satisfied is the range of intrinsic conduction. At lower temperatures, the range of impurity conduction follows. In the latter the electrical properties of semiconductors are sensitive to structural variations (cf. p. 129), that is, they are determined by the imperfections of the crystal structure.

The subdivision into intrinsic and impurity semiconductors refers not to substances but to imperfections and to the conduction mechanisms in a given semiconductor crystal, which depend on the nature of these imperfections. In principle, both conduction mechanisms are present simultaneously; the classification means only that one strongly dominates the other.

III. Impurity Semiconductors

Impurity conduction is generally due to imperfection (cf. pp. 106 ff.). In the case of semiconductors with chiefly covalent bonds (see below) the degree of impurity that is, chemical imperfection, is of importance. We shall not discuss the weaker influence of structural imperfections.

The influence of impurities is most easily understood when we consider an atomistic model of a semiconductor. So far we only characterized semiconductors by the occupation of the energy bands. The band structure, however, is a consequence of the form of the periodic potential and the type of bond in the crystal. We may therefore observe some correlation between type of bond and solid-state properties. There is, however, no sharp division between the various types of bond and the corresponding solid-state properties.

Many semiconductors are characterized by covalent bonds. The covalent bond is formed by an electron pair with antiparallel spins (electron pair binding). The mean number of valence electrons corresponds to the number of nearest neighbours (coordination number). Typical representatives of this form of bond are the elements of group IV of the periodic system (C as diamond, Si, Ge, α-Sn); silicon and germanium are the best-known semiconductors of this group. All these elements have the crystal structure of diamond (cf. Fig. 143): each atom has four nearest neighbours arranged at the corner points of a tetrahedron. The electron pairs are situated at about the midpoint of the line connecting two atoms.

The isoelectronic compounds of atoms of groups III and V, II and VI, or I and VII of the periodic system also possess semiconductor properties. The compounds InSb, ZnSe, or CuI are examples of such semiconductors. Because of the different electronegativity of the elements, the Coulomb forces contribute to the bond in such compounds. The type of bond is thus no longer purely covalent but has an ionic component. The centre of mass of the electron pair is displaced from the middle of the connection line towards the anion.

In an ideal semiconductor crystal only the valence electrons can be responsible for the electrical properties. At sufficiently low temperatures all valence

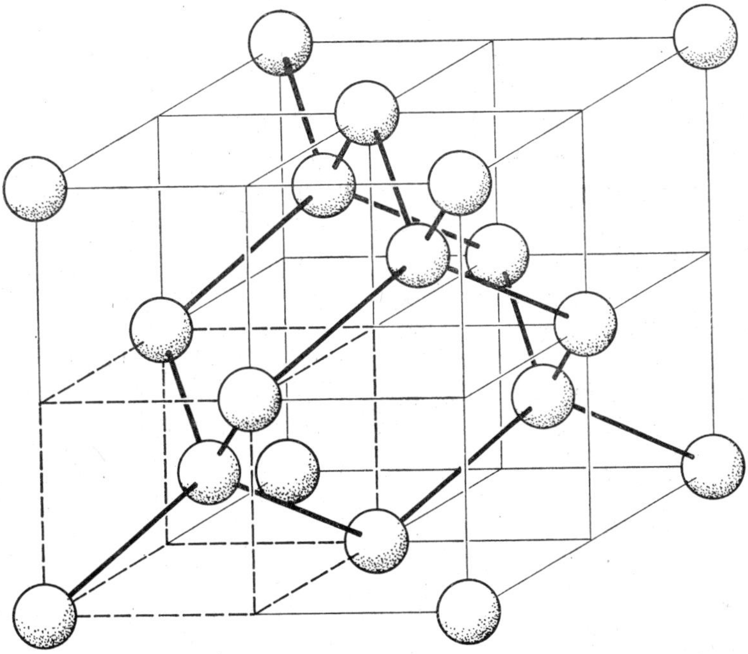

FIG. 143. Diamond structure.

electrons are involved in binding and are thus not available for conduction. This state corresponds to a completely filled valence band. Energy supply from without (temperature rise, irradiation by electromagnetic waves) may release electrons from the bonds, which corresponds to an excitation of electrons from the valence band to the conduction band; the energy necessary to break a bond is equal to the width of the energy gap.

Chemical imperfection is the result of an incorporation of foreign atoms at lattice sites or interstitial positions (cf. pp. 141 ff.). The way of incorporation depends on the ratio of the ion radii of host atoms and impurity atoms; together with the valence of the latter, it determines the mechanism of impurity conduction. The action of the defects is due to their ability to accept or give off electrons; the energy necessary for such a process is smaller than the energy gap between valence and conduction bands. Very many applications of semiconductors are based on the effect of impurities. Therefore, in order to achieve certain desired properties, semiconductors are "doped" with suitable impurity atoms. The concentration of random impurities is reduced by purification methods (e.g. zone melting) to such a degree that the subsequent doping is not affected.

In the following we shall consider the impurity conduction due to substitution sites in an element semiconductor. These considerations also apply to semiconducting compounds with predominantly covalent bonds.

As a typical example we choose silicon containing phosphorus or boron impurities. The phosphorus atom has five valence electrons, four of which are involved in electron pair bonds with four neighbour atoms, thus compensating four positive charges of the phosphorus core. The fifth electron of the phosphorus atomic shells is bound only by the Coulomb attraction of the non-compensated charge of the phosphorus core. The work necessary to release this electron (to excite it up to the conduction band) can be easily estimated following Bethe and Mott, on the basis of the hydrogen model. The phosphorus core with one positive charge and the fifth shell electron may be considered as a hydrogen atom. The energy states of the H atom *in vacuo* are, according to the elementary Bohr theory,

$$E_n = -\frac{1}{2}\frac{e^4 m}{(4\pi\varepsilon_0\hbar)^2}\frac{1}{n^2} \approx -\frac{13.6}{n^2}\,\text{eV}, \qquad \text{(G. 59)}$$

where ε_0 is the vacuum permeability, m is the electron rest mass and n is an integer.

From this we obtain the energy states of the impurity atom in the crystal if we replace the vacuum by a polarizable medium with dielectric constant ε, and the rest mass m by the mean effective mass m_n. Equation (G. 59) will then take the form

$$E_{n,\,\varepsilon} = -\frac{1}{2}\frac{e^4 m}{(4\pi e_0\hbar)^2}\frac{m_n}{m\varepsilon^2}\frac{1}{n^2} \approx -\frac{13.6}{n^2}\frac{m_n}{m}\frac{1}{\varepsilon^2}\,\text{eV}. \qquad \text{(G. 60)}$$

We see that the energy levels of the impurity atom are denser by the factor $m_n/m\varepsilon^2$ than in the case of the free H atom. Accordingly, the work necessary to separate the electron from the phosphorus core is much lower than the ionization energy of the free H atom. The separation work ΔE_D is given by

$$\Delta E_D = E_{\infty,\,\varepsilon} - E_{1,\,\varepsilon} = \frac{1}{2}\frac{e^4 m}{(4\pi\varepsilon_0\hbar)^2}\frac{m_n}{m}\frac{1}{\varepsilon^2} \approx 13.6\frac{m_n}{m}\frac{1}{\varepsilon^2}\,\text{eV}. \qquad \text{(G. 61)}$$

This energy is of the order of 10^{-2} eV (cf. Table 22) and comparable with the thermal energy kT of a phonon at room temperature. Thus the fifth shell electron is only loosely connected with the phosphorus core and, through a small energy supply (e.g. interaction with a phonon) it is excited into the conduction band. Impurity atoms which supply conduction electrons are called donors.

TABLE 22. Typical data for semiconductors (after R. A. Smith, *Semiconductors* [Cambridge University Press, London 1961], pp. 346, 347, 350; and O. Madelung, *Halbleiter*, in *Handbuch der Physik*, Vol. 20, Part 2 [Springer, Berlin 1957], pp. 21, 233)

	Energy gap	Static dielectric constant	Lattice constant	Effective masses†		Activation energies for			
						donors		acceptors	
	ΔE_{i0} [eV]	ε	a [Å]	$\frac{m_n}{m}$	$\frac{m_p}{m}$		ΔE_D [eV]		ΔE_A [eV]
Si	1.21	11.8	5.43	0.33	0.55	P	0.044	B	0.045
						As	0.049	Al	0.057
						Sb	0.039	Ga	0.065
						Li	0.033	In	0.16
Ge	0.78	15.6	5.66	0.22	0.39	P	0.012	B	0.010
						As	0.013	Al	0.010
						Sb	0.010	Ga	0.011
						Li	0.009	In	0.011

† The values for m_n and m_p are approximated by those of the density-of-states masses.

In an atomistic model it is justified to work with a macroscopic dielectric constant ε since the radius $r_{1,\varepsilon}$ of the first Bohr orbit in a polarizable medium is much larger than the lattice constant a. We have

$$r_{1,\varepsilon} = \frac{4\pi\varepsilon_0\hbar^2}{me^2}\frac{m}{m_n}\varepsilon \approx 0.53\frac{m}{m_n}\varepsilon \quad \text{Å}. \tag{G. 62}$$

The charge cloud of the fifth shell electron comprises a crystal volume of the order of $(4\pi/3)r_{1,\varepsilon}^3$. This volume contains $(8/a^3)(4\pi r_{1,\varepsilon}^3/3)$ atoms, that is, according to Table 22, about 10^3 atoms (the unit cell of the diamond structure contains 8 atoms, cf. Table 2). Such a crystal volume may be treated as approximately macroscopic.

When trivalent boron is incorporated in the silicon lattice, one of the four electron pair bonds remains incomplete. The neutral defect may be taken as a negative ion which binds a defect electron (a positively charged hole) by Coulomb attraction. Just as in the case of a pentavalent defect, the hole is not localized in the immediate neighbourhood of the defect as, for instance, in the incomplete bond between defect and neighbour atom. This gap is filled by a valence electron from a neighbouring bond so that a hole appears in the initially complete neighbouring bond. Through a supply of a small energy ΔE_A the hole is withdrawn from the range of attraction of the negative ion and becomes "freely" mobile. The work necessary to remove the hole is [by analogy to eqn. (G. 61)] given by

$$\Delta E_A = \frac{1}{2}\,\frac{e^4 m}{(4\pi\varepsilon_0\hbar)^2}\,\frac{m_p}{m}\,\frac{1}{\varepsilon^2} \approx 13.6\,\frac{m_p}{m}\,\frac{1}{\varepsilon^2}\ \text{eV}. \qquad \text{(G. 63)}$$

This energy differs from ΔE_D only by the mean effective mass m_p which, as to its magnitude, is to be attributed to an electron from the top of the valence band.

Thus the influence of a trivalent impurity atom on the semiconductor properties is due to the capture of a valence electron by a small energy supply, which is equivalent to the excitation of a hole to the valence band. Impurity atoms which accept electrons are generally called acceptors.

On the basis of the hydrogen model the ionization energies ΔE_D and ΔE_A are independent of the nature of the impurity atoms. Actually, however, there exist different types of donors and acceptors which are caused by different impurity atoms as well as structural imperfections, and whose ionization energies have characteristic values (cf. Table 22).

1. BAND SCHEME

Impurity atoms may supply conduction electrons or holes. The energies necessary for this are generally much smaller than the energy gap between valence and conduction bands. The energy spectrum of the electrons thus contains levels within the energy gap which are due to impurities. This can be shown strictly for simple cases by substituting a perturbed potential into the Schrödinger equation instead of the exact periodic potential, and determining the corresponding eigenvalues.

The presence of impurity atoms requires an extension of the scheme of the energy spectrum by the introduction of a position coordinate. The energy scale (cf. Fig. 99) is replaced by the so-called band scheme $E(\boldsymbol{r})$ (cf. Fig. 144) which must not be confused with the band structure $E_n(\boldsymbol{k})$. In an intrinsic semiconductor the electrons are nowhere localized so that their energy is independent of the position and the energy bands lie horizontal. In an impurity semiconductor, however, carriers may be bound to defects. The levels due to the presence of defects lie in the energy gap; their spatial position is given by the position of the defect, that is, the impurity atom.

The band scheme is a suitable means of describing the energy conditions in a crystal, in particular for the treatment of contact phenomena (cf. pp. 308 ff.).

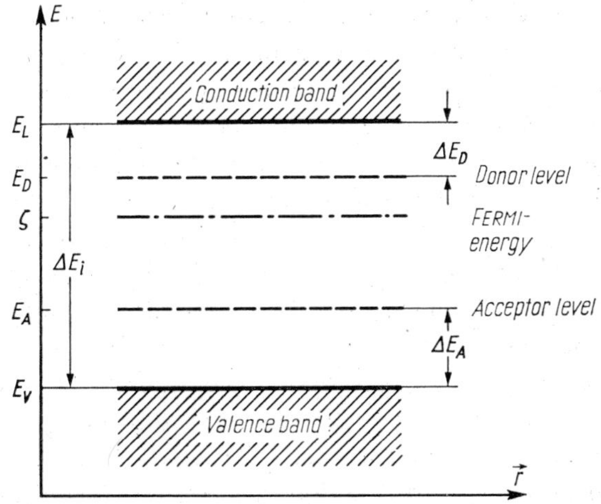

FIG. 144. Band scheme of an extrinsic semiconductor.

The impurity levels are localized discrete levels for not too high impurity concentrations. Above some critical concentration the impurity levels are split and form the so-called impurity bands. This is due to an overlapping of the electron or hole clouds of neighbouring neutral impurity atoms. The critical concentration N_{crit} is reached when the wave functions corresponding to the ground states of the impurity atoms come into contact. A rough estimate for N_{crit} can be obtained from the condition for hydrogenic impurity atoms that the spacing between them is equal to twice the Bohr radius:

$$N_{crit} \approx \frac{1}{(2r_{1,\,\varepsilon})^3}. \qquad \text{(G. 64)}$$

From (G. 62) this yields—depending on the values of m_n/m and ε—for N_{crit} values of the order of magnitude of 10^{18} to 10^{20} cm^{-3}. In the following we shall assume the donor and acceptor concentrations N_D and N_A to be always smaller than the critical values:

$$N_D \ll N_{crit}, \quad N_A \ll N_{crit}. \qquad \text{(G. 65)}$$

2. CARRIER CONCENTRATION IN AN IMPURITY SEMICONDUCTOR

Let us consider an impurity semiconductor which is characterized by the band scheme shown in Fig. 144. This semiconductor contains N_D donors and N_A acceptors per unit volume. The impurity levels caused by them have energies E_D and E_A and lie below the bottom of the conduction band by ΔE_D and above the top of the valence band by ΔE_A, respectively. The impurity atoms are assumed to be partly ionized at the temperature T. The concentrations of the *neutral* donors and acceptors are n_D and n_A. The donors are neutral if occupied by an electron, the acceptors are neutral if unoccupied, that is, if occupied by a hole. We shall determine the position of the Fermi energy and the carrier concentrations n and p in the conduction and valence bands as functions of the temperature.

The occupation probabilities of the impurity levels are generally given not by the Fermi–Dirac function (E. 16) but by a similar function which depends on the type of the model used. In the case of our model of the impurity semiconductor (cf. p. 294) each impurity level can be occupied by only one electron of arbitrary spin orientation. A population of two electrons with antiparallel spins is electrostatically forbidden since it relates to localized energy states. The occupation probability for the donor levels, i.e. the fraction of neutral donors, is given by

$$\frac{n_D}{N_D} = F_D(E_D) = \frac{1}{\frac{1}{2}\exp\left(\frac{E_D-\zeta}{kT}\right)+1} \qquad \text{(G. 66)}$$

and for the fraction of ionized donors we obtain

$$\frac{N_D-n_D}{N_D} = 1-F_D(E_D) = \frac{1}{2\exp\left(\frac{\zeta-E_D}{kT}\right)+1}. \qquad \text{(G. 67)}$$

The occupation probability for the acceptor levels, that is, the fraction of ionized acceptors, is

$$\frac{N_A-n_A}{N_A} = F_A(E_A) = \frac{1}{2\exp\left(\frac{E_A-\zeta}{kT}\right)+1}. \qquad \text{(G. 68)}$$

The occupation probabilities are thus identical with the Fermi–Dirac functions apart from the factors $\frac{1}{2}$ and 2, respectively.

The required carrier concentrations n and p and the impurity site populations n_D and n_A depend on the position of the Fermi energy ζ which results from the neutrality condition of the semiconductor. This reads

$$n+(N_A-n_A) = p+(N_D-n_D)$$

or (G. 69)

$$n-(N_D-n_D) = p-(N_A-n_A).$$

The carrier concentrations n and p are obtained in the same way as in (G. 2) and (G. 6), by integration over the conduction and valence bands

$$n = 2 \int_{E_c}^{\infty} D_n(E)\,F(E)\,dE, \tag{G. 70}$$

$$p = 2 \int_{-\infty}^{E_V} D_p(E)\,[1-F(E)]\,dE. \tag{G. 71}$$

Substitution of these relations and of eqns. (G. 67) and (G. 68) in the neutrality condition yields the basic equation determining the Fermi energy ζ in the case of thermodynamic equilibrium:

$$2 \int_{E_c}^{\infty} D_n(E)\,F(E)\,dE + N_A F_A(E_A) = 2 \int_{-\infty}^{E_V} D_p(E)\,[1-F(E)]\,dE + N_D[1-F_D(E_D)]. \tag{G. 72}$$

This equation applies to a semiconductor with only two types of impurities, that is, impurity levels with two different energies. It can in general not be solved analytically. So we restrict ourselves to a solution under simplifying assumptions, that is, we consider certain definite types of semiconductors.

IV. Types of Semiconductors

According to the impurity concentrations N_D and N_A we distinguish the following types of semiconductors with the corresponding conduction mechanisms:

1. *i*-type semiconductors (*i* stands for *i*ntrinsic) with intrinsic conduction,

$$N_D = N_A = 0 \rightarrow n = p = n_i.$$

2. *n*-type semiconductors (*n* stands for *n*egative, that is, the majority of carriers are electrons) with excess conduction,

$$N_D \neq 0; \qquad N_A = 0 \rightarrow n \gg p.$$

3. *p*-type semiconductors (*p* stands for *p*ositive, that is, the majority of carriers are holes) with deficit conduction,

$$N_A \neq 0; \qquad N_D = 0 \rightarrow p \gg n.$$

4. *c*-type semiconductors (*c* means that the actions of donors and acceptors partly *c*ompensate one another) with mixed conduction,

$$N_D \neq 0; \qquad N_A \neq 0 \rightarrow n \gtreqless p.$$

Actually the first three types can strictly speaking not be realized. As the temperature rises, the *i*-type properties become probable for all types of defect semiconductors (cf. p. 291).

The following classification of impurity semiconductors is also used:

2a. n-type semiconductors with $N_D > N_A$ and $n > p$,
3a. p-type semiconductors with $N_A > N_D$ and $p > n$,
4a. c-type semiconductors with $N_D \approx N_A \neq 0$ and $n \approx p$.

1. n-TYPE SEMICONDUCTORS

An evaluation of the neutrality condition (G. 69) may be based on the following assumptions:

$$N_D \neq 0; \qquad N_A = 0, \tag{G. 73}$$

$$\Delta E_i \gg \Delta E_D, \tag{G. 74}$$

$$n \gg p. \tag{G. 75}$$

The energy gap between valence and conduction bands is so broad and the temperature T is so low that, because of the steep drop of the Fermi–Dirac function, the hole concentration can be neglected (cf. Fig. 145). One can thus

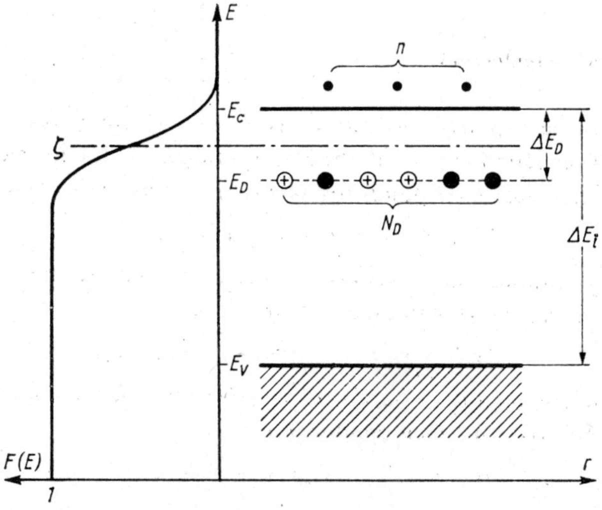

FIG. 145. Band scheme of an n-type semiconductor.

neglect the lower energy bands, including the valence band. Under these conditions the neutrality condition takes the form

$$n = N_D - n_D \tag{G. 76}$$

or, using eqn. (G. 72),

$$2\int_{E_c}^{\infty} D_n(E)\,F(E)\,dE = N_D[1-F_D(E_D)]. \tag{G. 77}$$

Substitution of (G. 10) and (G. 67) yields

$$\frac{1}{2\pi^2}\left(\frac{2m_n}{\hbar^2}\right)^{3/2}\int_{E_c}^{\infty}\frac{(E-E_c)^{1/2}\,dE}{\exp\left(\frac{E-\zeta}{kT}\right)+1} = \frac{N_D}{2\exp\left(\frac{\zeta-E_D}{kT}\right)+1}. \tag{G. 78}$$

With the substitutions (cf. p. 282)

$$\frac{E-E_c}{kT} = x, \qquad \frac{E_c-E_D}{kT} = \frac{\Delta E_D}{kT} = \delta, \tag{G. 79}$$

$$\frac{\zeta-E_c}{kT} = \alpha, \qquad \frac{\zeta-E_D}{kT} = \alpha+\delta \tag{G. 80}$$

we obtain from (G. 78)

$$n = \frac{1}{2\pi^2}\left(\frac{2m_n kT}{\hbar^2}\right)^{3/2} F_{1/2}(\alpha) = \frac{N_D}{2\exp(\alpha+\delta)+1}. \tag{G. 81}$$

From this equation we can always graphically determine the unknown value $\alpha(N_D, \Delta E_D, T)$ which depends on the concentration and the ionization energy of the impurity atoms as well as on the temperature. Knowing α we can determine the Fermi energy ζ and the conduction electron concentration. Using (G. 79) we obtain from (G. 81) the following parameter representation:

$$\frac{n}{N_D} = \frac{1}{2\pi^2}\left(\frac{2m_n \Delta E_D}{\hbar^2}\right)^{3/2}\frac{1}{N_D\,\delta^{3/2}}F_{1/2}(\alpha), \tag{G. 82}$$

$$\frac{n}{N_D} = \frac{1}{2\exp(\alpha+\delta)+1}. \tag{G. 83}$$

When we plot these two equations in the form $\ln(n/N_D)$ versus α (cf. Fig. 146), the points of intersection of the curves of equal parameters δ determine the required values of α and n/N_D. From the course of the curve (G. 83) we see immediately that no points of intersection exist for $n/N_D > 1$. Under our assumptions (G. 73) to (G. 75) the conduction electron concentration above a certain temperature (that is, below a certain value of the parameter δ) becomes practically independent of the temperature and equal to the donor concentration N_D. In this case all donors are ionized and the state is called donor exhaustion. Below this temperature $n < N_D$, that is, one has a reserve of donors.

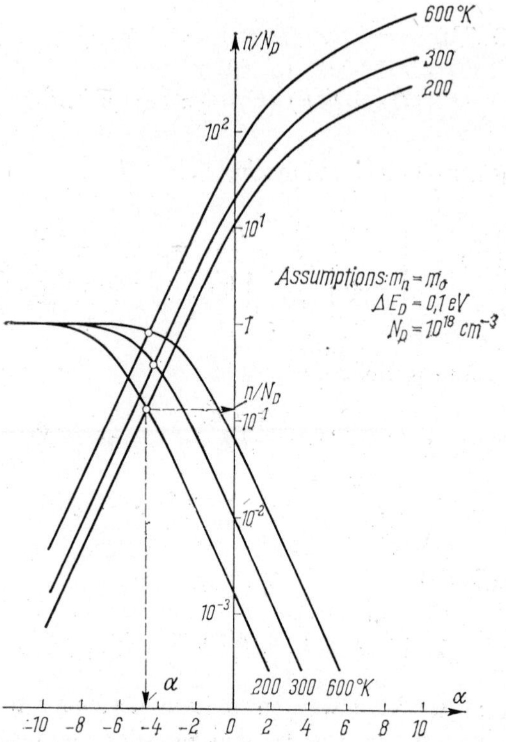

FIG. 146. Graphic determination of the Fermi energy and of the electron concentration of an n-type semiconductor using eqns. (G. 82) and (G. 83).

The same considerations apply to the case of p-type semiconductors. We speak generally of exhaustion and reserve of defects and denote the corresponding temperature ranges as the ranges of saturation and reserve.

Analytically the values of α and n can be calculated from (G. 81) only under the assumption of non-degeneracy for the two cases of reserve and saturation. Non-degeneracy means (cf. p. 287):

$$\frac{E_c-\zeta}{kT} \gg 1 \quad \text{or} \quad \alpha \ll -1. \tag{G. 84}$$

1. The *reserve region* is characterized by the condition

$$\alpha+\delta \gg 1, \tag{G. 85}$$

i.e. because of (G. 84),

$$|\alpha| \ll \delta, \tag{G. 86}$$

or, from eqns. (G. 79) and (G. 80),

$$\frac{E_c-\zeta}{kT} \ll \frac{E_c-E_D}{kT} \tag{G. 87}$$

or

$$\frac{\zeta}{kT} \gg \frac{E_D}{kT}. \tag{G. 88}$$

The Fermi energy ζ should lie so far above the donor term (cf. Fig. 145) that the fraction of ionized donors (G. 67) is very small. Taking into account eqns. (G. 84) and (G. 85) yields with (G. 17) and (G. 19) instead of (G. 81)

$$n = n_0 \exp(\alpha) = \tfrac{1}{2} N_D \exp[-(\alpha+\delta)] \tag{G. 89}$$

or

$$\exp(2\alpha) = \frac{1}{2}\frac{N_D}{n_0}\exp(-\delta). \tag{G. 90}$$

Hence we obtain

$$n = n_0 \exp(\alpha) = \left(\frac{1}{2} n_0 N_D\right)^{1/2} \exp\left(-\frac{\delta}{2}\right) \tag{G. 91}$$

or, using (G. 19) and (G. 79),

$$n = \left(\frac{m_n kT}{2\pi\hbar^2}\right)^{3/4} N_D^{1/2} \exp\left(-\frac{\Delta E_D}{2kT}\right). \tag{G. 92}$$

This relation, which, just as (G. 34), was derived by Wilson, is applicable only for electron concentrations far below the degeneracy concentration and far below the donor concentration, that is, if

$$n \ll n_{\text{crit}} \quad \text{and} \quad n \ll N_D. \tag{G. 93}$$

The curve of $\ln(n/T^{3/4})$ versus $1/T$ is a straight line whose slope is proportional to the distance between donor level and conduction band.

The position of the Fermi energy ζ is thus obtained by solving (G. 91) with respect to α:

$$\alpha = -\frac{\delta}{2}+\frac{1}{2}\ln\left(\frac{N_D}{2n_0}\right). \tag{G. 94}$$

Substitution of (G. 79) and (G. 80) yields

$$\zeta = \frac{E_c+E_D}{2}+\frac{1}{2}kT\ln\left(\frac{N_D}{2n_0(T)}\right). \tag{G. 95}$$

At $T = 0°K$ the Fermi energy lies in the middle between the bottom of the conduction band and the donor level [cf. eqn. (G. 30)]. The temperature dependence of ζ depends on the donor concentration and the effective mass m_n contained in the expression for the effective density of states, n_0. According to whether $[N_D/2n_0(T)] \gtrless 1$ the Fermi energy ζ increases or decreases with the temperature. Equation (G. 95) holds only if the simplified neutrality condition (G. 76) is satisfied.

2. For the *saturation region* we have the condition

$$\alpha+\delta \ll -1, \tag{G. 96}$$

that is, because of the assumption of non-degeneracy (G. 84)

$$|\alpha| \gg \delta \tag{G. 97}$$

or, using eqns. (G. 79) and (G. 80),

$$\frac{E_c-\zeta}{kT} \gg \frac{E_c-E_D}{kT} \tag{G. 98}$$

or

$$\frac{\zeta}{kT} \ll \frac{E_D}{kT}. \tag{G. 99}$$

In this case the Fermi energy lies far below the donor term so that, according to (G. 67), almost all donors are ionized. Because of (G. 96) the right-hand side of (G. 83) can be expanded and we obtained instead of (G. 89)

$$n = n_0 \exp(\alpha) = N_D[1-2\exp(\alpha+\delta)]. \tag{G. 100}$$

Hence follows from (G. 79) and (G. 80)

$$n = \frac{N_D}{1+2\dfrac{N_D}{n_0}\exp\left(\dfrac{\Delta E_D}{kT}\right)}. \tag{G. 101}$$

At high temperatures the exponential term in the denominator becomes small compared with unity and the electron concentration reaches the temperature-independent maximum value N_D:

$$n \approx N_D\left[1-2\frac{N_D}{n_0}\exp\left(\frac{\Delta E_D}{kT}\right)\right] \approx N_D. \tag{G. 102}$$

Combining eqns. (G. 100) and (G. 101) we obtain for the Fermi energy in the case of donor exhaustion

$$\exp(\alpha) = \frac{n}{n_0} = \frac{N_D}{n_0}\,\frac{1}{1+2\dfrac{N_D}{n_0}\exp\left(\dfrac{\Delta E_D}{kT}\right)} \tag{G. 103}$$

or, in the high-temperature approximation,

$$\zeta = E_c - kT \ln \left(\frac{n_0}{N_D}\right). \tag{G. 104}$$

The distance between the Fermi energy and the bottom of the conduction band increases linearly with the temperature.

For an n-type semiconductor ($N_A = 0$) the Fermi energy ζ is shown in Fig. 147 as a function of the temperature. In the reserve region the Fermi energy increases with temperature as long as the donor concentration N_D exceeds the temperature-dependent state density $n_0(T)$, cf. eqn. (G. 95). As soon as with rising temperature the state density becomes higher than the donor concentration, the Fermi energy drops and, at sufficiently high temperatures, the $\zeta(T)$ curve passes over to the saturation region (G. 104) and then to the intrinsic conduction region (G. 31).

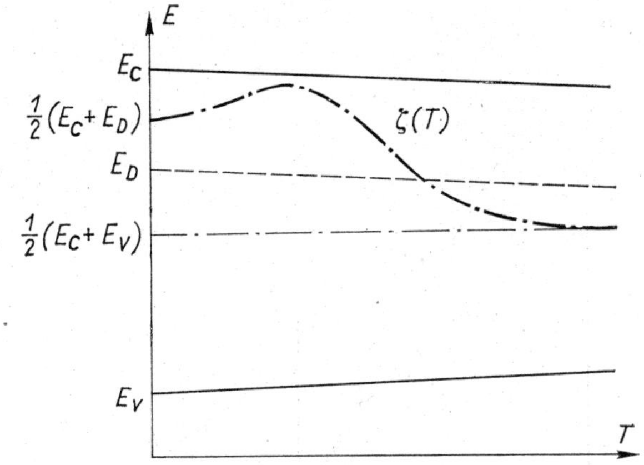

FIG. 147. Plot of the Fermi energy versus temperature for an n-type semiconductor.

2. INVERSION DENSITY

For each type of semiconductor the carrier concentrations n and p are given by eqns. (G. 70) and (G. 71). Under the assumption of non-degeneracy of both concentrations, that is, if

$$\alpha \ll -1 \quad \text{and} \quad \beta \ll -1 \tag{G. 105}$$

we have in the case of thermodynamic equilibrium

$$n = n_0 \exp(\alpha), \tag{G. 106}$$

$$p = p_0 \exp(\beta). \tag{G. 107}$$

Hence follows

$$np = n_0 p_0 \exp(\alpha+\beta) = n_0 p_0 \exp\left(-\frac{\Delta E_i}{kT}\right) \qquad \text{(G. 108)}$$

or, using eqns. (G. 33) and (G. 34),

$$np = n_i^2 . \qquad \text{(G. 109)}$$

Under the condition of non-degeneracy, in thermodynamic equilibrium, the product of electron and hole concentrations is for all semiconductors equal to the square of the inversion density n_i at the temperature T. Thus the electron and hole concentrations are not mutually independent. The electron concen-

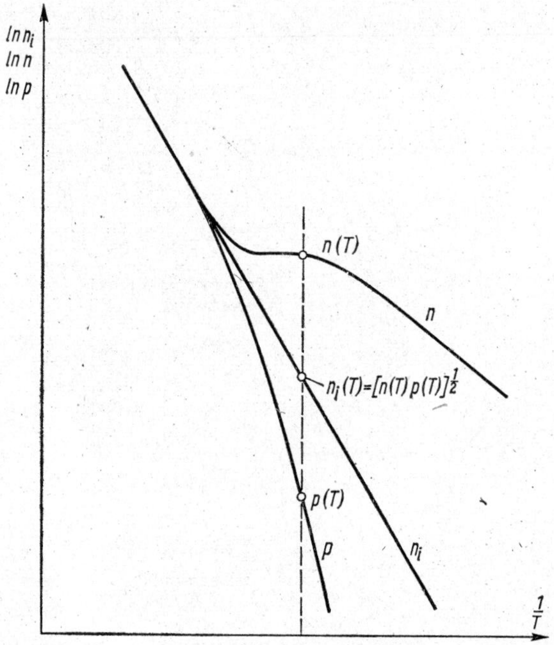

FIG. 148. Temperature dependence of the carrier concentrations in thermodynamic equilibrium.

tration may only grow at the expense of the hole concentration and vice versa. In the case of thermodynamic equilibrium we always have (cf. Fig. 148)

$$n \geqslant n_i \quad \text{and} \quad p \leqslant p_i$$

or (G. 110)

$$n < n_i \quad \text{and} \quad p > p_i .$$

The term inversion density can be explained from the fact that the conduction type undergoes an inversion (cf. p. 300 the cases 2a ⇄ 3a) from excess conduction to deficit conduction and vice versa as soon as the carrier concentration becomes higher or smaller than the inversion density.

The carriers are called majority or minority carriers, respectively, if their concentrations are higher or lower than the inversion density.

References

ADLER, R. B., SMITH, A. C., and LONGINI, R. L., *Introduction to Semiconductor Physics* (Wiley, New York 1964).

AIGRAIN, P., and BALKANSKI, P. (Editors), *Constantes sélectionées relatives aux semiconducteurs* (Pergamon, Oxford 1961).

ANSELM, A. I., *Einführung in die Halbleitertheorie* (Akademie-Verlag, Berlin 1964).

BEAM, W. R., *Electronics of Solids* (McGraw-Hill, New York 1965).

BLAKEMORE, J. S., *Semiconductor Statistics* (Pergamon, Oxford 1962).

HANNAY, N. B. (Editor), *Semiconductors* (Reinhold, New York 1959).

JOFFÉ, A. F., *Physik der Halbleiter* (Akademie-Verlag, Berlin 1958).

MADELUNG, O., *Halbleiter*, in *Handbuch der Physik*, Vol. 20 (Springer, Berlin 1957).

MADELUNG, O., *Grundlagen der Halbleiterphysik* (Springer, Berlin 1970 [Heidelberger Taschenbücher Vol. 71]).

MCKELVEY, J. P., *Solid State and Semiconductor Physics* (Harper and Row, New York 1966).

MOLL, J. L., *Physics of Semiconductors* (McGraw-Hill, New York 1964).

MÜSER, H. A., *Einführung in die Halbleiterphysik* (Steinkopff, Darmstadt 1960.)

RHODES, R. G., *Imperfections and Active Centers in Semiconductors* (Pergamon, Oxford 1964).

SHOCKLEY, W., *Electrons and Holes in Semiconductors* (Van Nostrand, Princeton 1950).

SMITH, R. A., *Semiconductors* (Cambridge University Press, London 1959).

SPENKE, E., *Elektronische Halbleiter*, 2nd ed. (Springer, Berlin 1965); also *Electronic Semiconductors* (McGraw-Hill, New York 1958).

Series

Halbleiterprobleme, SCHOTTKY, W., from Vol. 5 SAUTER, F. (Editor), continued as *Festkörperprobleme*, SAUTER, F., from Vol. 6 MADELUNG, O. (Editor) (Vieweg, Braunschweig 1954 ff.).

International Series of Monographs on Semiconductors HENISCH, H. K. (Editor) (Pergamon, Oxford 1961 ff.).

Progress in Semiconductors, GIBSON, A. F., AIGRAIN, P. and BURGESS, R. E. (Editors) (Heywood, London 1956 ff.).

Semiconductor Monographs, HOGARTH, C. A. (Editor) (Butterworth, London 1959 ff.).

Semiconductors and Semimetals, WILLARDSON, R. K., and BEER, A. C. (Editors) (Academic Press, New York 1966 ff.).

H. Contact Effects

Two substances are in electrical contact if charge carriers can pass through the interface of them. Contacts are:

1. Boundary surfaces in a crystal; for example, two-dimensional defects, grain boundaries (cf. p. 111), sudden transitions between different dopings in one and the same semiconductor ("homo-junctions" such as *pn* junctions, cf. pp. 334 ff.).
2. Interfaces between two or several substances; for example, between metal and vacuum, metal and semiconductor, or different semiconductors ("hetero-junctions"), between metal, insulator, and semiconductor ("MOS" junctions – *m*etal, *o*xide, *s*emiconductor).

The action of most electronic circuit and control elements is based on charge transport through contacts. This transport depends on the behaviour of the carrier potential energy which, in the neighbourhood of the contact, is position-dependent. One often speaks of the potential curve; the potential energy of electrons or holes is obtained from it by multiplication by $-e$ or $+e$. The basis for the explanation of a contact phenomenon is given by the band scheme in the environment of the boundary surface in the case of thermodynamic equilibrium. The potential curve will then depend only on the carrier concentrations on either side of the interface and the work functions (see below) of the substances in contact.

I. Thermodynamic Equilibrium

Thermodynamic equilibrium occurs in a crystal or between several crystals 1, 2, . . . if the Fermi energy in the system considered has everywhere the same value, so that ζ in the band scheme is horizontal:

$$\zeta_1 = \zeta_2 = \ldots = \text{const.} \qquad \text{(H. 1)}$$

All $\zeta_1, \zeta_2, \ldots$ must be reckoned from the same zero of the energy scale. Condition (H. 1) means, for instance, that in a crystal the net current of charge carriers of a given energy must be zero in all directions. This is possible only if all states of the same energy have the same occupation probabilities $f(E)$, since otherwise there would exist a net particle flux in the direction of higher occupation probability.

Condition (H. 1) follows in general from thermodynamic considerations: consider two systems of particles (e.g. electrons) which are brought into contact at a temperature T. Before contact the constant volumina V_1 and V_2 contain N_1 and N_2 particles, respectively. After establishment of the contact, the particle numbers in both volumina have changed until thermodynamic equilibrium between the two volumina has been established. The condition for this process is that the free energy F of the whole system reaches a minimum value. From

$$F = F_1 + F_2 \tag{H. 2}$$

and

$$(\delta F)_{T,\,V,\,N,} = 0 \tag{H. 3}$$

we obtain

$$\delta F = \left(\frac{\partial F_1}{\partial N_1}\right)_{T,\,V_1} \delta N_1 + \left(\frac{\partial F_2}{\partial N_2}\right)_{T,\,V_2} \delta N_2 = 0. \tag{H. 4}$$

With the additional condition of a constant total number of particles

$$N_1 + N_2 = N = \text{const.} \tag{H. 5}$$

we find

$$\delta N_1 + \delta N_2 = 0. \tag{H. 6}$$

Using this we obtain from (H. 4)

$$\left(\frac{\partial F_1}{\partial N_1}\right)_{T,\,V_1} = \left(\frac{\partial F_2}{\partial N_2}\right)_{T,\,V_2}. \tag{H. 7}$$

It can be confirmed from the derivation of the Fermi–Dirac function $f(E)$ that the chemical potential (i.e. the variation of the free energy with the particle number at constant temperature and constant volume) is equal to the Fermi energy ζ:

$$\left(\frac{\partial F}{\partial N}\right)_{T,\,V} \equiv \zeta. \tag{H. 8}$$

Therefore eqn. (H. 7) can be written in the form

$$\zeta_1 = \zeta_2. \tag{H. 9}$$

In this way we have shown that the Fermi energy is horizontal also in the band scheme of a contact in thermal equilibrium.

The Fermi energy is determined with respect to the energy bands by the carrier concentration on the one hand (cf. p. 298) and, on the other hand, with respect to the vacuum potential by the work function. In general the work function for metals (cf. p. 174) as well as for semiconductors is defined as the energy difference between the Fermi level ζ and the potential energy E_∞ of an electron *in vacuo* at an infinite distance from the crystal (vacuum potential). This definition of the work function applies also to semiconductors whose Fermi level is generally unoccupied, since, for the high energies necessary for emission, the electron distribution function can be replaced by a Maxwell–Boltzmann distribution. The electron gas then behaves like a Maxwell gas of concentration n_0 [cf. eqn. (G. 19)] and potential energy ζ.

1. VOLTA POTENTIAL

Considering the band model of a contact, we choose the vacuum potential or an energy fixed to it as the zero of the energy scale. When two crystals with different work functions come into contact, the Fermi levels must get established at the same position. The carrier concentrations in the interior of the crystals are not influenced by the contact, that is, the position of the Fermi levels relative to the energy bands remains unchanged. The adjusting of the Fermi levels to the same height is thus only possible if the band schemes of the materials in contact are shifted with respect to one another. This means that

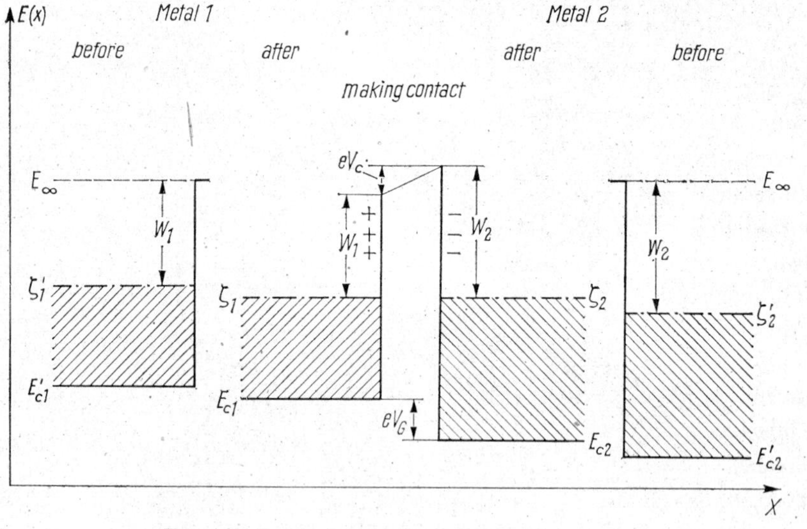

FIG. 149. Band scheme of a metal–metal contact.

in thermodynamic equilibrium a potential difference will arise between the contacting materials, whose value is equal to the difference of the work functions (cf. Fig. 149). The potential difference V_c is called the contact potential or Volta potential:

$$|V_c| = \frac{1}{e}|W_1 - W_2|. \tag{H. 10}$$

The contact potential causes an electric field whose spatial distribution depends on the carrier concentrations of the materials in contact.

In the following we shall describe the formation and the measurement of a contact potential through the example of a metal–metal contact (cf. Fig. 149). The contact is assumed to be made between the two metals 1 and 2, with the electron concentrations n_1 and n_2 and the work functions W_1 and W_2. The zero of the energy scale lies below the vacuum potential by an amount of

$$\zeta_1' + W_1 = \zeta_2' + W_2, \tag{H. 11}$$

where ζ_1', ζ_2' are the Fermi energies of the metals 1 and 2 before the contact has been made.

By analogy to (E. 24) we can neglect the temperature dependence of the Fermi energy and give the electron concentrations as

$$n_1 = \frac{1}{3\pi^2}\left[\frac{2m}{\hbar^2}(\zeta_1' - E_{c1}')\right]^{3/2}, \tag{H. 12}$$

$$n_2 = \frac{1}{3\pi^2}\left[\frac{2m}{\hbar^2}(\zeta_2' - E_{c2}')\right]^{3/2}. \tag{H. 13}$$

E_{c1}' and E_{c2}' are the energies of the bottoms of the conduction bands before contact (on p. 162 this energy was taken as the zero of the energy scale).

We assume both metals to be at the same temperature T and $W_1 < W_2$. When these two metals are brought from infinity to a small mutual distance, an exchange of electrons will set in because of the thermionic emission which always exists at $T > 0°K$; since $W_1 < W_2$, at first more electrons pass over from metal 1 to metal 2. This results in an excess of negative carriers on metal 2, an electric field is built up and a potential difference arises between the two metals. This potential difference is established at such a value that the net current between the metals is vanishing; in thermodynamic equilibrium it reaches the value V_c of the contact potential. Since no potential differences may exist in a metal, the energies of all electrons of the metals 1 and 2 are reduced by $\zeta_1' - \zeta_1$ and increased by $\zeta_2 - \zeta_2'$, respectively, and we have

$$\zeta_1' - \zeta_1 + \zeta_2 - \zeta_2' = eV_c; \tag{H. 14}$$

ζ_1 and ζ_2 are the Fermi energies of the metals 1 and 2 after the contact has been made. From this we obtain from (H. 1) and (H. 11)

$$\zeta_1' - \zeta_2' = W_2 - W_1 = eV_c. \qquad \text{(H. 15)}$$

Thus, after the contact has been made, the band schemes of the initially separated metals are mutually displaced by eV_c (cf. Fig. 149). Since physically only energy differences are significant, it is common practice to take the band scheme of one of the contact materials as fixed on the common energy scale (e.g. $\zeta_1' = \zeta_1$) so that the scheme of the other is shifted by the full amount of the contact potential (e.g. $\zeta_2 - \zeta_2' = eV_c$).

The carrier exchange between the two metals which causes the contact potential involves so few particles that the concentrations n_1 and n_2 are almost uninfluenced by the contact. Thus we can write in a good approximation, according to (H. 12) and (H. 13),

$$\zeta_1' - E_{c1}' = \zeta_1 - E_{c1} \qquad \text{(H. 16)}$$

and

$$\zeta_2' - E_{c2}' = \zeta_2 - E_{c2}. \qquad \text{(H. 17)}$$

Taking (H. 1) into account and subtracting the above two equations, we obtain

$$E_{c2} - E_{c1} = (\zeta_1' - E_{c1}') - (\zeta_2' - E_{c2}') = \zeta_1' - \zeta_2' - (E_{c1}' - E_{c2}') \qquad \text{(H. 18)}$$

or, with (H. 15),

$$eV_G \equiv E_{c2} - E_{c1} = eV_c - (E_{c1}' - E_{c2}'). \qquad \text{(H. 19)}$$

The quantity V_G is called the Galvani voltage. It corresponds to the energy difference between the bottoms of the conduction bands of the contact materials. The Galvani voltage is identical with the contact potential difference only if the electron affinities of the two metals are equal; then $E_{c1}' = E_{c2}'$. The Galvani voltage cannot be measured immediately; it can be determined from eqns. (H. 12), (H. 13), and (H. 18).

In contrast to this, the contact potential can be determined directly.

2. THE KELVIN METHOD FOR THE MEASUREMENT OF THE CONTACT POTENTIAL DIFFERENCE

In principle it should be possible to measure electrostatically the contact potential difference between the two metals 1 and 2, if the two poles of the electrostatic voltmeter consist of the same metals 1 and 2 and each is connected with the same metal. The establishment of thermodynamic equilibrium is then only based on thermionic emission; at temperatures reached in practice, however, this process takes too long a time.

One uses rather a retarding field method. The two metals form the plates of a plate capacitor, which are connected through a galvanometer with a potentiometer (cf. Fig. 150). In this way the thermodynamic equilibrium and thus the contact potential difference can be established very rapidly and independently of the plate separation. As long as the voltage applied is equal to zero, a change of the plate separation gives rise to a change in charge of the capacitor and a current may flow. If an external voltage U is applied to the capacitor,

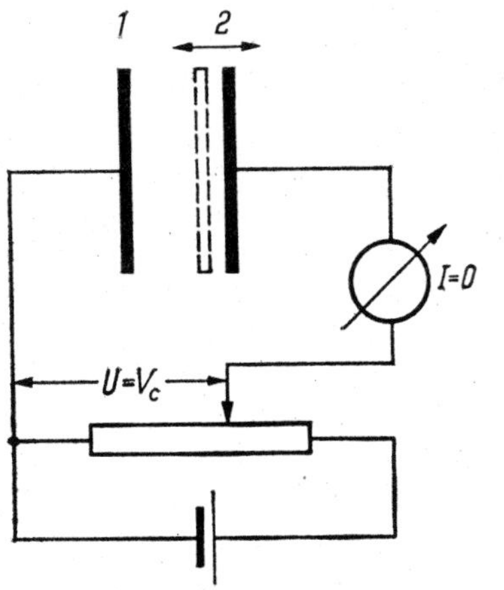

FIG. 150. Measurement of contact potential differences using the compensation method by Kelvin.

which is opposite to the contact potential difference, the charge of the two metals is reduced until it vanishes at the particular value $U = -V_c$. In this case a change of plate separation with zero charge will not give rise to a current flow. In practice the two metals 1 and 2 are used in the form of an oscillating capacitor and the alternating current is compensated by a variation of the direct voltage U. The accuracy with which the contact potential difference can be measured is $\Delta V_c \approx 10^{-4}$ V.

In the band scheme the measurement of the contact potential means the following. If $W_1 < W_2$ in thermodynamic equilibrium the metal 1 is charged positively relative to the metal 2 (cf. Fig. 149). The potential $-V_c$ of metal 2 is therefore lower than that of 1, and accordingly the potential energy eV_c of 2 is higher than that of 1 (the band scheme of 1 is assumed to be fixed, cf. p. 312). If by application of the external voltage U the potential of 2 with respect to 1 is increased, the potential energy of 2 drops by $-eU$. If $U = -V_c$ the metals in contact have the same band schemes as the metals at infinite separation.

II. Metal–Semiconductor Contacts

Contacts involving semiconductors are in principle different from metal–metal contacts since the semiconductor may have a space charge near the boundary surface. By virtue of the Poisson equation of electrostatics

$$\frac{d^2V}{dx^2} = -\frac{1}{\varepsilon\varepsilon_0}\varrho(x) \tag{H. 20}$$

(ε is the static dielectric constant of the semiconductor and ε_0 is the vacuum permittivity), a space charge $\varrho(x)$ means the existence of a position-dependent potential and thus a position-dependent variation of the potential energy $-eV(x)$ for the electron ensemble in the semiconductor. Therefore the energy bands are shifted by $-eV(x)$ in the space charge region (the so-called bending of energy bands, band curvature). In thermodynamic equilibrium the Fermi energy remains independent of position, according to (H. 1).

1. CONTACT BETWEEN METALS AND *n*-TYPE SEMICONDUCTORS

Let us consider a contact between a metal with the work function W_m and an *n*-type semiconductor with the work function W_n (cf. Fig. 151). We assume

$$W_m > W_n \tag{H. 21}$$

and

$$W_m - W_n \ll \Delta E_i. \tag{H. 22}$$

The assumption (H. 22) makes it unnecessary to take into account the presence of the valence band. We shall also assume the band scheme of the *insulated* semiconductor to be independent of position; the bands are thus horizontal up to the semiconductor surface, that is, the donor concentration is position-independent and the surface states (cf. p. 327) are neglected.

Thermodynamic equilibrium is established if electrons pass from the semiconductor to the metal until the charge Q produced in this way gives rise to the contact potential difference $eV_c = W_m - W_n$. The positive charge of the semiconductor is constituted by the ionized donors whose electrons represent the negative surface charge on the metal. As the distance between metal and semiconductor is reduced, the charge required by the thermodynamic equilibrium becomes higher and higher and cannot be produced any longer by the donors in the immediate vicinity of the semiconductor surface. For example, for $V_c = 10^{-1}$ V and a separation of 10^{-7} cm, the necessary charge per unit area is $Q \approx 10^{-7}$ A sec/cm^2 $\approx 10^{12}$ unit charges per cm^2. A strongly doped n-type semiconductor may have a donor concentration of about $N_D \approx 10^{18}$ cm^{-3}, that is, in a volume of 1 cm$^2 \times 10^{-6}$ cm at most 10^{12} ionized donors are available as positive charges. From this typical example it follows that the

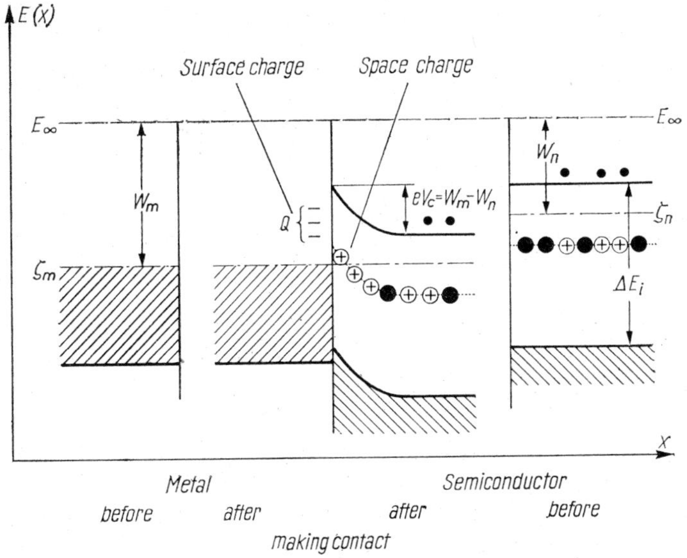

FIG. 151. Band scheme of an n-type semiconductor–metal contact. Band bending in the space charge region.

charge in the semiconductor must be distributed over a depth of at least 10^{-6} cm, that is, as to the order of magnitude, over at least 100 interatomic spacings. As the separation between metal and semiconductor vanishes, the surface charge of the semiconductor spreads to form a space charge.

The electric field penetrates into the semiconductor. The contact potential difference is displaced more and more into the interior of the semiconductor. The energy bands of the semiconductor become bent according to (H. 20). By virtue of the continuity condition of the dielectric displacement density the potential curve has a break at the transition from vacuum ($\varepsilon = 1$) to the semiconductor ($\varepsilon > 1$): the curved potential begins in the semiconductor with an inclined section whose increment is smaller by the factor $1/\varepsilon$ than *in vacuo*. If the surfaces of metal and semiconductor are in contact, the entire contact potential difference lies inside the semiconductor. The energy band curvature due to this fact yields a position-dependent electron concentration $n_s(x)$ in the space charge region, in the so-called boundary layer of the semiconductor. Positive volume charge means that the energies of the states have increased compared with the Fermi level ("upward" band curvature) and the occupation probability has decreased. The boundary layer will then have a deficit of electrons; for this "depletion" layer $n_s < n$. The opposite case of an "accumulation" layer with $n_s > n$ is encountered when a negative space charge raises the occupation probability of the states in the boundary layer ("downward" band curvature).

While the position-dependent carrier concentration in the boundary layer is determined by the two contact substances, the carrier concentration in the inner part of the semiconductor has the position-independent value $n \leq N_D$ given by the condition of neutrality.

2. THE SCHOTTKY BOUNDARY LAYER

In the following we shall calculate the potential curve and the thickness of the so-called Schottky boundary layer which is characterized by a position-independent space charge density ϱ (cf. Fig. 152):

$$\varrho(x) = eN_D = \text{const.} \tag{H. 23}$$

The space charge ϱd per unit area is contained in the boundary layer of thickness d; the value of d is to be calculated. In this layer all donors are ionized and no conduction electrons are present.

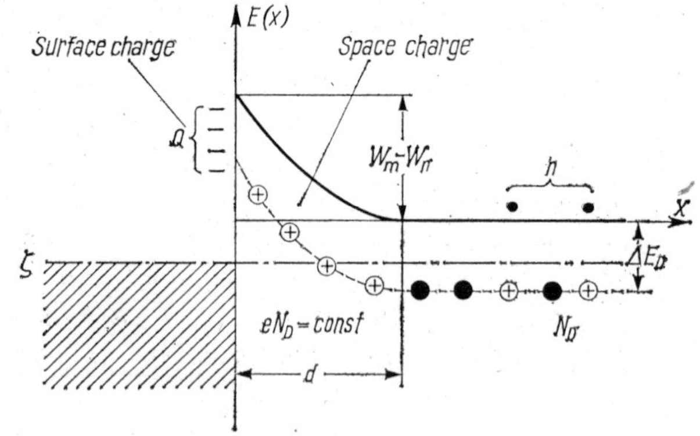

FIG. 152. Schottky layer in an n-type semiconductor.

The potential curve is obtained from a solution of Poisson's equation (H. 20) with the help of (H. 23):

$$\frac{d^2V}{dx^2} = -\frac{1}{\varepsilon\varepsilon_0}\varrho = -\frac{eN_D}{\varepsilon\varepsilon_0} = A. \tag{H. 24}$$

Here x is the position coordinate ($x=0$ denotes the metal–semiconductor interface, $x > 0$ is the interior of the semiconductor) and A is a constant. The solution of (H. 24) is a parabolic potential distribution in the form

$$V(x) = \frac{A}{2}x^2 + Bx + C. \tag{H. 25}$$

The constants B and C are obtained from the boundary conditions. At $x = 0$

$$-eV(0) = W_m - W_n, \tag{H. 26}$$

or

$$C = \frac{1}{e}(W_m - W_n). \tag{H. 27}$$

Then at $x = d$

$$-eV(d) = 0 \tag{H. 28}$$

and

$$\frac{dV}{dx}(d) = Ad + B = 0, \tag{H. 29}$$

that is,

$$B = -Ad. \tag{H. 30}$$

Substitution of (H. 30) in (H. 25) yields

$$V(x) = \frac{A}{2}(d-x)^2, \tag{H. 31}$$

where

$$d^2 = \frac{2C}{A} \tag{H. 32}$$

or, using (H. 24) and (H. 27),

$$V(x) = -\frac{eN_D}{2\varepsilon\varepsilon_0}(d-x)^2 \tag{H. 33}$$

and we have for the required thickness of the boundary layer

$$d = \left[\frac{2\varepsilon\varepsilon_0}{e^2N_D}(W_m - W_n)\right]^{1/2}. \tag{H. 34}$$

The thickness of this layer decreases as the donor concentration, that is, the space charge density, increases. Equation (H. 33), however, is valid only as long as the distance δ_D between the donors is much smaller than the boundary layer d, i.e. if

$$d \gg \delta_D \approx N_D^{-1/3}. \tag{H. 35}$$

For a metal–germanium contact, for example, under the assumption $W_m - W_n = 0.3$ eV, $\varepsilon_{Ge} = 16$ and $N_D = 10^{17}$ cm^{-3}, we obtain a value of $d \approx 7 \times 10^{-6}$ cm. For comparison, we have to point out that the mean distance of the donors is $\delta_D \approx 2 \times 10^{-6}$ cm so that eqn. (H. 35) is barely satisfied. This example shows that the condition must be checked in every case.

If a voltage is applied to the contact (e.g. by changing the potential of the semiconductor by $\pm U$ relative to the metal), the band scheme of the semiconductor is shifted with respect to that of the metal by $\mp eU$ (cf. Fig. 153). The

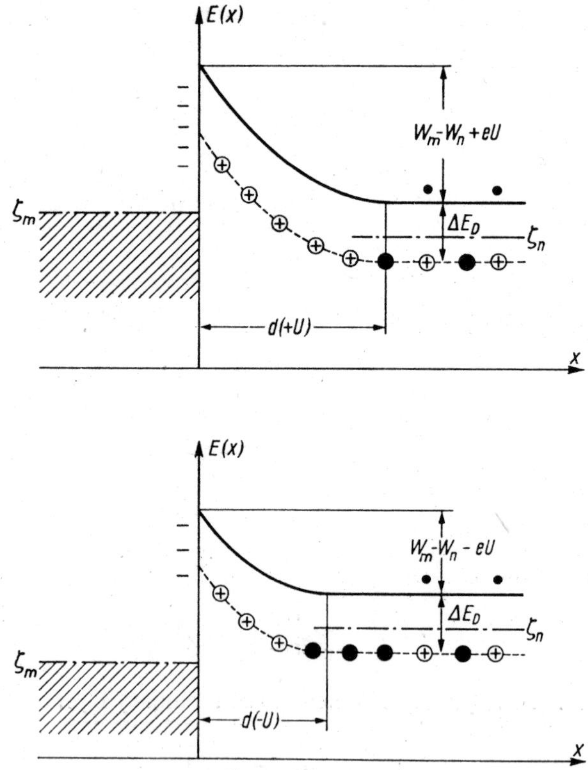

FIG. 153. Schottky layer in an n-type semiconductor with applied voltage $\pm U$.

potential curve at the contact behaves as if the work function of the metal, W_m, had been changed by $\pm eU$. The thickness of the Schottky layer is thus a function of the voltage; we have instead of (H. 34)

$$d(\pm U) = \left[\frac{2\varepsilon\varepsilon_0}{e^2 N_D}(W_m - W_n \pm eU)\right]^{1/2}. \qquad \text{(H. 36)}$$

This means that with constant space charge density ϱ the space charge ϱd, which is equal to the surface charge of the metal, varies with the voltage applied.

Since, according to the assumption (H. 23), the Schottky layer contains only immobile ionized donors and no conduction electrons, it represents an insulating layer between the metal ($x \leqslant 0$) and the semiconductor ($x \geqslant d$). Electrostatically, the contact between a metal and an n-type semiconductor behaves like a capacitor whose capacity per unit area is given by

$$C = \frac{\varepsilon\varepsilon_0}{d}, \qquad \text{(H. 37)}$$

or, using eqn. (H. 36), by

$$C = \left[\frac{\varepsilon\varepsilon_0 e^2 N_D}{2(\Delta W \pm eU)}\right]^{1/2}, \qquad \text{(H. 38)}$$

where

$$\Delta W = W_m - W_n . \qquad \text{(H. 39)}$$

We see that the capacity of the Schottky boundary layer depends on the voltage applied. Measurements of the capacitance plotted in the form of $1/C^2$ versus voltage U yield straight lines and are thus in good agreement with (H. 38). From the slope of this straight line we can determine the donor concentration or the dielectric constant of the semiconductor; the extrapolation $1/C^2 \to 0$ yields the value of ΔW. The values thus obtained for ΔW, however, are generally not in agreement with the contact potential difference $W_m - W_n$ obtained from other measurements, for instance, from determinations of the work functions W_m and W_n. This is due to the existence of surface states (cf. pp. 327 ff.) which also influence the potential curve.

3. CONTACT BETWEEN METALS AND p-TYPE SEMICONDUCTORS

The contact between a metal and a p-type semiconductor with $W_m < W_p$ can be treated in the same way as the contact between a metal and an n-type semiconductor with $W_m > W_n$. In the former case more electrons go over from the metal to the semiconductor. This results in the appearance of a negative

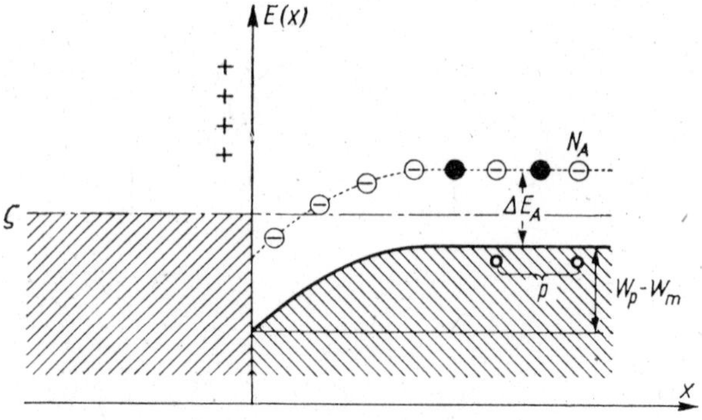

FIG. 154. Depletion layer for holes.

charge of the semiconductor which can be taken up by the holes or acceptors as a negative space charge. According to (H. 20) this gives rise to a downward curvature of the bands, i.e. if $W_p - W_m < \Delta E_i$, a depletion layer for holes will arise (cf. Fig. 154).

4. INJECTING CONTACTS

The assumption of a Schottky layer is justified only if

$$|W_m - W_{n,\,p}| \ll \Delta E_i. \tag{H. 40}$$

In this case the contact can be described by means of a one-band model, that is, we can restrict ourselves to considering the behaviour of a *single* type of carriers under the assumption that all carriers are excited only from impurity levels.

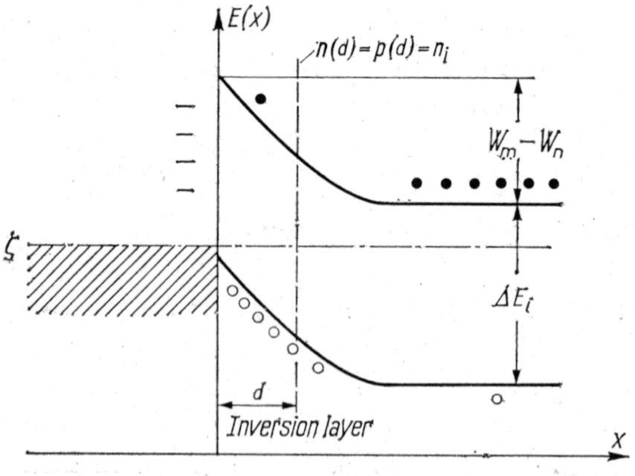

FIG. 155. Hole injection.

If the contact potential difference ΔW becomes comparable with the energy gap ΔE_i, it is qualitatively obvious that the initially neglected energy band must have approached the Fermi level near the surface layer (cf. Fig. 155). This means that the occupation probability for the corresponding other type of carriers can no longer be ignored. For example, the boundary layer near a contact between metal and p-type semiconductor can be of n-type conductivity, and vice versa. This part of the boundary layer is called inversion layer. Contacts with an inversion layer can be used for the injection of carriers: an inversion layer on a p-type semiconductor injects electrons, on an n-type semiconductor it injects holes. Such injecting contacts are basic for numerous semiconductor devices.

5. CONTACT BETWEEN METALS AND INTRINSIC SEMICONDUCTORS

Let us consider a contact which must be described by the two-band model (*both* types of carriers are taken into account) using the example of a metal of work function W_m in contact with an intrinsic semiconductor of work function W_i. We calculate the potential curve at the contact, the electric field at the interface, and the space charge. We assume

$$W_m > W_i \tag{H. 41}$$

and

$$W_m - W_i < \frac{\Delta E_i}{2}. \tag{H. 42}$$

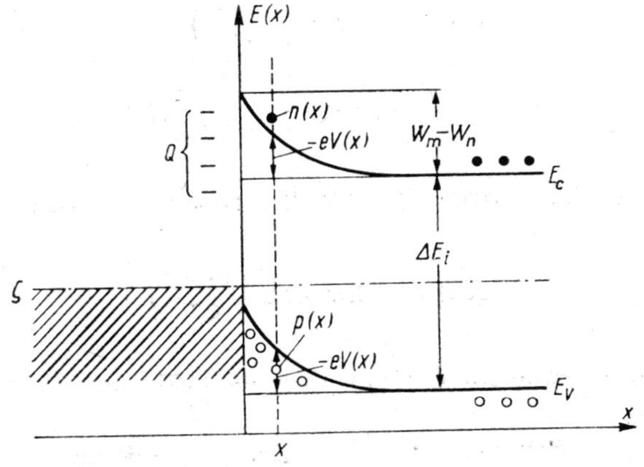

FIG. 156. Band scheme of a contact between a metal and an intrinsic semiconductor.

The assumption (H. 41) means that a positive space charge will arise in the semiconductor; according to (H. 42) the bands should display a weak upward curvature so that the Fermi energy at the interface $x = 0$ remains within the energy gap (cf. Fig. 156). The hole concentration near the interface is thus assumed to be non-degenerate (cf. p. 285).

In contrast to (H. 23) the space charge density depends on the position and is given by the difference of carrier concentrations which are position-dependent near the interface:

$$\varrho(x) = e[p(x) - n(x)]. \tag{H. 43}$$

In the case of non-degeneracy the position-dependent carrier concentrations are given by eqns. (G. 107) and (G. 108):

$$n(x) = n_0 \exp [\alpha(x)] = n_0 \exp \left(\frac{\zeta - E_c(x)}{kT}\right), \tag{H. 44}$$

$$p(x) = p_0 \exp [\beta(x)] = p_0 \exp \left(\frac{E_V(x) - \zeta}{kT}\right). \tag{H. 45}$$

In the space charge region the energy curves for the bottom of the conduction band and the top of the valence band are given by

$$E_c(x) = E_c + (-e)\, V(x), \tag{H. 46}$$
$$E_V(x) = E_V + (-e)\, V(x), \tag{H. 47}$$

where E_c and E_V are the energy values outside the barrier layer *after* the establishment of thermodynamic equilibrium. Using this and eqns. (G. 9), (G. 26) and (G. 27), we can rewrite eqns. (H. 44) and (H. 45) in the form

$$n(x) = n_i \exp \left(\frac{eV(x)}{kT}\right), \tag{H. 48}$$

$$p(x) = n_i \exp \left(-\frac{eV(x)}{kT}\right). \tag{H. 49}$$

Using this we obtain from (H. 43) the position-dependent space charge density

$$\varrho(x) = -2en_i \sinh \left(\frac{eV(x)}{kT}\right). \tag{H. 50}$$

The unknown potential curve $V(x)$ is then obtained from the solution of Poisson's equation:

$$\frac{d^2V}{dx^2} = -\frac{1}{\varepsilon\varepsilon_0}\,\varrho(x) = \frac{2en_i}{\varepsilon\varepsilon_0} \sinh \left(\frac{eV(x)}{kT}\right). \tag{H. 51}$$

The following method enables us to obtain an analytical solution of this differential equation. With the substitutions

$$y = -\frac{eV(x)}{kT} \tag{H. 52}$$

and

$$b = \left(\frac{2e^2 n_i}{\varepsilon\varepsilon_0 kT}\right)^{1/2} \tag{H. 53}$$

eqn. (H. 51) takes the form

$$\frac{d^2y}{dx^2} = -b^2 \sinh(-y) = b^2 \sinh y. \tag{H. 54}$$

Multiplication by dy/dx yields

$$\frac{dy}{dx}\frac{d^2y}{dx^2} = \frac{1}{2}\frac{d}{dx}\left(\frac{dy}{dx}\right)^2 = b^2\frac{dy}{dx}\sinh y \tag{H. 55}$$

and, after integrating with respect to x, we have

$$\left(\frac{dy}{dx}\right)^2 = 2b^2 \int \sinh y\, dy + C_1 \tag{H. 56}$$

or

$$\left(\frac{dy}{dx}\right)^2 = 2b^2 \cosh y + C_1. \tag{H. 57}$$

The integration constant is obtained from the condition that the derivative must vanish if $V = 0$, i.e. if

$$y = 0 \quad \rightarrow \quad \frac{dy}{dx} = 0; \tag{H. 58}$$

we obtain in this way

$$C_1 = -2b^2. \tag{H. 59}$$

Using this eqn. (H. 57) can be written as

$$\left(\frac{dy}{dx}\right)^2 = 2b^2(\cosh y - 1) = 4b^2 \sinh^2\left(\frac{y}{2}\right) \tag{H. 60}$$

or

$$\frac{dy}{dx} = \pm 2b \sinh\left(\frac{y}{2}\right). \tag{H. 61}$$

Hence follows

$$dx = \pm\frac{1}{2b}\,\frac{dy}{\sinh\left(\frac{y}{2}\right)} \tag{H. 62}$$

or

$$x = \pm\frac{1}{b}\ln\left(\tanh\frac{y}{4}\right)+C_2. \tag{H. 63}$$

The constant C_2 is obtained from the boundary condition at the interface $x = 0$,

$$y(0) = \frac{-eV(0)}{kT} = \frac{W_m - W_i}{kT}. \tag{H. 64}$$

Using this relation we obtain

$$C_2 = \mp\frac{1}{b}\ln\left(\tanh\frac{W_m - W_i}{4kT}\right). \tag{H. 65}$$

The choice of the physically reasonable sign is based on the fact that the x scale is given such that only positive x values are defined. According to (H. 53) the quantity b is always positive and we always have

$$-\infty < \frac{1}{b}\ln\left(\tanh\frac{y}{4}\right) < 0. \tag{H. 66}$$

Furthermore,

$$\left|\frac{1}{b}\ln\left(\tanh\frac{y}{4}\right)\right| \geqslant \left|\frac{1}{b}\ln\left(\tanh\frac{y(0)}{4}\right)\right|. \tag{H. 67}$$

Therefore, under the condition $x \geqslant 0$, eqn. (H. 63) takes the form

$$x = -\frac{1}{b}\ln\left(\tanh\frac{y}{4}\right)+\frac{1}{b}\ln\left(\tanh\frac{y(0)}{4}\right). \tag{H. 68}$$

Solution with respect to y yields the required curve for the potential energy at the contact:

$$y = \frac{(-e)\,V(x)}{kT} = 4\ \mathrm{ar}\tanh\left[a\exp\left(-bx\right)\right] \tag{H. 69}$$

where

$$a = \tanh\left(\frac{y(0)}{4}\right) = \tanh\left(\frac{W_m - W_i}{4kT}\right). \tag{H. 70}$$

In the range $x > 0$ the right-hand side of (H. 69) takes only positive values. Thus the change $-eV(x)$ in the potential energy of the electrons is positive, that is, the bands display an upward curvature according to the positive space charge.

Strictly speaking, the position dependence of the potential energy (the curvature of the energy bands) must exist up to $x \to \infty$. However, it can be shown that the potential energy at a distance of $x \approx 1/b$ amounts to only a small fraction $1/\alpha$ of its value at $x = 0$. The quantity $1/b$ (H. 53) is a measure for the effective thickness of the boundary layer, which is proportional to the square root of the dielectric constant and inversely proportional to the square root of the carrier concentration n_i. Let us calculate the depth $x = d_\alpha$ for which the potential energy is given by

$$y(d_\alpha) = \frac{1}{\alpha}\, y(0) \tag{H. 71}$$

where

$$\alpha > 1. \tag{H. 72}$$

Using (H. 68) we obtain

$$d_\alpha = \frac{1}{b} \ln \left\{ \frac{\tanh\left(\frac{y(0)}{4}\right)}{\tanh\left(\frac{y(d_\alpha)}{4}\right)} \right\} \tag{H. 73}$$

and from eqns. (H. 64) and (H. 71)

$$d_\alpha = \frac{1}{b} \ln \left\{ \frac{\tanh\left(\frac{W_m - W_i}{4kT}\right)}{\tanh\left(\frac{W_m - W_i}{4kT}\,\frac{1}{\alpha}\right)} \right\}. \tag{H. 74}$$

For a typical value $W_m - W_i \approx 0.3$ eV and for $T = 300°$K, the quotient $(W_m - W_i)/4kT \approx 3$. For the boundary layer in which the potential drops to $1/\alpha = 1/10$ of its value at $x = 0$ we obtain

$$d_\alpha = \frac{1}{b}\, 1.2 \approx \frac{1}{b}. \tag{H. 75}$$

For instance, we get for $1/b$ in the case of germanium at $T = 300°$K ($\Delta E_i =$ $= 0.75$ eV, $n_i \approx 10^{13}$ cm^{-3} and $\varepsilon = 16$):

$$\frac{1}{b} = \left(\frac{\varepsilon\varepsilon_0 kT}{2e^2 n_i}\right)^{1/2} \approx 1.1 \times 10^{-4}\ \text{cm}. \tag{H. 76}$$

When we differentiate (H. 69) with respect to x, we obtain the distribution of the electric field strength F in the boundary layer as well as the so-called barrier field strength at $x = 0$

$$F(x) = -\frac{dV}{dx} = \frac{kT}{e}\frac{dy}{dx} = -4b\frac{kT}{e}\frac{a\exp(-bx)}{1-a^2\exp(-2bx)}. \qquad \text{(H. 77)}$$

Hence we obtain for the boundary field strength

$$F(0) = -4b\frac{kT}{e}\frac{a}{1-a^2} \qquad \text{(H. 78)}$$

or, using (H. 53) and (H. 70),

$$F(0) = -2\left(\frac{2n_i kT}{\varepsilon\varepsilon_0}\right)^{1/2}\sinh\left(\frac{W_m - W_i}{2kT}\right). \qquad \text{(H. 79)}$$

The total space charge Q is obtained by integrating the space charge density ϱ from $x = 0$ to $x = \infty$. With the help of Poisson's equation (H. 20) we obtain an immediate relation with the boundary field strength $F(0)$:

$$Q = \int_0^\infty \varrho(x)\,dx = -\varepsilon\varepsilon_0 \left.\frac{dV}{dx}\right|_0^\infty = \varepsilon\varepsilon_0 F(0). \qquad \text{(H. 80)}$$

This relation represents a general connection between the surface charge Q of the semiconductor which is equal in magnitude to the space charge in the interior of the semiconductor and the boundary field strength in the semiconductor.

In the present case of a contact between a metal and an intrinsic semiconductor we obtain for the charge at the contact

$$Q = 2(2n_i\varepsilon\varepsilon_0 kT)^{1/2}\sinh\left(\frac{W_m - W_i}{2kT}\right), \qquad \text{(H. 81)}$$

which results from eqns. (H. 79) and (H. 80). In the case of germanium, for example, with the given data we obtain for the surface charge

$$Q \approx 1.2\times 10^{-7}\frac{\text{A sec}}{\text{cm}^2} \approx 10^{12}\frac{\text{elementary charges}}{\text{cm}^2} \qquad \text{(H. 82)}$$

and for the boundary field strength in the interior

$$F(0) = \frac{Q}{\varepsilon\varepsilon_0} \approx 2\times 10^4\frac{\text{V}}{\text{cm}}. \qquad \text{(H. 83)}$$

6. SURFACE STATES

According to what has been said so far we might expect the electrical properties of metal–semiconductor contacts to depend strongly on the relative values of the work functions of the junction materials. It has been observed, however, that, apart from the capacitor properties (cf. p. 319), also the rectifying action of such contacts (cf. pp. 330ff.) is essentially independent of the choice of the metal and thus of the work function W_m. Apart from the considerable experimental difficulties connected with the production of a contact without a "foreign" intermediate layer, this independence has a fundamental cause which results in the action of the so-called surface states or interface states.

The surface of a crystal, as well as any interface inside a crystal (e.g. a grain boundary), represents a perturbation of the regular structure, that is, of the strict periodicity of the crystal, which we assumed to exist in the derivation of the energy band spectrum. The surface atoms have a smaller coordination number than the atoms in the interior of the crystal. Because of the open bonds the atoms near the surface are displaced from the positions determined by the periodicity of the lattice inside the crystal. The unsaturated valence electrons of the surface atoms ("dangling bonds") can easily bind foreign atoms and are the cause of the strong adsorption of impurities on initially atomically pure surfaces.

Tamm was the first to show that the surface boundary of a crystal gives rise to the presence of new states. The corresponding energy values lie in the energy gap between valence and conduction bands. The pertinent wave functions are non-vanishing only near the surface and display an exponential decrease on either side, that is, the states are localized at the surface (surface states).

The surface states for perfectly pure surfaces, the so-called Tamm states, are due to the free, non-saturated valences of the surface atoms. The number of these states per unit area is of the same order of magnitude as the areal density of the surface atoms (about 10^{15} cm^{-2}).

The binding of foreign atoms also gives rise to the appearance of surface states the properties of which may differ essentially from those of the Tamm states.

The crystallographic orientation of the surface as well as the nature of the adsorbed impurity layer determine the number, the energetic position, and the character of the surface states (donor or acceptor states). In the following we shall ignore the origin of the surface states, that is, whether they are due to the finite size of the crystal (Tamm states) or to impurity adsorption. It is essential that the occupation of the surface states is governed by the Fermi statistics and depends on the energetic position of the states with respect to the band edges and the Fermi energy in the interior of the semiconductor. Since the surface states act as donors or acceptors, the carrier concentrations are changed near the surface and thus become position-dependent. The surface will then have a positive or negative charge which is compensated by an equally large negative or positive space charge near the surface. This means that in thermodynamic equilibrium with a horizontal Fermi energy the bands are bent upward or downward. Whereas far in the interior of the semiconductor the carrier concentrations are given by the condition of neutrality, in the surface region they are given by the surface state population. According to the character and the energy distribution of the states there are formed depletion, accumulation or inversion boundary layers.

In order to calculate the potential curve we shall again start from Poisson's equation (H. 20). The space charge density is position-dependent since in the boundary layer the carrier concentrations as well as the occupation of the imperfections depend on the potential curve $V(x)$ which enters into the expression for the occupation probability $f(E)$. The value $V(0)$ of the potential at the surface is not immediately given [as in the case of contacts neglecting surface states; cf. eqns. (H. 26) and (H. 64)], but is established such that the crystal as a whole is neutral. The charge of the surface states must be equal to the space charge in the crystal.

A closed analytical calculation of the potential curve is in general not possible. The results of graphical or numerical evaluations have been compiled in the form of graphs which are contained in the specialist literature.†

The difference between the potential $V(0)$ at the surface ($x = 0$) and the potential $V(\infty)$ far in the interior of the semiconductor ($x \to \infty$) is called the diffusion voltage V_d. The potential energy of the carriers at the surface and far in the interior of the semiconductor differs by an amount of eV_d which is a measure of the band bending.

The surface states give rise to the formation of a boundary layer in the semiconductor even if it is in contact with empty space. Bardeen was the first to take into account the presence of surface states in his theory of semiconductor–metal rectifiers. When passing over from a semiconductor–vacuum or –gas contact to a direct contact between two crystals the number and energetic position of the surface states do not necessarily remain unchanged, that is, the

† Cf. for example R. H. Kingston and S. F. Neustadter, *J. Appl. Phys.* **26**, 718 (1955); R. Seiwatz and M. Green, *J. Appl. Phys.* **29**, 1034 (1958); C. E. Young, *J. Appl. Phys.* **32**, 329 (1961).

spectrum of the surface states is generally different from the spectrum of the interface states, which are likewise a consequence of the perturbed periodicity of the lattice. It is, however, essential that the space charge boundary layer formed in the semiconductor does not merely arise from the contact with a metal. If the number of surface states or interface states is sufficiently high ($> 10^{13}$ cm^{-2}), that is, if the band bending eV_d caused by these states is large compared with the difference of the work functions of the contact materials, the changes of band bending eV_d caused by the contact formation are small. In this case the influence of the metal is screened by the high charge of the surface or interface states. The potential curve and thus the electrical properties of metal–semiconductor junctions may therefore be virtually independent of the work function of the metal.

The relation $eV_d = \Delta W$ between the band bending and the work function difference, which holds for a contact without accounting for interface states [cf. eqn. (H. 39)], is usually not satisfied in practical cases (cf. p. 314). In order to treat contact problems of semiconductor–metal junctions it is expedient to assume the presence of a direct, established contact. Since it is difficult to distinguish between the influences of the metal and the interface states on the

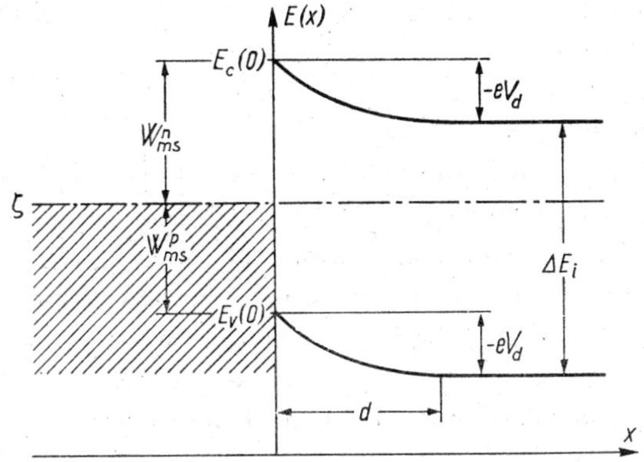

FIG. 157. Band scheme of a metal–semiconductor interface.

space charge boundary layer, new quantities are defined: the work function W^n_{ms} for the transition of electrons and the work function W^p_{ms} for the transition of holes from the metal to the semiconductor. We have (cf. Fig. 157)

$$W^n_{ms} = E_c(0) - \zeta \qquad \text{(H. 84)}$$

and

$$W^p_{ms} = \zeta - E_V(0) \qquad \text{(H. 85)}$$

or, using eqns. (H. 46) and (H. 47),

$$W^n_{ms} = E_c+(-e)V_d-\zeta \tag{H. 86}$$

and

$$W^p_{ms} = \zeta-E_V-(-e)V_d\,. \tag{H. 87}$$

In the case of non-degeneracy, in thermodynamic equilibrium, the electron and hole concentrations at the semiconductor boundary are denoted by n_b and p_b [cf. eqns. (H. 44) and (H. 45)] and given by

$$n_b = n(0) = n_0 \exp\left(-\frac{W^n_{ms}}{kT}\right) \tag{H. 88}$$

and

$$p_b = p(0) = p_0 \exp\left(-\frac{W^p_{ms}}{kT}\right) \tag{H. 89}$$

or, using eqns. (G. 13), (G. 106), (G. 107), (H. 86), and (H. 87),

$$n_b = n \exp\left(\frac{eV_d}{kT}\right) \tag{H. 90}$$

and

$$p_b = p \exp\left(-\frac{eV_d}{kT}\right). \tag{H. 91}$$

The boundary layer is an accumulation layer for electrons or holes if $n_b > n$ or $p_b > p$, and a depletion layer for electrons or holes if $n_b < n$ or $p_b < p$. In the case of inversion layers $n_b < n_i < n$ and $p_b > n_i > p$ (inversion layer of an n-type semiconductor) or $n_b > n_i > n$ and $p_b < n_i < p$ (inversion layer of a p-type semiconductor).

Equations (H. 88) and (H. 89) show that the boundary concentrations are independent of the carrier concentrations far in the interior of the semiconductor and depend only on the characteristic contact parameters W_{ms}. Adding eqns. (H. 86) and (H. 87) we see that the sum of the work functions W^n_{ms} and W^p_{ms} is equal to the energy gap of the semiconductor:

$$W^n_{ms}+W^p_{ms} = E_c-E_V = \Delta E_i\,. \tag{H. 92}$$

7. METAL–SEMICONDUCTOR CONTACT UNDER LOAD

When a voltage is applied to the contact, this results in a change of the potential curve (cf. p. 318) as well as in the flowing of a net current. According to whether the contact layer is enriched with or deprived of carriers, the electrical properties of the contact are different.

The resistance of an accumulation layer is smaller than that of a layer of like thickness in the interior of the semiconductor. The current through the contact is therefore chiefly determined by the properties in the interior of the semiconductor and not by the boundary layer. Contacts with accumulation boundary layers are very important for non-rectifying metal–semiconductor junctions.

In contrast to this, contacts with depletion boundary layers display marked rectifier properties. The current passing through the contact is determined by the boundary layer which is deprived of carriers and possesses a high resistivity. A depletion boundary layer represents an energetic wall (potential barrier) for the corresponding carriers, that is, the potential energy of the carriers in the boundary layer is higher than in the interior of the contact materials. In the following we shall assume the potential barrier to be so high and so deep that tunnelling can be neglected (cf. p. 187). In this case the entire transport of carriers through the contact is only due to thermionic emission over the barrier. The conduction mechanism also depends on whether the carriers in the boundary layer undergo many or few collisions with lattice defects and phonons, that is, whether the mean free path Λ of the carriers is small or large compared with the thickness d of the boundary layer.

Under the assumption of $\Lambda \ll d$ the current–voltage characteristic of the contact is calculated on the basis of diffusion theory. The number of collisions will then be so high that the carrier concentrations in the boundary layer depend on the local value of the potential $V(x)$. This results in the appearance of density gradients and diffusion currents which, together with the field-induced currents ($V(x) \neq$ const. $\rightarrow F \neq 0$), determine the conduction mechanism.

Under the assumption of $\Lambda \gg d$ we can use the so-called diode theory. The carrier transport is in this case determined by the unilateral thermal current densities, which depend on the carrier concentrations far in the interior of the contact substances and which reach the contact surface $x = 0$ between metal and semiconductor owing to thermal excitation.

The rectifier characteristics obtained on the basis of diffusion theory or on that of diode theory differ by a factor which, in the case of diffusion theory, depends slightly on the external voltage U. The relation between current and voltage, which is characteristic for all rectifiers, can be derived most simply with the help of diode theory.

In calculations of the current-voltage characteristics the current per junction area, that is, the current density j, is usually employed; experimental current–voltage characteristics, however, often show immediately the current I through a given rectifier as a function of the external voltage U. When we speak of the current in the following, we mean the current density.

THE DIODE THEORY

Let us consider the contact between a metal and an n-type semiconductor. In thermodynamic equilibrium the currents flowing from the two contact materials towards the junction are equal as to their magnitudes:

$$j_{ms}^0 = j_{sm}^0. \qquad \text{(H. 93)}$$

In the case of $\Lambda \gg d$ they depend only on the height of the potential barrier: by analogy with (E. 90) we have (cf. Fig. 157):

$$j_{ms}^0 = C(T) \exp\left(-\frac{W_{ms}}{kT}\right) \qquad \text{(H. 94)}$$

and

$$j_{sm}^0 = C(T) \exp\left(-\frac{E_c + |eV_d| - \zeta}{kT}\right). \qquad \text{(H. 95)}$$

Because of (H. 86) and (H. 93) the constant $C(T)$ is the same for the two currents.

The application of a voltage U changes the potential of the metal with respect to the grounded semiconductor by $\pm U$ and thus shifts the band scheme of the metal with respect to that of the semiconductor by $\mp eU$ (cf. p. 318). The potential drop over the boundary layer is then not V_d but $V_d \mp U$; the value of W_{ms} remains unchanged. The height of the potential barrier for the semiconductor carriers has thus been changed by $\mp eU$, which strongly increases or reduces the current j_{sm} from the semiconductor:

$$j_{sm} = j_{sm}^0 \exp\left(\pm\frac{eU}{kT}\right). \qquad \text{(H. 96)}$$

Independent of the applied voltage U the metal electrons "see" the same potential barrier W_{ms}; the current j_{ms} from the metal can therefore be considered to be a saturation current, j_s.

$$j_{ms} = j_{ms}^0 = j_s\,. \qquad \text{(H. 97)}$$

The net current j through the contact is obtained from the difference of the currents j_{ms} and j_{sm},

$$j = j_s\left[\exp\left(\pm\frac{eU}{kT}\right) - 1\right]. \qquad \text{(H. 98)}$$

When the positive pole is connected to the metal, the exponent in eqn. (H. 98) is positive and the current j increases exponentially with the voltage U;

the electrons flow in a "forward" direction through the metal–n-type rectifier, from the semiconductor to the metal.

When the metal is connected with the negative pole, the exponent is negative and the electrons flow in a "reverse" direction, from the metal to the semicon-

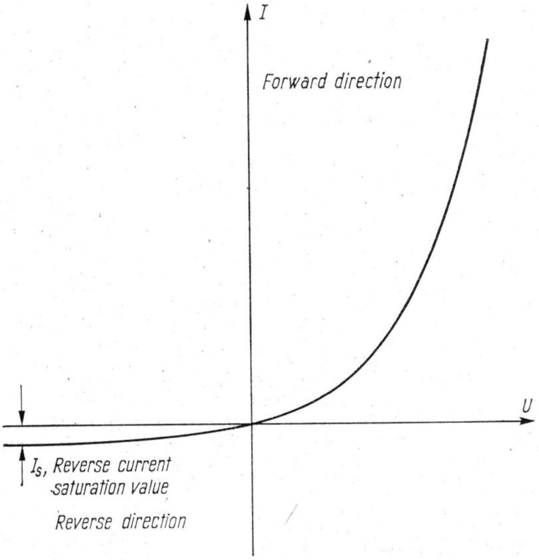

FIG. 158. Rectifier characteristic.

ductor; the reverse current reaches the voltage-independent value j_s for $eU > kT$ (cf. Fig. 158).

For small voltages ($eU < kT$), Ohm's law is valid independently of the polarity, since in this approximation we have

$$j \approx j_s\left[1 \pm \frac{eU}{kT} - 1\right] = \pm j_s \frac{e}{kT} U. \tag{H. 99}$$

The same current–voltage characteristic (H. 98) is obtained for the contact between a metal and a p-type semiconductor. In this case, however, the "forward" and "reverse" directions have been reversed, that is, negative polarity of the metal corresponds to passage (forward direction), positive polarity of the metal to blocking (reverse direction).

The measured current–voltage characteristics of metal–semiconductor contacts are given by (H. 98) in a qualitatively correct form. As to the quantitative results, the measurements agree—apart from a few special cases—neither with the diode theory nor with the diffusion theory. They are often described by an equation of the form

$$j = j_s\left[\exp\left(\pm\frac{eU}{\beta kT}\right) - 1\right], \tag{H. 100}$$

where $\beta > 1$, β being a phenomenological correction factor.

The deviations from eqn. (H. 98) are due to the simplifying assumptions which are necessary for a calculation that avoids substantial mathematical difficulties. These simple theories are based on the assumption of a single type of carriers (one-band model) and ignore the minority carriers as well as the possible appearance of an inversion layer. Moreover, it is uncertain whether the boundary layer is the only factor determining the rectifier properties or whether the diffusion of defects between metal and semiconductor induces an additional *pn* junction (see below) which also influences the charge transport.

III. Semiconductor–Semiconductor Contacts

In the case of a contact between two different semiconductors, we must take into account the interface states just as with the metal–semiconductor contact. This makes the theoretical treatment complex and requires the use of unverifiable assumptions as to the energy distribution of the interface states.

Here we restrict ourselves to the treatment of *pn* junctions within one and the same semiconductor single crystal. With the help of suitable diffusion techniques a single crystal is produced, one part of which is doped with acceptors and the other by donors. Since the junction of these parts lies in one and the same crystal, no interface states arise in the band scheme at the position of the interface. The type of doping alone gives rise to the impurity levels which are different on either side of the interface.

1. THE *pn* JUNCTION

In thermodynamic equilibrium the two differently doped parts of the crystal are separated by a potential step which is produced in the following way. At the interface between *n*-type and *p*-type both carrier concentrations have a strong gradient. Electrons pass over from the *n*-type to the *p*-type part where they recombine with the holes. At the same time holes go over from the *p*-type to the *n*-type section and there they recombine with electrons. This has the consequence that the charge of the ionized, immobile impurities near the interface remains uncompensated and that a negative space charge arises in the *p*-type part and a positive one in the n-type part. These space charges produce an electric field and thus a potential difference (the diffusion voltage V_d) at the pn junction. In thermodynamic equilibrium the diffusion currents due to the concentration gradients are compensated, just as are the field-induced currents of the two types of carriers.

The formation of a potential step means that the band schemes of the p-type and n-type semiconductors (assumed to be initially separated) are shifted

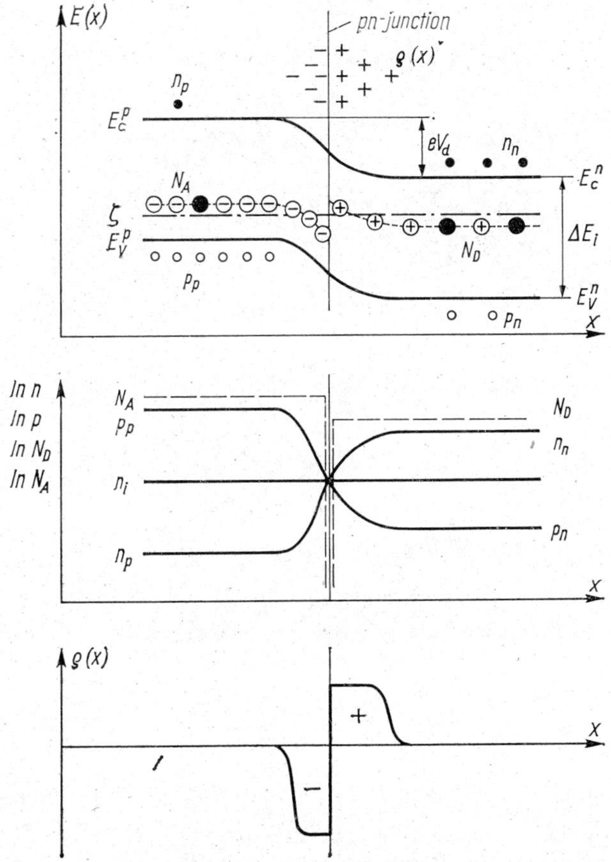

FIG. 159. Band scheme, carrier concentration, impurity concentration, and charge density for a pn junction.

with respect to one another until their Fermi energies have reached the same value ζ (cf. Fig. 159). Owing to the transition of electrons from the n-type to the p-type semiconductor the energy of the former decreases and the energy of the latter increases.

The height eV_d of the potential energy step is equal to the difference of the initial Fermi energies and is related to the carrier concentrations of the n-type and p-type sections outside the space-charge regions: n_n, p_n and n_p, p_p.

In thermodynamic equilibrium and in the case of non-degeneracy we have by virtue of eqns. (G. 106) and (G. 107)

$$n_n = n_0 \exp\left(\frac{\zeta - E_c^n}{kT}\right), \tag{H. 101}$$

$$p_n = p_0 \exp\left(\frac{E_V^n - \zeta}{kT}\right), \tag{H. 102}$$

$$n_p = n_0 \exp\left(\frac{\zeta - E_c^p}{kT}\right), \tag{H. 103}$$

$$p_p = p_0 \exp\left(\frac{E_V^p - \zeta}{kT}\right), \tag{H. 104}$$

where E_c^n, E_c^p, E_V^n, E_V^p are the band edge energies in the n-type and p-type sections outside the space charge regions. Moreover, according to (G. 109) we obtain for the inversion density n_i

$$n_i^2 = n_n p_n = n_p p_p = n_0 p_0 \exp\left(-\frac{\Delta E_i}{kT}\right) \tag{H. 105}$$

where

$$\Delta E_i = E_c^n - E_V^n = E_c^p - E_V^p. \tag{H. 106}$$

Dividing (H. 101) by (H. 103) and (H. 102) by (H. 104), respectively, we obtain for the diffusion voltage V_d the relation

$$eV_d = kT \ln\left(\frac{n_n}{n_p}\right) = kT \ln\left(\frac{p_p}{p_n}\right) = kT \ln\left(\frac{n_n p_p}{n_i^2}\right) \tag{H. 107}$$

where (cf. Fig. 159)

$$eV_d = E_c^p - E_c^n = E_V^p - E_V^n. \tag{H. 108}$$

It is easy to show that in the case of strong doping, that is, if $n_n \gg n_i$ and $p_p \gg n_i$, the diffusion voltage is given by the magnitude of the energy gap. Applying eqns. (H. 101), (H. 104), and (H. 105), we obtain

$$n_0 \exp\left(\frac{\zeta - E_c^n}{kT}\right) \gg (n_0 p_0)^{1/2} \exp\left(-\frac{\Delta E_i}{2kT}\right) \tag{H. 109}$$

and

$$p_0 \exp\left(\frac{E_V^p - \zeta}{kT}\right) \gg (n_0 p_0)^{1/2} \exp\left(-\frac{\Delta E_i}{2kT}\right). \tag{H. 110}$$

The effective state densities n_0 and p_0 are different only because of the different effective masses of electrons and holes and in many cases they are of the same order of magnitude, that is, we have approximately

$$n_0 \approx p_0 \approx (n_0 p_0)^{1/2}, \tag{H. 111}$$

and from eqns. (H. 109) and (H. 110) we obtain

$$E_c^n - \zeta \ll \tfrac{1}{2}\,\Delta E_i \tag{H. 112}$$

and

$$\zeta - E_V^p \ll \tfrac{1}{2}\,\Delta E_i\,. \tag{H. 113}$$

Addition of these two relations yields

$$E_c^n - E_V^p \ll \Delta E_i\,, \tag{H. 114}$$

and a combination of (H. 106) and (H. 108) yields

$$eV_d = \Delta E_i - (E_c^n - E_V^p); \tag{H. 115}$$

with the assumptions $n_n \gg n_i$ and $p_p \gg n_i$, that is, using eqn. (H. 114), we obtain

$$eV_d \approx \Delta E_i\,. \tag{H. 116}$$

While the height V_d of the potential step depends only on the carrier concentrations of the n-type and p-type substances, the potential curve $V(x)$ at the *pn* junction depends on the local distribution of the space charges, that is, on the distribution of impurities $N_A(x)$ and $N_D(x)$ near the boundary surface. These distributions are called impurity profiles. Electrical properties of *pn* junctions adapt themselves to the impurity profiles and thus to the potential curve. Both in metal–semiconductor contacts and in *pn* junctions the carrier transport depends on the width d of the space charge layer as compared with an average mean free path of the charge carriers. In *pn* junctions this mean free path is the so-called diffusion length L of the minority carriers. It is determined by the mean free path traversed by a minority carrier before it recombines with a majority carrier. We distinguish *pn* junctions with high recombination rates ($d \gg L$) from such with low recombination rates ($d \ll L$).

The most important electrical property of a *pn* junction, namely its rectifying action, can be understood without exact knowledge of the impurity profiles and the potential curve.

2. THE *pn* JUNCTION UNDER LOAD

The application of a voltage U at the pn junction changes the height of the potential step, i.e. the diffusion voltage V_d. If the p-type part is positive with respect to the n-type part, the diffusion voltage is reduced to $eV_d - eU$; in the case of inverse polarity it is raised to $eV_d + eU$. The unilateral thermal currents of minority carriers, $j(n_p)$ and $j(p_n)$, flowing towards the interface, are independent of the voltage applied. They are proportional to the equilibrium concentrations n_p and p_n:

$$j(n_p) = C_1 n_p, \tag{H. 117}$$

$$j(p_n) = C_3 p_n. \tag{H. 118}$$

C_1 and C_3 are constants.

The unilateral thermal currents of majority carriers, $j(n_n)$ and $j(p_p)$, are proportional to the fraction of carriers able to surmount the potential step of height $eV_d \pm eU$, i.e.

$$j(n_n) = C_2 n_n \exp\left(-\frac{eV_d \pm eU}{kT}\right), \tag{H. 119}$$

$$j(p_p) = C_4 p_p \exp\left(-\frac{eV_d \pm eU}{kT}\right). \tag{H. 120}$$

From the condition of thermodynamic equilibrium ($U = 0$)

$$j_0(n_n) = j_0(n_p) = j(n_p) \tag{H. 121}$$

and

$$j_0(p_p) = j_0(p_n) = j(p_n) \tag{H. 122}$$

we obtain from (H. 107)

$$C_1 = C_2 \tag{H. 123}$$

and

$$C_3 = C_4. \tag{H. 124}$$

The net current through the pn junction is then obtained as the sum of the electron current and the hole current:

$$j = j(n_n) - j(n_p) + j(p_p) - j(p_n). \tag{H. 125}$$

Substitution of eqns. (H. 117) and (H. 118) as well as (H. 119) and (H. 120) yields with (H. 123) and (H. 124) the current–voltage characteristic of the pn junction

$$j = [j(n_p) + j(p_n)]\left[\exp\left(\pm\frac{eU}{kT}\right) - 1\right]. \tag{H. 126}$$

In this way we have obtained a typical rectifier characteristic, just as in the case of metal–semiconductor contacts. Reduction of the diffusion voltage, that is, application of a negative potential to the n-type section, corresponds to the forward direction. The magnitude of the saturation current densities $j(n_p)$ and $j(p_n)$ depends on the potential curve at the pn junction and on the recombination mechanism.

Measurements of the current–voltage characteristics of pn junctions are essentially in agreement with the theoretical relation (H. 126). In many cases, as with the metal–semiconductor contacts [cf. eqn. (H. 100)] a correction factor $\beta > 1$ must be added to the denominator of the exponent.

If very high voltages are applied in the reverse direction of the rectifier (n-type section positive, negative sign of the exponent), eqn. (H. 126) loses its validity. Above the so-called breakdown voltage the saturation current grows rapidly, that is, the number of carriers passing through the junction increases strongly. There are two mechanisms which are responsible for this effect:

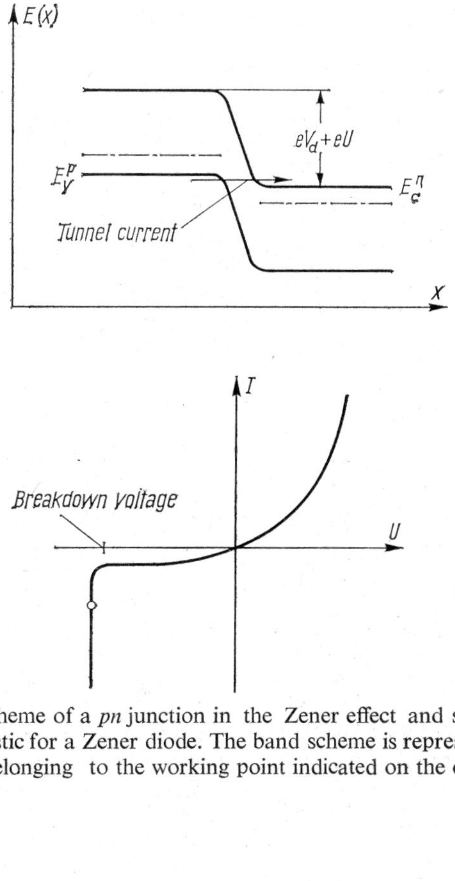

FIG. 160. Band scheme of a pn junction in the Zener effect and schematic current–voltage characteristic for a Zener diode. The band scheme is represented for a blocking voltage belonging to the working point indicated on the characteristic.

1. *Production of secondary electrons by impact ionization* (*carrier multiplication*). With increasing blocking voltage the electric field strength at the junction may reach very high values. The carriers are then strongly accelerated in these high fields so that their energies are sufficient for an excitation of additional electrons from the valence band. The holes and electrons supplied in this way contribute to the reverse current.

2. *Internal field emission* (*Zener effect*). When there is a reverse voltage at sufficiently high electric field strengths in the semiconductor the energy bands in the band scheme are so strongly inclined that there exists an energy range ($E_V^p - E_c^n$; see Fig. 160) which is common to both the valence and the conduction bands (this case must not be confused with the case of band overlapping, cf. p. 209). This energy range of the valence band is spatially separated from that of the conduction band by an energetically forbidden region. If the spatial separation between valence and conduction bands is sufficiently small, that is, if under the action of high electric fields the bands have a sufficient inclination, tunnelling between valence and conduction bands becomes probable. Zener was the first to calculate the corresponding tunnelling probability. At high reverse voltages the conditions for internal field emission can be satisfied in the case of narrow *pn* junctions (*pn* junctions with steep impurity profiles) so that electrons from the valence band of the *p*-type section may tunnel to the *n*-type conduction band. This will also cause a strong increase of the reverse current with increasing voltage.

3. TUNNEL DIODES (ZENER AND ESAKI DIODES)

The strong increase of the reverse current above the breakdown voltage (see Fig. 160) is fundamental for various interesting device applications (stabilizations, reference voltage sources, digital techniques, etc.). Such *pn* junctions which are operated below and above the breakdown voltage are called Zener diodes although their action is based not only on the Zener effect but also on the carrier multiplication by impact ionization.

According to Esaki the charge transport in extremely high-doped *pn* junctions is determined by the tunnel effect for both polarities, even at very low voltages. The mechanism of operation of the Esaki diodes is immediately understood from the band scheme (cf. Fig. 161).

The *p*-type and *n*-type doping must be so high that the Fermi energy lies within the valence band of the *p*-type part as well as within the conduction band of the *n*-type part. The carrier concentrations are thus strongly degenerate

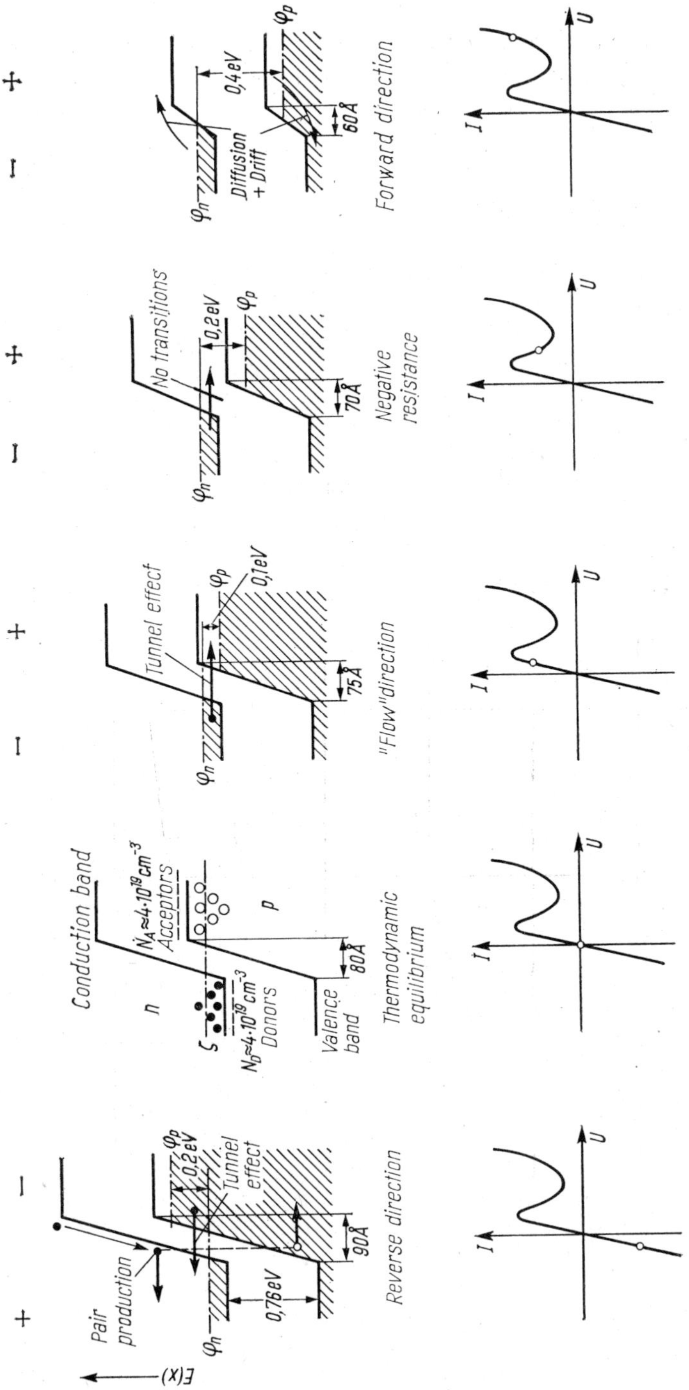

FIG. 161. Mechanism of operation of the Esaki diode (the numbers in the figure represent typical values for a germanium diode).

(cf. p. 285). In this case, even for $U = 0$ the valence band of the p-type part and the conduction band of the n-type part have a common range of energy. If, moreover, the impurity profiles are sufficiently steep, the forbidden region at the pn junction will be so narrow that the tunnelling probability is high. In thermodynamic equilibrium the tunnel currents between the valence band of the p-type part and the conduction band of the n-type part are of equal magnitude.

When a voltage is applied in the reverse direction (positive n-type section) a net tunnel current from the p-type valence band to the n-type conduction band will result, which grows strongly as the voltage is raised. Thus the resistance of the pn junction is small even with a polarity corresponding to the reverse direction.

Also for small voltages in the forward direction (negative n-type section) the charge transport is primarily determined by the tunnel effect. Electrons from the conduction band of the n-type part tunnel to the holes of the valence band of the p-type part. This tunnel current grows with increasing voltage until the fastest electrons from the n-type section have reached the energy of the top of the valence band of the p-type section. The tunnel current then drops again

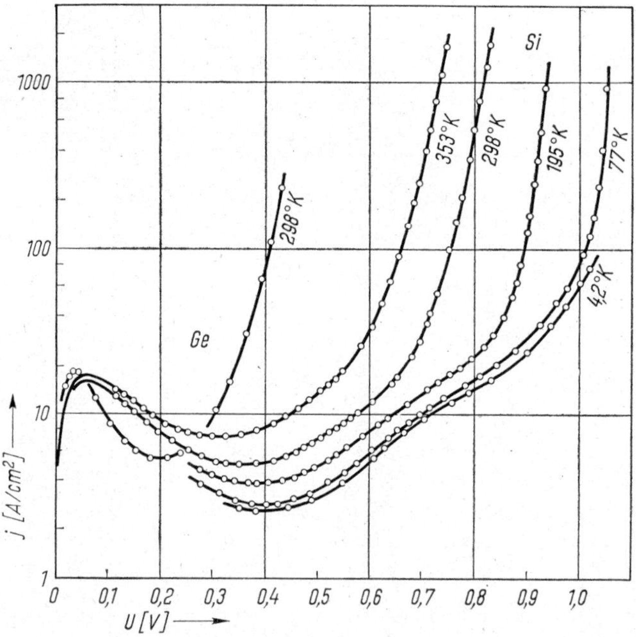

FIG. 162. Current–voltage characteristics in the forward direction for germanium and silicon tunnel diodes for various temperatures [after L. Esaki and Y. Miyahara, *Solid State Electronics* **1,** 13 (1960)].

with further rising voltage since, because of the relative displacement of the energy bands, the electrons from the n-type sections find less and less vacant energy states in the p-type section. The tunnel current is zero as soon as the bottom of the conduction band of the n-type section has reached the same energy as the top of the valence band of the p-type section. The electrons are then confronted with an infinitely thick potential hill (potential step) which can only be surmounted by means of thermal excitation. The forward current is then given by the diffusion mechanism (cf. p. 338).

The current–voltage characteristic in the forward direction has in this case a section of negative inclination (cf. Fig. 162) which makes the Esaki diode particularly important for circuit technical applications (e.g. for oscillators of extremely high frequencies).

4. RADIATIVE TRANSITIONS NEAR *pn* JUNCTIONS

Semiconductor contacts are not only important because of their electrical properties, they also play an important role in the conversion of radiative energy into electrical energy and vice versa. Semiconductor junctions which convert radiative energy to electrical energy are called photocells, photoelectric diodes, or solar cells. Semiconductor junctions converting electrical energy to radiation are called light diodes or luminescent diodes; under certain conditions light diodes may act as injection lasers (cf. pp. 345 ff.). Both cases of energy conversion through *pn* junctions will be treated in the following sections.

(a) *PHOTOEFFECT IN pn JUNCTIONS*

If in a homogeneous semiconductor exposed to light or some other radiation electrons are excited from the valence band to the conduction band (generation of electron–hole pairs), the carrier concentrations and thus the conductivity of the semiconductor will grow (photoconduction). The concentration growth is counteracted by increased recombination and ambipolar diffusion, that is, by diffusion of electron–hole pairs without charge transport.

If, however, electron–hole pairs are produced in the region of a *pn* junction, they are separated from one another by the electric field of the space charge region. This gives rise to a charge transport (an electric current) through the *pn* junction.

Assume the excess concentrations due to light absorption to be given by Δn and Δp; if the wavelength of the light corresponds to the energy gap of the semiconductor,

$$\Delta n = \Delta p. \qquad \text{(H. 127)}$$

In impurity semiconductors the excess concentrations are usually much lower than the concentrations of the majority carriers. For the two sections of a *pn* junction we can therefore write in a good approximation

$$\Delta n \ll n_n, \tag{H. 128}$$

$$\Delta p \ll p_p. \tag{H. 129}$$

Thus the concentrations of majority carriers are virtually unchanged by irradiation of light. The minority carrier concentrations, however, reach the values $n_p+\Delta n$ and $p_n+\Delta p$. Under the assumptions of eqns. (H. 128) and (H. 129) the current–voltage characteristic of a *pn* junction under suitable irradiation is obtained analogously to the case derived on p. 338. The minority carrier currents are increased by irradiation; instead of eqns. (H. 117) and (H. 118) we have

$$j_{h\nu}(n_p) = j(n_p)+j(\Delta n) = C_1(n_p+\Delta n) \tag{H. 130}$$

and

$$j_{h\nu}(p_n) = j(p_n)+j(\Delta p) = C_3(p_n+\Delta p). \tag{H. 131}$$

Because of eqns. (H. 128) and (H. 129) the majority carrier currents remain unchanged, that is,

$$j_{h\nu}(n_n) \approx j(n_n) \tag{H. 132}$$

and

$$j_{h\nu}(p_p) \approx j(p_p). \tag{H. 133}$$

In the case of thermodynamic equilibrium, i.e. if $U = 0$ and $h\nu = 0$, eqns. (H. 121) and (H. 122) are again valid. Instead of (H. 125) we obtain for the total net current from (H. 132) and (H. 133)

$$j = j(n_n)-j_{h\nu}(n_p)+j(p_p)-j_{h\nu}(p_n) \tag{H. 134}$$

or, using eqns. (H. 130) and (H. 131),

$$j = j(n_n)-j(n_p)+j(p_p)-j(p_n)-j_{h\nu} \tag{H. 135}$$

where

$$j_{h\nu} = j(\Delta n)+j(\Delta p). \tag{H. 136}$$

Hence it follows by analogy with (H. 126) that

$$j = j_s\left[\exp\left(\pm\frac{eU}{kT}\right)-1\right]-j_{h\nu} \tag{H. 137}$$

with the saturation current for $h\nu = 0$:

$$j_s = j(n_p)+j(p_n). \tag{H. 138}$$

In a closed circuit with $U=0$ the photocurrent flows as a short-circuit current in the reverse direction of the *pn* junction. From eqn. (H. 137) we have

$$j = -j_{h\nu}. \tag{H. 139}$$

When a *pn* junction in an open circuit is irradiated an external photoelectric voltage arises as open-circuit voltage; the photodiode acts as a photocell. For $j = 0$ we obtain from eqn. (H. 137)

$$U = \frac{kT}{e} \ln\left(1+\frac{j_{h\nu}}{j_s}\right). \tag{H. 140}$$

For a symmetrical *pn* junction, i.e. if $n_p = p_n$, we obtain from eqns. (H. 117), (H. 127), (H. 130), (H. 136), and (H. 138)

$$U = \frac{kT}{e} \ln\left(1+\frac{\Delta n}{n_{\min}}\right) \tag{H. 141}$$

with the minority carrier concentration

$$n_{\min} = n_p = p_n. \tag{H. 142}$$

The photoelectric voltage causes a reduction of the potential step between the *p*-type and *n*-type sections. In the case of irradiation of the *pn* junction in open circuit the height of the potential step drops to $eV_d - eU$ (this means positive polarity of the *p*-type part for $U > 0$; cf. p. 338). From the band scheme it is easy to see that the photoelectric voltage amounts to at most $U_{\max} = V_d$. The potential step and thus the electric field at the *pn* junction becomes smaller as the radiation intensity is raised. As soon as the field vanishes, the produced electron–hole pairs are no longer separated and the charge variation at the *pn* junction becomes equal to zero. According to eqn. (H. 116) the diffusion voltage at highly doped *pn* junctions, and thus also the maximum photoelectric voltage, is given by the value ΔE_i of the energy gap; the maximum photoelectric voltage is then of the order of 0.1–1 V.

(b) *INJECTION LASERS* (*LASER DIODES*)

When the polarity of a *pn* junction corresponds to the forward direction, additional electrons are injected into the *p*-type section and additional holes are injected into the *n*-type section. If recombination takes place with emission of radiation, electrical energy is converted to radiation in the *pn* junction. Under certain conditions, which will be discussed briefly in the following, *pn* junctions can be used for the production and amplification of coherent and thus monochromatic radiation in the optical range of the spectrum. Apart from spatial and temporal coherence (equality of phase), the light emitted is highly collimated and of high intensity. Devices which may produce light of the above properties are generally called lasers (*l*ight *a*mplification by *s*timulated *e*mission of *r*adiation). Suitable *pn* junctions are called injection lasers or laser diodes. However, not every *pn* junction with forward polarity is a laser diode, even if all recombination processes are accompanied by light emission.

Generally, optical transitions between two energy levels $E_1 > E_2$ mean spontaneous and stimulated emission ($E_1 \rightarrow E_2$) or absorption ($E_2 \rightarrow E_1$). The number of transitions per unit time is in any case proportional to the particle concentration in the initial state. Spontaneous emission is a statistical process as regards the direction, phase, and polarization of the emitted radiation. Absorption and stimulated emission occur under the action of an exciting radiation field of suitable frequency, which, for example, can also be produced by spontaneous emission. The stimulated quanta form a wave of the same phase, direction and polarization as the exciting wave. Because of the finite widths of the two energy levels, however, stimulated emission alone cannot yield coherent radiation.

The following three conditions must be satisfied for a junction to act as a laser:

1. The number of stimulated emissions must exceed the number of absorptions as well as the number of spontaneous emissions. This condition is fulfilled in the case of an "inverted population" of the two energy levels, that is, if the higher energy level E_1 is populated by more particles than the lower level E_2 (the so-called laser condition).

In a semiconductor the optical transitions are mainly band-to-band transitions (conduction band $\rightarrow$ valence band) as well as band-to-impurity or impurity-to-band transitions. When we consider only band-to-band transitions, the laser condition is satisfied if the bottom of the conduction band has a higher electron population than the top of the valence band. The number of electrons with energy E_1 at the bottom of the conduction band is given by the distribution function (cf. p. 226)

$$N_1(E_1)\,dE = 2D_n(E_1)\,F_n(E_1)\,dE. \qquad \text{(H. 143)}$$

Similarly the number of electrons of energy E_2 at the top of the valence band is given by

$$N_2(E_2)\,dE = 2D_p(E_2)\,F_p(E_2)\,dE. \qquad \text{(H. 144)}$$

The functions $F_n(E)$ and $F_p(E)$ give the occupation probabilities for electrons with energies in the conduction band and the valence band, respectively, if the system is not in thermal equilibrium. We have

$$F_n(E) = \frac{1}{\exp\left(\frac{E-\varphi_n}{kT}\right)+1} \qquad \text{(H. 145)}$$

and

$$F_p(E) = \frac{1}{\exp\left(\frac{E-\varphi_p}{kT}\right)+1}. \qquad \text{(H. 146)}$$

In the case of non-equilibrium these two equations are determined by the so-called quasi-Fermi levels for electrons and holes, φ_n and φ_p. The occupation probabilities F_n and F_p are mutually independent and refer only to the valence band and the conduction band, respectively. Hence, in non-equilibrium, the electron and hole concentrations are no longer related through the inversion density (cf. p. 305). Only in thermodynamic equilibrium we have

$$F(E) = F_n(E) = F_p(E) \qquad \text{(H. 147)}$$

and

$$\zeta = \varphi_n = \varphi_p. \qquad \text{(H. 148)}$$

The laser condition reads

$$N_1(E_1) > N_2(E_2). \qquad \text{(H. 149)}$$

Under the assumption of almost equal values $D_n(E_1)$ and $D_p(E_2)$ of the eigenvalue densities [cf. eqns. (G. 5) and (G. 7)], we obtain from (H. 149), with the help of (H. 143) and (H. 144),

$$\varphi_n - \varphi_p > E_1 - E_2, \qquad \text{(H. 150)}$$

that is, for band–band transitions,

$$\varphi_n - \varphi_p > \Delta E_i. \qquad \text{(H. 151)}$$

For injection lasers the laser condition (H. 149) means the following: the maximum electron concentration that can be injected into the p-type section by the application of a voltage in the forward direction is equal to the electron concentration in the n-type section. The necessary voltage is equal to the diffusion voltage, the potential jump between p-type and n-type sections has then been levelled out (see Fig. 163). Far away from the junction, the electron and hole concentrations in the n-type and p-type sections are determined as in thermodynamic equilibrium by a single Fermi energy, φ_n and φ_p, respectively. Because

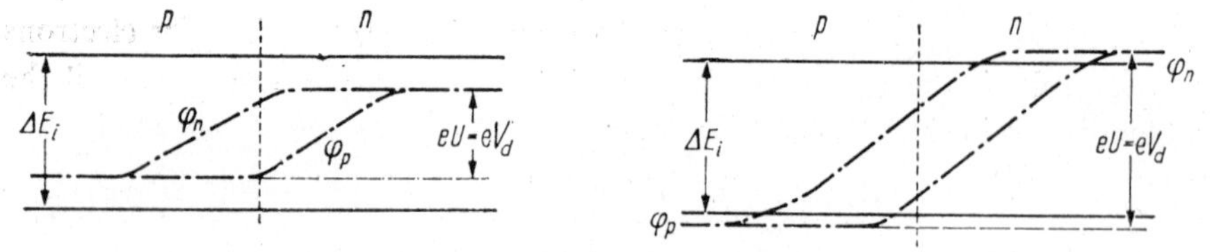

FIG. 163. The laser condition, $\varphi_n - \varphi_p > \Delta E_i$.

of the carrier injection the thermodynamic equilibrium is disturbed near the *pn* junction, that is, on either side of the junction the electron and hole concentrations are mutually independent. The Fermi energies φ_p or φ_n extend to the *n*-type and *p*-type sections, respectively, and increase or decrease according to the local reductions of the excess concentrations to the values φ_n and φ_p, respectively. Figure 163 shows that the laser condition (H. 151) is not satisfied for normally doped *pn* junctions. It can only be fulfilled if at least one side of the *pn* junction is doped so strongly that there the Fermi energy lies within an allowed energy band. For symmetrical *pn* junctions both the *p*-type and the *n*-type sections must be strongly degenerate, i.e. for lasers with band–band transitions only Esaki diodes (cf. p. 340) can be used.

The laser condition is necessary but not sufficient for laser action.

2. The stimulated emission is coherent only if a standing wave can exist in the active material. By choosing a suitable geometry optical feedback to the stimulated emission can be achieved, which means selection of frequency from the broad spectrum of spontaneous emission. For this purpose the crystal is split at right angles to the plane of the *pn* junction crystal in a plane-parallel way. The crystal then represents a Fabry–Pérot interferometer acting as optical cavity with different oscillatory states ("modes"). The condition for standing waves reads

$$q\frac{\lambda}{2} = l, \tag{H. 152}$$

where q is an integer, λ is the wavelength of light in the laser material, and l is the distance between the plane-parallel faces. Using the refractive index $n(\nu)$ of the laser material we can write for the light frequency ν

$$\nu = \frac{c}{\lambda n(\nu)}, \tag{H. 153}$$

where c is the vacuum light velocity. Using this we obtain from eqn. (H. 152)

$$q = \frac{2l}{c}\,\nu n(\nu) \tag{H. 154}$$

or

$$\Delta q = \frac{2l}{c}\left[n(\nu)+\nu\frac{dn}{d\nu}\right]\Delta\nu. \qquad \text{(H. 155)}$$

Hence we obtain for $\Delta q = 1$ the difference $\Delta\nu$ of the frequencies ν of the different oscillatory states of the Fabry–Pérot interferometer,

$$\Delta\nu = \frac{c}{2l}\,\frac{1}{n(\nu)+\nu\dfrac{dn}{d\nu}}. \qquad \text{(H. 156)}$$

The light emitted by the laser diode has the preferred frequency (H. 154), which lies closest to the maximum of the spectral distribution of normal recombination luminescence (cf. Fig. 164). The corresponding oscillation state is called the "laser mode".

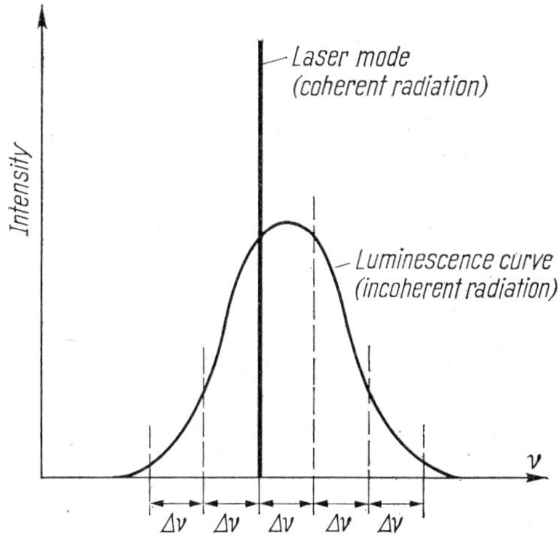

FIG. 164. Laser mode and luminescence curve for a laser diode.

3. Light of the preferred frequency is amplified if the energy loss of the standing wave (mainly due to absorption and non-radiative transitions) is smaller than the energy gain from stimulated emission. The number of light quanta produced per unit time must exceed the number of lost light quanta. This condition is satisfied if the forward current density exceeds a certain threshold. Only then will the semiconductor diode emit laser radiation.

We have assumed that the recombination of the injected carriers is radiative. Radiative transitions are chiefly direct ones. They are most probable if the minimum of the conduction band and the maximum of the valence band pertain to the same wave vector ***k*** (cf. p. 246). This is the case, for example, for III–V compounds such as GaAs, InAs, InP, and so on, which are the principal materials for injection lasers.

References

Contact Phenomena

HARPER, W. R., *Contact and Frictional Electrification* (Oxford University Press, London 1968).
HOLM, R., *Electric Contacts*, 4th ed. (Springer, Berlin 1967).
MYAMLIN, V. A., and PLESKOV, Y. V., *Electrochemistry of Semiconductors* (Plenum, New York 1967).
SPENKE, E., *Elektronische Halbleiter*, 2nd ed. (Springer, Berlin 1965).
SZE, S. M., *Physics of Semiconductor Devices* (Wiley, New York 1969).

Surface Phenomena

FRANKL, D. R., *Electrical Properties of Semiconductor Surfaces* (Pergamon Press, Oxford 1967).
MANY, A., GOLDSTEIN, Y., and GROVER, N. B., *Semiconductor Surfaces* (North-Holland, Amsterdam 1965).

J. Transport Phenomena

I. The Problem

The motion of carriers under the influence of external forces or density gradients gives rise to the so-called transport phenomena. The carriers transport electric charge (→ electric current density, electrical conductivity) as well as energy (→ heat current density, thermal conductivity).

Generally, the problem can be stated as follows. Assume a given inhomogeneous and anisotropic single crystal, in which, in certain directions with respect to the crystallographic orientation, an electric field $\boldsymbol{E}$, a magnetic induction $\boldsymbol{B}$, and a temperature gradient grad $\boldsymbol{T}$ act as exciting quantities. Determine the net electric current density $\boldsymbol{j}$ and the net heat current density $\boldsymbol{w}$.

In the following we shall consider only the contributions of electrons and holes to the transport of charge and heat. Charge transport by ions is mainly important for ionic crystals (cf. pp. 126 ff.). The heat transport through lattice vibrations (phonons) is mainly important for insulators and semiconductors or metals at low temperatures.

The electric current density $\boldsymbol{j}$ is given by the number of carriers passing per unit time through a unit area perpendicular to the current direction:

$$\boldsymbol{j} = e \int_0^\infty \boldsymbol{v}(E)\, N(E)\, dE \tag{J. 1}$$

or, using eqn. (E. 15),

$$\boldsymbol{j} = 2e \int_0^\infty \boldsymbol{v}(E)\, D(E)\, F(E)\, dE; \tag{J. 2}$$

$\boldsymbol{v}(E)$ is the group velocity (cf. p. 215). When we use the wave vector $\boldsymbol{k}$ instead of the energy E as the independent variable, we obtain instead of (J. 2)

$$\boldsymbol{j} = 2e \int_{-\infty}^{+\infty} \boldsymbol{v}(\boldsymbol{k})\, D(\boldsymbol{k})\, F(\boldsymbol{k})\, d^3\boldsymbol{k} \tag{J. 3}$$

where [cf. eqn. (F. 122)]

$$dZ_k = D(\boldsymbol{k})\, d^3\boldsymbol{k} = \left(\frac{L}{2\pi}\right)^3 d^3\boldsymbol{k}; \tag{J. 4}$$

$D(\boldsymbol{k})$ is the density of states in $\boldsymbol{k}$-space and $F(\boldsymbol{k})$ is the occupation probability for a state with wave vector $\boldsymbol{k}$. The heat current density $\boldsymbol{w}$ is given by the number of particles which per unit time transport the energy difference between their total energy and the chemical potential through a unit area. The total energy is the sum of potential and kinetic energy; the chemical potential of an electron is equal to the Fermi energy ζ. The heat current is thus given by

$$w = \int_0^\infty (E-\zeta)\, v(E)\, N(E)\, dE \tag{J. 5}$$

or, using eqn. (E. 15),

$$w = 2\int_0^\infty (E-\zeta)\, v(E)\, D(E)\, F(E)\, dE, \tag{J. 6}$$

where E is the total energy of the particle.

In analogy to eqn. (J. 3) we have from (J. 4) for $L^3 = 1$

$$\boldsymbol{w} = \frac{1}{4\pi^3} \int_{-\infty}^{+\infty} (E(\boldsymbol{k})-\zeta)\, \boldsymbol{v}(\boldsymbol{k})\, F(\boldsymbol{k})\, d^3\boldsymbol{k}. \tag{J. 7}$$

Under the assumption that the distribution function $N(E)$, or $N(\boldsymbol{k})$, is not influenced by external factors (forces, temperature gradients), the particle currents $\boldsymbol{j}$ and $\boldsymbol{w}$ are exactly equal to zero. In this case, since $E(\boldsymbol{k}) = E(-\boldsymbol{k})$ [cf. eqn. (F. 20)], the number $N(\boldsymbol{k})$ of particles with velocity $\boldsymbol{v}(\boldsymbol{k})$ is equal to the number $N(-\boldsymbol{k})$ of particles with velocity $-\boldsymbol{v}(-\boldsymbol{k})$.

II. The Boltzmann Equation

A discussion of the transport phenomena makes it necessary to know a *perturbed* distribution function. We assume that the eigenvalue density $D(E)$, or $D(\boldsymbol{k})$, remains unchanged (compare in contrast to this p. 259). In this case a perturbation of the distribution function is only due to a change of the occupation probabilities $F(E)$ or $F(\boldsymbol{k})$ (cf. Fig. 165):

$$F(E) = F_0(E)+g(E) \tag{J. 8}$$

or

$$F(\boldsymbol{k}) = F_0(\boldsymbol{k})+g(\boldsymbol{k}), \tag{J. 9}$$

where $F_0(E)$ or $F_0(\boldsymbol{k})$ is the unperturbed occupation probability (Fermi–Dirac function) and $g(E)$ or $g(\boldsymbol{k})$ denotes its perturbation.

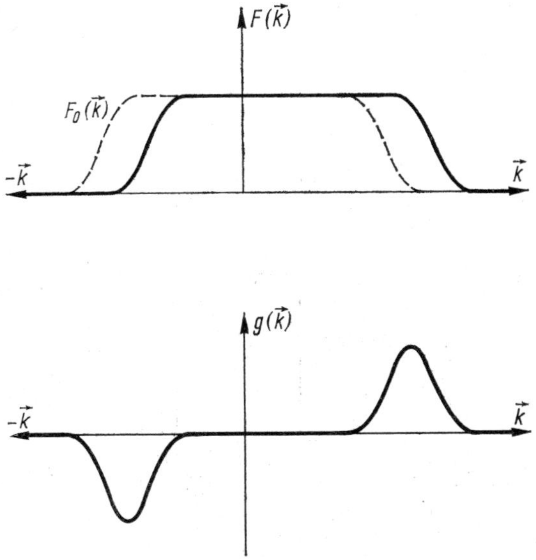

FIG. 165. Perturbation of the occupation probability by a constant force.

The probability of finding an electron with wave vector $\boldsymbol{k}$ in the crystal near the position $\boldsymbol{r}$ at the time t is given by

$$F = F(\boldsymbol{r}, \boldsymbol{k}, t). \tag{J. 10}$$

The position dependence is due to chemical inhomogeneities, that is, to the position dependence of the Fermi energy $\zeta(\boldsymbol{r})$, on the one hand, and temperature gradients, that is, $\zeta(T(\boldsymbol{r}))$, on the other hand. The wave vector dependence is equivalent to an energy dependence. The time dependence is due to the action of external forces which make the wave vector time dependent [cf. eqn. (F. 104)].

If not all energy bands are completely filled, external forces change the mean momentum and thus induce a current (cf. p. 228). The occupation probability then loses its symmetry with respect to the wave vector $\boldsymbol{k}$ since

$$F(\boldsymbol{k}, t+\Delta t) = F(\boldsymbol{k}-\dot{\boldsymbol{k}}\,\Delta t, t) \tag{J. 11}$$

and

$$F(-\boldsymbol{k}, t+\Delta t) = F(-\boldsymbol{k}-\dot{\boldsymbol{k}}\,\Delta t, t) \tag{J. 12}$$

where

$$\boldsymbol{k}(t+\Delta t) = \boldsymbol{k}(t)-\dot{\boldsymbol{k}}\Delta t \tag{J. 13}$$

and

$$\boldsymbol{F} = \hbar\dot{\boldsymbol{k}}, \tag{J. 14}$$

$\boldsymbol{F}$ being the external force [cf. eqn. (F. 104)] which must not be confused with the occupation probability F.

From eqns. (J. 11) and (J. 12) we obtain immediately (cf. Fig. 166)

$$F(\boldsymbol{k}, t+\Delta t) \neq F(-\boldsymbol{k}, t+\Delta t). \tag{J. 15}$$

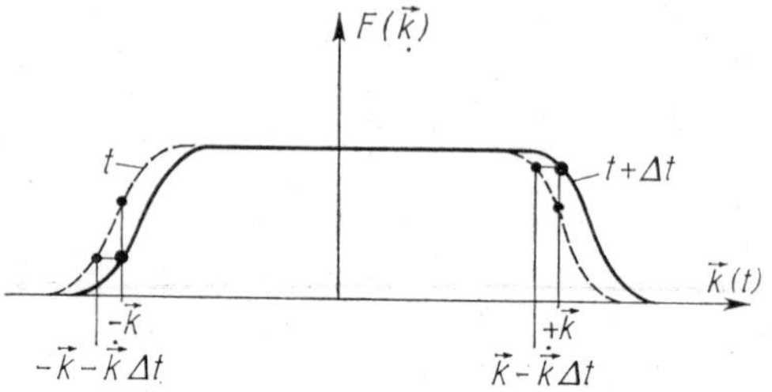

FIG. 166. Variation of the occupation probability with time under the influence of a constant force.

So far we have neglected all interactions of electrons with lattice imperfections (phonons, defects) as well as the electron–electron interactions. Under these assumptions (cf. p. 195) the momentum change caused by external fields is conserved when the fields are switched off. This means that, even without the action of a force, the particle flux would be continued, that is, the resistivity of such an ideal crystal is equal to zero (cf. p. 216).

An explanation of the electrical resistance and the thermal resistance makes it necessary to take into account the interactions of electrons with phonons or lattice imperfections and, in certain cases, with other electrons. All these interactions (collisions with the corresponding interaction partners) also result in a change of the occupation probability.

The variation with time of the occupation probability has two constituents: a component $(\partial F/\partial t)_{\text{drift}}$ describing the translational processes (particle drift and diffusion) and a component $(\partial F/\partial t)_{\text{coll}}$, which is due to collisions with the interaction partners:

$$\frac{dF}{dt} = \left(\frac{\partial F}{\partial t}\right)_{\text{drift}} + \left(\frac{\partial F}{\partial t}\right)_{\text{coll}}. \tag{J. 16}$$

We are only interested in stationary processes, that is, we assume that

$$\frac{dF}{dt} = 0 \tag{J. 17}$$

and neglect switching-on or switching-off processes. Here we must not ignore the difference between stationarity defined by (J. 17) and equilibrium defined by the Fermi–Dirac function

$$F_0(E(\boldsymbol{k})) = \frac{1}{\exp\left(\dfrac{E(\boldsymbol{k})-\zeta}{kT}\right)+1}. \tag{J. 18}$$

A consideration of the transport phenomena makes it necessary to know both components, $(\partial F/\partial t)_{\text{drift}}$ and $(\partial F/\partial t)_{\text{coll}}$, which we shall discuss in the following.

1. *The drift component.* Let at time t the fraction $F(t)$ of the total number of electrons be in the phase-space element $d\Omega$:

$$F(t)\, d\Omega = F(x, y, z; k_x, k_y, k_z; t)\, d\Omega \tag{J. 19}$$

where

$$d\Omega = dx\, dy\, dz\, dk_x\, dk_y\, dk_z. \tag{J. 20}$$

Under the influence of diffusion and the action of forces the occupation probability in coordinate and momentum spaces is changed. At time $t+dt$ the phase-space element $d\Omega$ will contain those electrons which, at the time t, had the coordinates $x-v_x dt$, $y-v_y dt$, $z-v_z dt$ and the wave vector components $k_x-\dot{k}_x dt$, $k_y-\dot{k}_y dt$, $k_z-\dot{k}_z dt$, that is, at time $t+dt$,

$$F(t+dt)\, d\Omega = F(x-v_x dt, \ldots, k_x-\dot{k}_x dt, \ldots, t)\, d\Omega. \tag{J. 21}$$

We thus obtain from eqns. (J. 19) and (J. 21) the time variation of the occupation probability due to external influences:

$$\left(\frac{\partial F}{\partial t}\right)_{\text{drift}} = -\left(\frac{\partial F}{\partial x} v_x + \frac{\partial F}{\partial y} v_y + \frac{\partial F}{\partial z} v_z + \frac{\partial F}{\partial k_x}\dot{k}_x + \frac{\partial F}{\partial k_y}\dot{k}_y + \frac{\partial F}{\partial k_z}\dot{k}_z\right) \tag{J. 22}$$

or

$$\left(\frac{\partial F}{\partial t}\right)_{\text{drift}} = -(\boldsymbol{v}\cdot\text{grad}_r F) - (\dot{\boldsymbol{k}}\cdot\text{grad}_k F). \tag{J. 23}$$

2. *The interaction component.* An accurate calculation of $(\partial F/\partial t)_{\text{coll}}$ is possible only if the individual collision mechanisms are known. It is possible, however, to simplify the problem by introducing a collision probability $W(\boldsymbol{k})$ or a relaxation time $\tau(\boldsymbol{k}) = W^{-1}$ as a phenomenological quantity which characterizes the individual interactions. The relaxation time is defined by the assump-

tion that the perturbation $g(\boldsymbol{k})$ of the occupation probability vanishes exponentially with time after the establishment of equilibrium conditions, that is,

$$g(\boldsymbol{k}, t) = g_0(\boldsymbol{k}) \exp\left(-\frac{t}{\tau(\boldsymbol{k})}\right) \tag{J. 24}$$

or

$$\left(\frac{\partial g}{\partial t}\right)_{\text{coll}} = -\frac{g}{\tau}. \tag{J. 25}$$

It can be shown that in thermal equilibrium the occupation probability of the electron states is not changed by the action of phonons, that is, the Fermi–Dirac distribution of the electrons is in equilibrium with the Bose–Einstein distribution of the phonons (cf. p. 65). Therefore

$$\left(\frac{\partial F_0}{\partial t}\right)_{\text{coll}} = 0 \tag{J. 26}$$

and using (J. 9)

$$\left(\frac{\partial F}{\partial t}\right)_{\text{coll}} = \left(\frac{\partial g}{\partial t}\right)_{\text{coll}} = -\frac{g}{\tau}. \tag{J. 27}$$

Using eqns. (J. 16), (J. 17), (J. 23), and (J. 27), we obtain the steady-state condition (the so-called Boltzmann transport equation)

$$\frac{F-F_0}{\tau} = -(\boldsymbol{v}\cdot\text{grad}_r\, F)-(\dot{\boldsymbol{k}}\cdot\text{grad}_k\, F) \tag{J. 28}$$

or

$$F = F_0-\tau(\boldsymbol{v}\cdot\text{grad}_r\, F)-\tau(\dot{\boldsymbol{k}}\cdot\text{grad}_k\, F). \tag{J. 29}$$

This equation is basic to all transport phenomena. It is very difficult to solve it in this form since both sides of the equation contain the perturbed occupation probability F. In all practical cases, however, the variation of the occupation probability through external influences is very small so that

$$F(\boldsymbol{k}) \approx F_0(\boldsymbol{k}) \gg g(\boldsymbol{k}). \tag{J. 30}$$

An estimate shows that in metals, for example, the electron velocity due to external fields is smaller by a factor of 10^8 than the Fermi velocity at $T = 0°\text{K}$. With a typical value of $\zeta = 5$ eV for the Fermi energy (cf. Table 20) we obtain from the relation $\zeta = mv_F^2/2$ a Fermi velocity of $v_F \approx 10^6$ m/sec. On the other hand, a very high current density of $j = 10^8\ \text{A/m}^2$ corresponds to a superimposed drift velocity of only $v_d \approx 10^{-2}$ m/sec.

Because of (J. 30) we can, to a good approximation, therefore use in (J. 29) the gradients of the unperturbed occupation probability, that is, the Fermi–Dirac function. Thus we have instead of (J. 29)

$$F = F_0 - \tau(\boldsymbol{v}\cdot\text{grad}_r\, F_0) - \tau(\dot{\boldsymbol{k}}\cdot\text{grad}_k\, F_0). \tag{J. 31}$$

This linearized Boltzmann equation gives an explicit expression for the perturbed occupation probability F which is necessary for a calculation of the carrier fluxes [eqns. (J. 3) and (J. 7)].

III. Electrical Conductivity and Thermal Conductivity

We shall first calculate the electric current density for a homogeneous crystal under the assumption of time- and position-independent temperature, that is,

$$F(\boldsymbol{r}, \boldsymbol{k}) = F(\boldsymbol{k}) \tag{J. 32}$$

or

$$\text{grad}_r\, F = \text{grad}_r\, F_0 = 0. \tag{J. 33}$$

The action of an external force on the carriers is assumed to be due only to the electric field strength $\boldsymbol{E}$, i.e.

$$\boldsymbol{F} = e\boldsymbol{E} = \hbar\dot{\boldsymbol{k}}. \tag{J. 34}$$

With these conditions [(J. 33) and (J. 34)] the Boltzmann equation (J. 31) yields the perturbed occupation probability

$$F = F_0 - \frac{e}{\hbar}\tau(\boldsymbol{k})\,(\boldsymbol{E}\cdot\text{grad}_k\, F_0). \tag{J. 35}$$

Hence we obtain from eqns. (F. 100) and (J. 18)

$$F = F_0 - e\tau(\boldsymbol{k})\,(\boldsymbol{v}(\boldsymbol{k})\cdot\boldsymbol{E})\,\frac{\partial F_0}{\partial E}. \tag{J. 36}$$

Substitution in eqn. (J. 3) yields for the electric current density (the volume of the crystal considered is $L^3 = 1$)

$$\boldsymbol{j} = -\frac{2e^2}{(2\pi)^3}\int_{-\infty}^{+\infty} \tau(\boldsymbol{k})\,\boldsymbol{v}(\boldsymbol{k})\,(\boldsymbol{v}(\boldsymbol{k})\cdot\boldsymbol{E})\,\frac{\partial F_0}{\partial E}\, d^3\boldsymbol{k}. \tag{J. 37}$$

The integration over $\boldsymbol{k}$-space can be replaced by means of (F. 125) by an integration over isoenergetic surfaces. We obtain

$$\boldsymbol{j} = -\frac{e^2}{4\pi^3}\int\limits_0^\infty \int\limits_{E=\text{const}} \tau(\boldsymbol{k})\boldsymbol{v}(\boldsymbol{k})(\boldsymbol{v}(\boldsymbol{k})\cdot\boldsymbol{E})\frac{\partial F_0}{\partial E}\,dS\,\frac{dE}{|\text{grad}_k\,E(\boldsymbol{k})|} \tag{J. 38}$$

or, from (F. 100),

$$\boldsymbol{j} = -\frac{e^2}{4\pi^3}\int\limits_0^\infty \int\limits_{E=\text{const}} \tau(\boldsymbol{k})\,\frac{\boldsymbol{v}(\boldsymbol{k})\,(\boldsymbol{v}(\boldsymbol{k})\cdot\boldsymbol{E})}{\hbar\,|\boldsymbol{v}(\boldsymbol{k})|}\,\frac{\partial F_0}{\partial E}\,dS\,dE. \tag{J. 39}$$

This relation holds in general for a homogeneous crystal. Further calculations are based on the use of approximations which are justified for special models (cf. pp. 360 and 366).

The heat current density is macroscopically given by

$$\boldsymbol{w} = -\varkappa\text{ grad }T. \tag{J. 40}$$

In the general case the thermal conductivity $\varkappa$ is a symmetrical tensor of rank two, that is,

$$\varkappa_{ij} = \varkappa_{ji}. \tag{J. 41}$$

The thermal conductivity of a crystal has two components: the contribution from lattice vibrations (phonons) and the contribution from the carriers. Both contributions are additive:

$$\varkappa = \varkappa_{Ph} + \varkappa_c. \tag{J. 42}$$

In the following we shall only consider the heat conduction by carriers.

Because of the temperature dependence of the occupation probability, a temperature gradient will in principle cause a carrier-concentration gradient. This results in a diffusion current which transports both energy and electric charge. Since the thermal conductivity is always measured in an open circuit, the net total electric current is equal to zero. This is only possible if an electric field exists which induces a current opposite to the diffusion current. Thus we have

$$\text{grad }T \neq 0 \tag{J. 43}$$

as well as

$$\boldsymbol{E} \neq 0. \tag{J. 44}$$

The linearized Boltzmann equation (J. 31) will therefore be more complicated than in the calculation of the electrical conductivity at constant temperature:

$$F = F_0 - \tau\left[\left(\boldsymbol{v}\cdot\left\{\frac{\partial F_0}{\partial \zeta}\,\mathrm{grad}_r\,\zeta + \frac{\partial F_0}{\partial T}\,\mathrm{grad}_r\,T\right\}\right) + \frac{e}{\hbar}\left(\boldsymbol{E}\cdot\frac{\partial F_0}{\partial E}\,\mathrm{grad}_k\,E\right)\right]. \quad \text{(J. 45)}$$

Evaluating the appropriate derivatives of the Fermi–Dirac function F_0 we obtain, using eqn. (F. 100),

$$F = F_0 - \tau\left(\boldsymbol{v}\cdot\frac{\partial F_0}{\partial E}\left[e\boldsymbol{E} - \mathrm{grad}_r\,\zeta - \frac{E-\zeta}{T}\,\mathrm{grad}_r\,T\right]\right). \quad \text{(J. 46)}$$

The term with $\mathrm{grad}_r\,\zeta$ takes into account both the chemical inhomogeneities, that is, $\zeta(\boldsymbol{r})$, and the temperature dependence of the Fermi energy, that is, $\zeta(T(\boldsymbol{r}))$. In the case of a chemically uniform crystal we have

$$\mathrm{grad}_r\,\zeta = \frac{\partial \zeta}{\partial T}\,\mathrm{grad}_r\,T. \quad \text{(J. 47)}$$

Substitution of (J. 46) in (J. 3) and (J. 7), using (F. 125) and (J. 4), yields the general transport equations which apply in the presence of an electric field and a temperature gradient:

$$\boldsymbol{j} = -\frac{e}{4\pi^3\hbar}\int_0^\infty \int_{E=\mathrm{const}} \tau\,\frac{\partial F_0}{\partial E}\,\frac{\boldsymbol{v}}{|\boldsymbol{v}|}\left(\boldsymbol{v}\cdot\left[e\boldsymbol{E} - \mathrm{grad}_r\,\zeta - \frac{E-\zeta}{T}\,\mathrm{grad}_r\,T\right]\right) dS\,dE, \quad \text{(J. 48)}$$

$$w = -\frac{1}{4\pi^3\hbar}\int_0^\infty \int_{E=\mathrm{const}} (E-\zeta)\tau\,\frac{\partial F_0}{\partial E}\,\frac{\boldsymbol{v}}{|\boldsymbol{v}|}\left(\boldsymbol{v}\cdot\left[e\boldsymbol{E} - \mathrm{grad}_r\zeta - \frac{E-\zeta}{T}\,\mathrm{grad}_r\,T\right]\right) dS\,dE. \quad \text{(J. 49)}$$

With the substitutions

$$L_s = \frac{1}{4\pi^3\hbar}\int_0^\infty \int_{E=\mathrm{const}} \tau\,\frac{v_i v_j}{|\boldsymbol{v}|}\left(-\frac{\partial F_0}{\partial E}\right) E^s\,dS\,dE \quad \text{(J. 50)}$$

for $s = 0, 1, 2$, the transport equations can be written in the form

$$\boldsymbol{j} = eL_0\left(e\boldsymbol{E} - \mathrm{grad}_r\,\zeta + \frac{\zeta}{T}\,\mathrm{grad}_r\,T\right) - eL_1\,\frac{1}{T}\,\mathrm{grad}_r\,T \quad \text{(J. 51)}$$

and

$$w = (L_1 - \zeta L_0)\left(e\boldsymbol{E} - \operatorname{grad}_r \zeta + \frac{\zeta}{T}\operatorname{grad}_r T\right) - (L_2 - \zeta L_1)\frac{1}{T}\operatorname{grad}_r T. \quad \text{(J. 52)}$$

These equations are basic for a calculation of the thermoelectrical effects (cf. pp. 371ff.). It should be mentioned that in the literature one may find instead of L_s (J. 50) an integral which is defined in a slightly different way: instead of E^s the term $(E-\zeta)^s$ is contained in the integrand. In this case the transport equations (J. 51) and (J. 52) contain coefficients which differ from those given here.†

Let us first consider the thermal conductivity which is measured in the case of zero current. Under the condition that $j = 0$ we obtain a relation between the field strength $\boldsymbol{E}$ and the temperature gradient $\operatorname{grad}_r T$, i.e.

$$L_0\left(e\boldsymbol{E} - \operatorname{grad}_r \zeta + \frac{\zeta}{T}\operatorname{grad}_r T\right) = L_1\frac{1}{T}\operatorname{grad}_r T. \quad \text{(J. 53)}$$

Using this we obtain from eqn. (J. 52) for the heat flux

$$w = \left(\frac{L_1^2}{L_0} - L_2\right)\frac{1}{T}\operatorname{grad}_r T \quad \text{(J. 54)}$$

and, from eqn. (J. 40), for the carrier contribution to the thermal conductivity

$$\varkappa_c = -\frac{1}{T}\left(\frac{L_1^2}{L_0} - L_2\right). \quad \text{(J. 55)}$$

This relation holds generally, for any crystalline solid, as long as, after the establishment of equilibrium, the scattering processes cause an exponential decay of the perturbation of the occupation probability [cf. eqn. (J. 24)]. The integrals L_s can be calculated only under special assumptions.

1. DEGENERATE ONE-BAND MODEL

The transport equations (J. 39), (J. 48), and (J. 49) all contain the energy derivative of the Fermi–Dirac function. This derivative is essentially different from zero only near the Fermi energy (cf. Fig. 167). In the case of metals and strongly degenerate semiconductors, for which particle velocities at $E = \zeta$ are defined, it is therefore possible to expand the factors of $\partial F_0/\partial E$ at $E = \zeta$.

† A comparison of both definitions may be found for example in H. Jones, *Handbuch der Physik*, **19**, 271 (1956).

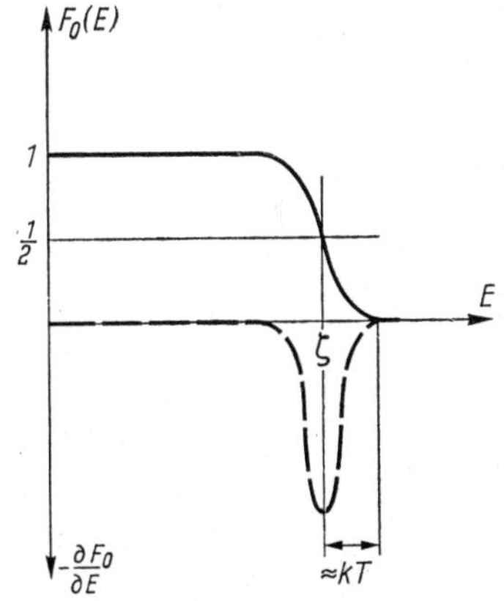

FIG. 167. Fermi–Dirac function and its derivative with respect to energy.

(a) *ELECTRICAL CONDUCTIVITY*

To obtain the electric current density in a homogeneous crystal it is sufficient to place the factor of $\partial F_0/\partial E$ with its value at $E = \zeta$ in front of the integral over the energy; doing this, we can write eqn. (J. 39) as

$$\boldsymbol{j} = -\frac{e^2}{4\pi^3\hbar}\int\limits_{E=\zeta}\tau(\boldsymbol{k})\,\frac{\boldsymbol{v}(\boldsymbol{k})(\boldsymbol{v}(\boldsymbol{k})\cdot\boldsymbol{E})}{v_F}\,dS_F\int\limits_0^\infty\frac{\partial F_0}{\partial E}\,dE, \qquad \text{(J. 56)}$$

where v_F is the Fermi velocity and dS_F is a surface element on the Fermi surface $E = \zeta$. Integration over the energy yields

$$\int\limits_0^\infty\frac{\partial F_0}{\partial E}\,dE = F_0(\infty) - F_0(0) = -1. \qquad \text{(J. 57)}$$

The electric current density in a metal is thus given by an integral over the Fermi surface:

$$\boldsymbol{j} = \frac{e^2}{4\pi^3\hbar}\int\limits_{E=\zeta}\tau(\boldsymbol{k})\,\frac{\boldsymbol{v}(\boldsymbol{k})(\boldsymbol{v}(\boldsymbol{k})\cdot\boldsymbol{E})}{v_F}\,dS_F. \qquad \text{(J. 58)}$$

From a comparison with the macroscopic relation

$$\boldsymbol{j} = \sigma\boldsymbol{E} \qquad \text{(J. 59)}$$

we obtain the conductivity tensor

$$\sigma_{ij} = \frac{1}{4\pi^3}\frac{e^2}{\hbar}\int\limits_{E=\zeta} \tau(\boldsymbol{k})\frac{v_i(\boldsymbol{k})\,v_j(\boldsymbol{k})}{v_F}\,dS_F. \qquad \text{(J. 60)}$$

This equation shows immediately that the conductivity in metals depends only on the electron properties at the Fermi energy. The high conductivity of metals is, strictly speaking, due to the high velocities of a few electrons possessing the Fermi energy, and not to the high concentration of free or quasifree electrons [cf. eqn. (J. 65)].

In crystals with cubic symmetry the conductivity tensor is reduced to a scalar. For $\boldsymbol{E}(E_x, 0, 0)$ we then obtain instead of eqn. (J. 58)

$$j_x = \frac{1}{4\pi^3}\frac{e^2}{\hbar}\int\limits_{E=\zeta} \tau(\boldsymbol{k})\frac{v_x^2}{v_F}\,dS_F E_x. \qquad \text{(J. 61)}$$

For cubic symmetry we obtain

$$\overline{v_x^2} = \tfrac{1}{3}v_F^2 \qquad \text{(J. 62)}$$

and thus instead of (J. 60)

$$\sigma = \frac{1}{12\pi^3}\frac{e^2}{\hbar}\int\limits_{E=\zeta} \tau(\boldsymbol{k})\,v_F\,dS_F. \qquad \text{(J. 63)}$$

In the particular case of spherical energy surfaces we have

$$\sigma = \frac{1}{12\pi^3}\frac{e^2}{\hbar}\tau(k_\zeta)\,v_F 4\pi k_\zeta^2 \qquad \text{(J. 64)}$$

or, using (F. 101) and (F. 146),

$$\sigma = ebn, \qquad \text{(J. 65)}$$

with the electron mobility [cf. eqns. (G. 49) or (J. 105)]

$$b = \frac{e}{m^*}\tau(k_\zeta), \qquad \text{(J. 66)}$$

where k_ζ is the radius of the Fermi sphere and n is the electron concentration. The temperature dependence of the electrical conductivity of metals depends on the temperature dependence of the relaxation time τ; all other quantities are independent of temperature.

(b) *THERMAL CONDUCTIVITY*

We shall show that—in contrast to the case of the electrical conductivity (cf. p. 361)—a non-vanishing value of the thermal conductivity is only obtained in the second approximation. The factors of $\partial F_0/\partial E$ contained in the transport equations (J. 48) and (J. 49) are the following:

$$E^s\phi(E) \equiv \frac{1}{4\pi^3\hbar} \int\limits_{E=\text{const}} \tau \frac{v_i v_j}{|v|} E^s \, dS \tag{J. 67}$$

where $s = 0, 1, 2$. An expansion of these factors at $E = \zeta$ yields

$$E^s\phi(E) \approx \zeta^s\phi(\zeta) + (E-\zeta)\left[\frac{d}{dE}(E^s\phi(E))\right]_\zeta + \frac{1}{2}(E-\zeta)^2\left[\frac{d^2}{dE^2}(E^s\phi(E))\right]_\zeta + \dots . \tag{J. 68}$$

Substitution in (J. 50) yields

$$L_s \approx M_1\zeta^s\phi(\zeta) + M_2\left[\frac{d}{dE}(E^s\phi(E))\right]_\zeta + M_3\left[\frac{d^2}{dE^2}(E^s\phi(E))\right]_\zeta \tag{J. 69}$$

with the integrals

$$M_1 = -\int_0^\infty \frac{\partial F_0}{\partial E}\, dE, \tag{J. 70}$$

$$M_2 = -\int_0^\infty (E-\zeta)\frac{\partial F_0}{\partial E}\, dE, \tag{J. 71}$$

$$M_3 = -\frac{1}{2}\int_0^\infty (E-\zeta)^2\frac{\partial F_0}{\partial E}\, dE. \tag{J. 72}$$

Since $\partial F_0/\partial E$ is a symmetrical function and $(E-\zeta)(\partial F_0/\partial E)$ is antisymmetric (cf. Fig. 167) we obtain immediately for the first two integrals

$$M_1 = 1, \tag{J. 73}$$

$$M_2 = 0. \tag{J. 74}$$

For the third integral we obtain using (J. 18)

$$M_3 = +\frac{1}{2}(kT)^2 \int\limits_{-(\zeta/kT)}^{\infty} \frac{x^2 e^x}{(1+e^x)^2}\, dx = \frac{1}{2}kT^2 \int\limits_{-(\zeta/kT)}^{\infty} \frac{x^2\, dx}{(e^{x/2}+e^{-(x/2)})^2} \tag{J. 75}$$

where

$$x = \frac{E-\zeta}{kT}. \tag{J. 76}$$

It is possible to a good approximation to replace the lower limit of the integral by $-\infty$ since $\partial F_0/\partial E$ and $\partial F_0/\partial x$ are essentially different from zero only near $E = \zeta$ and $x = 0$, respectively. The integration then yields

$$M_3 = \frac{1}{2}(kT)^2 \frac{\pi^2}{3} = \frac{\pi^2}{6}(kT)^2. \tag{J. 77}$$

Using eqns. (J. 73), (J. 74), and (J. 77), eqn. (J. 69) can be written in the form

$$L_s \approx \zeta^s\phi(\zeta) + \frac{\pi^2}{6}(kT)^2 \left[\frac{d^2}{dE^2}(E^s\phi(E))\right]_\zeta. \tag{J. 78}$$

Neglecting consistently $d\phi(E)/dE$ and $d^2\phi(E)/dE^2$, we hence obtain

$$L_0 \approx \phi(\zeta), \tag{J. 79}$$

$$L_1 \approx \zeta\phi(\zeta), \tag{J. 80}$$

$$L_2 \approx \zeta^2\phi(\zeta) + \frac{\pi^2}{3}(kT)^2\,\phi(\zeta). \tag{J. 81}$$

Substitution of these relations into eqn. (J. 55) yields

$$\varkappa_c = \frac{\pi^2}{3} k^2 T\phi(\zeta) \tag{J. 82}$$

for the thermal conductivity, where [cf. eqn. (J. 67)]

$$\phi(\zeta) = \frac{1}{4\pi^3\hbar} \int\limits_{E=\zeta} \tau \frac{v_i v_j}{|v_F|}\, dS_F. \tag{J. 83}$$

By analogy with the electrical conductivity [cf. eqn. (J. 60)] the electronic contribution to the thermal conductivity of metals depends on the electron properties at the Fermi energy. For spherical energy surfaces in particular we obtain [cf. eqn. (J. 65)]

$$\varkappa_c = \frac{\pi^2}{3} \frac{k^2}{m^*} T\tau(\zeta)n. \tag{J. 84}$$

When, just as in the derivation of the electrical conductivity, we consider only the first term in the approximation (J. 68), the thermal conductivity becomes equal to zero.

Passing over to the second approximation, we see that the electrical conductivity remains unchanged. Substituting $\text{grad}_r \zeta = \text{grad}_r T = 0$ in (J. 51) we obtain the electrical conductivity of a crystalline solid:

$$\frac{j}{E} \equiv \sigma = e^2 L_0 . \qquad \text{(J. 85)}$$

Using eqns. (J. 78) and (J. 79) we obtain for the electrical conductivity of a metal or of a strongly degenerate semiconductor [cf. eqn. (J. 60)]:

$$\sigma = e^2 \phi(\zeta) = \frac{e^2}{4\pi^3 \hbar} \int\limits_{E=\zeta} \tau \frac{v_i v_j}{|v_F|} dS_F . \qquad \text{(J. 86)}$$

Dividing equations (J. 82) and (J. 86) yields the Wiedemann–Franz law, according to which the electrical conductivity and the electron contribution to the thermal conductivity are proportional to one another:

$$\frac{\varkappa_c}{\sigma} = \frac{\pi^2}{3} \left(\frac{k}{e} \right)^2 T = LT, \qquad \text{(J. 87)}$$

where L is the universal Lorenz number

$$L = \frac{\pi^2}{3} \left(\frac{k}{e} \right)^2 = 2.45 \times 10^{-8} \, (\text{V}/^\circ\text{K})^2 . \qquad \text{(J. 88)}$$

We might have derived the Wiedemann–Franz law in this form from Sommerfeld's free electron theory. It must be mentioned that in this theory the electron scattering mechanism is taken into account through the mean free path. The dependence on this quantity is the same in the two equations for electrical and thermal conductivities and therefore drops out of the Wiedemann–Franz law. As will be discussed in the following, the range of validity of the Wiedemann–Franz law, however, depends on the scattering mechanisms.

The Wiedemann–Franz law is independent of the shape of the energy surfaces and the presence of cubic symmetry. The only essential fact is that the relaxation time $\tau(\boldsymbol{k})$ must remain unchanged when in the direction of $\boldsymbol{k}$ an electric field is applied or a temperature gradient is acting, or even both. This means that the same scattering processes must be responsible for both the electrical resistance and the thermal resistance. This condition is satisfied for temperatures above the Debye temperature, that is, for $T > \Theta_D$, when elastic scattering is dominant. In this temperature range $\tau \propto T^{-1}$, that is, according to eqn. (J. 86), the electrical conductivity is inversely proportional to the temperature, and, according to eqn. (J. 82), the thermal conductivity is independent of temperature. In this range the Wiedemann–Franz law is valid (cf. Table 23). In good conductors the phonon contribution to the thermal conductivity above $T > \Theta_D$ is small compared with the electron contribution. In this case the measured thermal conductivity is approximately equal to the contribution from the electrons.

TABLE 23. Electrical conductivities, thermal conductivities and Lorenz numbers of various metals (after J. L. Olsen, *Electron Transport in Metals* [Interscience, New York 1962], p. 5)

Metal	$\sigma \times 10^{-5}$ $[\Omega\ \text{cm}]^{-1}$ 273°K	$\varkappa$ [W/cm·°K] 273°K	$L \times 10^8$ $[\text{V}/°\text{K}]^2$ 273°K	373°K
Na	2.34	1.35	2.10	—
Cu	6.45	3.85	2.18	2.30
Ag	6.6	4.18	2.31	2.37
Be	3.6	2.3	2.34	—
Mg	2.54	1.5	2.16	2.32
Al	4.0	2.38	2.18	2.22
Pb	0.52	0.35	2.46	2.57
Bi	0.093	0.085	3.30	2.88
Pt	1.02	0.69	2.47	2.56

At lower temperatures ($T < \Theta_D$) inelastic scattering processes become important which influence the electrical conductivity only slightly but the thermal conductivity strongly. The two conductivities are then determined by relaxation times which have different temperature dependences. Without derivation we give the results (for $T < \Theta_D$):

$$\tau_\sigma \propto T^{-5}, \qquad \text{(J. 89)}$$

$$\tau_\varkappa \propto T^{-3}. \qquad \text{(J. 90)}$$

In this temperature range the Wiedemann–Franz law has lost its validity.

2. NON-DEGENERATE ISOTROPIC TWO-BAND MODEL

The transport equations (J. 51) and (J. 52) as well as the transport coefficients, the electrical conductivity, and the thermal conductivity derived from them [(J. 85) and (J. 55)] hold generally for a crystalline solid if the same scattering processes apply to the charge and energy transport in them. The explicit calculation of the integrals L_s for non-degenerate semiconductors differs from that for metals and degenerate semiconductors, since the eigenvalue density at the Fermi energy is in principle different for these two cases. In non-degenerate semiconductors the eigenvalue density at $E = \zeta$ is equal to zero, that is, there are no electrons with the Fermi energy. It is therefore not meaningful to expand the factor of $\partial F_0/\partial E$ at $E = \zeta$ (cf. p. 361).

As regards the further calculation of the integrals, we restrict ourselves to isotropic and non-degenerate semiconductors, when the effective masses are independent of direction and energy. For this case of quasi-free electrons (cf. p. 224) we can use the following relations:

$$E - E_R = \frac{\hbar^2 k^2}{2m^*}, \tag{J. 91}$$

$$v = \frac{\hbar k}{m^*}. \tag{J. 92}$$

In the isotropic case the velocities and relaxation times pertaining to $E =$ const. are also independent of direction. The squares of the velocity components in a given direction, averaged over $E =$ const., are equal to one-third of the square of the total velocity [cf. eqn. (J. 62)]:

$$\overline{v_i^2} = \tfrac{1}{3} v^2. \tag{J. 93}$$

In the case of isotropy the integrals L_s take the form

$$L_s = -\frac{1}{12\pi^3 \hbar} \int_0^\infty \tau(E)\, v(E) \frac{\partial F_0}{\partial E} E^s 4\pi k^2 \, dE, \tag{J. 94}$$

or, using eqns. (J. 91), (J. 92), and (F. 135),

$$L_s = -\frac{4}{3m^*} \int_0^\infty \tau(E)\, E^s (E - E_R)\, D(E) \frac{\partial F_0}{\partial E} \, dE. \tag{J. 95}$$

In non-degenerate semiconductors the two uppermost energy bands contribute to the transport of charge and energy. The integrals are therefore the result of a contribution from the conduction band and one from the valence band:

$$L_s = L_{sn} + L_{sp} \tag{J. 96}$$

where

$$L_{sn} = -\frac{4}{3m_n} \int_{E_c}^\infty \tau_n(E)\, E^s (E - E_c)\, D_n(E) \frac{\partial F_0}{\partial E} \, dE \tag{J. 97}$$

and

$$L_{sp} = -\frac{4}{3m_p} \int_{-\infty}^{E_V} \tau_p(E)\, E^s (E - E_V)\, D_p(E) \frac{\partial}{\partial E} (1 - F_0) \, dE. \tag{J. 98}$$

(a) *ELECTRICAL CONDUCTIVITY*

The electrical conductivity of an isotropic non-degenerate semiconductor is given by [cf. eqns. (J. 85) and (J. 96)]

$$\sigma = e^2(L_{on}+L_{op}) \tag{J. 99}$$

where

$$L_{on} = -\frac{4}{3m_n}\int_{E_c}^{\infty} \tau_n(E-E_c)\, D_n(E)\,\frac{\partial F_0}{\partial E}\,dE \tag{J. 100}$$

and

$$L_{op} = -\frac{4}{3m_p}\int_{E_V}^{-\infty} \tau_p(E_V-E)\, D_p(E)\,\frac{\partial}{\partial E}(1-F_0)\,dE. \tag{J. 101}$$

The integrals can be calculated when averaged relaxation times $\bar{\tau}_n$ and $\bar{\tau}_p$ are introduced.

We define

$$\bar{\tau}_n = \frac{\displaystyle\int_{E_c}^{\infty} \tau_n(E-E_c)\, D_n(E)\frac{\partial F_0}{\partial E}\,dE}{\displaystyle\int_{E_c}^{\infty} (E-E_c)\, D_n(E)\frac{\partial F_0}{\partial E}\,dE} \tag{J. 102}$$

and

$$\bar{\tau}_p = \frac{\displaystyle\int_{E_V}^{-\infty} \tau_p(E_V-E)\, D_p(E)\frac{\partial}{\partial E}(1-F_0)\,dE}{\displaystyle\int_{E_V}^{-\infty} (E_V-E)\, D_p(E)\frac{\partial}{\partial E}(1-F_0)\,dE}. \tag{J. 103}$$

Using this and eqn. (G. 5) we obtain from (J. 100)

$$L_{on} = -\frac{4}{3m_n}\,\frac{1}{4\pi^2}\left(\frac{2m_n}{\hbar^2}\right)^{3/2}\bar{\tau}_n\int_{E_c}^{\infty} (E-E_c)^{3/2}\,\frac{\partial F_0}{\partial E}\,dE. \tag{J. 104}$$

Integration by parts yields

$$L_{on} = -\frac{4}{3m_n}\,\frac{1}{4\pi^2}\left(\frac{2m_n}{\hbar^2}\right)^{3/2}\bar{\tau}_n\left\{\Big[(E-E_c)^{3/2}F_0\Big]_{E_c}^{\infty} - \frac{3}{2}\int_{E_c}^{\infty}(E-E_c)^{1/2}F_0\,dE\right\} \tag{J. 105}$$

or, with the help of eqns. (G. 5) and (G. 17),

$$L_{on} = \frac{\bar{\tau}_n n}{m_n}. \tag{J. 106}$$

Similarly we obtain

$$L_{op} = \frac{\bar{\tau}_p p}{m_p}. \tag{J. 107}$$

Hence we get from eqns. (J. 99), (J. 106), and (J. 107) for the electrical conductivity

$$\sigma = eb_n n + |e| b_p p \tag{J. 108}$$

with the so-called carrier mobilities for electrons and holes [cf. eqn. (G. 49)]:

$$b_n = \frac{e}{m_n} \bar{\tau}_n \tag{J. 109}$$

and

$$b_p = \frac{|e|}{m_p} \bar{\tau}_p. \tag{J. 110}$$

The mobilities are mainly determined by the scattering processes for the carriers, which are characterized by relaxation times. A further calculation of the mobilities and the electrical conductivity requires a knowledge of the scattering processes and thus of the relaxation times $\tau_{n,p}(E)$ (cf. pp. 396 ff.).

(b) *THERMAL CONDUCTIVITY*

The carrier contribution to the thermal conductivity of an isotropic non-degenerate semiconductor is given by [cf. eqn. (J. 55)]

$$\varkappa_c = -\frac{1}{T}\left(\frac{L_1^2}{L_0} - L_2\right). \tag{J. 111}$$

The integrals L_1 and L_2 can be calculated in the same way as L_0 when the mean values $(\overline{\tau E})_{n,p}$ and $(\overline{\tau E^2})_{n,p}$ corresponding to eqns. (J. 102) and (J. 103), respectively, are introduced. We obtain

$$L_1 = \frac{n}{m_n}(\overline{\tau E})_n + \frac{p}{m_p}(\overline{\tau E})_p \tag{J. 112}$$

and

$$L_2 = \frac{n}{m_n}(\overline{\tau E^2})_n + \frac{p}{m_p}(\overline{\tau E^2})_p. \tag{J. 113}$$

In a good approximation we can write

$$(\overline{\tau E})_{n,p} \approx \bar{\tau}_{n,p} \bar{E}_{n,p} \tag{J. 114}$$

and

$$(\overline{\tau E^2})_{n,p} \approx \bar{\tau}_{n,p} \overline{E^2_{n,p}} \tag{J. 115}$$

where $\bar{E}_{n,p}$ is the mean energy of the electrons in the conduction band and the holes in the valence band, respectively. We also set

$$\overline{E^2_{n,p}} \approx \bar{E}^2_{n,p}. \tag{J. 116}$$

Substitution of eqns. (J. 106), (J. 107), (J. 112), and (J. 113) in (J. 111) yields in the above approximation

$$\varkappa_c = \frac{1}{e^2 T} \frac{\sigma_n \sigma_p}{\sigma} (\bar{E}_n - \bar{E}_p)^2 \tag{J. 117}$$

with the electron and hole conductivities

$$\sigma_n = \frac{e^2}{m_n} \bar{\tau}_n n, \tag{J. 118}$$

$$\sigma_p = \frac{e^2}{m_p} \bar{\tau}_p p. \tag{J. 119}$$

The mean energies are approximately given by

$$\bar{E}_n \approx E_c + \tfrac{3}{2} kT \tag{J. 120}$$

and

$$\bar{E}_p \approx E_V - \tfrac{3}{2} kT. \tag{J. 121}$$

Thus we obtain for the carrier contribution to the thermal conductivity of isotropic non-degenerate semiconductors

$$\varkappa_c = \frac{1}{e^2 T} \frac{\sigma_n \sigma_p}{\sigma} (\Delta E_i + 3kT)^2, \tag{J. 122}$$

where the width of the energy gap is given by

$$\Delta E_i = E_c - E_V. \tag{J. 123}$$

Just as in the case of the Wiedemann–Franz law we obtain from eqn. (J. 122)

$$\frac{\varkappa_c}{\sigma} = L^* T \tag{J. 124}$$

with the modified Lorenz number

$$L^* = 2\left(\frac{k}{e}\right)^2 \frac{\sigma_n \sigma_p}{2\sigma^2}\left(\frac{\Delta E_i}{kT}+3\right)^2. \qquad \text{(J. 125)}$$

Under the assumption of pure thermal scattering (cf. p. 396) and an exact calculation of the integrals L_s we obtain instead of (J. 125) for the modified Lorenz number

$$L^* = 2\left(\frac{k}{e}\right)^2 \left[\frac{\sigma_n \sigma_p}{2\sigma^2}\left(\frac{\Delta E_i}{kT}+4\right)^2+1\right]. \qquad \text{(J. 126)}$$

A comparison with the universal Lorenz number (J. 88) shows that the ratio of the carrier contribution to the thermal conductivity and the electrical conductivity can be much higher in semiconductors than in metals. The factor $\sigma_n\sigma_p/2\sigma^2$ has its maximum value when the electron and hole conductivities are equal, that is, when $\sigma_n = \sigma_p = \sigma/2$. This is satisfied for an intrinsic semiconductor if the mobilities b_n and b_p are not too different from one another. An estimation shows that with $\Delta E_i \approx 1$ eV and $\sigma_n = \sigma_p$ at $T = 500°$K the modified Lorenz number L^* is about 100 times the universal Lorenz number. The relatively high thermal conductivity of the carriers is due to the so-called ambipolar diffusion, that is, the diffusion of electron–hole pairs. At the warmer end of the semiconductor more electron–hole pairs are produced than at the colder end. These pairs diffuse to the colder end and transport besides their kinetic energy also their excitation energy ΔE_i. In the subsequent recombination process ΔE_i is liberated as recombination energy. The influence of ambipolar diffusion can be observed if the samples are sufficiently pure and the temperature is sufficiently high (cf. Fig. 168), that is, if the concentration of the electron–hole pairs is high compared with the concentrations of the carriers excited from impurities. The effect of ambipolar diffusion therefore decreases with increasing impurity conduction. The ambipolar diffusion does not contribute to the electrical conductivity.

It should furthermore be mentioned that in semiconductors, besides the carrier contribution which has been considered here, the contribution of phonons to the thermal conductivity cannot be neglected [cf. eqn. (J. 42)].

IV. Thermoelectrical Effects

The transport equations (J. 51) and (J. 52) show that both an electric field and a temperature gradient can give rise to an electric current and a heat flux. According to the choice of the experimental conditions it is possible to measure separately the electrical, thermal, or thermoelectrical properties.

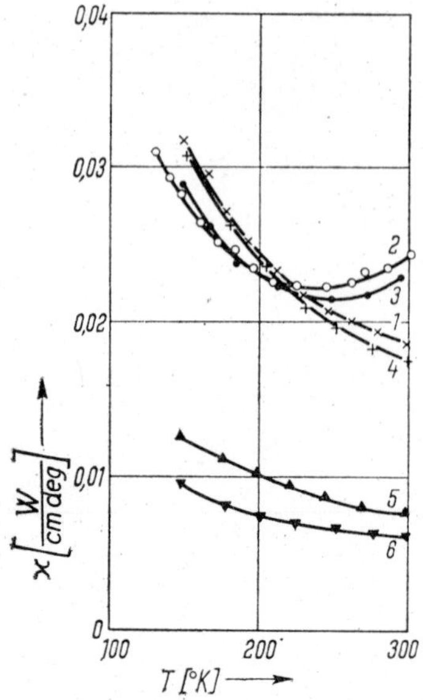

FIG. 168. Temperature dependence of the thermal conductivity of Bi_2Te_3, the samples 2 and 3 which are nearly intrinsic show *ambipolar diffusion*. The other curves are valid for extrinsic samples of various doping and orientation [after H. J. Goldsmid, *Proc. Phys. Soc.* B **69**, 203 (1956)].

1. THE SEEBECK EFFECT

We have shown already that in an open circuit ($j = 0$) a temperature gradient produces an electric field. Its magnitude follows from (J. 53) as

$$\boldsymbol{E} = \frac{1}{e} \operatorname{grad} \zeta + \frac{1}{eT} \left(\frac{L_1}{L_0} - \zeta \right) \operatorname{grad} T. \tag{J. 127}$$

This field strength consists of two components: a component which is due to the chemical inhomogeneity and contacts between different materials (the so-called contact field, cf. p. 310), and a component due to the temperature gradient (the thermoelectric field).

In practice it is impossible to measure this field strength directly with a single definite material; the measuring device would then have to display the same temperature gradient. One rather uses a so-called open thermocouple series. In the simplest case it consists of a single couple, that is, of two substances A and B which are connected as in Fig. 169. The two junctions are at

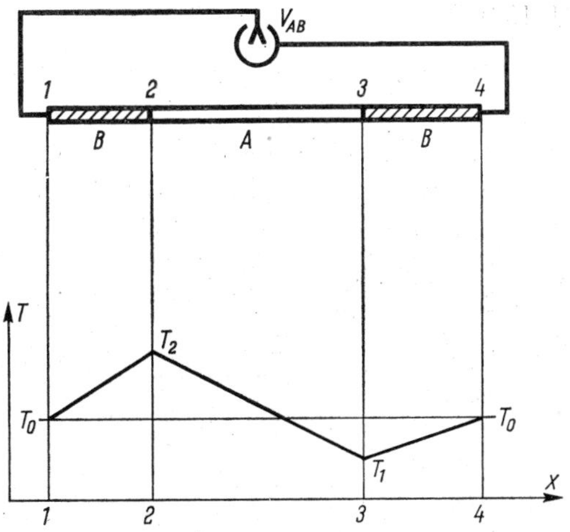

FIG. 169. Open thermocouple (Seebeck effect).

different temperatures T_1 and T_2. The series is interrupted at the temperature $T_0(T_2 > T_0 > T_1)$. The so-called integral thermoelectric voltage between the two ends is given by the integral over the field strength along the thermocouple series (x direction):

$$V_{AB} = -\int E_x \, dx \tag{J. 128}$$

where [cf. eqn. (J. 127)]

$$E_x = \frac{1}{e}\frac{\partial \zeta}{\partial x} + \frac{1}{eT}\left(\frac{L_1}{L_0} - \zeta\right)\frac{dT}{dx}. \tag{J. 129}$$

When we take into account the different properties of the materials A and B, we obtain from (J. 128)

$$V_{AB} = -\frac{1}{e}\int \frac{\partial \zeta}{\partial x} dx - \int_{T_0}^{T_2} S_B \, dT - \int_{T_2}^{T_1} S_A \, dT - \int_{T_1}^{T_0} S_B \, dT \tag{J. 130}$$

where the absolute thermoelectric power (the Seebeck coefficient) of the material A or B is given by

$$S_{A,B} = \frac{1}{eT}\left(\frac{L_1}{L_0} - \zeta\right)_{A,B}. \tag{J. 131}$$

The first term in (J. 130), which represents the contact potential, vanishes, since material and temperature are identical by definition, at points 1 and 4. Thus we obtain from (J. 130)

$$V_{AB} = \int_{T_1}^{T_2} (S_A - S_B)\, dT. \tag{J. 132}$$

Thus the thermoelectric voltage is obtained as the difference of the thermoelectric powers of two substances. It is a function of the temperature difference between the two junctions. It must be stressed that the electric field exists along the entire volume of the thermocouple series and not only at the junctions of two different materials. It is therefore reasonable to define the integral absolute thermoelectric voltage as a material property:

$$V_{A,B}(T) = \int_0^T S_{A,B}\, dT. \tag{J. 133}$$

We also define a so-called differential thermoelectric power ϕ_{AB} as the difference of the absolute thermoelectric powers of the materials A and B at the temperature T:

$$\phi_{AB}(T) = \frac{dV_A}{dT} - \frac{dV_B}{dT} = S_A - S_B. \tag{J. 134}$$

Just as the electrical conductivity, the thermoelectric power is a transport property since it contains the integrals L_s and thus the relaxation time τ which is characteristic of transport phenomena.

The further calculation of the thermoelectric power again depends on the special approximation for metals or semiconductors.

(a) *THERMOELECTRIC POWER OF METALS*

In the case of metals the integrals L_s can be given approximately by eqn. (J. 78). It can be seen immediately that, just as the thermal conductivity, the thermoelectric power is non-vanishing only in the second approximation. Substitution of eqn. (J. 78) in (J. 131) and neglecting the higher-order terms yields

$$S_m = \frac{1}{eT}\frac{\pi^2}{3}(kT)^2 \left\{\frac{d[\ln \phi(E)]}{dE}\right\}_{E=\zeta} \tag{J. 135}$$

or, using (J. 86),

$$S_m = \frac{\pi^2}{3}\frac{k^2}{e}T \left\{\frac{d[\ln \sigma(E)]}{dE}\right\}_{E=\zeta}. \tag{J. 136}$$

The quantity $\sigma(E)$ is the electrical conductivity calculated in terms of hypothetic Fermi energies E. The variation of σ with E determined at the point of the true Fermi energy ζ of the material considered defines the thermoelectric power.

It depends, on the one hand, on the electron-scattering mechanism, that is, on the energy dependence of the relaxation time $\tau(E)$, and, on the other hand, on the form variation of the energy surfaces near the Fermi surface [cf. eqn. (J. 67) for $s = 0$].

Let us estimate the value of the thermoelectric power in the case of quasi-free electrons, under the assumption of an energy dependence of the relaxation time of the form (cf. pp. 397 ff.)

$$\tau(E) = \text{const.}\, E^j. \tag{J. 137}$$

According to eqn. (J. 65) the electrical conductivity is given by

$$\sigma = \frac{e^2}{m^*}\tau(\zeta)\, n. \tag{J. 138}$$

Using the hypothetical energy dependence of the electron concentration (E. 24) and (J. 137), we obtain for the energy dependence of the conductivity (in the case of quasi-free electrons the effective mass is independent of the energy)

$$\sigma(E) = \text{const.}\, E^{(3/2)+j}. \tag{J. 139}$$

Using this we obtain from (J. 136) for the thermoelectric power of a metal with spherical energy surfaces

$$S_m = \frac{dV}{dT} = \frac{\pi^2}{3}\frac{k^2}{e}\frac{T}{\zeta}\left(\frac{3}{2}+j\right). \tag{J. 140}$$

In the case of scattering from acoustic phonons $j = -1/2$. Taking $\zeta \approx 5$ eV (cf. Table 16) and $T = 300$°K we obtain an order-of-magnitude value of 10^{-6} V/°K for the thermoelectric power, which is in agreement with experimental results.

(b) *THERMOELECTRIC POWER OF NON-DEGENERATE SEMICONDUCTORS*

Using eqns. (J. 106), (J. 107), and (J. 112), we obtain from (J. 131)

$$S_s = \frac{1}{eT}\,\frac{\dfrac{n}{m_n}\bar{\tau}_n(\bar{E}_n-\zeta)-\dfrac{p}{m_p}\bar{\tau}_p(\zeta-\bar{E}_p)}{\dfrac{n}{m_n}\bar{\tau}_n+\dfrac{p}{m_p}\bar{\tau}_p}. \tag{J. 141}$$

Hence we obtain from (J. 120) and (J. 121)

$$S_s = \frac{k}{e}\left\{\frac{\dfrac{n}{m_n}\bar{\tau}_n\dfrac{E_c-\zeta}{kT}-\dfrac{p}{m_p}\bar{\tau}_p\dfrac{\zeta-E_V}{kT}}{\dfrac{n}{m_n}\bar{\tau}_n+\dfrac{p}{m_p}\bar{\tau}_p}+\frac{3}{2}\,\frac{\dfrac{n}{m_n}\bar{\tau}_n-\dfrac{p}{m_p}\bar{\tau}_p}{\dfrac{n}{m_n}\bar{\tau}_n+\dfrac{p}{m_p}\bar{\tau}_p}\right\}. \tag{J. 142}$$

This relation shows that the sign of the thermoelectric power depends on the typical semiconductor properties. For n-type semiconductors, that is, for $n \gg p$, eqn. (J. 142) takes the form

$$S_n = \frac{k}{e}\left(\frac{E_c-\zeta}{kT}+\frac{3}{2}\right). \qquad \text{(J. 143)}$$

Similarly, we obtain for p-type semiconductors, i.e. for $p \gg n$,

$$S_p = -\frac{k}{e}\left(\frac{\zeta-E_V}{kT}+\frac{3}{2}\right). \qquad \text{(J. 144)}$$

The absolute thermoelectric power of a semiconductor is generally much higher than that of a metal. For a typical distance of the Fermi energy from the band edge of the order of 10^{-1} eV, we obtain from eqn. (J. 143) or (J. 144) an order of 10^{-3} V/°K, compared with 10^{-6} V/°K for metals. For a metal–semiconductor thermoelement we obtain to a good approximation

$$V_{s-m} = \int_{T_1}^{T_2} (S_s - S_m)\, dT \approx \int_{T_1}^{T_2} S_s\, dT \qquad \text{(J. 145)}$$

or

$$\frac{dV_{s-m}}{dT} \approx S_s. \qquad \text{(J. 146)}$$

The determination of the sign of the thermoelectric voltage represents a simple method of determining the conduction type of a semiconductor. The semiconductor crystal is contacted with two metal electrodes at different temperatures and the voltage is measured. The sign of this voltage is essentially determined by the type of the majority carriers; its value is because of (J. 145) essentially independent of the metals used.

The value of the thermoelectric power of non-degenerate impurity semiconductors depends only on the position of the Fermi energy which is given by the energy distribution of the impurities. In the simplest case of a single impurity level we obtain from (J. 143) with the help of (G. 95) for the case of donor reserve

$$S_n = \frac{k}{e}\left[\frac{\Delta E_D}{2kT} - \ln\left(\frac{N_D}{2n_0}\right)^{1/2} + \frac{3}{2}\right]. \qquad \text{(J. 147)}$$

This shows that the differential thermoelectric power is proportional to the activation energy ΔE_D of the donors and inversely proportional to the temperature. At low temperatures, however, this relation (J. 147) loses its validity since then the transport equations can no longer be described in terms of a single relaxation time.

2. THE PELTIER EFFECT

Under the condition grad $T = 0$ we can show from the transport equations (J. 51) and (J. 52) that the electric current is always accompanied by a heat flux

$$w = \Pi j \tag{J. 148}$$

with the so-called Peltier coefficient

$$\Pi = \frac{1}{e}\left(\frac{L_1}{L_0} - \zeta\right). \tag{J. 149}$$

Comparing this with eqn. (J. 131), we obtain a relation between the Seebeck and the Peltier coefficients, which can also be obtained from thermodynamic calculations:

$$\Pi = TS. \tag{J. 150}$$

Thus the theory of the Peltier coefficient is contained in the theory of the thermoelectric power.

The presence of a heat flux is proved with the help of a closed thermocouple series in which a current j is produced by means of a battery (cf. Fig. 170). In the two materials A and B the heat flux is different as Π_A and Π_B are different [cf. eqn. (J. 148)]. The difference between the heat fluxes must be produced

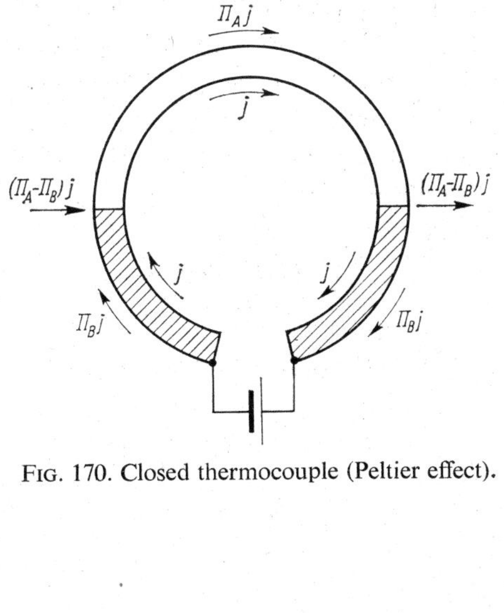

FIG. 170. Closed thermocouple (Peltier effect).

at the two junctions. In the ideal case (grad $T = 0$) the difference at one junction would be obtained from an infinitely large heat reservoir and discharged at the other to another infinitely large reservoir so that the temperature of the whole system would remain unchanged. In practice, however, the temperature of the finite heat reservoirs is changed and the heat flux difference is brought about by additional temperature gradients. Therefore one junction will be cooled while the other is heated. This effect is called the Peltier effect. A measure for the temperature change is the quantity

$$\Pi_{AB} = \Pi_A - \Pi_B. \tag{J. 151}$$

It represents the quantity of heat supplied to or withdrawn from each junction per unit time and current density. The value of the temperature change at the junctions, moreover, depends on the external conditions, especially on the magnitude of the heat reservoirs and thus on the geometry of the thermocouple series.

By virtue of the relation (J. 150) and the order of magnitude given for S (cf. p. 376) we may expect strong Peltier effects particularly for semiconductor–metal elements.

The Peltier effect may be considered the inverse of the Seebeck effect.

3. THE THOMSON EFFECT

In the volume of a conductor in which an electric current and a heat current exist at the same time, the entire energy balance consists of the following components: (1) Joule heat, (2) heat due to heat conduction, and (3) the so-called Thomson heat. We shall show in the following that the Thomson heat is proportional to the scalar product of the current density and the temperature gradient; the proportionality factor, called the Thomson coefficient, is related to the Seebeck and Peltier coefficients.

The heat produced per unit volume and unit time is given by the electric power and the heat flux absorbed:

$$\frac{dQ}{dt} = (\boldsymbol{j} \cdot \boldsymbol{E}) - \operatorname{div} \boldsymbol{w}. \tag{J. 152}$$

Using eqns. (J. 85) and (J. 131), we obtain from (J. 51)

$$\boldsymbol{E} = \frac{\boldsymbol{j}}{\sigma} + S \operatorname{grad} T + \frac{1}{e} \operatorname{grad} \zeta. \tag{J. 153}$$

For the heat flux we obtain from (J. 52), using (J. 51), (J. 55), and (J. 149),

$$\boldsymbol{w} = \Pi \boldsymbol{j} - \varkappa_c \operatorname{grad} T. \tag{J. 154}$$

Using eqns. (J. 153) and (J. 154), eqn. (J. 152) yields

$$\frac{dQ}{dt} = \frac{j^2}{\sigma} + \left(\frac{\boldsymbol{j}}{e} \cdot \operatorname{grad} \zeta\right) + \operatorname{div}(\varkappa_c \operatorname{grad} T) + S(\boldsymbol{j} \cdot \operatorname{grad} T) - (\boldsymbol{j} \cdot \operatorname{grad} \Pi). \tag{J. 155}$$

For a homogeneous crystal we have

$$\operatorname{grad} \Pi = \frac{d\Pi}{dT} \operatorname{grad} T. \tag{J. 156}$$

Using this and (J. 150) we obtain

$$\frac{dQ}{dt} = \frac{j^2}{\sigma} + \left(\frac{\boldsymbol{j}}{e} \cdot \operatorname{grad} \zeta\right) + \operatorname{div}(\varkappa_c \operatorname{grad} T) - \mu(\boldsymbol{j} \cdot \operatorname{grad} T), \tag{J. 157}$$

where the Thomson coefficient is given by

$$\mu = \frac{d\Pi}{dT} - S = T \frac{dS}{dT}. \tag{J. 158}$$

In eqn. (J. 157) the first two terms stand for the Joule heat (the current density is due to the external field $\boldsymbol{E}$ and the position dependence of the Fermi energy ζ); the third term represents the heat which has been brought into the volume by the heat conduction for the case $\boldsymbol{j} = 0$; the last term is the Thomson heat. The latter depends on the angle made by the directions of the current and the temperature gradient and can therefore be determined by means of a simple measurement. One determines the power at constant temperature gradient with parallel as well as antiparallel current directions. The difference of the power in these two cases is equal to the Thomson heat. Since the Thomson coefficient depends only on the thermoelectric power, the *absolute* thermoelectric power is obtained immediately if this coefficient is measured as a function of temperature.

The equations (J. 150) and (J. 158) can also be obtained from thermodynamic considerations (the so-called Thomson relations), independently of model representations; we have

$$S = \frac{dV}{dT}, \tag{J. 159}$$

$$\Pi = TS = T \frac{dV}{dT}, \tag{J. 160}$$

$$\mu = T \frac{dS}{dT} = T \frac{d^2V}{dT^2}. \tag{J. 161}$$

V. Galvanomagnetic Effects

Galvanomagnetic effects occur under the simultaneous action of electric and magnetic fields. According to the field directions we distinguish between transverse ($\boldsymbol{E} \perp \boldsymbol{B}$) and longitudinal ($\boldsymbol{E} \parallel \boldsymbol{B}$) effects.

In the following we restrict ourselves to the consideration of transverse effects, namely the Hall effect and the transverse magnetoresistance. The Hall effect consists in the appearance of an electric field E_y which is perpendicular to the initially applied field E_x and to the magnetic induction B_z (cf. Fig. 171). The Hall coefficient R_H is defined as

$$R_H \equiv \frac{E_y}{j_x B_z}. \qquad \text{(J. 162)}$$

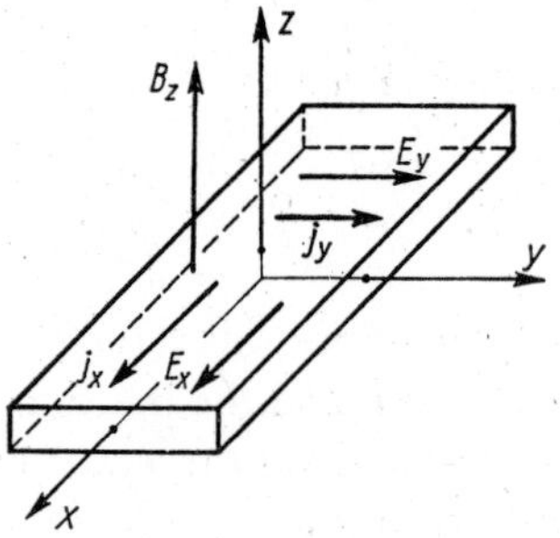

FIG. 171. Usual arrangement of the electric field and magnetic induction for the isothermal Hall effect.

Its value depends on the experimental conditions. The current density j_x and the induction B_z cause not only the appearance of the Hall field E_y but also a temperature gradient $\partial T/\partial y$ (the Ettingshausen effect). According to whether this permits a heat flux in the y direction or not, we speak of an isothermal or an adiabatic Hall effect.

The magneto-resistance consists in an increase of the electrical resistance in a magnetic field. It is a consequence of the Lorentz force which curves the carrier trajectories. This reduces the mean free path in the direction of the electric field and thus also the electrical conductivity.

In the following derivations we consider a homogeneous, isotropic single crystal which, at constant temperature ($\rightarrow$ isothermal effects), is under the action of an electric field $\boldsymbol{E}(E_x, E_y, 0)$ and a magnetic induction $\boldsymbol{B}(0, 0, B_z)$. Under the assumption of sufficiently weak magnetic fields the eigenvalue density of the carriers is in a good approximation given by its value at $B = 0$ [cf. eqn. (F. 135)]; the perturbation of the distribution function will then be given by the variation of the occupation probability only (cf. p. 352). The condition of weak magnetic fields means that the relaxation time τ is small compared with the reciprocal cyclotron frequency, that is (cf. p. 256),

$$|\omega_c \tau| < 1. \tag{J. 163}$$

In the case of spherical energy surfaces we obtain from this using eqn. (F. 193)

$$\left| \frac{e}{m^*} \tau B_z \right| < 1 \tag{J. 164}$$

or, in terms of the carrier mobility [cf. (J. 66) or (J. 109)],

$$|bB_z| < 1. \tag{J. 165}$$

An estimate shows that this condition is well satisfied for metals up to magnetic flux densities of the order of 10^{-4} V sec/cm^2 ($b_{\text{metal}} \approx 10^2$ cm^2/V sec). The carrier mobilities of certain semiconductors, however, may be of a higher order of magnitude than that of metals (e.g. for InSb the electron mobility $b_n \approx 10^4$ cm^2/V sec) so that eqn. (J. 165) is only satisfied for correspondingly smaller values of B_z.

Since we have assumed homogeneity, we can write for the isothermal case

$$\text{grad}_r\, F = 0. \tag{J. 166}$$

The change in the occupation probability due to external forces [cf. eqn. (J. 23)] is then given by

$$\left(\frac{\partial F}{\partial t}\right)_{\text{drift}} = -(\dot{\boldsymbol{k}} \cdot \text{grad}_k\, F) \tag{J. 167}$$

where

$$\hbar \dot{\boldsymbol{k}} = e(\boldsymbol{E} + [\boldsymbol{v} \wedge \boldsymbol{B}]). \tag{J. 168}$$

With the given fields, we have

$$\left(\frac{\partial F}{\partial t}\right)_{\text{drift}} = -\frac{e}{\hbar}\left[(E_x + v_y B_z)\frac{\partial F}{\partial k_x} + (E_y - v_x B_z)\frac{\partial F}{\partial k_y}\right]. \tag{J. 169}$$

The scattering processes are described by a single relaxation time $\tau(E)$ which, in the isotropic case, depends only on the energy and not on the direction of $\boldsymbol{k}$. Using eqns. (J. 27) and (J. 169) we obtain by analogy with the Boltzmann equation (J. 29) the condition of stationarity

$$g = -\frac{e}{\hbar}\tau\left[(E_x + v_y B_z)\frac{\partial F}{\partial k_x} + (E_y - v_x B_z)\frac{\partial F}{\partial k_y}\right]. \tag{J. 170}$$

When we again replace the derivatives on the right-hand side by the derivatives of the unperturbed occupation probabilities [under the assumption of eqn. (J. 30)], the terms which contain the magnetic induction are cancelled. In this way it would in principle be impossible to consider the galvanomagnetic effect. It is therefore necessary, at least in the terms with B_z, to take into account the perturbation of the occupation probability. Using eqns. (F. 97) and (J. 8) and the derivatives

$$\frac{\partial F}{\partial k_x} = \hbar v_x \frac{\partial F_0}{\partial E} + \frac{\partial g}{\partial k_x} \qquad \text{(J. 171)}$$

and

$$\frac{\partial F}{\partial k_y} = \hbar v_y \frac{\partial F_0}{\partial E} + \frac{\partial g}{\partial k_y} \qquad \text{(J. 172)}$$

we obtain from (J. 170) a differential equation for the perturbation g of the occupation probability

$$g = -e\tau \frac{\partial F_0}{\partial E}(v_x E_x + v_y E_y) - \frac{e}{\hbar}\tau B_z\left(v_y \frac{\partial g}{\partial k_x} - v_x \frac{\partial g}{\partial k_y}\right). \qquad \text{(J. 173)}$$

According to Gans, the equation can be solved with the help of the following ansatz:

$$g(E) = v_x \varphi_x(E) + v_y \varphi_y(E), \qquad \text{(J. 174)}$$

where $\varphi_x(E)$ and $\varphi_y(E)$ are functions which depend only on the energy. This assumes the perturbation to consist of two mutually independent components acting in the x and y directions. The substitution of (J. 174) in the differential equation (J. 173) yields

$$v_x\varphi_x + v_y\varphi_y + e\tau \frac{\partial F_0}{\partial E}(v_x E_x + v_y E_y) + \frac{e}{\hbar}\tau B_z\left(v_y\varphi_x \frac{\partial v_x}{\partial k_x} - v_x\varphi_y \frac{\partial v_y}{\partial k_y}\right) = 0. \qquad \text{(J. 175)}$$

The velocity components are independent of one another. Therefore the factor of v_x as well as that of v_y in (J. 175) must be equal to zero. In this way we obtain two equations for the functions φ_x and φ_y:

$$\varphi_x + e\tau \frac{\partial F_0}{\partial E} E_x - \frac{e}{\hbar}\tau B_z \varphi_y \frac{\partial v_y}{\partial k_y} = 0, \qquad \text{(J. 176)}$$

$$\varphi_y + e\tau \frac{\partial F_0}{\partial E} E_y + \frac{e}{\hbar}\tau B_z \varphi_x \frac{\partial v_x}{\partial k_x} = 0. \qquad \text{(J. 177)}$$

The further calculation makes it necessary to know explicitly the relation $E(\boldsymbol{k})$ which is contained in the derivatives of the velocity components with respect to the wave number components. Under the assumption of isotropy we have spherical energy surfaces. Using eqns. (F. 100) and (F. 113), we obtain from (F. 136)

$$\frac{\partial v_x}{\partial k_x} = \frac{1}{\hbar}\frac{\partial^2 E}{\partial k_x^2} = \frac{\hbar}{m^*} \tag{J. 178}$$

and

$$\frac{\partial v_y}{\partial k_y} = \frac{1}{\hbar}\frac{\partial^2 E}{\partial k_y^2} = \frac{\hbar}{m^*}. \tag{J. 179}$$

Using this eqns. (J. 176) and (J. 177) take the form

$$\varphi_x - \frac{e}{m^*}\tau B_z \varphi_y = -e\tau \frac{\partial F_0}{\partial E} E_x \tag{J. 180}$$

and

$$\varphi_y + \frac{e}{m^*}\tau B_z \varphi_x = -e\tau \frac{\partial F_0}{\partial E} E_y . \tag{J. 181}$$

Their solutions read

$$\varphi_x = -\frac{e\tau \dfrac{\partial F_0}{\partial E}}{1+(\omega_c\tau)^2}(E_x + \omega_c \tau E_y) \tag{J. 182}$$

and

$$\varphi_y = -\frac{e\tau \dfrac{\partial F_0}{\partial E}}{1+(\omega_c\tau)^2}(E_y - \omega_c \tau E_x), \tag{J. 183}$$

where

$$\omega_c = \frac{e}{m^*} B_z . \tag{J. 184}$$

Substitution into eqn. (J. 174) yields the perturbation of the occupation probability [instead of the linearized Boltzmann equation (J. 31)]:

$$g = -\frac{e\tau \dfrac{\partial F_0}{\partial E}}{1+(\omega_c\tau)^2}[v_x(E_x + \omega_c \tau E_y) + v_y(E_y - \omega_c \tau E_x)]. \tag{J. 185}$$

Using this we obtain from (J. 3) the electric current components

$$j_x = -\frac{2e^2}{(2\pi)^3}\int\limits_{-\infty}^{+\infty} \frac{\tau}{1+(\omega_c\tau)^2}[v_x^2(E_x + \omega_c \tau E_y) + v_x v_y(E_y - \omega_c \tau E_x)]\frac{\partial F_0}{\partial E}\, d^3\mathbf{k}, \tag{J. 186}$$

$$j_y = -\frac{2e^2}{(2\pi)^3}\int\limits_{-\infty}^{+\infty} \frac{\tau}{1+(\omega_c\tau)^2}[v_x v_y(E_x + \omega_c \tau E_y) + v_y^2(E_y - \omega_c \tau E_x)]\frac{\partial F_0}{\partial E}\, d^3\mathbf{k}. \tag{J. 187}$$

Since no external forces act in the z direction, $j_z = 0$.

The integration over $\boldsymbol{k}$-space is again replaced with the help of (F. 125) and (F. 126) by the integration over the surfaces of constant energy [cf. eqn. (J. 38)]. The integration over spherical energy surfaces (as well as over energy surfaces with cubic symmetry) yields

$$\int_{E=\text{const}} v_{x,y}^2 \, dS = \tfrac{1}{3} v^2(E) \int_{E=\text{const}} dS \tag{J. 188}$$

and

$$\int_{E=\text{const}} v_x v_y \, dS = 0. \tag{J. 189}$$

Equations (J. 186) and (J. 187) can then be rewritten with the help of (F. 100), (F. 127), and (F. 136) in the form

$$j_x = e^2(K_1 E_x + K_2 E_y), \tag{J. 190}$$

$$j_y = e^2(K_1 E_y - K_2 E_x) \tag{J. 191}$$

with the integrals

$$K_1 = -\frac{4}{3m^*} \int_{E_R}^{\infty} \frac{\tau}{1+(\omega_c\tau)^2} (E-E_R)\, D(E) \frac{\partial F_0}{\partial E}\, dE \tag{J. 192}$$

and

$$K_2 = -\frac{4}{3m^*} \int_{E_R}^{\infty} \frac{\omega_c\tau^2}{1+(\omega_c\tau)^2} (E-E_R)\, D(E) \frac{\partial F_0}{\partial E}\, dE. \tag{J. 193}$$

In the case of zero magnetic induction, i.e. $\omega_c = 0$, eqns. (J. 190) and (J. 191), can be replaced by (J. 85); it is easy to see that, in the isotropic case with $\omega_c = 0$, a comparison of eqns. (J. 95) and (J. 192) yields

$$K_1 = L_0\,. \tag{J. 194}$$

According to the geometrical form of the crystals, the further calculation of the galvanomagnetic effects is different. Let us consider a crystalline rod, that is, a crystal whose dimension in the x direction (the direction of the primary field E_x) is much larger than that in the y direction. It is then virtually impossible that the Lorentz force induces a y component of current. Instead of this it produces the Hall field E_y whose action compensates the Lorentz force and forces the carriers to move along the x direction. When the Hall field is measured under zero current conditions, we have

$$j_y = 0. \tag{J. 195}$$

Under this experimental condition we obtain from (J. 190) and (J. 191) the following quantities:

The Hall angle γ_H which is defined as the ratio of Hall field E_y and primary field E_x, that is, from (J. 191) and (J. 195), we have

$$\tan \gamma_H \equiv \frac{E_y}{E_x} = \frac{K_2}{K_1}. \tag{J. 196}$$

The substitution of this relation in (J. 190) yields the Hall field strength in terms of the current density j_x:

$$E_y = \frac{1}{e^2} \frac{K_2}{K_1^2+K_2^2} j_x. \tag{J. 197}$$

By means of a comparison with (J. 162) we obtain the Hall coefficient

$$R_H = \frac{1}{e^2} \frac{K_2}{K_1^2+K_2^2} \frac{1}{B_z}. \tag{J. 198}$$

A combination of eqns. (J. 190) and (J. 196) yields the transverse magneto-resistance, that is, the dependence of the electrical conductivity on the magnetic induction,

$$\sigma(B_z) = \frac{j_x}{E_x} = e^2 \frac{K_1^2+K_2^2}{K_1}. \tag{J. 199}$$

The quantities defined by eqns. (J. 196), (J. 198), and (J. 199) refer generally to homogeneous isotropic crystals, metals as well as semiconductors with spherical energy surfaces. They apply, however, only to weak magnetic fields, that is, the condition $|\omega_c \tau| < 1$.

The calculation of the integrals K_1 and K_2, (J. 192) and (J. 193), that is, of the Hall coefficient and the magneto-resistance, again depends on the chosen models and the approximations connected with them.

1. DEGENERATE ONE-BAND MODEL

(a) *THE HALL EFFECT*

Considering metals and degenerate semiconductors we take into account that only near $E = \zeta$ the derivative $\partial F_0/\partial E$ is essentially different from zero. In the case of isotropy it is sufficient to put in front of the integrals the factors $\tau/(1+(\omega_c\tau)^2)$ and $\omega_c\tau^2/(1+(\omega_c\tau)^2)$, respectively, with their values at $E = \zeta$. The integrals modified in this way can then be solved exactly by integration by parts, just as for the case of (J. 105), and one obtains

$$K_1 = \frac{\tau(\zeta)}{1+(\omega_c\tau(\zeta))^2} \frac{n}{m^*}, \tag{J. 200}$$

$$K_2 = \frac{\omega_c\tau^2(\zeta)}{1+(\omega_c\tau(\zeta))^2} \frac{n}{m^*} = \omega_c\tau(\zeta) K_1. \tag{J. 201}$$

From these equations we obtain for the Hall angle (J. 196)

$$\tan \gamma_H = \omega_c \tau(\zeta) = \frac{e}{m^*} \tau(\zeta) B_z \tag{J. 202}$$

or, in terms of the carrier mobility,

$$\tan \gamma_H = bB_z . \tag{J. 203}$$

Using eqns. (J. 200) and (J. 201) we can write for the Hall coefficient (J. 198)

$$R_H = \frac{1}{en} . \tag{J. 204}$$

According to this relation the Hall coefficient of a metal with spherical energy surfaces is inversely proportional to the electron concentration, which can be given approximately in terms of the concentration n_{at} of atoms and the valence z:

$$n \approx zn_{at} . \tag{J. 205}$$

The Hall coefficient is thus, according to eqns. (J. 204) and (J. 205), independent of the temperature, the effective mass, and the relaxation time, that is, of the scattering mechanism. We would have arrived at the same result by way of an elementary calculation for perfectly free electrons. The negative sign of the electron charge lets us expect negative Hall coefficients (normal Hall effect). This is in agreement with experimental results mainly for monovalent metals, as regards the sign as well as the value (cf. Table 24). We also observe agreement with experimental results for liquid metals in which the electrons behave like free or quasi-free particles.

A great number of multivalent metals, such as Cd, Zn, Pb, Mo, and W, however, have positive Hall coefficients (anomalous Hall effect). These metals display a behaviour as if holes were responsible for the charge transport. The anomalous Hall effect can be explained when the corresponding $E(\boldsymbol{k})$ relations are taken into account. For the derivations (cf. pp. 382 ff.) obtained so far we used the $E(\boldsymbol{k})$ relation (F. 136) which is valid only near the *bottom* of a band. The opposite signs are obtained for both the Hall coefficient and the Hall angle, if instead of (F. 136) the $E(\boldsymbol{k})$ relation is used which holds for spherical energy surfaces near the *top* of a band:

$$E = E_R - \frac{\hbar^2}{2\,|m^*|} k^2 . \tag{J. 206}$$

This assumes a high population of the corresponding energy band. Since near the top of a band the electron effective mass is always negative (cf. p. 219), the quantity ω_c will be negative. However, in the integrals K_1' and K_2' which are obtained similarly as (J. 192) and (J. 193) by using (J. 206), instead of

TABLE 24. Hall coefficients of several metals, measured at room temperature and calculated according to eqns. (J. 204) and (J. 205) (after C. Kittel, *Introduction to Solid State Physics*, 2nd ed. [Wiley, New York 1956], p. 298; and J. L. Olsen, *Electron Transport in Metals* [Interscience, New York 1962], p. 64)

Metal	R_H	$1/en_{at}$	$z_{exp} =$	z
	$\left[10^{-5} \dfrac{\text{cm}^3}{\text{A sec}}\right]$		$1/en_{at}R_H$	
Li	−17.0	−13.1	0.77	1
Na	−25.0	−24.5	0.98	1
K	−42	−47	1.12	1
Cs	−78	−73	0.94	1
Cu	− 5.5	− 7.4	1.34	1
Ag	− 9.0	−10.4	1.15	1
Au	− 7.2	−10.5	1.45	1
Be	+24.4	− 5.1	—	2
Zn	+ 3.3	− 4.6	—	2
Cd	+ 6.0	− 6.5	—	2
Al	− 3.5	−10.0	2.86	3
Bi	$\approx -10^4$	− 4.1	$\approx 10^4$	5

m^* only the absolute value $|m^*|$ of the effective mass is contained. Therefore the sign of K_1' remains the same as that of K_1 while that of K_2' is opposite to that of K_2. The Hall angle (J. 196) and the Hall coefficient (J. 198) change their signs when the integrals $K_{1,2}'$ are used. As we assumed a heavily populated energy band, we have in analogy to (J. 204)

$$R_H = \frac{|m^*|}{m^*}\,\frac{1}{en}, \tag{J. 207}$$

where the pertinent carrier concentration n is equal to the hole concentration in the energy band considered. We see that within the framework of the one-band model we have a normal Hall effect for low population and an anomalous Hall effect for high population of the appropriate energy band.

In many cases the one-band model may explain the sign, but not the magnitude of the Hall coefficient. In this case several energy bands have to be taken into account for the calculation of galvanomagnetic effects (such as in the case of band overlapping). The Hall coefficient will then be the result of the contributions of all these bands [see, for example, eqn. (J. 224)].

(b) *MAGNETO-RESISTANCE*

The transverse magneto-resistance vanishes only in a first approximation within the isotropic one-band model of metals. With (J. 200) and (J. 201) we obtain from (J. 199)

$$\sigma(B_z) = e^2K_1[1+(\omega_c\tau(\zeta))^2] = \frac{e^2}{m^*}\tau(\zeta)n \tag{J. 208}$$

or, from (J. 65),

$$\sigma(B_z) = \sigma. \tag{J. 209}$$

This result means that, owing to the action of the Hall field, the deviation due to the Lorentz force is vanishing for all carriers. Thus the trajectories remain the same as in the case of zero magnetic field. This behaviour is a consequence of the assumption that only electrons with the Fermi velocity are involved in the charge transport (cf. p. 362). Thus the Lorentz force exerts the same action on all these electrons.

However, when we take into account that the electrons which are essentially involved in the transport phenomena have energies in the range $\zeta \pm kT$, and that the relaxation time is energy-dependent, we see that the Lorentz force is no longer the same for all carriers. Only on average can the Hall field compensate the Lorentz force so that the trajectories of part of the carriers are changed. Hence results a magneto-resistance. Theoretically, this effect can be considered when, as in the case of heat conduction (cf. p. 363), we pass over to the second approximation, that is, when in the isotropic case the energy dependence of the factors $\tau/(1+(\omega_c\tau)^2)$ and $\omega_c\tau^2/(1+(\omega_c\tau)^2)$ in the integrals K_1 and K_2 is taken into account. This calculation carried out for weak magnetic fields yields the magneto-resistance

$$\frac{\Delta\sigma}{\sigma} \equiv \frac{\sigma-\sigma(B_z)}{\sigma} = \frac{c_1B_z^2}{1+c_2B_z^2}, \tag{J. 210}$$

where

$$c_1 = \frac{\pi^2}{3}\left(\frac{ekT}{2m^*\zeta}\tau(\zeta)\right)^2 \tag{J. 211}$$

and

$$c_2 = \sigma^2R_H^2 = \left(\frac{e}{m^*}\tau(\zeta)\right)^2. \tag{J. 212}$$

Thus the magneto-resistance varies as the square of magnetic induction and is, as would be expected, independent of the magnetic field direction. This agrees qualitatively with the experimental facts; quantitatively, however, the theoretical values are smaller than the measured ones by several orders of magnitude (about 10^3-10^4 times). This is true even for the alkali metals whose electronic properties are otherwise in good agreement with the free or quasi-free electron model.

Generally, the magneto-resistance is mainly due to the following two causes (which may exist alone or simultaneously):

1. Even in the simplest cases (e.g. for the alkali metals), the energy surfaces including the Fermi surface are not strictly spherical. The $E(\boldsymbol{k})$ relation will then depend on the direction of motion of the electrons and accordingly the velocity and the effective mass become dependent on the direction and the energy. In the anisotropic case the balance between the Lorentz force and the force due to the Hall field is satisfied for fewer carriers than in the isotropic case. This means that the trajectories of more carriers are changed and the magneto-resistance is thus stronger. In the case of anisotropic energy surfaces the derivation of the magneto-resistance is complicated and makes it necessary to know the corresponding $E(\boldsymbol{k})$ relations.

2. The transport properties are determined by several incompletely populated energy bands. In the simplest case one can assume within the framework of a two-band model that both bands are isotropic. One has then two groups of charge carriers with different effective masses and different velocities, which are differently influenced by the magnetic induction. The Hall field is established at some average value which depends on the relative contribution of the bands to the charge transport, that is, on the corresponding $E(\boldsymbol{k})$ relations (→ eigenvalue density, effective mass). Thus the Lorentz force is but incompletely compensated for both groups of carriers, and the magnetic field deflects the charge carriers and thus changes the conductivity. For an isotropic two-band model this is given by

$$\frac{\Delta\sigma}{\sigma} = \frac{\sigma_1\sigma_2(b_1-b_2)^2 B^2}{(\sigma_1+\sigma_2)^2+(b_1\sigma_2+b_2\sigma_1)^2 B^2}, \qquad \text{(J. 213)}$$

where

$$b_{1,2} = \frac{e}{m^*_{1,2}}\tau_{1,2} \qquad \text{(J. 214)}$$

and

$$\sigma_{1,2} = |eb_{1,2}|\, n_{1,2}. \qquad \text{(J. 215)}$$

The subscripts 1 and 2 refer to the two energy bands; n_1 and n_2 are the charge carrier densities which determine the charge transport. The magneto-resistance is particularly large when the two mobilities b_1 and b_2 have opposite signs. This is the case when the charge transport is brought about by the electrons of a weakly occupied band and of a highly occupied band, or by electrons and holes.

Apart from the transverse magneto-resistance one can also observe a longitudinal one (this means, in the case of $\boldsymbol{E} \parallel \boldsymbol{B}$). From the above considerations one sees easily that this effect can, as a matter of principal, not be explained by an isotropic two-band model.

2. NON-DEGENERATE ISOTROPIC TWO-BAND MODEL

In order to calculate transport phenomena in crystals in which several energy bands are incompletely filled, we have to add the contributions from these bands. In particular for semiconductors the integrals K_1 and K_2 consist of the contribution from the valence and conduction bands, just as the integrals L_s [cf. eqn. (J. 96)]:

$$K_{1,\,2} = K_{1,\,2n} + K_{1,\,2p}\,. \tag{J. 216}$$

In the approximation of weak magnetic fields, that is, for $|\omega_c \tau| \ll 1$, the denominators of the integrals K_1 and K_2 [cf. eqns. (J. 192) and (J. 193)] can be expanded. Neglecting terms of higher orders we obtain (for *one* type of carrier)

$$K_1 = -\frac{4}{3m^*}\int\limits_{E_R}^{\infty} \tau(1-(\omega_c\tau)^2)\,(E-E_R)\,D(E)\frac{\partial F_0}{\partial E}\,dE, \tag{J. 217}$$

$$K_2 = -\frac{4}{3m^*}\int\limits_{E_R}^{\infty} \omega_c\tau^2(E-E_R)\,D(E)\frac{\partial F_0}{\partial E}\,dE. \tag{J. 218}$$

Just as in eqn. (J. 102) we introduce mean values for the various powers of the relaxation times and, using the derivation of (J. 106), we obtain

$$K_1 = \frac{n}{m_n}\,\overline{\tau}_n + \frac{p}{m_p}\,\overline{\tau}_p - e^2B_z^2\left(\frac{n}{m_n^3}\,\overline{\tau_n^3} + \frac{p}{m_p^3}\,\overline{\tau_p^3}\right), \tag{J. 219}$$

$$K_2 = -|e|\,B_z\left(\frac{n}{m_n^2}\,\overline{\tau_n^2} - \frac{p}{m_p^2}\,\overline{\tau_p^2}\right). \tag{J. 220}$$

(a) *THE HALL EFFECT*

Using eqns. (J. 219) and (J. 220) and neglecting quadratic terms in B_z, we obtain from (J. 198) to a first approximation

$$R_H = -\frac{1}{|e|}\frac{\frac{n}{m_n^2}\overline{\tau_n^2}-\frac{p}{m_p^2}\overline{\tau_p^2}}{\left(\frac{n}{m_n}\bar{\tau}_n+\frac{p}{m_p}\bar{\tau}_p\right)^2} \qquad \text{(J. 221)}$$

for the Hall coefficient of a non-degenerate semiconductor; or, in terms of the carrier mobilities b_n and b_p [eqns. (J. 109) and (J. 110)] we have

$$R_H = -\frac{1}{|e|}\frac{\frac{\overline{\tau_n^2}}{\bar{\tau}_n^2}b_n^2 n-\frac{\overline{\tau_p^2}}{\bar{\tau}_p^2}b_p^2 p}{(|b_n|\,n+b_p p)^2}. \qquad \text{(J. 222)}$$

The factors $\left(\overline{\tau_{n,p}^2}/\bar{\tau}_{n,p}^2\right)$ are of the order of unity (cf. p. 401). They are determined by the energy dependences of the relaxation times, that is, they depend on the characteristic scattering processes. If they are the same for electrons and holes,

$$\frac{\overline{\tau_n^2}}{\bar{\tau}_n^2}=\frac{\overline{\tau_p^2}}{\bar{\tau}_p^2}\equiv r \qquad \text{(J. 223)}$$

and therefore

$$R_H = -\frac{r}{|e|}\frac{b_n^2 n-b_p^2 p}{(|b_n|\,n+b_p p)^2}. \qquad \text{(J. 224)}$$

Thus, depending on the carrier concentrations and mobilities, the Hall effect of a semiconductor is positive or negative. The Hall coefficient is zero, that is, the Hall field vanishes, if

$$b_n^2 n = b_p^2 p. \qquad \text{(J. 225)}$$

Under this condition no net charge is transported in the y direction by the electrons and holes deflected by a Lorentz force. It should be pointed out that in a magnetic field the electrons and holes are deflected in the same direction since both the charge and the mean velocity component parallel to the primary field E_x have different signs for electrons and holes.

In the general case, when eqn. (J. 225) is not satisfied, the Hall field prevents a charge transport in the y direction. The Hall field accelerates electrons and decelerates holes, or vice versa, and is established in such a way that the currents of electrons and holes are equal in the y direction and no net charge is transported. The electrons and holes which reach the surface of the crystal undergo recombination, while at the opposite surface electron–hole pairs are produced; in this way the carrier density remains constant in the whole crystal.

The sign of the Hall effect tells us which type of carrier is mainly responsible for the charge transport. Electron conduction is predominant if

$$b_n^2 n > b_p^2 p, \quad \text{i.e.} \quad R_H < 0 \tag{J. 226}$$

and hole conduction is predominant if

$$b_n^2 n < b_p^2 p, \quad \text{i.e.} \quad R_H > 0. \tag{J. 227}$$

Because of the dependence on the square of the mobilities this need not mean that electron or hole conduction is due to predominant electron or hole concentrations. Therefore, in general, the Hall coefficient changes sign not precisely at equal concentrations ($n = p$) but at

$$\beta^2 \equiv \left(\frac{b_n}{b_p}\right)^2 = \frac{p}{n}, \tag{J. 228}$$

where β is the mobility ratio.

In the particular case of intrinsic semiconductors with $n_i = n = p$ we obtain from (J. 222)

$$R_i = -\frac{1}{|e|\, n_i} \frac{\frac{\overline{\tau_n^2}}{\bar{\tau}_n^2} b_n^2 - \frac{\overline{\tau_p^2}}{\bar{\tau}_p^2} b_p^2}{(|b_n| + b_p)^2}. \tag{J. 229}$$

The electron mobility is in most cases higher than the hole mobility and the Hall coefficient is therefore negative. An exception is, for instance, Mg_3Sb_2 with $R_i > 0$.

For an n-type semiconductor with $\beta^2 n \gg p$ we obtain from eqn. (J. 222)

$$R_n = -\frac{\overline{\tau_n^2}}{\bar{\tau}_n^2} \frac{1}{|e|\, n} < 0, \tag{J. 230}$$

and, similarly, for a p-type semiconductor with $p \gg \beta^2 n$

$$R_p = +\frac{\overline{\tau_p^2}}{\bar{\tau}_p^2} \frac{1}{|e|\, p} > 0. \tag{J. 231}$$

It is thus possible to distinguish n-type and p-type semiconductors according to the sign of their Hall coefficients if the ratio of the hole and electron concentrations is much smaller or much larger than the square of the mobility ratio.

(α) *Hall mobility*. The electrical conductivity of an n-type or p-type semiconductor is given by

$$\sigma_n = |eb_n|n \tag{J. 232}$$

or

$$\sigma_p = |eb_p|p. \tag{J. 233}$$

Combining this with eqn. (J. 230) or (J. 231) we obtain

$$|R_n\sigma_n| = \frac{\overline{\tau_n^2}}{\overline{\tau_n}^2}|b_n| \tag{J. 234}$$

or

$$|R_p\sigma_p| = \frac{\overline{\tau_p^2}}{\overline{\tau_p}^2}b_p. \tag{J. 235}$$

We generally define the Hall mobility as

$$b_H \equiv |R_H\sigma|. \tag{J. 236}$$

A comparison with eqns. (J. 234) and (J. 235) shows that this definition is reasonable for predominant impurity conduction. In this case the carrier mobility and the Hall mobility are equal except for a factor r which is of the order of unity (cf. p. 401). In the case of impurity semiconductors it is thus possible to obtain information about the carrier mobility from measurements of the electrical conductivity and the Hall coefficient. Moreover, a measurement of the temperature dependences of the conductivity and the Hall coefficient informs us about the principal scattering mechanisms (cf. p. 400).

If, however, the products $b_n^2 n$ and $b_p^2 p$ are of the same order of magnitude, the concept of the Hall mobility loses its physical significance.

(β) *Temperature dependence of the Hall coefficient*. The temperature dependence of the Hall coefficient is essentially determined by the temperature dependence of the predominant carrier concentration.

In the impurity conduction range the Hall coefficient is inversely proportional to the carrier concentration according to eqn. (J. 230) or (J. 231). In the reserve range the concentration increases exponentially with the temperature (G. 92). Thus we obtain in the impurity reserve range (assuming a single donor level of energy E_D) for the Hall coefficient

$$R_n = \frac{1}{e}\left(\frac{m_n k}{2\pi\hbar^2}\right)^{-3/4} N_D^{-1/2} T^{-3/4} \exp\left(\frac{\Delta E_D}{2kT}\right). \tag{J. 237}$$

Plotting $\ln(R_n T^{3/4})$ versus $1/T$ we obtain a straight line whose slope is a measure of the activation energy of the impurity concerned.

As the temperature rises all impurity atoms become ionized and thus exhausted. The carrier concentration is then temperature-independent and given by the impurity concentration [cf. eqn. (G. 102)]. The Hall coefficient in the impurity exhaustion range is thus temperature-independent and inversely proportional to the determining impurity concentration. Such a behaviour is observed only when the impurity activation energy is small compared with the width of the energy gap and if the impurity concentration is sufficiently large so that the thermal generation of electron–hole pairs, and thus the intrinsic conduction, can be neglected.

At higher temperatures intrinsic conduction will dominate. The Hall coefficient is then given by (J. 229). Its temperature dependence is essentially determined by the intrinsic carrier concentration n_i. The factor of $1/en_i$ contains only the mobility ratio β whose temperature dependence can be neglected in most cases. From eqns. (G. 34) and (J. 229) we obtain for the Hall coefficient in the intrinsic conduction range

$$R_i = \frac{4}{e}\left(\frac{2k}{\pi\hbar^2}\right)^{-3/2}(m_n m_p)^{-3/4}T^{-3/2}\exp\left(\frac{\Delta E_i}{2kT}\right). \qquad \text{(J. 238)}$$

A plot of $\ln(R_i T^{3/2})$ versus $1/T$ again yields a straight line whose slope is a measure of the energy gap ΔE_i. When we take into account the linear temperature dependence of the energy gap [cf. eqn. (G. 57)], we obtain from the slope of the straight line the width ΔE_{i0} of the energy gap at $T = 0$°K.

The intrinsic lines of the Hall coefficient, just as those of the electrical conductivity, are identical for all samples of the same semiconductor (cf. p. 291). In contrast to the temperature dependence of the electrical conductivity, the curve of the Hall coefficient does not additionally depend on the temperature dependence of the mobility and is therefore better suited for the determination of energy differences by means of (J. 237) and (J. 238).

When the sign of the Hall coefficient is taken into account, the temperature dependence is that shown in Fig. 172. Since in general the intrinsic Hall coefficient is negative (cf. p. 392), we observe a change of sign of the Hall coefficient for p-type semiconductors.

(b) *TRANSVERSE MAGNETO-RESISTANCE*

The transverse magneto-resistance can be explained by means of an isotropic two-band model, even if the relaxation time is assumed to be energy-independent (cf. p. 389). This is true for degeneracy (mainly metals) as well as

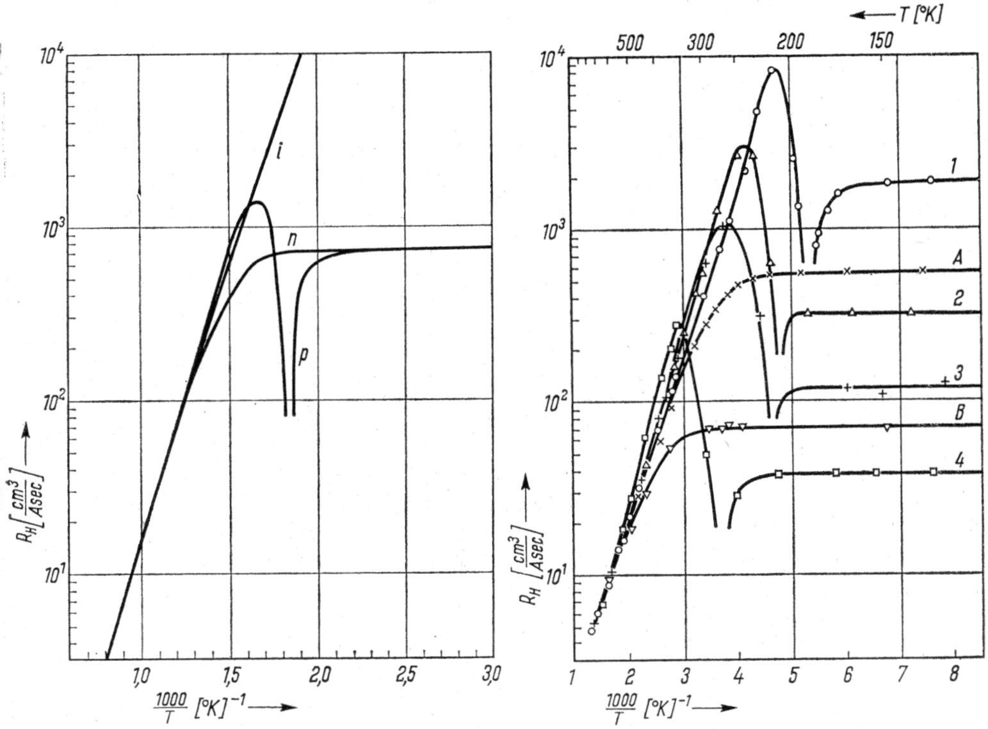

FIG. 172. Hall coefficient as a function of temperature. (a) Calculated for $\Delta E_i = 1$ eV, $\beta = 10$. (b) Measured for InSb (after O. Madelung and H. Weiss).

non-degeneracy (semiconductors). An expression of the form of (J. 213) is then obtained.

In the following we shall calculate the magneto-resistance of a non-degenerate isotropic n-type semiconductor. To solve this problem it is sufficient to use a *non-degenerate* isotropic one-band model. In this case the magneto-resistance is due only to the energy dependence of the relaxation time, and a consistent application of eqns. (J. 199), (J. 219), and (J. 220) yields values which are correct as to order of magnitude. As already mentioned (cf. p. 389), it is not sufficient in the case of a *degenerate* isotropic one-band model to take into account only the energy dependence of the relaxation time in order to explain the measured magneto-resistance of metals.

According to eqn. (J. 199), the magneto-resistance or change in conductivity of a rod-shaped sample (cf. p. 384) is given by

$$\frac{\Delta\sigma}{\sigma} = \frac{\sigma - \sigma(B_z)}{\sigma} = 1 - \frac{e^2(K_1^2 + K_2^2)}{\sigma K_1}. \qquad \text{(J. 239)}$$

The electrical conductivity of an n-type semiconductor at $B = 0$ amounts to

$$\sigma_n = \frac{e^2}{m_n} \bar{\tau}_n n. \tag{J. 240}$$

From eqns. (J. 219) and (J. 220) it follows that

$$K_{1n} = \frac{n}{m_n} \tau_n - e^2 B_z^2 \frac{n}{m_n^3} \overline{\tau_n^3} \tag{J. 241}$$

and

$$K_{2n} = -|e| B_z \frac{n}{m_n^2} \overline{\tau_n^2}. \tag{J. 242}$$

In the approximation of small magnetic fields $\left(\frac{|e|}{m_n} \cdot \tau_n B_z < 1\right)$ we obtain

$$\frac{1}{K_{1n}} \approx \frac{m_n}{n\bar{\tau}_n} \left(1 + \frac{e^2 B_z^2}{m_n^2} \frac{\overline{\tau_n^3}}{\bar{\tau}_n}\right). \tag{J. 243}$$

Substitution of eqns. (J. 240), (J. 241), (J. 242), and (J. 243) yields for the transverse conductivity change in a non-degenerate isotropic n-type semiconductor

$$\frac{\Delta\sigma}{\sigma} = \frac{e^2 B_z^2}{m_n^2} \left(\frac{\overline{\tau_n^2}^2 - \overline{\tau_n^3}\bar{\tau}_n}{\bar{\tau}_n^2}\right) \tag{J. 244}$$

or, in terms of the mobility (J. 109),

$$\frac{\Delta\sigma}{\sigma} = b_n^2 B_z^2 \left[\left(\frac{\overline{\tau_n^2}}{\bar{\tau}_n^2}\right)^2 - \frac{\overline{\tau_n^3}}{\bar{\tau}_n^3}\right]. \tag{J. 245}$$

The various mean values can be calculated with the help of eqn. (J. 263) [cf. also (J. 276) and (J. 277)]. We see immediately that the magneto-resistance is vanishing within the framework of this isotropic one-band model, if the relaxation time is independent of the energy. Since eqn. (J. 244) contains only the ratios and not the absolute magnitudes of the averaged relaxation times, it is possible to draw conclusions from the magneto-resistance about the value of the effective mass. The magneto-resistance is large if the effective mass of the dominant carriers is small (and their mobility is high). A typical example is the III–V compound InSb with $m_n/m \approx 10^{-2}$ and $|b_n| \approx 10^4$ cm^2/V sec.

VI. Scattering Processes

As a rule the transport coefficients are linked with a relaxation time; the latter is determined by the scattering mechanisms in which the carriers are involved, and is essentially the time between two collisions of a carrier. Hence it follows that the relaxation time must be inversely proportional to a collision probability or collision frequency. In this connection one often uses the mean free path Λ of a carrier which is defined as

$$\Lambda(\boldsymbol{k}) = \tau(\boldsymbol{k})\, v(\boldsymbol{k}). \tag{J. 246}$$

Both the relaxation time τ and the mean free path Λ depend on the wave vector $\boldsymbol{k}$, i.e. on energy and direction. In an ideal crystal the relaxation time as well as the mean free path is infinitely large (cf. p. 354). Both quantities, however, are limited by any perturbation of the periodic potential. In particular, the following perturbations may cause scattering processes:

1. *Thermal scattering* (*lattice scattering*), that is, scattering by acoustic or optical phonons. Owing to the lattice vibrations the carriers suffer collisions with the lattice particles which occupy their regular sites only over long time averages.
2. *Defect scattering.* All lattice defects, structural as well as chemical (cf. pp. 106 ff.), can act as scattering centres. We distinguish between scattering by ionized and neutral impurities or defects as well as scattering by dislocations.
3. *Mutual carrier scattering*, that is, electron–electron scattering as well as electron–hole scattering. Electron–electron scattering is especially important in the case of high electric fields, when the electrons acquire by acceleration in the electric field more energy per unit time than they can lose to the lattice. We talk about "hot" electrons. Electron–hole scattering can become noticeable at high temperatures in addition to thermal scattering, if the carrier concentrations are sufficiently large.

The calculation of relaxation times as a function of the wave vector is basically possible only through quantum-mechanical scattering theory. These calculations go far beyond the framework of the transport theory discussed here.

We therefore restrict ourselves to the energy dependences of the relaxation times for the most important and most frequent scattering processes. In this way we can for *isotropic* solids obtain information about the temperature dependence of the mobility [cf. eqns. (J. 66), (J. 109), (J. 110)] as well as the values of the ratios of the mean relaxation times [cf. eqns. (J. 222) and (J. 244)].

1. *Scattering by acoustic phonons.* According to Bardeen and Shockley the mean free path is obtained as

$$\Lambda_{\text{ac}} = \text{const.} \frac{1}{T}. \tag{J. 247}$$

In the case of isotropy we hence obtain from (J. 91), (J. 92), and (J. 246)

$$\tau_{ac} = a_{ac}\frac{1}{T}|E-E_R|^{-1/2}, \qquad (J. 248)$$

where a_{ac} is a proportionality factor independent of T and E.

2. *Scattering by ionized impurities.* On the basis of Rutherford's scattering formula Conwell and Weisskopf obtained for the mean free path

$$\Lambda_{ion} = \text{const}\frac{1}{N_{ion}}|E-E_R|^2, \qquad (J. 249)$$

and for the corresponding relaxation time

$$\tau_{ion} = \frac{a_{ion}}{N_{ion}}|E-E_R|^{3/2}, \qquad (J. 250)$$

where N_{ion} is the concentration of ionized impurities and a_{ion} is a T- and E-independent proportionality factor. The corresponding theory can be easily modified to apply to electron–hole scattering: the value of the factor a_{ion} must be changed and N_{ion} is to be replaced by the hole concentration.

3. *Scattering by neutral impurities.* Following Erginsoy we obtain for the mean free path

$$\Lambda_{neutral} = \text{const.}\frac{1}{N_{neutral}}|E-E_R|^{1/2}, \qquad (J. 251)$$

and for the corresponding relaxation time

$$\tau_{neutral} = \frac{a_{neutral}}{N_{neutral}}, \qquad (J. 252)$$

where $N_{neutral}$ is the concentration of the neutral impurities and $a_{neutral}$ is a proportionality factor independent of T and E.

Knowing the energy-dependence of the relaxation time we can calculate the mean values $\bar{\tau}$ as well as the mean powers $\overline{\tau^l}$ for isotropic semiconductors, according to eqn. (J. 102) for electrons and holes:

$$\overline{\tau_j^l} = \frac{a_j^l \int\limits_{E_R}^{\infty} |E-E_R|^{(lj+1)} D(E)\frac{\partial F_0}{\partial E}\,dE}{\int\limits_{E_R}^{\infty} |E-E_R|\,D(E)\frac{\partial F_0}{\partial E}\,dE} \qquad (J. 253)$$

or, using eqn. (F. 135), we obtain

$$\overline{\tau_j^l} = \frac{a_j^l \int\limits_{E_R}^{\infty} |E-E_R|^{[lj+(3/2)]} \frac{\partial F_0}{\partial E}\, dE}{\int\limits_{E_R}^{\infty} |E-E_R|^{3/2} \frac{\partial F_0}{\partial E}\, dE}. \tag{J. 254}$$

The energy dependence of the relaxation time, that is, $\tau_j = a_j |E-E_R|^j$, depends on the nature of the scattering mechanism, which is characterized by the value of j: $j = -\frac{1}{2}$ for scattering by acoustic phonons, $j = +\frac{3}{2}$ for scattering by ionized impurities, and $j = 0$ for scattering by neutral impurities. In the last case the relaxation time is independent of the energy.

Integration by parts in numerator and denominator of eqn. (J. 254) yields by analogy with (J. 105)

$$\overline{\tau_j^l} = \overline{a_j^l}\, \frac{lj+\frac{3}{2}}{\frac{3}{2}}\, \frac{\int_{E_R}^{\infty} |E-E_R|^{[lj+(1/2)]} F_0\, dE}{\int_{E_R}^{\infty} |E-E_R|^{1/2} F_0\, dE}. \tag{J. 255}$$

Introducing the substitutions

$$x = \frac{E-E_R}{kT}, \tag{J. 256}$$

$$\alpha = \frac{\zeta-E_R}{kT} \tag{J. 257}$$

and applying (J. 18), we obtain from (J. 255)

$$\overline{\tau_j^l} = \frac{2}{3}\left(lj+\frac{3}{2}\right) a_j^l (kT)^{lj}\, \frac{\int\limits_0^{\infty} \frac{x^{[lj+(1/2)]}}{\exp(x-\alpha)+1}\, dx}{\int\limits_0^{\infty} \frac{x^{1/2}}{\exp(x-\alpha)+1}\, dx}. \tag{J. 258}$$

The integrals in numerator and denominator are Fermi integrals [cf. eqns. (E. 38) and (G. 21)], that is, independent of the degree of degeneracy we can write

$$\overline{\tau_j^l} = \frac{2}{3}\left(lj+\frac{3}{2}\right) a_j^l (kT)^{lj} \frac{F_{[lj+(1/2)]}(\alpha)}{F_{1/2}(\alpha)}. \tag{J. 259}$$

In the special case of non-degeneracy, that is, for $\alpha < 0$, the Fermi integrals can be expressed in terms of the gamma function, provided the subscript q of the Fermi integral is not a negative integer:

$$F_q(\alpha) = e^{\alpha} \int_0^{\infty} x^q e^{-x}\, dx = e^{\alpha}\Gamma(q+1). \tag{J. 260}$$

We have

$$\Gamma(q+1) = q\Gamma(q) \tag{J. 261}$$

and

$$\Gamma(\tfrac{3}{2}) = \tfrac{1}{2}\Gamma(\tfrac{1}{2}) = \tfrac{1}{2}\pi^{1/2}. \tag{J. 262}$$

For non-degenerate semiconductors (J. 259) can be rewritten in the form

$$\overline{\tau_j^l} = \frac{4}{3\pi^{1/2}} a_j^l (kT)^{lj}\, \Gamma\left(lj+\frac{5}{2}\right). \tag{J. 263}$$

From this we obtain the following temperature dependence of the mean relaxation time $\bar{\tau}$ ($l = 1$) and thus of the carrier mobilities [eqns. (J. 109) and (J. 110)]. For thermal scattering by acoustic phonons ($j = -\frac{1}{2}$; recall that $a_{-1/2} = a_{ac}/T$)

$$\bar{\tau}_{ac} = \text{const. } T^{-3/2}, \tag{J. 264}$$

that is, the carrier mobilities decrease with increasing temperature:

$$b_{ac} = \text{const. } T^{-3/2}. \tag{J. 265}$$

In the case of scattering by ionized impurities or electron–hole scattering ($j = +\frac{3}{2}$) we have

$$\bar{\tau}_{ion} = \text{const. } T^{+3/2}, \tag{J. 266}$$

that is, the carrier mobilities increase as the temperature rises:

$$b_{ion} = \text{const. } T^{+3/2}. \tag{J. 267}$$

The scattering by neutral impurities ($j = 0$) is temperature independent according to (J. 263). It is observed only at low temperatures when the concentrations of neutral and ionized impurities are comparable or when the concentration of the neutral impurities exceeds that of the ionized ones.

With the help of eqn. (J. 263) we can obtain values for the various ratios of mean relaxation times contained in the formulas for the Hall effect and the magneto-resistance. Using eqns. (J. 261) and (J. 262) we obtain for the scattering by acoustic phonons ($j=-\frac{1}{2}$)

$$\frac{\overline{\tau^2}}{\bar{\tau}^2}=\frac{3}{8}\pi=1.18, \qquad \text{(J. 268)}$$

$$\frac{\overline{\tau^3}}{\bar{\tau}^3}=\frac{9}{16}\pi=0.56 \qquad \text{(J. 269)}$$

and for the scattering by ionized impurities ($j=+\frac{3}{2}$)

$$\frac{\overline{\tau^2}}{\bar{\tau}^2}=\frac{315}{512}\pi=1.93, \qquad \text{(J. 270)}$$

$$\frac{\overline{\tau^3}}{\bar{\tau}^3}=\frac{15}{8}\pi=5.89. \qquad \text{(J. 271)}$$

For an impurity semiconductor in particular we obtain for the Hall coefficients [cf. eqns. (J. 230) and (J. 231)] in the case of acoustic scattering

$$(R_n)_{\text{ac}}=-\frac{3\pi}{8}\frac{1}{|e|\,n} \qquad \text{(J. 272)}$$

and

$$(R_p)_{\text{ac}}=+\frac{3\pi}{8}\frac{1}{|e|\,p} \qquad \text{(J. 273)}$$

and in the case of scattering by ionized impurities

$$(R_n)_{\text{ion}}=-\frac{315}{512}\pi\frac{1}{|e|\,n}, \qquad \text{(J. 274)}$$

$$(R_p)_{\text{ion}}=+\frac{315}{512}\pi\frac{1}{|e|\,p}. \qquad \text{(J. 275)}$$

In the case of acoustic scattering the magneto-resistance of an n-type semiconductor is obtained from (J. 245) as

$$\left(\frac{\Delta\sigma}{\sigma}\right)_{\text{ac}}=\frac{\pi-4}{\pi}\left(\frac{3\pi}{8}\right)^2(b_nB_z)^2=-0.379(b_nB_z)^2 \qquad \text{(J. 276)}$$

and in the case of ionic scattering it is

$$\left(\frac{\Delta\sigma}{\sigma}\right)_{\text{ion}} = \left[\left(\frac{315}{512}\pi\right)^2 - \frac{15}{8}\pi\right](b_n B_z)^2 = -2.15(b_n B_z)^2. \quad \text{(J. 277)}$$

The numerical factors given here are strictly valid only for a definite scattering mechanism. Because of the different temperature dependences of the relaxation times [cf. eqns. (J. 264) and (J. 266)] it is often the case that in certain temperature ranges a single scattering mechanism predominates so that the corresponding relations for the Hall coefficient and the magneto-resistance [e.g. eqns. (J. 274) and (J. 277)] are decisive for these temperature ranges.

In principle, however, several scattering mechanisms will be simultaneously active at a given temperature. The resulting collision frequency of a carrier of energy E is then obtained from an addition of the collision frequencies characteristic for the various processes, that is,

$$\frac{1}{\tau(E)} = \sum_j \frac{1}{\tau_j(E)}. \quad \text{(J. 278)}$$

The relaxation time (averaged over the energy), which is contained in the expressions of the transport coefficients, is obtained when (J. 278) is substituted in (J. 102). This is a very complex calculation; to a good approximation, however, we can add instead of the collision frequencies pertaining to the energy E the *individually* averaged collision frequencies, that is,

$$\frac{1}{\bar{\tau}} \approx \sum \frac{1}{\bar{\tau}}. \quad \text{(J. 279)}$$

Thus we obtain for the net carrier mobility in the case of the so-called mixed scattering (i.e. the simultaneous action of several scattering mechanisms):

$$\frac{1}{b_{\text{total}}} \approx \sum_j \frac{1}{b_j}. \quad \text{(J. 280)}$$

A comparison with (J. 264) and (J. 266) shows that the resulting mobility is at low temperatures mainly due to scattering by ionized impurities while at higher temperatures thermal scattering becomes decisive (cf. Fig. 173). The net mobility as a function of the temperature can be obtained immediately from the measurement of the temperature dependence of the Hall mobility, provided that $b_n^2 n \lesssim b_p^2 p$ (cf. p. 393). This behaviour also manifests itself in the temperature dependence of the electrical conductivity (cf. Fig. 174).

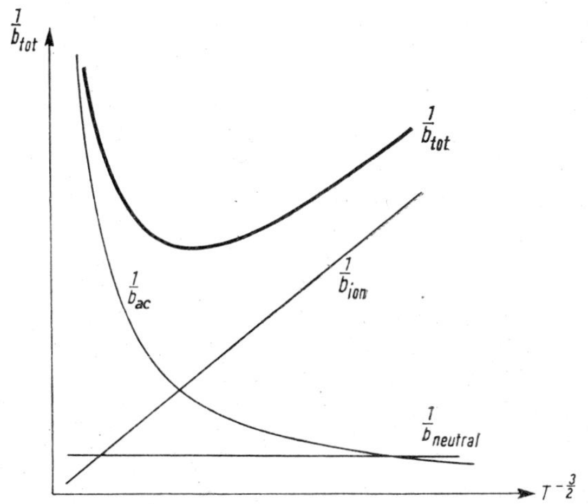

FIG. 173. Mobility contributions from various scattering mechanisms under the assumption of additivity of the inverse mobility contributions.

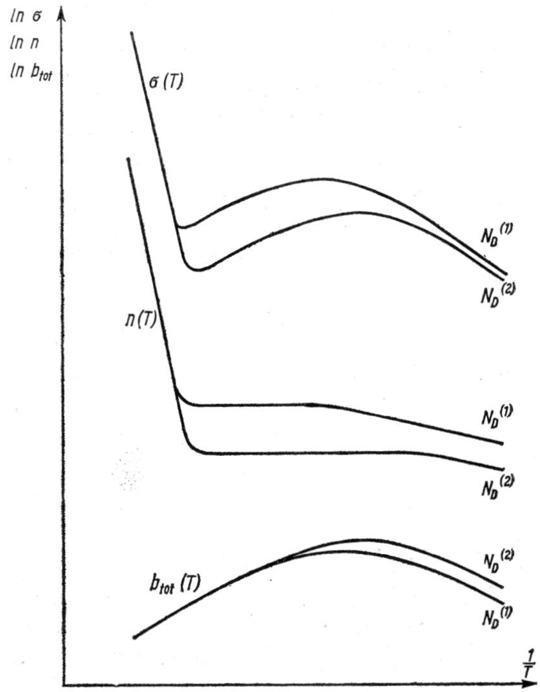

FIG. 174. Temperature dependence of mobility, carrier concentration, and electrical conductivity of a non-degenerate n-type semiconductor for two different doping concentrations, $N_D^{(1)} > N_D^{(2)}$.

References

BEER, A. C., *Galvanomagnetic Effects in Semiconductors*, in *Solid State Physics*, Suppl. 4 (Academic Press, New York 1963).

BLATT, F. J., *Physics of Electronic Conduction in Solids* (McGraw-Hill, New York 1968).

BORELIUS, G., *Grundlagen des metallischen Zustandes, Physikalische Eigenschaften der Metalle*, in *Handbuch der Metallphysik*, edited by MASING, G., Vol. 1/1 (Akademische Verlagsgesellschaft, Leipzig 1935).

BUBE, R. H., *Photoconductivity of Solids* (Wiley, New York 1960).

CUSAK, N., *The Electric and Magnetic Properties of Solids* (Longmans, London 1958).

DRABBLE, J. R., and GOLDSMID, H. J., *Thermal Conduction in Semiconductors* (Pergamon, Oxford 1961).

EHRENBERG, W., *Electrical Conductivity in Semiconductors and Metals* (Oxford University Press, London 1958).

GERRITSEN, A. N., *Metallic Conductivity, Experimental Part*, in *Handbuch der Physik*, Vol. 19 (Springer, Berlin 1956).

HEIKES, R. R. and URE, R. W., *Thermoelectricity* (Interscience, New York 1961).

JONES, H., *Theory of Electrical and Thermal Conductivity of Metals*, in *Handbuch der Physik*, Vol. 19 (Springer, Berlin 1956).

JUSTI, E., *Leitungsmechanismus und Energieumwandlung in Festkörpern* (Vandenhoek, Göttingen 1965).

MCDONALD, D. K. C., *Thermoelectricity* (Wiley, New York 1962).

MEADEN, G. T., *Electrical Resistance of Metals* (Heywood, London 1966).

OLSEN, J. L., *Electron Transport in Metals* (Interscience, New York 1962).

PUTLEY, E. H., *The Hall Effect and Related Phenomena* (Butterworth, London 1960).

ROSE, A., *Concepts in Photoconductivity* (Interscience, New York 1963).

RYVKIN, S. M., *Photoelectric Effects in Semiconductors* (Consultants Bureau, New York 1964).

SMITH, A. C., JANAK, J. F., and ADLER, R. B., *Electronic Conduction in Solids* (McGraw-Hill, New York 1967).

TAUC, J., *Photo- and Thermoelectric Effects in Semiconductors* (Pergamon, Oxford 1962).

TSIDIL'KOVSKII, L. M., *Thermomagnetic Effects in Semiconductors* (Infosearch, London 1962).

ZIMAN, J. M., *Electrons and Phonons* (Oxford University Press, London 1960).

K. Magnetism

BESIDES mechanical, electrical, and optical properties a solid also possesses magnetic properties. The model which we have satisfactorily used to derive various properties of the solid took into account only the electrostatic interaction between the electrons and the ions in the crystal. This was based on the essential assumption of a strict periodicity of the crystal (electrons in a periodic potential). The introduction of lattice imperfections permitted an explanation of further solid-state properties. Whereas so far the only properties of electrons and ions to be taken into account have been the charges and the ion radii, new basic quantities come into play when we consider magnetism: the magnetic moments of the electrons as well as of the lattice particles play a decisive part. The magnetic moment (cf. pp. 421 ff.) either is due to a magnetic field (→ diamagnetism) or is a permanent property of the particle (→ paramagnetism). The interactions of the magnetic moments with one another as well as with a magnetic excitation give rise to the magnetic phenomena which will be dealt with in the following.

I. Phenomenological Description of Magnetic Properties

The electrical and magnetic macroscopic quantities are related to one another by the Maxwell equations:

$$\operatorname{curl} \boldsymbol{E} = -\frac{\partial \boldsymbol{B}}{\partial t}, \tag{K. 1}$$

$$\operatorname{curl} \boldsymbol{H} = \boldsymbol{j} + \frac{\partial \boldsymbol{D}}{\partial t}, \tag{K. 2}$$

$$\operatorname{div} \boldsymbol{B} = 0, \tag{K. 3}$$

$$\operatorname{div} \boldsymbol{D} = \varrho, \tag{K. 4}$$

where $\boldsymbol{E}$ is the electric field strength, $\boldsymbol{D}$ is the electric displacement, $\boldsymbol{j}$ is the electric current density, ϱ is the space charge density, $\boldsymbol{H}$ is the magnetic field strength, and $\boldsymbol{B}$ is the magnetic induction. From eqns. (K. 2) and (K. 4) we obtain the equation of continuity

$$\frac{\partial \varrho}{\partial t}+\operatorname{div} \boldsymbol{j} = 0. \tag{K. 5}$$

The Maxwell equations are supplemented by so-called material equations which introduce specific properties of the material such as the electrical conductivity σ, the dielectric permeability ε, and the permeability μ:

$$\boldsymbol{j} = \sigma \boldsymbol{E}, \tag{K. 6}$$

$$\boldsymbol{D} = \varepsilon_0 \varepsilon \boldsymbol{E}, \tag{K. 7}$$

$$\boldsymbol{B} = \mu_0 \mu \boldsymbol{H}. \tag{K. 8}$$

Here ε_0 is the dielectric vacuum permittivity and μ_0 is the magnetic vacuum permeability. In the general case of anisotropic materials the material quantities σ, ε, and μ are symmetrical tensors which, after transformation to principal axes, have three components each. We shall restrict ourselves to isotropic materials for which σ, ε, and μ are scalars, that is, the related vectors are always parallel to one another.

According to Mie the magnetization $\boldsymbol{M}$ of a material is defined by the relation

$$\boldsymbol{B} = \mu_0(\boldsymbol{H}+\boldsymbol{M}). \tag{K. 9}$$

Thus the total induction has two components: one, $\mu_0 \boldsymbol{H}$, which also exists *in vacuo*, and one, $\mu_0 \boldsymbol{M}$, $\boldsymbol{M}$ being the contribution of the substance to the magnetic field. According to eqn. (K. 9) $\boldsymbol{M}$ and $\boldsymbol{H}$ have the same dimensions. Using the formulation of (K. 8) and (K. 9), we have introduced the Giorgi–Mie system of units.

The magnetization $\boldsymbol{M}$ is the sum of the magnetic moments $\boldsymbol{m}_i$ per unit volume:

$$\boldsymbol{M} = \sum_{\text{Volume}} \boldsymbol{m}_i. \tag{K. 10}$$

The magnetic properties of a substance are thus the result of the simultaneous action of elementary magnetic moments which, according to Ampère, are equivalent to molecular circular currents, that is,

$$\boldsymbol{m} = \lim_{\substack{I \to \infty \\ F \to 0}} (I\boldsymbol{A}), \tag{K. 11}$$

where I is the circular current which encircles an area $|\boldsymbol{A}|$ while the vector $\boldsymbol{A}$ is normal to this area in a direction such that, seen in the direction of $\boldsymbol{A}$, the circular current flows in a clockwise direction.

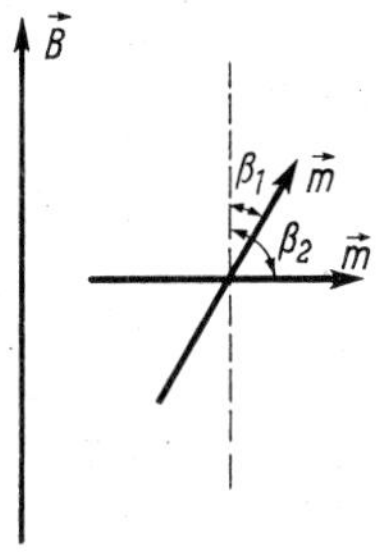

FIG. 175. Definition of the potential energy of a magnetic moment subject to a magnetic induction **B**.

From this definition we obtain immediately the dimension of the magnetic moment in the Giorgi–Mie system as A m^2; the magnetization per unit volume, **M**, has then the dimension A/m, that is, the same dimension as the magnetic field **H**.

Under the influence of the induction **B** a mechanical torque will act on the magnetic moment **m**, which is given by

$$\mathbf{G} = [\mathbf{m} \wedge \mathbf{B}]. \tag{K. 12}$$

When the dipole is rotated over an angle $\beta_2 - \beta_1$, the torque will perform the work (see Fig. 175)

$$E = \int_{\beta_2}^{\beta_1} (\mathbf{G} \cdot d\boldsymbol{\beta}), \tag{K. 13}$$

where $\boldsymbol{\beta}$ is the angle between the direction of the magnetic moment and the induction ($d\boldsymbol{\beta}$ and **G** have the same direction). Applying (K. 12) we obtain then

$$E = mB(\cos\beta_2 - \cos\beta_1). \tag{K. 14}$$

When we choose the zero of energy to correspond to the case of perpendicular vectors **m** and **B** ($\beta_2 = \pi/2$), the energy variation E_p caused by a rotation through $\frac{1}{2}\pi - \beta_1$ is given by

$$E_p = -(\mathbf{m} \cdot \mathbf{B}). \tag{K. 15}$$

The energy E_p is called the energy of orientation or the potential energy of the dipole **m** under the action of the induction **B**.

If the magnetic induction depends on the position **r**, a translatory force **F** will act on the magnetic moment in addition to the torque; it is given by

$$\mathbf{F} = (\mathbf{m} \cdot \mathrm{grad}_r)\, \mathbf{B}, \tag{K. 16}$$

or, resolved into components, by

$$F_x = m_x \frac{\partial B_x}{\partial x} + m_y \frac{\partial B_x}{\partial y} + m_z \frac{\partial B_x}{\partial z},$$

$$F_y = m_x \frac{\partial B_y}{\partial x} + m_y \frac{\partial B_y}{\partial y} + m_z \frac{\partial B_y}{\partial z}, \quad \text{(K. 17)}$$

$$F_z = m_x \frac{\partial B_z}{\partial x} + m_y \frac{\partial B_z}{\partial y} + m_z \frac{\partial B_z}{\partial z}.$$

The equations (K. 12), (K. 13) and (K. 16) represent the basis for measurements of magnetizations or magnetic susceptibilities.

1. MAGNETIC SUSCEPTIBILITY

Combining eqns. (K. 8) and (K. 9) we obtain

$$\boldsymbol{M} = (\mu - 1)\boldsymbol{H} = \varkappa \boldsymbol{H} \quad \text{(K. 18)}$$

where

$$\varkappa = \mu - 1. \quad \text{(K. 19)}$$

Like μ, the quantity $\varkappa$ is a dimensionless material property, the so-called volume susceptibility. When the magnetization is referred not to unit volume but to unit mass, the volume susceptibility must be divided by the density ϱ of the material. The so-called mass susceptibility (specific susceptibility) χ is given by

$$\chi = \frac{\varkappa}{\varrho}. \quad \text{(K. 20)}$$

The atomic and molar susceptibilities (χ_A and χ_M) are obtained by multiplying χ by the atomic weight A and the molecular weight M, respectively:

$$\chi_A = \chi A \quad \text{and} \quad \chi_M = \chi M. \quad \text{(K. 21)}$$

According to the value of $\varkappa$ or μ we distinguish the following magnetic properties:

1. Diamagnetism, i.e. $\varkappa < 0$ or $\mu < 1$: the resulting moment $\boldsymbol{M}$ per unit volume and the magnetic field $\boldsymbol{H}$ have opposite directions according to eqn. (K. 18).
2. Paramagnetism, i.e. $\varkappa > 0$ or $\mu > 1$: $\boldsymbol{M}$ and $\boldsymbol{H}$ have the same directions.

In diamagnetic and paramagnetic substances the susceptibility is usually independent of the magnetic field, that is,

$$\varkappa = \frac{M}{H} = \text{const.} \qquad \text{(K. 22)}$$

As a rule, diamagnetism is independent of the temperature while paramagnetism is temperature dependent.

Below a certain temperature paramagnetic substances may become ferrimagnetic, ferromagnetic, or antiferromagnetic (cf. pp. 459 ff.). In these cases the susceptibility is always positive and depends on the magnetic field $\boldsymbol{H}$, that is

$$\boldsymbol{M} = \varkappa(\boldsymbol{H})\boldsymbol{H} \qquad \text{(K. 23)}$$

where

$$\varkappa(\boldsymbol{H}) > 0.$$

Ferrimagnetism or ferromagnetism occurs if $\varkappa(\boldsymbol{H}) \approx \mu(\boldsymbol{H}) \gg 1$. The interrelation of $\boldsymbol{M}$ and $\boldsymbol{H}$ is generally not single-valued and has the form of a so-called hysteresis loop. In this case a substance is characterized not by the field-dependent susceptibility $\varkappa(\boldsymbol{H})$ but directly by the magnetization curve $\boldsymbol{M}(\boldsymbol{H})$.

Finally, substances are said to be metamagnetic when they display an antiferromagnetic behaviour in weak magnetic fields and a ferromagnetic behaviour in strong fields.

With this we have sketched the variety of magnetic phenomena. We have thus the problem of calculating the susceptibility or the magnetization curve $\boldsymbol{M}(\boldsymbol{H})$ in terms of the specific properties of the material, the history of the material, and the temperature (see pp. 433 ff.).

2. MAGNETIZATION WORK

The magnetization work is defined as the energy per unit volume necessary to change the magnetization of a substance by a certain amount. The change of magnetization at constant temperature is a consequence of a change in magnetic field [cf. eqn. (K. 18)] and can be achieved in two ways:

1. The body is brought into a coil and the current through the coil is changed.
2. The distance between the pole of a permanent magnet and the body is changed.

The body together with the device producing the magnetic field represents a system. The work A_1 to be performed by this system in order to change the magnetization by a certain amount depends on the type of the device. In order to obtain the true magnetization work A of the body, a thought experiment is necessary in which the device is separated from the magnetized body. The work A_2 is necessary to bring the device back to its original state in such a way that the change of magnetization of the body achieved previously is not influenced. In this case the magnetization work is independent of the way of magnetization and equal to

$$A = A_1 + A_2. \tag{K. 24}$$

Let us now calculate the magnetization work for the two different ways of magnetization:

1. *Magnetization with the help of the coil.* Assume the body to be a closed ring which completely fills the interior of the coil. When the ohmic resistance of the coil is neglected, the applied voltage U is equal to the induced voltage U_{ind} which is given by the time variation of the magnetic induction:

$$U = -U_{\text{ind}} = nq\frac{\partial B}{\partial t}, \tag{K. 25}$$

where q is the coil cross-section and n is the number of windings. The coil current I produces the magnetic field

$$H = nI/l, \tag{K. 26}$$

where l is the mean circumference of the coil. In the time dt the current source performs the work

$$dE_1 = UI\,dt, \tag{K. 27}$$

which using (K. 25) and (K. 26) can be written as

$$dE_1 = qlH\,dB. \tag{K. 28}$$

In order to change the magnetic induction from $B = 0$ to $B = B_1$, the system coil plus body must perform the work A_1 per unit volume:

$$A_1 = \frac{1}{ql}\int dE_1 = \int_0^{B_1} H\,dB. \tag{K. 29}$$

The magnetic field H_1 produces the magnetic induction B_1 and the magnetization M_1. With the help of (K. 9) we obtain

$$A_1 = \mu_0\left(\int_0^{H_1} H\,dH + \int_0^{M_1} H\,dM\right) \tag{K. 30}$$

or

$$A_1 = \tfrac{1}{2}\mu_0 H_1^2 + \mu_0 \int_0^{M_1} H\, dM. \tag{K. 31}$$

The first term in eqn. (K. 31) represents the energy density of the magnetic field in empty space; this energy, per unit volume of the coil, is needed to produce the same magnetic field H_1 as when no body is inside the coil. The second term is the energy per unit volume necessary to produce the magnetization M_1. When we assume the achieved magnetization M_1 to remain unchanged when the coil current is switched off, that is, when the magnetization device is brought back to its original state, then the work A_2 is gained per coil volume. By analogy with (K. 29) we have

$$A_2 = \int_{B_1}^{0} H\, dB. \tag{K. 32}$$

For the case of constant magnetization we obtain by analogy to (K. 30)

$$A_2 = -\tfrac{1}{2}\mu_0 H_1^2. \tag{K. 33}$$

According to (K. 24) the magnetization work amounts to

$$A = \mu_0 \int_0^{M_1} H\, dM \tag{K. 34}$$

or, for small changes of the magnetization,

$$\delta A = \mu_0 H\, dM. \tag{K. 35}$$

2. *Magnetization with the help of a permanent magnet.* The pole of a magnetic rod produces a position-dependent magnetic field $\boldsymbol{H}(\boldsymbol{r})$, and thus according to (K. 8) the vacuum magnetic induction

$$\boldsymbol{B}(\boldsymbol{r}) = \mu_0 \boldsymbol{H}(\boldsymbol{r}). \tag{K. 36}$$

According to (K. 10) and (K. 16), the force

$$\boldsymbol{F} = (\boldsymbol{M}\cdot\text{grad}_r)\, \boldsymbol{B} \tag{K. 37}$$

acts per unit volume on a body with the magnetization $\boldsymbol{M}$; $\boldsymbol{B}(\boldsymbol{r})$ is the magnetic induction at the point $\boldsymbol{r}$ *in vacuo* without a body. The body is assumed to be so small that $\boldsymbol{B}(\boldsymbol{r})$ may be taken as constant over its volume. The action of the force results in a motion of the body whereby the energy of the system body plus permanent magnet is changed by A_1'.

For the sake of simplicity we consider only the motion in the direction of the axis of the bar magnet (x-axis) and assume the body to be magnetizable only in the x direction. Equation (K. 37) then takes the form

$$F_x = M \frac{\partial B}{\partial x}. \tag{K. 38}$$

When $M > 0$ the body is attracted by the magnetic pole. In the motion over distance dx the mechanical work δA_1 is set free, that is, using (K. 38), we have

$$\delta A_1' = -M \, dB \tag{K. 39}$$

or using (K. 36)

$$\delta A_1' = -\mu_0 M \, dH. \tag{K. 40}$$

The total work connected with a displacement from a point where $H = 0$ to a point where $H = H_1$ amounts to

$$A_1' = -\mu_0 \int_0^{H_1} M \, dH \tag{K. 41}$$

or

$$A_1' = -\mu_0 \int_0^{M_1 H_1} d(MH) + \mu_0 \int_0^{M} H \, dM = -\mu_0 M_1 H_1 + \mu_0 \int_0^{M_1} H \, dM. \tag{K. 42}$$

In this form (K. 42) can be compared with (K. 31). The first term represents the reduction of the potential energy of the body [cf. eqn. (K. 15)] when it is brought from a place where $H = 0$ to one where $H = H_1$. The second term is the actual magnetization work. When we again imagine the body to be brought back to its original position so that the initial distance between body and pole is restored while the magnetization M_1 of the body should remain unchanged, the work A_2' must be performed,

$$A_2' = -\mu_0 \int_{H_1}^{0} M \, dH, \tag{K. 43}$$

that is, for $M = M_1$

$$A_2' = +\mu_0 M_1 H_1. \tag{K. 44}$$

Using (K. 42) and (K. 44) we obtain by virtue of (K. 24) the same magnetization work

$$A = \mu_0 \int_0^{M_1} H \, dM \tag{K. 45}$$

or

$$\delta A = \mu_0 H \, dM. \tag{K. 46}$$

3. DEMAGNETIZATION FACTOR

The magnetic field H is in general *not* identical with the externally applied magnetic field, but rather is the internal or effective magnetic field. It is this field which determines the magnetization of the material, cf. eqn. (K. 18). The relation between the magnetic field H and the externally applied magnetic field H_{ext} depends on the geometry and the chemical and structural homogeneity of the magnetic sample. We have

$$H = H_{ext} - NM, \qquad \text{(K. 47)}$$

where N is the demagnetization factor. This numerical factor N is a measure for the attenuation of the external field by the presence of the magnetized matter and is therefore called the demagnetization factor.

Strictly speaking, the relation (K. 47) is only meaningful when a uniform external magnetic field induces a uniform effective magnetic field and thus a uniform magnetization. This is the case for ellipsoidal samples whose major axes are parallel to the external field H_{ext}, or for a ring core in a coil. Generally, however, a uniform external field is distorted by the insertion of an arbitrarily shaped sample and thus the effective magnetic field is non-uniform. The influence of chemical and structural inhomogeneities of the sample cannot be calculated in the general case.

Let us calculate the demagnetization factor for a ring core with an air gap (cf. Fig. 176) and show that there exists a connection between the external magnetic field and the effective magnetic field in the form of eqn. (K. 47). If

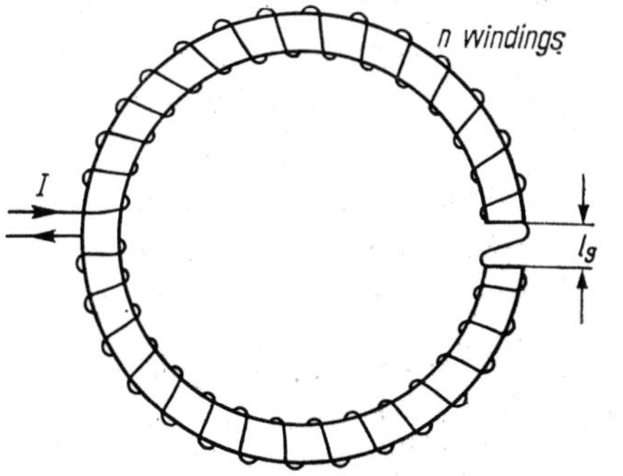

FIG. 176. Calculation of the demagnetization factor of magnetic cores with a gap.

the ring radius is large compared with the radius of the ring cross-section, the coil field H_{ext} can be assumed to be uniform. The coil field is identical with the magnetic field in the coil, in the absence of the ring core, that is, in air or vacuum. It represents the external magnetic field the ring core is exposed to, and it is given by

$$H_{\text{ext}} = \frac{n}{l} I, \tag{K. 48}$$

where n is the number of turns of the coil, l is its mean circumference, and I is the current intensity in the coil. If the ring core has a gap of the width l_g, its "length" will be given by $l_c = l - l_g$. Since no sources exist for the magnetic induction B [cf. eqn. (K. 3)], the inductions in the gap (B_g) and in the core (B) are equal, i.e.

$$B_g = B. \tag{K. 49}$$

This equation results from (K. 3) under the assumption of such a small gap width l_g that the magnetic lines of force cross the gap uniformly. In this case the induction has a single component in the core as well as in the gap, which is perpendicular to the core cross-section.

Using (K. 9) we obtain from (K. 49) a relation linking the magnetic fields in the gap, H_g, and in the core, H:

$$\mu_0 H_g = \mu_0 (H + M). \tag{K. 50}$$

The vacuum magnetization is identically equal to zero.

The relation to the external magnetic field (coil field) H_{ext} results from Laplace's law which represents an integral form of the Maxwell equation (K. 2):

$$\oint H_l \, dl = \int_F j \, df. \tag{K. 51}$$

Here H_l is the component of the magnetic excitation along a closed curve l, dl is a line element along it, df is a surface element of the area F bounded by the closed curve l, and j is the current density perpendicular to the surface element df.

The displacement current is here equal to zero. For a coil with n turns and a coil current I eqn. (K. 51) can be rewritten in the form

$$\oint H_l \, dl = nI. \tag{K. 52}$$

The contour integral can easily be calculated for piecewise constant magnetic fields. Using (K. 48) we obtain from (K. 52)

$$H l_c + H_g l_g = H_{\text{ext}} l. \tag{K. 53}$$

Combining eqns. (K. 50) and (K. 53) we obtain for the effective magnetic excitation in the core

$$H = H_{\text{ext}} - \frac{l_g}{l} M. \tag{K. 54}$$

That is, using eqn. (K. 47),

$$N = \frac{l_g}{l}. \tag{K. 55}$$

In particular for a closed ring core, that is, when $l_g = 0$, we have

$$H = H_{\text{ext}}. \tag{K. 56}$$

Only in this particular case is the effective magnetic field equal to the applied coil field, as has been assumed in our calculation of the magnetization work (cf. p. 410).

Substituting the expression for the magnetization, we obtain from (K. 18) instead of (K. 54)

$$H = H_{\text{ext}} - N\varkappa H$$

or

$$H = \frac{H_{\text{ext}}}{1+N\varkappa}. \tag{K. 57}$$

For ellipsoids of revolution as well as circular cylinders the values of N are compiled in Table 25. In the case of a sphere $N = \frac{1}{3}$.

TABLE 25. Demagnetization factors for ellipsoids of revolution and circular cylinders; magnetic field parallel to the axis of rotation and axis of cylinder, respectively; a and b are the dimensions of the sample parallel and perpendicular to the magnetic field, respectively (after W. F. Brown, *Magnetostatic Principles in Ferromagnetism* [North-Holland, Amsterdam 1962], p. 192)

Ratio of dimensions a/b	Demagnetization factor N: Ellipsoid of revolution	Demagnetization factor N: Circular cylinder	Ratio of dimensions a/b	Demagnetization factor N: Ellipsoid of revolution	Demagnetization factor N: Circular cylinder
0.0	1.000	1.000	20	0.006 749	0.020 91
0.2	0.750 5	0.680 2	30	0.003 444	0.014 01
0.4	0.588 2	0.528 1	40	0.002 116	0.010 53
0.8	0.394 4	0.361 9	60	0.001 053	0.007 039
1.0	0.333 3	0.311 6	80	0.000 637	0.005 286
1.5	0.233 0	0.230 1	100	0.000 430	0.004 232
2.0	0.173 6	0.181 9	150	0.000 209	0.002 824
3.0	0.108 7	0.127 8	200	0.000 125	0.002 119
4.0	0.075 41	0.098 35	300	0.000 060	0.001 413
6.0	0.043 23	0.067 28	400	0.000 036	0.001 060
8.0	0.028 42	0.051 10	600	0.000 017	0.000 707
10	0.020 29	0.041 19	800	0.000 010	0.000 530
15	0.010 75	0.027 74	1000	0.000 007	0.000 424

The demagnetizing field is particularly strong for ferrimagnetic and ferromagnetic samples with high susceptibilities. In the case of paramagnetic and diamagnetic samples we can write generally to a good approximation

$$H \approx H_{\text{ext}}. \quad \text{(K. 58)}$$

II. Thermodynamics of Magnetization

The magnetic properties of matter are generally not independent of the thermal and mechanical properties. The phenomenological relationship between these properties can generally be described with the help of thermodynamics.

A body possesses an internal energy U which depends only on its state. According to the first law of thermodynamics the increment of internal energy is equal to the sum of heat and work supplied to the body, that is,

$$dU = \delta Q + \delta A. \quad \text{(K. 59)}$$

The quantities U, Q, and A refer to unit volume.

We have shown that the change of magnetization is related to the magnetization work. Under the assumption that magnetization does not entail a change in volume ($dV = 0$) and applying eqn. (K. 35), the first law can be written in the form

$$dU = \delta Q + \mu_0 H\, dM. \quad \text{(K. 60)}$$

While for diamagnetic and paramagnetic substances the assumption of $dV = 0$ is justified, this is not the case for ferromagnetic substances which, in general, display a magnetization-dependent volume variation: magneto-striction, $(\partial V/\partial M)_{p,T} \neq 0$.

The internal energy U is a quantity of state which, according to eqn. (K. 60), depends only on the temperature and the magnetization. The total differential of $U(T, M)$ consists of a thermal component corresponding to the temperature variation and a magnetic component corresponding to the magnetization variation:

$$dU = \left(\frac{\partial U}{\partial T}\right)_M dT + \left(\frac{\partial U}{\partial M}\right)_T dM. \quad \text{(K. 61)}$$

1. THE MAGNETO-CALORIC EFFECT

The temperature variation connected with a change of the magnetic state only is called the magneto-caloric effect. Let us calculate the temperature variation due to an adiabatic variation of the magnetization, i.e. the body is assumed not to exchange heat with its environment, $\delta Q = 0$. We then obtain from (K. 60) and (K. 61)

$$\mu_0 H\, dM = \left(\frac{\partial U}{\partial T}\right)_M dT + \left(\frac{\partial U}{\partial M}\right)_T dM. \tag{K. 62}$$

By analogy to (C. 40) the specific heat per unit mass at constant magnetization is given by

$$c_M = \frac{1}{\varrho}\left(\frac{\partial U}{\partial T}\right)_M, \tag{K. 63}$$

where ϱ is the matter density. Using this we obtain from (K. 62)

$$dT = \frac{1}{\varrho c_M}\left[\mu_0 H - \left(\frac{\partial U}{\partial M}\right)_T\right] dM. \tag{K. 64}$$

In order to calculate the change $(\partial U/\partial M)_T$, in the internal energy with magnetization, we must apply the second law of thermodynamics, according to which the entropy S is a function of state. The differential dS is defined by the heat δQ exchanged in a reversible process, and the absolute temperature:

$$dS = \frac{\delta Q_{\text{rev}}}{T}. \tag{K. 65}$$

A change of state is reversible if, with the help of a finite number of variables of state, the state of a system can be defined unambiguously at any point within the range of the variation of state. For ferrimagnetic and ferromagnetic substances for example there are irreversible changes of state since there exists no unambiguous relation between magnetization and magnetic field (hysteresis loop $M(H)$).

Substitution of (K. 65) in (K. 60) yields

$$dS = \frac{1}{T}\,dU - \mu_0 \frac{H}{T}\,dM \tag{K. 66}$$

or, using (K. 66),

$$dS = \frac{1}{T}\left(\frac{\partial U}{\partial T}\right)_M dT + \frac{1}{T}\left[\left(\frac{\partial U}{\partial M}\right)_T - \mu_0 H\right] dM. \tag{K. 67}$$

According to this, the quantity of state S is a function of T and M; the total differential is given by

$$dS = \left(\frac{\partial S}{\partial T}\right)_M dT + \left(\frac{\partial S}{\partial M}\right)_T dM. \tag{K. 68}$$

A comparison of (K. 67) and (K. 68) shows us that

$$\left(\frac{\partial S}{\partial T}\right)_M = \frac{1}{T}\left(\frac{\partial U}{\partial T}\right)_M \tag{K. 69}$$

and

$$\left(\frac{\partial S}{\partial M}\right)_T = \frac{1}{T}\left[\left(\frac{\partial U}{\partial M}\right)_T - \mu_0 H\right]. \tag{K. 70}$$

For total differentials the mixed second derivatives with respect to the pertinent independent variables are identical, that is,

$$\frac{\partial}{\partial M}\left(\frac{\partial S}{\partial T}\right)_M = \frac{\partial}{\partial T}\left(\frac{\partial S}{\partial M}\right)_T \tag{K. 71}$$

or, from (K. 69) and (K. 70),

$$\frac{1}{T}\frac{\partial^2 U}{\partial T\,\partial M} = -\frac{1}{T^2}\left[\left(\frac{\partial U}{\partial M}\right)_T - \mu_0 H\right] + \frac{1}{T}\left[\frac{\partial^2 U}{\partial M\,\partial T} - \mu_0\frac{\partial H}{\partial T}\right] \tag{K. 72}$$

or

$$\left(\frac{\partial U}{\partial M}\right)_T = \mu_0 H - \mu_0 T\left(\frac{\partial H}{\partial T}\right)_M. \tag{K. 73}$$

This is the variation of the internal energy with magnetization, at constant temperature, in terms of the magnetic field, the temperature T, and the temperature variation of the magnetic field at constant magnetization.

Substitution of (K. 73) in (K. 64) yields for the magneto-caloric effect

$$dT = \frac{\mu_0}{\varrho c_M} T\left(\frac{\partial H}{\partial T}\right)_M dM. \tag{K. 74}$$

The temperature variation in the magneto-caloric effect is, for practical reasons, suitably given in terms of the variation of magnetic field, dH. In order to recalculate (K. 74) we have to make use of the relation between the specific heats at constant magnetization (c_M) and at constant magnetic field (c_H) which will be derived on p. 419. According to it, we have

$$c_H - c_M = -\frac{\mu_0}{\varrho} T\left(\frac{\partial H}{\partial T}\right)_M\left(\frac{\partial M}{\partial T}\right)_H. \tag{K. 75}$$

The magnetization M is a function of temperature T and magnetic field H, that is,

$$M = M(T, H) \tag{K. 76}$$

and

$$dM = \left(\frac{\partial M}{\partial T}\right)_H dT + \left(\frac{\partial M}{\partial H}\right)_T dH. \tag{K. 77}$$

In the special case of M = const. we obtain from this

$$\left(\frac{\partial M}{\partial H}\right)_T \left(\frac{\partial H}{\partial T}\right)_M = -\left(\frac{\partial M}{\partial T}\right)_H. \qquad \text{(K. 78)}$$

Substitution of (K. 77) in (K. 74) yields

$$dT = \frac{\mu_0}{\varrho c_M} T \left(\frac{\partial H}{\partial T}\right)_M \left[\left(\frac{\partial M}{\partial H}\right)_T dH + \left(\frac{\partial M}{\partial T}\right)_H dT\right]. \qquad \text{(K. 79)}$$

Hence follows with (K. 75) and (K. 78) the magneto-caloric effect in terms of dH:

$$dT = -\frac{\mu_0}{\varrho c_H} T \left(\frac{\partial M}{\partial T}\right)_H dH. \qquad \text{(K. 80)}$$

Equations (K. 74) and (K. 80) hold generally if the variations of state can be taken to be reversible. In diamagnetic substances the magnetization is temperature independent, that is, no magneto-caloric effect occurs. In paramagnetic substances $(\partial M/\partial T)_H < 0$ so that, according to (K. 80), a reduction of magnetic field ($dH < 0$) will entail a reduction of temperature ($dT < 0$). The adiabatic demagnetization of paramagnetic substances is an important method used to reach low temperatures.

2. SPECIFIC HEATS

We define here the specific heat as the ratio of the heat supplied per unit mass and the resultant temperature rise:

$$c \equiv \frac{1}{\varrho} \frac{\delta Q}{dT}. \qquad \text{(K. 81)}$$

Just as the specific heat at constant volume (c_V) is different from that at constant pressure (c_p), one obtains different values for the specific heats in the case of heat supply at constant magnetic field (c_H) and in the case of heat supply at constant magnetization (c_M). Actually, a magnetic substance has four different specific heats, namely c_{VM}, c_{VH}, c_{pM}, and c_{pH}. We again neglect the volume variations and do not distinguish between c_{VM} and c_{pM} or c_{VH} and c_{pH}.

Using eqns. (K. 60) and (K. 61) we obtain the specific heat at constant magnetization from (K. 81) as

$$c_M = \frac{1}{\varrho} \left(\frac{\partial U}{\partial T}\right)_M. \qquad \text{(K. 82)}$$

For the specific heat at constant magnetic field we find with the help of (K. 65), (K. 67), and (K. 73) from (K. 81)

$$c_H = \frac{1}{\varrho}\left(\frac{\delta Q}{dT}\right)_H = \frac{1}{\varrho}\left(\frac{\partial U}{\partial T}\right)_M - \frac{\mu_0}{\varrho}T\left(\frac{\partial H}{\partial T}\right)_M\left(\frac{\partial M}{\partial T}\right)_H \qquad \text{(K. 83)}$$

or, using eqn. (K. 82),

$$c_H = c_M - \frac{\mu_0}{\varrho}T\left(\frac{\partial H}{\partial T}\right)_M\left(\frac{\partial M}{\partial T}\right)_H. \qquad \text{(K. 84)}$$

Experimentally one determines the specific heat c_H at constant magnetic field generally as $H \to 0$.

3. SPECIFIC HEAT ANOMALIES

The internal energy, and thus also the specific heats c_M and c_H, consist of two components:

1. A contribution U_0 which is independent of the magnetic properties, which is determined by the vibrational energy U_v (cf. p. 53), the imperfection energy nW (cf. p. 118), and the energy U of the electron gas (cf. p. 169).
2. A contribution U_M due to magnetization.

Thus the internal energy is the sum

$$U(T, M) = U_0(T) + U_M(T, M). \qquad \text{(K. 85)}$$

From this we obtain instead of (K. 82)

$$c_M = c_0 + \frac{1}{\varrho}\left(\frac{\partial U_M}{\partial T}\right)_M \qquad \text{(K. 86)}$$

and instead of (K. 84)

$$c_H = c_0 + \frac{1}{\varrho}\left(\frac{\partial U_M}{\partial T}\right)_M - \frac{\mu_0}{\varrho}T\left(\frac{\partial H}{\partial T}\right)_M\left(\frac{\partial M}{\partial T}\right)_H \qquad \text{(K. 87)}$$

where

$$c_0 = \frac{1}{\varrho}\left(\frac{\partial U_0}{\partial T}\right). \qquad \text{(K. 88)}$$

The difference $c_H - c_0$ is called the specific heat anomaly. Its value is always positive since $(\partial H/\partial T)_M > 0$ and $(\partial M/\partial T)_H < 0$ (cf. pp. 439 ff.). The specific heat anomaly is due to the energy which is necessary to overcome the magnetic interactions. It is particularly marked in the case of ferrimagnetic, ferromagnetic, and antiferromagnetic substances (cf. pp. 459 ff.).

III. Atomistic Description of the Magnetic Moments

The lattice particles are the carriers of the magnetic moments. According to the theory of atoms, the magnetic moments are caused by:

1. the motion of the electrons around the nuclei (orbital moments);
2. the electrons' angular momenta (spin moments); and
3. the nuclear angular momenta (nuclear moments).

1. ORBITAL MOMENT

According to Bohr's classical model an atom consists of a positive nucleus surrounded by electrons which move along certain stationary orbits. Each orbiting electron represents a circular current which, according to Ampère, is equivalent to a magnetic moment.

Let us first calculate the magnetic orbital moment of an isolated hydrogen atom. Its single electron moves with the angular velocity ω along a circular orbit of radius r around the positive nucleus. The resulting circular current is

$$I = -\frac{|e|}{2\pi}\omega. \qquad \text{(K. 89)}$$

According to eqn. (K. 11) this corresponds to a magnetic moment

$$\mu_l = -\frac{|e|}{2} r^2\omega. \qquad \text{(K. 90)}$$

When we introduce the orbital angular momentum $\boldsymbol{l}$ of the atom as

$$\boldsymbol{l} = [\boldsymbol{r} \wedge m\boldsymbol{v}] = mr^2\omega, \qquad \text{(K. 91)}$$

we obtain from (K. 90)

$$\mu_l = -\frac{|e|}{2m}\boldsymbol{l}, \qquad \text{(K. 92)}$$

where m is the rest mass of the electron and $\boldsymbol{v}$ is the orbital velocity of the electron.

The relation (K. 92) is called the magnetomechanical parallelism; it relates the magnetic moment to the mechanical orbital angular momentum. Because of the negative electron charge the magnetic moment and the orbital angular momentum have opposite directions. The relation in (K. 92) applies generally to the motion of an electron under the influence of a central force, even if the closed orbit of the electron is not circular. To the coupling between the intrinsic angular momentum of the electron (or nucleus) and the pertinent spin (or nuclear) moment it applies however only qualitatively (magnetomechanical anomaly, cf. p. 426).

It follows already from the classical theory of the atom that the orbital angular momentum can assume only discrete values. According to Sommerfeld's quantum condition the phase integral is given by

$$\oint p\,dx = n\,2\pi\hbar, \tag{K. 93}$$

n being an integer. Hence there follows for the periodic motion of an electron along a circular orbit

$$\oint mv\,ds = 2\pi mvr = n\,2\pi\hbar \tag{K. 94}$$

or, from (K. 91),

$$|\boldsymbol{l}| = n\hbar. \tag{K. 95}$$

The orbital angular momentum and thus also the magnetic moment are quantized. Equation (K. 95), however, is not an exact relation as the following quantum-mechanical calculation will show. In order to calculate the orbital momentum we start from the Schrödinger equation for the system considered. It can be solved exactly only for the one-particle problem; so we first consider the motion of a single electron in the Coulomb field of the positive charge Ze of the nucleus. The electron has the potential energy

$$V(r) = -\frac{1}{4\pi\varepsilon_0}\,\frac{Ze^2}{r}; \tag{K. 96}$$

r is the distance between electron and nucleus.

The time-independent Schrödinger equation reads

$$\nabla^2\psi + \frac{2m}{\hbar^2}\left(E - \frac{1}{4\pi\varepsilon_0}\,\frac{Ze^2}{r}\right)\psi = 0. \tag{K. 97}$$

Since the potential energy is a spherical-symmetric function of position, we introduce the polar coordinates r, ϑ, φ. This permits a separation of variables and we obtain solutions of the form

$$\psi(r, \vartheta, \varphi) = R(r)\,\Theta(\vartheta)\,\Phi(\varphi). \tag{K. 98}$$

Apart from continuity and finiteness of the solutions, the boundary conditions require that ψ be single-valued and vanishing as $r \to \infty$. Single-valuedness means that ψ is a periodic function of the angles so that a change of an angle by 2π must not change the value of ψ. Taking these conditions into account, we obtain the various solutions $\psi_{n,\,l,\,m_l}(r, \vartheta, \varphi)$ which are characterized by the three integers n, l, m_l. It follows from the mathematical method of solution that the numerical ranges of these three numbers depend on one another, that is,

$$\begin{aligned} n &= 1, 2, \ldots, \\ l &= 0, 1, \ldots, n-1, \\ m_l &= 0, \pm 1, \pm 2, \ldots, \pm l. \end{aligned} \tag{K. 99}$$

The integers n, l, m_l are called quantum numbers which are attributed to the eigenfunctions $\psi_{n,\,l,\,m_l}$ of the system considered. In the first instance, the quantum numbers are merely a consequence of the mathematical problem. However, they determine the discrete values of physical quantities which characterize the system.

A physical quantity has precise definite values if the pertinent operator applied to the eigenfunction $\psi_{n,\,l,\,m_l}$ yields the same eigenfunction multiplied by a constant factor. The operator of a physical quantity $Q(x, y, z; p_x, p_y, p_z)$ can be obtained when the momentum components are replaced by the operators $-i\hbar\, \partial/\partial x$, $-i\hbar\, \partial/\partial y$, $-i\hbar\, \partial/\partial z$. The quantity Q has discrete values q (eigenvalues) if

$$Q_{\text{op}}\psi_{n,\,l,\,m_l} = q\psi_{n,\,l,\,m}, \tag{K. 100}$$

where q = const.

In our problem of a hydrogenic atom the relation (K. 100) is satisfied for the energy of the system as well as for the magnitude and one component of the angular momentum.

We obtain the energy eigenvalues E_n from

$$H\psi_{n,\,l,\,m_l} = E_n\psi_{n,\,l,\,m_l}, \tag{K. 101}$$

where

$$E_n = -\frac{me^4Z^2}{32\pi^2\varepsilon_0^2\hbar^2}\,\frac{1}{n^2}, \tag{K. 102}$$

H being the energy Hamiltonian.

The quantum number n is called the principal quantum number. It determines the energy of the electron. The K, L, M, N, O, P, Q electron shells correspond to the values n = 1, 2, 3, 4, 5, 6, 7. We obtain the eigenvalues of the square of the angular momentum from

$$l^2_{\text{op}}\psi_{n,\,l,\,m_l} = l_l^2\psi_{n,\,l,\,m_l} \tag{K. 103}$$

with

$$l_l^2 = l(l+1)\hbar^2. \tag{K. 104}$$

The operator $\boldsymbol{l}_{\text{op}}$ of the angular momentum is obtained when in the classical expression for the angular momentum

$$\boldsymbol{l} = [\boldsymbol{r} \wedge \boldsymbol{p}] \tag{K. 105}$$

the momentum $\boldsymbol{p}$ is replaced by the operator $(\hbar/i)\,\text{grad}_r$:

$$\boldsymbol{l}_{\text{op}} = \frac{\hbar}{i}[\boldsymbol{r} \wedge \text{grad}_r], \tag{K. 106}$$

or, resolved into components,

$$\begin{aligned} l_x &= yp_z - zp_y \quad \rightarrow \quad (l_x)_{\text{op}} = \frac{\hbar}{i}\left(y\frac{\partial}{\partial z} - z\frac{\partial}{\partial y}\right), \\ l_y &= zp_x - xp_z \quad \rightarrow \quad (l_y)_{\text{op}} = \frac{\hbar}{i}\left(z\frac{\partial}{\partial x} - x\frac{\partial}{\partial z}\right), \\ l_z &= xp_y - yp_x \quad \rightarrow \quad (l_z)_{\text{op}} = \frac{\hbar}{i}\left(x\frac{\partial}{\partial y} - y\frac{\partial}{\partial x}\right). \end{aligned} \tag{K. 107}$$

The integer l is the orbital angular momentum or azimuthal quantum number and characterizes the shape of the electron orbit. Electrons with the states $l = 0, 1, 2, \ldots (n-1)$ are called $s, p, d, f, g, \ldots$ electrons. The s electrons have in principle no orbital angular momentum and therefore, according to eqn. (K. 92), no orbital magnetic moment.

When by applying a magnetic field a certain spatial direction is preferred (e.g. the z direction), we can write

$$(l_z)_{\text{op}}\psi_{n,\,l,\,m_l} = (l_z)_{m_l}\psi_{n,\,l,\,m} \tag{K. 108}$$

with

$$(l_z)_{m_l} = m_l\hbar. \tag{K. 109}$$

Condition (K. 100) for discrete values is not satisfied for the other components. The magnetic quantum number m_l determines the value of the angular momentum component parallel to a preferred direction; according to (K. 109) these values are quantized. The possible directions of the total angular momentum are therefore not arbitrary; only such directions are allowed whose z component satisfies (K. 109) (the so-called directional quantization; cf. Fig. 177). According to eqn. (K. 99) $2l+1$ possible z components pertain to a definite value of the orbital quantum number. It is worth mentioning that the vector of the total angular momentum can never lie precisely in the z direction since the value of $\hbar[l(l+1)]^{1/2}$ always slightly exceeds the maximum z component $\hbar l$.

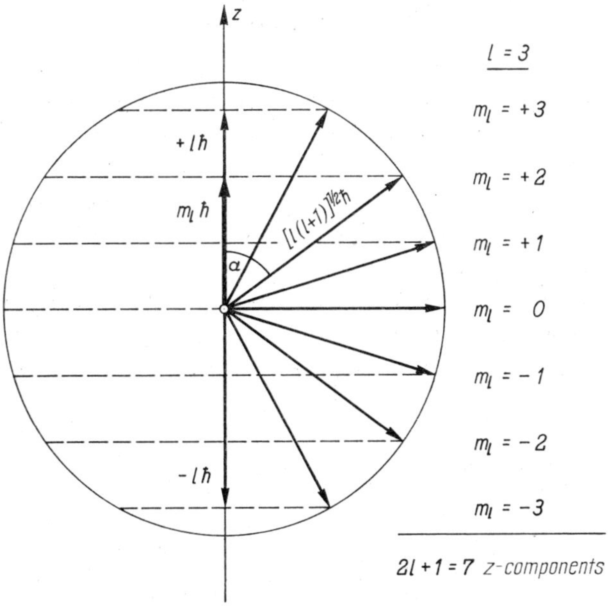

FIG. 177. Quantization of the z-component of the orbital angular momentum.

When we apply these quantum-mechanical results to eqn. (K. 92), we see that the magnitude of the orbital magnetic moment can only assume discrete values. Using (K. 92) we obtain

$$\mu_l = [l(l+1)]^{1/2}\,\mu_B \tag{K. 110}$$

with

$$\mu_B = \frac{|e|\,\hbar}{2m} = 0.927\times 10^{-23}\ \mathrm{A\ m^2}, \tag{K. 111}$$

where μ_B is Bohr's magneton, the basic unit for the magnetic moments of electrons.

From eqns. (K. 92) and (K. 109) we see that the component of the orbital magnetic moment parallel to the magnetic field (and not the magnitude of the total orbital moment) is equal to integral multiples of Bohr's magneton:

$$\mu_{l_z} = m_l \mu_B\,. \tag{K. 112}$$

The direction of the total orbital moment is not fixed but precesses about the direction of the magnetic field; according to eqns. (K. 110) and (K. 112), however, only certain definite precession angles α are allowed (cf. Fig. 177) which are given by

$$\cos\alpha = \frac{m_l}{[l(l+1)]^{1/2}}\,. \tag{K. 113}$$

2. SPIN MOMENT

Independent of its energy and orbital motion the electron rotates about its own axis and thus possesses an intrinsic angular momentum $\boldsymbol{s}$ which is called the spin. The classical calculation of this angular momentum and the magnetic moment caused by it (the so-called spin moment) again leads us to eqn. (K. 92). From measurements (spectroscopic line splitting in the magnetic field, Stern–Gerlach experiment, gyromagnetic ratio), however, it follows that eqn. (K. 92) does not describe the relationship between the intrinsic angular momentum and the spin moment: the ratio of spin moment to angular momentum is larger by a factor $g_e = 2$ than according to (K. 92). The spin moment $\boldsymbol{\mu}_s$ is given by

$$\boldsymbol{\mu}_s = -g_e \frac{|e|}{2m} \boldsymbol{s}, \qquad \text{(K. 114)}$$

where g_e is the gyromagnetic factor. This effect which is called the magneto-mechanical anomaly of the electron spin can only be explained by means of Dirac's relativistic wave-mechanical theory of the electron and cannot be given within the framework of this book. According to Dirac's theory $g_e = 2$; improved theories and accurate measurements yield the value

$$g_e = 2 \times 1.0011596. \qquad \text{(K. 115)}$$

By analogy with the orbital angular momentum we obtain for the eigenvalues of the spin

$$|\boldsymbol{s}| = [s(s+1)]^{1/2} \hbar \qquad \text{(K. 116)}$$

with the half-odd-integral spin quantum number

$$s = \tfrac{1}{2}. \qquad \text{(K. 117)}$$

In the presence of an external magnetic field (e.g. in the z direction) the magnetic spin quantum numbers m_s are also half-odd-integral,

$$m_s = \pm s = \pm \tfrac{1}{2}. \qquad \text{(K. 118)}$$

As in (K. 109) the z component of the spin is given by

$$s_z = \pm \tfrac{1}{2} \hbar. \qquad \text{(K. 119)}$$

Thus we obtain for the directional quantization of the spin moment

$$\mu_{sz} = \pm g_e \frac{e}{2m} \frac{\hbar}{2} \qquad \text{(K. 120)}$$

or, using eqns. (K. 111) and (K. 115),

$$\mu_{sz} = \pm 1.0011596\mu_B \approx \pm 1\mu_B. \qquad \text{(K. 121)}$$

The spin component parallel to a preferred direction is thus equal to $\pm\hbar/2$, and the corresponding component of the spin moment is virtually equal to one Bohr magneton. According to eqns. (K. 116) and (K. 117), the magnitude of the spin is given by

$$|\boldsymbol{s}| = \frac{3^{1/2}}{2}\hbar \qquad \text{(K. 122)}$$

and that of the spin moment [cf. (K. 111), (K. 114), and (K. 116)] by

$$|\boldsymbol{\mu}_s| = g_e[s(s+1)]^{1/2}\,\mu_B \qquad \text{(K. 123)}$$

or, using eqns. (K. 115) and (K. 117),

$$|\boldsymbol{\mu}_s| = g_e\frac{3^{1/2}}{2}\,\mu_B \approx 3^{1/2}\mu_B. \qquad \text{(K. 124)}$$

3. SPIN–ORBIT COUPLING

The superposition of several angular momenta yields a resulting angular momentum and a resulting magnetic moment. In particular the total angular momentum $\boldsymbol{j}$ of an electron, which consists of the orbital angular momentum and the intrinsic angular momentum, is given as

$$\boldsymbol{j} = \boldsymbol{l}+\boldsymbol{s}. \qquad \text{(K. 125)}$$

By analogy with eqns. (K. 104) and (K. 116) we can write for the magnitude of the total angular momentum

$$|\boldsymbol{j}| = [j(j+1)]^{1/2}\,\hbar \qquad \text{(K. 126)}$$

where

$$j = l \pm s = l \pm \tfrac{1}{2}, \qquad \text{(K. 127)}$$

j being the quantum number of angular momentum. For an atom with a single electron there exist for each orbital quantum number l two values of angular momentum which are determined by eqn. (K. 126). In this case the quantum numbers j are all half-odd-integral.

Because of the directional quantization of the orbital and intrinsic angular momenta the total angular momentum is also quantized. From eqns. (K. 109),

(K. 120), and (K. 125) we obtain for the component j_z of the total angular momentum parallel to the z direction

$$j_z = m_j\hbar \tag{K. 128}$$

where

$$m_j = m_l + m_s. \tag{K. 129}$$

According to (K. 99) m_j can assume the following values:

$$j, j-1, \ldots, -(j-1), -j. \tag{K. 130}$$

To each total angular momentum given by the quantum number j there exist $2j+1$ possible components parallel to an external magnetic field.

The total angular momentum has a resulting magnetic moment μ whose direction, because of the magneto-mechanical anomaly, is not parallel to the total angular momentum. When we draw the vectors of the orbital and spin moments in units of the corresponding angular momenta (cf. Fig. 178), the vector of the orbital moment is of equal magnitude but in the opposite direction of the orbital angular momentum; the vector of the spin moment, however, is twice as long as that of the intrinsic angular momentum and oppositely

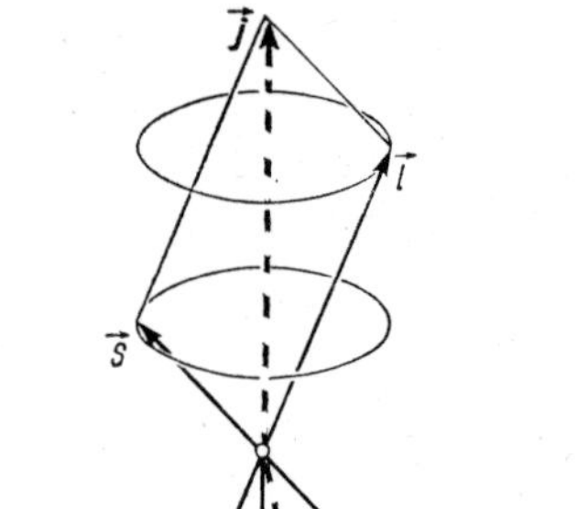

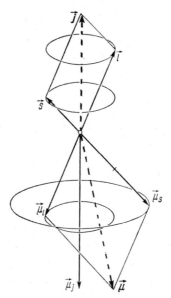

FIG. 178. Vector addition of the angular momenta $\boldsymbol{s}$ and $\boldsymbol{l}$ and coupling of the respective magnetic moments. Observe the magneto-mechanic anomaly of the electron spin.

directed [cf. eqns. (K. 92) and (K. 114)]. Considered from the classical viewpoint, the vectors $\boldsymbol{l}$ and $\boldsymbol{s}$ precess about the vector $\boldsymbol{j}$. The vectors μ_l and μ_s will then also precess about the direction of $\boldsymbol{j}$. This means that only the component μ_j parallel to $\boldsymbol{j}$ determines the magnetic moment. The component of μ perpendicular to $\boldsymbol{j}$, averaged over time, is equal to zero because of the precession of the vectors μ_l and μ_s.

The relationship between the total angular momentum and the effective component μ_j of the magnetic moment can be derived from Fig. 178 with the help of the cosine law. We have

$$\mu_j = \mu_l \cos(\boldsymbol{l}, \boldsymbol{j}) + \mu_s \cos(\boldsymbol{s}, \boldsymbol{j}) \tag{K. 131}$$

with

$$\cos(\boldsymbol{l}, \boldsymbol{j}) = \frac{\boldsymbol{l}^2 + \boldsymbol{j}^2 - \boldsymbol{s}^2}{2\,|\boldsymbol{l}|\,|\boldsymbol{j}|} = \frac{l(l+1)+j(j+1)-s(s+1)}{2[l(l+1)\,j(j+1)]^{1/2}} \tag{K. 132}$$

and

$$\cos(\boldsymbol{s}, \boldsymbol{j}) = \frac{\boldsymbol{s}^2 + \boldsymbol{j}^2 - \boldsymbol{l}^2}{2\,|\boldsymbol{s}|\,|\boldsymbol{j}|} = \frac{s(s+1)+j(j+1)-l(l+1)}{2[s(s+1)\,j(j+1)]^{1/2}}. \tag{K. 133}$$

Substitution of eqns. (K. 110), (K. 124), (K. 132), and (K. 133) in (K. 131) yields

$$\mu_j = g[j(j+1)]^{1/2}\,\mu_B \tag{K. 134}$$

with the Landé factor

$$g = 1 + \frac{j(j+1)+s(s+1)-l(l+1)}{2j(j+1)}. \tag{K. 135}$$

When we take into account only the orbital moment or the spin moment $s = 0$ and $j = l$ or $l = 0$ and $j = \pm s$ [cf. eqn. (K. 127)]. In this case the Landé factor becomes $g = g_l = 1$ or $g = g_e = 2$. We then obtain from (K. 134) either (K. 110) or (K. 124).

Because of the directional quantization of the total angular momentum, only certain components μ_{jz} of the magnetic moment are allowed, that is, we obtain, using (K. 128) and (K. 134),

$$\mu_{jz} = g m_j \mu_B. \tag{K. 136}$$

4. RUSSELL–SAUNDERS COUPLING

So far we have restricted our considerations to hydrogenic atoms with a single electron. In the case of atoms with several electrons, the orbital and intrinsic angular momenta of all electrons interact with one another. We are first confronted with the problem of calculating the quantum states of a neutral atom with nuclear charge $+Ze$. For the solution of such a many-body problem the so-called central-field method is a suitable approximation method, which can also explain the entire structure of the periodic system of elements. We consider one electron in the field of the nucleus and the spherically averaged charge distribution of the other $Z-1$ electrons. In this way we have reduced the many-body problem to a one-body problem; the way of solution for such a problem has been indicated on p. 422. The state of each electron is characterized by the quantum numbers n_i, l_i, m_{li}, m_{si}. The relations linking these quantum numbers are the same as in (K. 99).

The values of the resulting total angular momentum and the resulting magnetic moment of the atom considered depend on the kind of interaction between the various moments. Spectroscopic measurements have verified the assumption that normally the interaction between the orbital moments as well as between the spin moments is much stronger than that between the individual spin and orbital moments. One therefore adds up all the orbital angular momenta and all the intrinsic angular momenta separately and determines the total angular momentum from the resulting orbital and intrinsic angular momenta (the Russell–Saunders or LS coupling). We obtain for the resulting orbital angular momentum

$$\boldsymbol{L} = \sum_i \boldsymbol{l}_i, \tag{K. 137}$$

for the resulting intrinsic angular momentum

$$\boldsymbol{S} = \sum_i \boldsymbol{s}_i \tag{K. 138}$$

and for the total angular momentum

$$\boldsymbol{J} = \boldsymbol{L} + \boldsymbol{S}. \tag{K. 139}$$

According to the conditions of the one-body problem the magnitudes of these three vectors are quantized; the eigenvalues are

$$|\boldsymbol{L}| = [L(L+1)]^{1/2}\hbar \tag{K. 140}$$

with

$$L = 0, 1, 2, \ldots, \tag{K. 141}$$

$$|\boldsymbol{S}| = [S(S+1)]^{1/2}\hbar \tag{K. 142}$$

with

$$S = 0, \tfrac{1}{2}, 1, \tfrac{3}{2}, \ldots, \tag{K. 143}$$

and

$$|\boldsymbol{J}| = [J(J+1)]^{1/2}\hbar \tag{K. 144}$$

with

$$J = L+S, L+S-1, \ldots |L-S| \tag{K. 145}$$

or

$$J = 0, \tfrac{1}{2}, 1, \tfrac{3}{2}, \ldots .$$

Moreover, we have the following z components:

$$L_z = M_L \hbar \tag{K. 146}$$

with

$$M_L = L, L-1, \ldots -(L-1), -L, \tag{K. 147}$$

$$S_z = M_S \hbar \tag{K. 148}$$

with

$$M_S = S, S-1, \ldots -(S-1), -S, \tag{K. 149}$$

and

$$J_z = M_J \hbar \tag{K. 150}$$

with

$$M_J = J, J-1, \ldots -(J-1), -J. \tag{K. 151}$$

The atomic magnetic moments corresponding to the angular momenta are obtained as in (K. 92), (K. 114), and (K. 134). We obtain for the orbital moment

$$\boldsymbol{M}_L = -\frac{|e|}{2m}\boldsymbol{L} \tag{K. 152}$$

and

$$|\boldsymbol{M}_L| = [L(L+1)]^{1/2}\,\mu_B, \tag{K. 153}$$

for the spin moment

$$\boldsymbol{M}_S = -g_e\frac{|e|}{2m}\boldsymbol{S} \tag{K. 154}$$

and

$$|\boldsymbol{M}_S| = g_e[S(S+1)]^{1/2}\,\mu_B, \tag{K. 155}$$

for the effective component of the magnetic moment

$$\boldsymbol{M}_J = -g\frac{|e|}{2m}\boldsymbol{J} \tag{K. 156}$$

and

$$|\boldsymbol{M}_J| = g[J(J+1)]^{1/2}\,\mu_B \tag{K. 157}$$

with

$$g = 1+\frac{J(J+1)+S(S+1)-L(L+1)}{2J(J+1)}. \tag{K. 158}$$

From (K. 158) we obtain $g = 1$ for pure orbital moments ($S = 0$, $J = L$) and $g = 2$ for pure spin moments ($L = 0$, $J = S$). In the general case of Russell–Saunders coupling the Landé factor can also be higher than two since the spin quantum number S can assume higher values than $\frac{1}{2}$ [cf. eqns. (K. 117) and (K. 143)].

The magnetic moment observed in an external magnetic field is in analogy to (K. 136) given by

$$M_{Jz} = gM_J\mu_B. \tag{K. 159}$$

Besides the quantum rules (K. 140) to (K. 151) the Pauli principle and Hund's rules determine the resulting angular momentum and thus the resulting magnetic moment of an atom. According to the Pauli principle a state specified by the quantum numbers n, l, m_l, s can be occupied by only one electron. Hence it follows immediately that a completely filled electron shell defined by the quantum numbers n and l has no magnetic moment. In an external magnetic field the effective orbital moment as well as the effective spin moment, and thus according to (K. 139) also the effective total moment, are equal to zero ($L = S = J = 0$).

A permanent magnetic moment of an atom or lattice particle depends only on the electrons of the incomplete shells. The magnitude of this moment depends on Hund's rules, which were derived by Hund on the basis of spectroscopic studies of the atom. According to them, for the ground state of an atom,

1. the sum of the spin quantum numbers, $S = \sum_i m_{si}$, has the highest value consistent with the Pauli principle;
2. the sum of the orbital quantum numbers, $L = \sum_i m_{li}$, has the highest value possible compatible with the first rule;
3. for electron shells less than half filled

$$J = L - S \tag{K. 160}$$

and for shells which are more than half filled,

$$J = L + S. \tag{K. 161}$$

5. NUCLEAR MOMENT

Like the electrons, the protons and neutrons possess also angular momenta which lead to a resulting nuclear spin $\boldsymbol{I}$. The magnitude and the component of the nuclear spin parallel to a preferred direction are again quantized. We have

$$|\boldsymbol{I}| = [I(I+1)]^{1/2}\hbar \tag{K. 162}$$

with

$$I = 0, \tfrac{1}{2}, 1, \ldots \tag{K. 163}$$

and

$$I_z = M_I \hbar \tag{K. 164}$$

with

$$M_I = I, I-1, \ldots, -(I-1), -I. \tag{K. 165}$$

By analogy with the magnetic properties of the electron spin we obtain for the nuclear magnetic moment

$$\boldsymbol{M}_{\text{nucl}} = g_{\text{nucl}} \frac{|e|}{2M} \boldsymbol{I} \tag{K. 166}$$

and

$$|\boldsymbol{M}_{\text{nucl}}| = g_{\text{nucl}}[I(I+1)]^{1/2} \mu_{\text{nucl}} \tag{K. 167}$$

with the nuclear magneton

$$\mu_{\text{nucl}} = \frac{|e|\hbar}{2M} = \frac{1}{1836} \mu_B, \tag{K. 168}$$

where g_{nucl} is the nuclear g factor and M is the proton mass.

The nuclear moments are thus by three orders of magnitude smaller than the magnetic moments of the electron shell [cf. eqn. (K. 157) as well as (K. 167) and (K. 168)]. We can therefore neglect nuclear magnetism when we explain the macroscopic magnetic properties of solids.

In this context it should be mentioned that nuclear magnetism can be measured with high accuracy by means of the methods of nuclear resonance (NMR = *N*uclear *M*agnetic *R*esonance). In solid-state physics these methods are valuable aids in the investigation of the various interactions in a crystal, e.g. the mutual interaction of nuclear moments (in structural investigations, diffusion), the interactions between nuclear moments and shell electrons (shown in hyperfine splitting of spectral lines), and those between nuclear moments and conduction electrons (in the Knight shift).

IV. Diamagnetism and Paramagnetism

The classification of the various magnetic properties is phenomenologically based on the relationship between magnetization and magnetic field, or the values of magnetic susceptibility (cf. p. 408). This classification of the substances is in agreement with the atomic properties of the molecules, atoms, or ions comprising the substance. It depends essentially on their magnetic nature, that is, whether the structural particles of the substance had a net magnetic moment or not before the application of a magnetic field. Also the type of coupling between the orbital moments and the spin moments as well as the type of chemical bond is essential for the magnetic properties. The Russell–Saunders coupling mentioned above cannot be applied in the case of heavy atoms.

Apart from complicated exceptional cases, the atomistic viewpoint leads us to the following characterization of the various magnetic properties:

Diamagnetism. In diamagnetic substances there exist no net magnetic moments before the application of a magnetic field. We shall show (cf. pp. 435 ff.) that in principle a magnetic field induces magnetic moments in all substances (induced moments). These moments are opposite to the direction of the external field and give rise to diamagnetism; all substances display diamagnetism which, however, can be observed only in the case of materials whose atoms or ions have closed electron shells ($L = S = J = 0$). Typical representatives of this group are the noble gases as well as ionic crystals (e.g. alkali halides).

Moreover, the formation of chemical bonds may result in a mutual compensation of the orbital and spin moments of neighbouring atoms so that there are no resulting moments. Atomic hydrogen, for example, is paramagnetic by virtue of the spin moment of its 1*s* electron, while molecular hydrogen is diamagnetic since the effective spin components of the two H atoms of the H_2 molecule are antiparallel. Further examples of chemical-bond-induced diamagnetism are diamond, silicon, germanium, phosphorus, and sulphur.

Paramagnetism. Paramagnetism is observed when the atoms or ions possess, even when there is no magnetic field (so-called permanent moments), net magnetic moments which are not compensated by chemical binding. In the absence of a magnetic field only the magnitude (not the direction) of the magnetic moments is defined. Typical paramagnetic substances are the salts of elements with incomplete inner shells, that is, substances containing ions of transition elements, rare earth elements, or elements of the actinide group. Paramagnetic substances may become ferrimagnetic, ferromagnetic, or antiferromagnetic when the net magnetic moments interact via the so-called exchange forces (cf. pp. 459 ff.).

The calculation of the magnetic behaviour of a solid is based on the following considerations. The magnetic properties are due to the permanent and induced moments of all electrons. In a crystal the electrons are more or less strongly bound to the nuclei. Particularly those magnetic properties which are due to the valence electrons can therefore be quite different, according to whether the atom considered is free or bound in a crystal lattice. In the case of metals the valence electrons are virtually free, that is, they are not bound to a certain nucleus (→ free electrons, cf. p. 157), and therefore have no stationary orbits. Because of their random motions the free valence electrons (conduction electrons) will have no net orbital moment; because of their spin, however, they will contribute to the magnetism of the substance.

Generally, the magnetic properties of a solid are due to two components: the magnetism of the electrons bound to the nuclei, and the magnetism of the free or quasi-free conduction electrons. It is particularly easy to explain the diamagnetic and paramagnetic properties if one knows the limiting cases of the diamagnetism and paramagnetism of free atoms and the diamagnetism and paramagnetism of free electrons. In insulators, for instance, which contain no conduction electrons, only the magnetism of the atoms will thus play a role, while in metals, in addition to the magnetism of the atomic cores, one has to take into account also the magnetism of the conduction electrons.

1. DIAMAGNETISM OF FREE ATOMS

(a) *LARMOR–LANGEVIN THEORY*

The phenomenon of the diamagnetism of free atoms can be explained within the framework of classical electrodynamics. According to eqn. (K. 1) an electrical field is generated by the variation with time of the magnetic induction (induction law). In this way a circular current of electrons appears in the atom which, according to eqn. (K. 11), corresponds to a magnetic moment. Since the electrons move in their orbits with zero resistance, the circular current and the induced moment exist as long as the magnetic field is non-zero. Under the assumption that the electron wave functions remain unchanged when a magnetic field is applied, we shall calculate in the following the induced magnetic moment of an atom, that is, the corresponding diamagnetic susceptibility.

We write eqn. (K. 1) in an integral form:

$$\oint (\boldsymbol{E}\cdot d\boldsymbol{s}) = -\int \left(\frac{\partial \boldsymbol{B}}{\partial t}\cdot d\boldsymbol{f}\right). \qquad \text{(K. 169)}$$

The electric field strength E along a circular orbit of radius r in a plane perpendicular to the magnetic induction will then be given by

$$E = -\frac{1}{2} r \frac{\partial B}{\partial t}. \qquad \text{(K. 170)}$$

If we choose an axis through the centre of the nucleus, parallel to the magnetic field, in addition to the Coulomb force on each electron at a distance r from this axis there will act a force which is given by

$$eE = m_e \frac{dv}{dt} = -\frac{e}{2} r \frac{\partial B}{\partial t}, \tag{K. 171}$$

where m_e is the electron rest mass and v is the additional velocity of the electron which is due to the field E. Integrating this expression, we obtain the electron velocity due to the induction:

$$v = -\frac{e}{2m_e} rB. \tag{K. 172}$$

Independent of their distances r_i from the preferred axis, all electrons ($i=1, \ldots Z$) of the atom will have a superimposed angular velocity

$$\omega_L = \frac{v_i}{r_i} = -\frac{e}{2m_e} B, \tag{K. 173}$$

which is called the Larmor frequency. Assume the origin of a coordinate frame (x, y, z) to coincide with the centre of the nucleus and apply a magnetic induction in the z direction, then all electrons will precess with the Larmor frequency about the z direction. Under the assumption that the electron wave functions remain unchanged when $\boldsymbol{B}$ is applied, the charge distribution in the coordinate frame which rotates with the Larmor frequency about the z-axis is the same as in a system at rest with $\boldsymbol{B} = 0$ (Larmor's theorem).

Each electron produces a circular current given by

$$I = e \frac{\omega_L}{2\pi}. \tag{K. 174}$$

Averaging over the possible distances r_i of the ith electron from the z-axis, we obtain from (K. 11)

$$\mu_i = -\frac{e^2}{4m_e} \overline{r_i^2} B. \tag{K. 175}$$

Under the assumption of spherical atoms we have

$$\overline{x_i^2} = \overline{y_i^2} = \overline{z_i^2}. \tag{K. 176}$$

Moreover,

$$\overline{r_i^2} = \overline{x_i^2} + \overline{y_i^2} \tag{K. 177}$$

and the mean square orbital radius of the ith electron is given by

$$\overline{R_i^2} = \overline{x_i^2} + \overline{y_i^2} + \overline{z_i^2}. \tag{K. 178}$$

From the last three equations we obtain

$$\overline{r_i^2} = \tfrac{2}{3}\overline{R_i^2}. \tag{K. 179}$$

Using this, eqn. (K. 175) takes the form

$$\mu_i = -\frac{e^2}{6m_e}\overline{R_i^2}B. \tag{K. 180}$$

Summation over all mean-square orbital radii for the electrons $i = 1, \ldots Z$ yields the induced magnetic moment per atom

$$\mu = -\frac{e^2}{6m_e}B\sum_{i=1}^{Z}\overline{R_i^2}. \tag{K. 181}$$

For N atoms per unit volume the induced magnetization [cf. eqn. (K. 10)] is obtained as

$$M = -\frac{e^2}{6m_e}NB\sum_{i=1}^{Z}\overline{R_i^2}. \tag{K. 182}$$

Applying (K. 9) and (K. 18) we obtain

$$\varkappa_{\text{dia}} = \frac{M}{H} = \frac{-\dfrac{\mu_0 e^2}{6m_e}N\sum\limits_{i=1}^{Z}\overline{R_i^2}}{1+\dfrac{\mu_0 e^2}{6m_e}N\sum\limits_{i=1}^{Z}\overline{R_i^2}}. \tag{K. 183}$$

An estimate shows that the second term in the denominator is much smaller than unity; thus we have for the diamagnetic volume susceptibility of free atoms

$$\varkappa_{\text{dia}} = -\frac{\mu_0 e^2}{6m_e}N\sum_{i=1}^{Z}\overline{R_i^2}. \tag{K. 184}$$

The diamagnetic susceptibility is therefore always negative. Hence it follows that the induced moment is opposite to the direction of the magnetic field. This is in agreement with Lenz's rule.

The diamagnetic susceptibility is proportional to the sum of the square orbital radii, that is, $\varkappa_{\text{dia}}$ increases with the number Z of the electrons in the atom (see below). Such a diamagnetic susceptibility is essentially independent of the temperature.

A more accurate calculation of the susceptibility makes it necessary to know $\overline{R_i^2}$ and thus the wave functions of all electrons in the atom. From eqn. (K. 184) we see immediately that the outermost electrons make the highest contribution to the diamagnetic susceptibility. Apart from the case of hydrogenic atoms, the wave functions and thus the charge distribution in the atom can only be determined by means of approximations.† A comparison between the measured and calculated diamagnetic susceptibilities is a sensitive control on the validity of the wave-mechanical approximation used.

It must be stressed that diamagnetism exists in principle in every substance but it may be masked by other forms of magnetism.

(b) *COMPARISON WITH EXPERIMENTS*

Noble gases are purely diamagnetic; for them $J = 0$ and the model of spherical atoms is well suited (cf. Table 26; note the increase of diamagnetic susceptibility with the number of electrons, Z).

TABLE 26. Diamagnetic susceptibility of some noble gases (after A. H. Morrish, *The Physical Principles of Magnetism* [Wiley, New York 1965], p. 41)

Gas	$\chi_M \cdot 10^6/4\pi$ [cm³/mole] experimental	theoretical
He	− 2.02	− 1.86
Ne	− 6.96	− 7.48
Ar	−19.23	−18.8

The assumption of spherical atoms is also well suited to alkali halides. The molar susceptibility is equal to the sum of the susceptibilities of cations and anions,

$$\chi_M = \chi_{\text{cation}} + \chi_{\text{anion}}. \tag{K. 185}$$

The validity of this relation is supported by the fact that the changes in susceptibility are equal when one passes over from, for example, chlorides to the corresponding bromides [$\Delta\chi_M$(NaCl → NaBr) = $\Delta\chi_M$(KCl → KBr]. The same applies to an exchange of cations, e.g. in the series Na(–Cl, –Br, –J): $\Delta\chi_M$(NaCl →KCl) = $\Delta\chi_M$(NaBr → KBr).

† See, e.g., D. R. Hartree, *The Calculation of Atomic Structure* (Chapman and Hall, London 1957).

The values of χ_{cation} and χ_{anion} calculated for free ions are compiled in Table 27. The experimental values of χ_M agree essentially with the corresponding sums. The table also shows that the susceptibilities of solutions and of the corresponding crystals are almost equal. From this we may conclude that the ions have a spherical electron distribution also in the crystal (→ ionic binding).

TABLE 27. Diamagnetic susceptibility of the ions, solutions, and salts of alkali halides (after C. Kittel, *Introduction to Solid State Physics*, 2nd ed. [Wiley, New York 1956], p. 209; A. H. Morrish, *The Physical Principles of Magnetism* [Wiley, New York 1965], p. 41)

	$\chi_M \cdot 10^6/4\pi$ [cm³/mole]	Cl^- −24.2	Br^- −34.5	J^- −50.6	
Li^+	−0.7	−25.2	−35.8	−53.7	Solution
		−24.3	−34.7	−50.0	Salt
Na^+	−6.1	−30.3	−42.0	−59.9	Solution
		−30.2	−41.1	−57.0	Salt
K^+	−14.6	−40.2	−51.9	−68.2	Solution
		−38.7	−49.6	−65.5	Salt

The values for crystals are as a rule a little smaller since the diameters of the electron clouds and thus $\sum \overline{R_i^2}$ are smaller than in the case of solutions or free ions as the distances to the neighbouring ions are shorter.

2. PARAMAGNETISM OF FREE ATOMS

(a) *BRILLOUIN–LANGEVIN THEORY*

Atoms with non-zero angular momentum $\boldsymbol{J}$ possess a permanent magnetic moment whose magnitude is given by

$$\mu \equiv \boldsymbol{M}_J = -g \frac{|e|}{2m} \boldsymbol{J} = -g \frac{\boldsymbol{J}}{\hbar} \mu_B. \tag{K. 186}$$

Under the influence of a magnetic induction B_z the magnetic moment can take $2J+1$ different directions, that is, the possible z components of the magnetic moment are

$$\mu_z \equiv M_{Jz} = gM_J\mu_B \tag{K. 187}$$

with

$$M_J = J, (J-1), \ldots -(J-1), -J. \tag{K. 188}$$

We consider an ideal gas of N identical atoms with permanent magnetic moments $\boldsymbol{\mu}$ and calculate the magnetization M_z caused by the magnetic induction B_z at temperature T. The "ideal gas" is understood to be a gas whose atoms do not interact. The net magnetization consists of the always present induced diamagnetic magnetization plus the magnetization due to an alignment of the permanent moments in the magnetic field. We can write

$$M_z = \sum_{M_J=-J}^{+J} n_{\mu_z}\mu_z = N\bar{\mu}_z. \qquad \text{(K. 189)}$$

Here n_{μ_z} is the number of atoms per unit volume whose magnetic moments have the z component $\mu_z = gM_J\mu_B$, and $\bar{\mu}_z$ is the z component of the magnetic moments $\boldsymbol{\mu}$ averaged over N atoms per unit volume. The number n_{μ_z} is obtained with the help of statistical mechanics. It is determined by the ratio of the potential energy of the moment $\boldsymbol{\mu}$ in the magnetic field $\boldsymbol{B}$ and the thermal energy kT of the atom. According to eqn. (K. 15) the potential energy amounts to

$$E_{\text{pot}} = -(\boldsymbol{\mu}\cdot\boldsymbol{B}) = -\mu_z B_z = E_{\mu_z}. \qquad \text{(K. 190)}$$

The probability that n_{μ_z} atoms have the z component μ_z is equal to

$$\frac{n_{\mu_z}}{N} = A\exp\left(-\frac{E_{\mu_z}}{kT}\right). \qquad \text{(K. 191)}$$

The constant A is defined by the normalization

$$\sum_{M_J}\frac{n_{\mu_z}}{N} = 1 = A\sum_{M_J}\exp\left(-\frac{E_{\mu_z}}{kT}\right). \qquad \text{(K. 192)}$$

A substitution of (K. 191) and (K. 192) in (K. 189) yields the net paramagnetic moment

$$M_z = N\frac{\sum\limits_{M_J=-J}^{+J}\mu_z\exp\left(-\frac{E_{\mu_z}}{kT}\right)}{\sum\limits_{M_J=-J}^{+J}\exp\left(-\frac{E_{\mu_z}}{kT}\right)}. \qquad \text{(K. 193)}$$

Using eqns. (K. 187) and (K. 190) and the substitution

$$a = \frac{gJ\mu_B B_z}{kT} \qquad \text{(K. 194)}$$

we can rewrite eqn. (K. 193) as

$$M_z = NgJ\mu_B \frac{\sum_{M_J=-J}^{+J} \frac{M_J}{J} \exp\left(a\frac{M_J}{J}\right)}{\sum_{M_J=-J}^{J} \exp\left(a\frac{M_J}{J}\right)}. \tag{K. 195}$$

This expression can be calculated by elementary means if we take into account that the numerator is equal to the derivative of the denominator with respect to a:

$$M_z = N_g J\mu_B \frac{d}{da}\left\{\ln\left[\sum_{M_J=-J}^{+J} \exp\left(a\frac{M_J}{J}\right)\right]\right\}. \tag{K. 196}$$

We have to sum over the $2J+1$ members of a geometrical series with the first term $\exp(aJ/J)$ and the ratio $\exp(-a/J)$. In this way we obtain

$$M_z = NgJ\mu_B \frac{d}{da}\left\{\ln\left[\frac{\exp(a)\left[1-\exp\left(-\frac{a}{J}(2J+1)\right)\right]}{1-\exp\left(-\frac{a}{J}\right)}\right]\right\}. \tag{K. 197}$$

We hence get easily

$$M_z = NJ\mu_B \frac{d}{da}\left\{\ln\left[\frac{\sinh\left(\frac{a}{2J}(2J+1)\right)}{\sinh\left(\frac{a}{2J}\right)}\right]\right\}, \tag{K. 198}$$

and, after the differentiation,

$$M_z = NgJ\mu_B B_J(a) \tag{K. 199}$$

with the Brillouin function

$$B_J(a) \equiv \frac{2J+1}{2J} \coth\left(\frac{2J+1}{2J}a\right) - \frac{1}{2J}\coth\left(\frac{a}{2J}\right). \tag{K. 200}$$

Let us discuss the behaviour of the Brillouin function (cf. Fig. 179) with the help of the limiting cases of weak and strong magnetic induction.

1. In weak magnetic fields and at high temperatures $a \ll 1$ [cf. eqn. (K. 194)]. When we expand the coth we obtain for the Brillouin function

$$B_J(a \ll 1) \approx \frac{J+1}{3J} a \tag{K. 201}$$

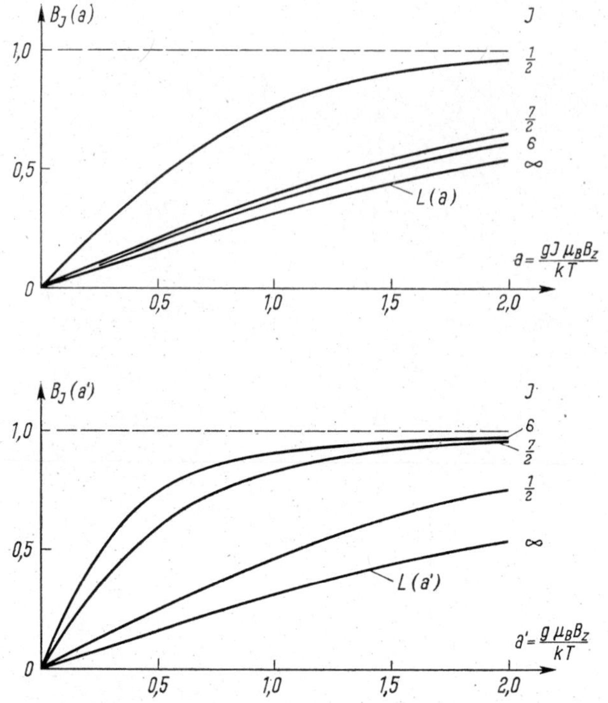

FIG. 179. The Brillouin function shown for different variables a and a' [cf. eqns. (K. 200) and (K. 215)]. For $J \to \infty$ and $\mu_B \propto \hbar \to 0$ the Brillouin function goes over into the Langevin function $L(a)$ with $a = a' = \mu_{\text{eff}} B_z/kT$ [see eqn. (K. 216)].

and when we substitute (K. 194) we obtain the net paramagnetic moment as

$$M_z = Ng^2J(J+1)\frac{\mu_B^2}{3k}\frac{B_z}{T}. \tag{K. 202}$$

Using (K. 9) and (K. 18) we obtain from this an expression for the paramagnetic susceptibility

$$\varkappa_{\text{para}} = \frac{Ng^2J(J+1)\,\mu_0\,\dfrac{\mu_B^2}{3kT}}{1-Ng^2J(J+1)\mu_0\,\dfrac{\mu_B^2}{3kT}}. \tag{K. 203}$$

An estimate shows that the second term in the denominator is small compared with unity so that we can write to a good approximation

$$\varkappa_{\text{para}} = Ng^2J(J+1)\mu_0\frac{\mu_B^2}{3kT} \tag{K. 204}$$

or

$$\varkappa_{\text{para}} = \frac{C}{T} \tag{K. 205}$$

with the Curie constant

$$C = Ng^2 J(J+1)\mu_0 \frac{\mu_B^2}{3k}. \tag{K. 206}$$

The $1/T$ dependence of the paramagnetic susceptibility in the limit of weak magnetic fields and high temperatures is called the Curie law. Substances for which (K. 205) is satisfied are called ideal paramagnetic substances. The reciprocal susceptibility plotted versus the temperature yields a straight line whose extrapolation passes through the origin. Its slope gives us the effective magnetic moments μ_{eff} of the atoms and the number n_B of Bohr magnetons of the atoms. We can write eqn. (K. 204) in the form

$$\varkappa_{\text{para}} = N\mu_0 \frac{\mu_{\text{eff}}^2}{3kT} = N\mu_0 \frac{(n_B\mu_B)^2}{3kT} \tag{K. 207}$$

with

$$\mu_{\text{eff}} = n_B \mu_B \tag{K. 208}$$

and

$$n_B = g[J(J+1)]^{1/2}. \tag{K. 209}$$

The range of validity ($a \ll 1$) of the Curie law is also determined by the total angular momenta of the atoms (besides magnetic induction and temperature). For example, with $g = 1$ and $J = 1$, and a typical value of the magnetic induction of 1 V s/m^2 ($= 10^4$ gauss) a substance behaves according to the Curie law at $T > 1$°K. At higher values of J the lower limit of the temperature is correspondingly higher. In general, however, the Curie law is valid in typical magnetic fields down to very low temperatures.

2. While in the range of validity of the Curie law the net magnetization is proportional to the magnetic induction, in the case of sufficiently high magnetic induction or sufficiently low temperatures we observe paramagnetic saturation. In the limiting case of $a \gg 1$ the Brillouin function (K. 200) takes the form

$$B_J(a \gg 1) \approx 1. \tag{K. 210}$$

Using this we obtain from (K. 199) the saturation value of the paramagnetic magnetization

$$M_z(a \gg 1) \equiv M_{\max} = NgJ\mu_B. \tag{K. 211}$$

In sufficiently strong magnetic fields and at sufficiently low temperatures in a paramagnetic substance all permanent moments are aligned, that is, all moments have the maximum z component in the direction of the field [cf. eqn. (K. 187)]. Recall that, by virtue of the directional quantization,

$$gJ\mu_B < g[J(J+1)]^{1/2}\mu_B = n_B\mu_B\,. \tag{K. 212}$$

According to (K. 199) and (K. 211) the Brillouin function is the ratio of magnetization to saturation magnetization:

$$B_J(a) = \frac{M_z}{M_{\text{max}}}\,. \tag{K. 213}$$

The total curve of the Brillouin function can be obtained on the basis of a discussion of the two limiting cases. Within the range of validity of the Curie law the slope of the linear part depends on the quantum number of angular momentum, J [cf. eqn. (K. 201)]. The total range of J values is limited by the two extreme slopes. The slope is a maximum for $J = \frac{1}{2}$; in this case eqn. (K. 201) takes the form

$$B_{1/2}(a \ll 1) \approx a. \tag{K. 214}$$

This case applies to atoms, for example, with vanishing orbital moment ($L = 0$) and non-saturated spin moment ($S = \frac{1}{2}$). According to (K. 158) the Landé factor $g = 2$ for $J = S = \frac{1}{2}$ and $L = 0$. Using this and eqns. (K. 208) and (K. 209) the effective magnetic moment is obtained as $\mu_{\text{eff}} = \sqrt{3}\mu_B$. In an external magnetic field there are only two possibilities of orientation, $M_J = +\frac{1}{2}$ and $M_J = -\frac{1}{2}$, that is, according to (K. 187), $\mu_z = \pm 1\ \mu_B$.

If $J = \frac{1}{2}$ the Brillouin function becomes

$$B_{1/2}(a) = 2\coth(2a) - \coth a = \tanh a. \tag{K. 215}$$

The other extremum corresponds to $J \to \infty$. The transition to very high quantum numbers means a transition from quantum mechanics to classical mechanics. This is coupled with the formal transition $\hbar \to 0$, that is, Planck's constant becomes small compared with the other quantities of this dimension. In the limit the effective magnetic moment has still a certain finite value μ_{eff} and the number $2J+1$ of possibilities of orientation in the magnetic field becomes infinitely large. The Brillouin function is replaced by the Langevin function which would have been obtained immediately from a classical derivation of paramagnetism (infinitely many possibilities of orientation of the permanent magnetic moments μ_{eff}):

$$B_\infty(a_\infty) \equiv L(a_\infty) = \coth a_\infty - \frac{1}{a_\infty} \tag{K. 216}$$

with

$$a_\infty = \frac{\mu_{\text{eff}} B_z}{kT}. \tag{K. 217}$$

For low magnetic inductions and high temperatures ($a_\infty \ll 1$) the function coth a_∞ can be expanded and one obtains [cf. eqn. (K. 201)]

$$B_\infty(a_\infty \ll 1) \approx \frac{a_\infty}{3}. \tag{K. 218}$$

The Brillouin function $B_J(a)$ is shown in Fig. 179 for various values of J. The influence of J becomes more obvious when instead of $B_J(a)$ the representation

$$B_J(a') = \frac{2J+1}{2J} \coth\left(\frac{2J+1}{2}a'\right) - \frac{1}{2J}\coth\left(\frac{a'}{2}\right) \tag{K. 219}$$

is used where

$$a' = \frac{g\mu_B B_z}{kT}. \tag{K. 220}$$

We see that as the number of magnetons increases [cf. (K. 209)], magnetic saturation is reached at smaller and smaller values of B_z/T.

(b) *COMPARISON WITH EXPERIMENTS*

The assumptions of the theory of paramagnetism of free atoms are best satisfied in the case of monatomic gases whose atoms have incomplete electron shells. Sodium vapour, for example, is an ideal paramagnetic substance (in contrast to solid sodium whose paramagnetism is temperature independent, cf. p. 458). Molecules with odd electron numbers are likewise paramagnetic since for them the resulting angular momentum and thus the magnetic moment is always non-vanishing. Gaseous nitrogen monoxide represents a typical example.

The theory of the paramagnetism of free atoms is also well applicable to many liquids and solids, in cases where the atoms or ions have incomplete electron shells and the magnetic interaction of neighbouring particles is negligible so that the assumption of "free atoms" is justified. This is true for salts with large quantities of crystal water and large spacings between the ions. Figure 180 shows as an example the results of a measurement with $CuSO_4 \cdot K_2SO_4 \cdot 6H_2O$ whose paramagnetism is due to the presence of the Cu^{2+} ions with incomplete $3d$ shells.

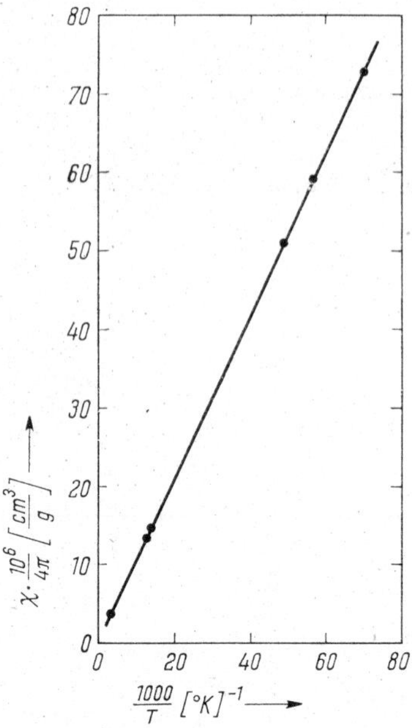

FIG. 180. Paramagnetic susceptibility as a function of the inverse temperature for $CuSO_4 \cdot K_2SO_4 \cdot 6H_2O$; the Curie law is valid [after J. C. Hupse, *Physica* **9**, 633 (1942)].

Also for the paramagnetic salts of the rare earth elements the theory of the paramagnetism of free atoms is essentially satisfied (cf. p. 449). The rare earth atoms have an incomplete inner shell; the number of the $4f$ electrons is incomplete and different for different elements of this group. As salts the rare earth elements occur as a rule in the form of trivalent positive ions, having lost two $6s$ electrons and one $5d$ or one $4f$ electron. Here the remaining $4f$ electrons in the incomplete inner shell of the atom are alone responsible for the paramagnetism. The incomplete $4f$ shell is surrounded by closed $5s$ and $5p$ shells and is thus well screened from the influence of the neighbouring ions. So the assumption of "free atoms" is often justified. It is also supported by measurements carried out with different salts and solutions containing one and the same rare-earth ion, in which almost the same values were obtained for the magneton number of this ion.

The effective magneton numbers of the rare earth ions are compiled in Table 28; Fig. 181 shows them as a function of the atomic number of the element. The experimental values which have been determined with the help of eqns. (K. 207) and (K. 208) agree as a rule with the theoretical values calculated from eqn. (K. 209). Exceptions are the Eu^{3+} and Sm^{3+} ions, which possess

TABLE 28. Number of magnetons of rare earths; theoretical values from eqn. (K. 209), experimental values from eqns. (K. 207) and (K. 208) (after J. S. Smart, *Effective Field Theories of Magnetism* [Saunders, Philadelphia 1966], p. 9)

Ion	Electron configuration	Ground state	L	S	J	g	n_B (theor.)	n_B (experim.)
La^{3+}, Ce^{4+}	$4f^0 5s^2 p^6$	$^1S_0$	0	0	0	—	0	diamagn.
Ce^{3+}, Pr^{4+}	$4f^1 5s^2 p^6$	$^2F_{5/2}$	3	$\frac{1}{2}$	$\frac{5}{2}$	$\frac{6}{7}$	2.54	2.6
Pr^{3+}	$4f^2 5s^2 p^6$	$^3H_4$	5	1	4	$\frac{4}{5}$	3.58	3.5
Nd^{3+}	$4f^3 5s^2 p^6$	$^4I_{9/2}$	6	$\frac{3}{2}$	$\frac{9}{2}$	$\frac{8}{11}$	3.62	3.5
Pm^{3+}	$4f^4 5s^2 p^6$	$^5I_4$	6	2	4	$\frac{3}{5}$	2.68	
Sm^{3+}	$4f^5 5s^2 p^6$	$^6H_{5/2}$	5	$\frac{5}{2}$	$\frac{5}{2}$	$\frac{2}{7}$	0.84	1.5
Sm^{2+}, Eu^{3+}	$4f^6 5s^2 p^6$	$^7F_0$	3	3	0	—	0	3.4
Eu^{2+}, Gd^{3+}	$4f^7 5s^2 p^6$	$^8S_{7/2}$	0	$\frac{7}{2}$	$\frac{7}{2}$	2	7.94	8.0
Tb^{3+}	$4f^8 5s^2 p^6$	$^7F_6$	3	3	6	$\frac{3}{2}$	9.72	9.5
Dy^{3+}	$4f^9 5s^2 p^6$	$^6H_{15/2}$	5	$\frac{5}{2}$	$\frac{15}{2}$	$\frac{4}{3}$	10.65	10.5
Ho^{3+}	$4f^{10} 5s^2 p^6$	$^5I_8$	6	2	8	$\frac{5}{4}$	10.61	10.4
Er^{3+}	$4f^{11} 5s^2 p^6$	$^4I_{15/2}$	6	$\frac{3}{2}$	$\frac{15}{2}$	$\frac{6}{5}$	9.58	9.5
Tm^{3+}	$4f^{12} 5s^2 p^6$	$^3H_6$	5	1	6	$\frac{7}{6}$	7.56	7.3
Yb^{3+}	$4f^{13} 5s^2 p^6$	$^2F_{7/2}$	3	$\frac{1}{2}$	$\frac{7}{2}$	$\frac{8}{7}$	4.54	4.5
Yb^{2+}, Lu^{3+}	$4f^{14} 5s^2 p^6$	$^1S_0$	0	0	0	—	0	diamagn.

excited states whose energies at room temperature lie only a few kT above the ground state. In these cases the magnetic moment is not exclusively determined by the ground state, that is, by Hund's rules (cf. p. 432). Only when the excited states are taken into account according to a model by Van Vleck do we get agreement with experimental values (cf. Fig. 181).

(c) *DEVIATIONS FROM THE PARAMAGNETISM OF FREE ATOMS*

In the following we shall briefly discuss some influences which give rise to deviations from the above-described paramagnetism of free atoms or ions. Such deviations are due to changes of the eigenfunctions of atoms caused by the effect of the magnetic field itself or by the way the atoms or ions are incorporated in the lattice.

In the approximate calculations of the paramagnetism of free atoms or ions the effect of the magnetic field on the eigenfunctions of the atoms is neglected. In fact, however, the eigenfunctions for $B = 0$ are slightly different from those

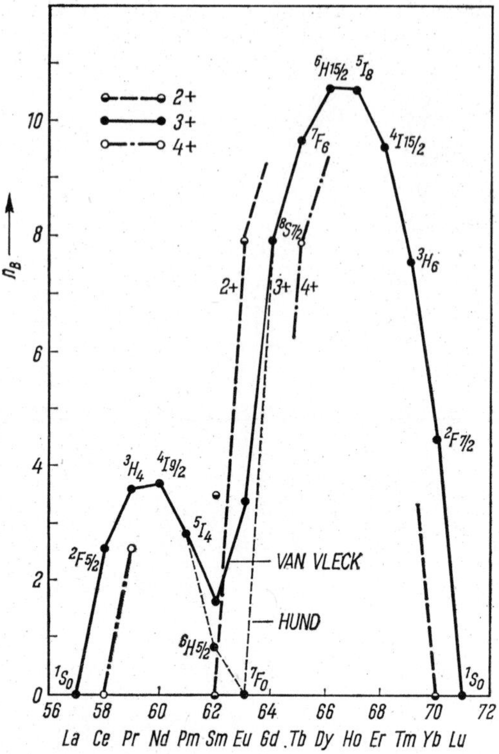

FIG. 181. Effective magnetic moments of rare earth ions versus atomic number.

for $B \neq 0$. In the latter case the eigenfunctions are a combination of the unperturbed eigenfunctions for $B = 0$. This results in a change of the magnetic moments of the unperturbed states. The contribution to the susceptibility due to this effect has been first calculated by Van Vleck with the help of perturbation theory, and is therefore called the Van Vleck paramagnetism.

The magnetic moment of the ions in paramagnetic salts is in most cases more or less different from that of free ions, since the eigenstates of the electrons of a magnetic ion are influenced by the presence of the neighbouring ions in the crystal. In paramagnetic salts the cations are as a rule the carriers of the magnetic moments while the anions, also called ligand ions, are non-magnetic. The influence of the neighbouring ions on a magnetic ion is often described by means of an electrostatic "ligand field" or "csystal field" acting on the electrons of the magnetic ion. One should take into account that in this approximation the electrons of an ion are assumed to be localized in its environment, in contrast to the assumption for Bloch electrons used in the energy band approximation (cf. p. 196).

In the case of ions with $L \neq 0$, $S \neq 0$, and $J \neq 0$, the ligand field tends to break the Russell–Saunders coupling (cf. pp. 429 ff.) between the spin moment and the orbital moment of the ion. According to the strength of the two opposite actions we distinguish the following two cases:

1. In the paramagnetic rare earth salts the incomplete $4f$ shells of the rare earth ions which cause the paramagnetism are screened by the filled $5s$ and $5p$ shells in such a way that the influence of the neighbouring ions on the $4f$ electrons is relatively weak (cf. p. 446). In this case J is a good quantum number and the ligand field splits the otherwise $(2J+1)$-fold degenerate levels.

2. In the paramagnetic salts of the transition element ions, however, the ligand field makes the Russell–Saunders coupling ineffective. In these cases the electron wave functions in the incomplete d-shells are strongly perturbed by the ligand field. This perturbation has the consequence that in the crystal the orbital moment cannot assume the orientation as in a free atom or ion so that the orbital moment does virtually not contribute to the magnetization. This effect is called the quenching of the orbital angular momentum L_z. The effective magneton number is then given approximately by

$$n_B = g[S(S+1)]^{1/2} \tag{K. 221}$$

instead of (K. 209). It should be mentioned that, under the action of the ligand field, J and L_z cannot be considered good quantum numbers.

3. LANDAU DIAMAGNETISM OF FREE ELECTRONS

Let us consider a gas of N electrons in a volume $V = L^3$, under the action of a magnetic induction $\boldsymbol{B}$. Besides the magnetization which is due to the electron spin moment (spin paramagnetism, cf. pp. 453 ff.), the electron gas suffers a diamagnetic magnetization which is caused by a change of the orbits in the magnetic field. The electrons move under the influence of a Lorentz force along helical trajectories; the axes of the spirals are parallel to the induction $\boldsymbol{B}$. The projection of the motion onto planes perpendicular to $\boldsymbol{B}$ yields circles which are traversed with the cyclotron frequency

$$\omega_c = \frac{e}{m} B \tag{K. 222}$$

(cf. p. 256). Note that the cyclotron frequency of free electrons is equal to twice the Larmor frequency [cf. eqns. (K. 173) and (K. 222)].

Each electron has an orbital moment. One can, however, show that from the viewpoint of classical mechanics the sum of all orbital moments is precisely equal to zero. This would mean zero diamagnetism of free electrons. This also results immediately from eqn. (K. 35) if the Lorentz force is assumed to be the only consequence of the magnetic induction. The Lorentz force performs no work (cf. p. 255), that is, in the classical case, for $B \neq 0$, the energy variation dA of the system, and thus the change of magnetization, is equal to zero.

Considered from the viewpoint of quantum theory, however, the action of the Lorentz force entails a quantization of the electron orbits in the magnetic field. This is connected with a change in the mean energy of the electrons and thus with a change in magnetization. According to Landau the electrons have the following energies under the influence of a magnetic induction $\boldsymbol{B}(0, 0, B_z)$:

$$E_{l,\,k_z} = \left(l+\frac{1}{2}\right)\hbar\omega_c + \frac{\hbar^2}{2\,m}\,k_z^2 \qquad \text{(K. 223)}$$

[cf. eqn. (F. 195)]; here $l = 0, 1, 2, \ldots$.

The motion in the planes perpendicular to B_z is determined by the transverse energy

$$E_\perp = (l+\tfrac{1}{2})\hbar\omega_c\,. \qquad \text{(K. 224)}$$

From the energy balance

$$E_\perp = \frac{m}{2}\,r^2\omega_c^2 \qquad \text{(K. 225)}$$

we obtain the selected allowed radii of orbits, and using eqns. (K. 90) and (K. 111) we obtain the following allowed orbital moments:

$$\mu_l = (2l+1)\mu_B\,. \qquad \text{(K. 226)}$$

In principle, the resulting diamagnetic moment would be obtained by a summation over all moments. In practice, however, the calculation is rendered difficult owing to the finite extension of the electron gas. The diamagnetic moment is therefore determined from the change of free energy in the magnetic field. We can generally use the thermodynamic relation

$$M = -\frac{1}{V}\left(\frac{\partial F}{\partial B}\right)_{T,\,V} \qquad \text{(K. 227)}$$

with

$$F = U - TS, \qquad \text{(K. 228)}$$

where F is the free energy, U is the internal energy, S is the entropy, and V is the volume.

The free energy depends on the statistics used. In the case of a non-degenerate electron gas, that is, Boltzmann statistics, we have

$$F = -kTNV \ln Z \qquad \text{(K. 229)}$$

with the partition function

$$Z = \sum_{l,\, k_z} Z_{l,\, k_z} \exp\left(-\frac{E_{l,\, k_z}}{kT}\right), \qquad \text{(K. 230)}$$

where $Z_{l,\, k_z}$ is the number of states with the quantum numbers l and k_z.

Summation is to be taken over all states of the system. The Landau levels are degenerate, that is, a number of states is attributed to each energy $E_{l,\, k_z}$ which, according to eqn. (F. 203), is given by

$$D(l, k_z)\, dk_z = \frac{L^3}{(2\pi)^2} \frac{m\omega_c}{\hbar} dk_z. \qquad \text{(K. 231)}$$

Since the quasi-continuous distribution of the k_z values is maintained also in the presence of a magnetic field, the summation over k_z in (K. 230) can be replaced by an integration. Using eqns. (K. 223) and (K. 231), we can rewrite (K. 230) in the form

$$Z = \frac{L^3}{(2\pi)^2} \frac{m\omega_c}{\hbar} \sum_{l=0}^{\infty} \int_{-\infty}^{+\infty} \exp\left[-\left(l+\frac{1}{2}\right)\frac{\hbar\omega_c}{kT}\right] \exp\left[-\frac{\hbar^2 k_z^2}{2mkT}\right] dk_z\,. \qquad \text{(K. 232)}$$

After integration and substitution of (K. 111) and (K. 222) we have

$$Z = A\mu_B B \sum_{l=0}^{\infty} \exp\left[-\left(l+\frac{1}{2}\right) 2\frac{\mu_B B}{kT}\right] \qquad \text{(K. 233)}$$

with the constant

$$A = \frac{L^3}{(2\pi)^2} \frac{2m}{\hbar^3} (2\pi mkT)^{1/2}. \qquad \text{(K. 234)}$$

The summation over l is a summation over an infinite geometrical series; we obtain

$$Z = \frac{A}{2} kT \frac{\mu_B B}{kT} \frac{1}{\sinh\left(\frac{\mu_B B}{kT}\right)}. \qquad \text{(K. 235)}$$

Substitution of this relation in (K. 229) yields the free energy of the non-degenerate electron gas in terms of the magnetic induction; so we obtain with the help of (K. 227) the resulting magnetization

$$M = kTN \frac{\partial}{\partial B} (\ln Z) = -N\mu_B\left[\coth\left(\frac{\mu_B B}{kT}\right) - \frac{kT}{\mu_B B}\right] \qquad \text{(K. 236)}$$

or, using eqn. (K. 216),

$$M = -N\mu_B L\left(\frac{\mu_B B}{kT}\right). \tag{K. 237}$$

The diamagnetic magnetization of the non-degenerate electron gas is thus determined by the Langevin function. We can see also here that the diamagnetic magnetization of the electron gas vanishes when we pass over to classical mechanics ($\hbar \to 0$ or $\mu_B \to 0$).

At high temperatures and low magnetic induction we can use the approximation [cf. eqn. (K. 218)]

$$M \approx -\frac{M\mu_B^2}{3}\frac{B}{kT}. \tag{K. 238}$$

In this approximation the diamagnetic susceptibility is given by

$$\varkappa_{\text{dia}} = -N\mu_0 \frac{\mu_B^2}{3k}\frac{1}{T}. \tag{K. 239}$$

Equation (K. 237) has been derived under the assumption of non-degeneracy and is therefore as a rule inapplicable to low temperatures. When we use Fermi–Dirac statistics instead of Boltzmann statistics, we obtain instead of (K. 229) the following relation for the free energy of the degenerate electron gas:

$$F = NV\zeta - kT\sum_{l,\,k_z} Z_{l,\,k_z} \ln\left[1+\exp\left(-\frac{E_{l,\,k_z}-\zeta}{kT}\right)\right]. \tag{K. 240}$$

Using eqns. (K. 111), (K. 222), (K. 223), and (K. 231) we obtain from this

$$F = NV\zeta - \frac{L^3}{(2\pi)^2}\frac{2m}{\hbar^2}\mu_B BkT\sum_{l=0}^{\infty}\int_{-\infty}^{+\infty}\ln\left[1+\exp\left(-\frac{E_{l,\,k_z}-\zeta}{kT}\right)\right]dk_z, \tag{K. 241}$$

where

$$E_{l,\,k_z} = \left(l+\frac{1}{2}\right)2\mu_B B + \frac{\hbar^2}{2m}k_z^2. \tag{K. 242}$$

This relation is connected with great mathematical difficulties and can only be calculated in the approximation of low magnetic induction and high temperature, that is, with $\mu_B B \ll kT$; in this approximation we obtain for the diamagnetic susceptibility of the degenerate electron gas

$$\varkappa_{\text{dia}} = -\tfrac{2}{3}\mu_0\mu_B^2 D(\zeta_0), \tag{K. 243}$$

where $D(\zeta_0)$ is the eigenvalue density at the energy ζ_0; ζ_0 is the Fermi energy for $B = 0$; generally $\zeta(B)$ is obtained from the relation $\delta N = 0$. When only terms linear in B are taken into account, $\zeta(B) = \zeta_0$ (cf. p. 455).

So far we have ignored the permanent magnetic moments which are coupled with the spin (cf. p. 426). These moments give rise to a paramagnetic susceptibility which will be calculated in the following section. Here we conclude that the magnetization of the electron gas has a paramagnetic and a diamagnetic component. A comparison between theory and experiment is only possible when both components are known (cf. p. 458).

4. PAULI PARAMAGNETISM OF FREE ELECTRONS

Independent of its energy state and its orbital motion, an electron possesses a magnetic spin moment which amounts to [cf. eqn. (K. 124)]

$$|\mu_s| = g_e \frac{3^{1/2}}{2} \mu_B = 3^{1/2} \mu_B . \qquad \text{(K. 244)}$$

Under the influence of a magnetic induction B_z the spin moment has the two possible z components

$$\mu_{sz} = \pm 1 \mu_B . \qquad \text{(K. 245)}$$

When we first consider the case of a non-degenerate electron gas, that is, apply Boltzmann statistics, we can immediately use the results from pp. 439 ff. For $J = S = \frac{1}{2}$ we obtain the magnetization from (K. 199), using (K. 194) and (K. 215):

$$M_z = N \mu_B \tanh\left(\frac{\mu_B B}{kT}\right). \qquad \text{(K. 246)}$$

In the approximation $\mu_B B \ll kT$ we obtain from this

$$M_z \approx N \frac{\mu_B^2}{k} \frac{B}{T} . \qquad \text{(K. 247)}$$

In this approximation the paramagnetic susceptibility of the non-degenerate electron gas amounts to

$$\varkappa_{\text{para}} = N \mu_0 \frac{\mu_B^2}{k} \frac{1}{T} . \qquad \text{(K. 248)}$$

A comparison with (K. 239) yields the following relation between the diamagnetic and the paramagnetic susceptibilities:

$$\varkappa_{\text{para}} = -3 \varkappa_{\text{dia}} . \qquad \text{(K. 249)}$$

This relation holds in the approximation $\mu_B B \ll kT$ for both the non-degenerate and the degenerate electron gas.

Let us now calculate the paramagnetic magnetization of the degenerate electron gas. This theory of the spin paramagnetism was developed by Pauli. He could show from it that the conduction electrons in metals behave like a free electron gas governed by Fermi–Dirac statistics (cf. p. 192). The Fermi–Dirac statistics are the reason for the fact that, in spite of the non-saturated spin moments of the electrons, many metals have only a weak and almost temperature-independent paramagnetism.

When E is the energy of an electron at $\boldsymbol{B} = 0$, its energy at $\boldsymbol{B} \neq 0$ is given either by

$$E_{\uparrow} = E - \mu_B B \tag{K. 250}$$

or by

$$E_{\downarrow} = E + \mu_B B, \tag{K. 251}$$

according to whether its spin moment is parallel or antiparallel to the direction of the magnetic induction. Because of the Pauli principle the electron gas in a magnetic field consists of two partial systems whose eigenvalue densities are shifted by $\pm \mu_B B$ along the energy scale, compared with the case of $\boldsymbol{B} = 0$. Since here only the influence of the magnetic field on the spin moments (and not on the orbital motion) is taken into account, the form of the eigenvalue density remains unchanged (the appearance of the Landau levels is a con-

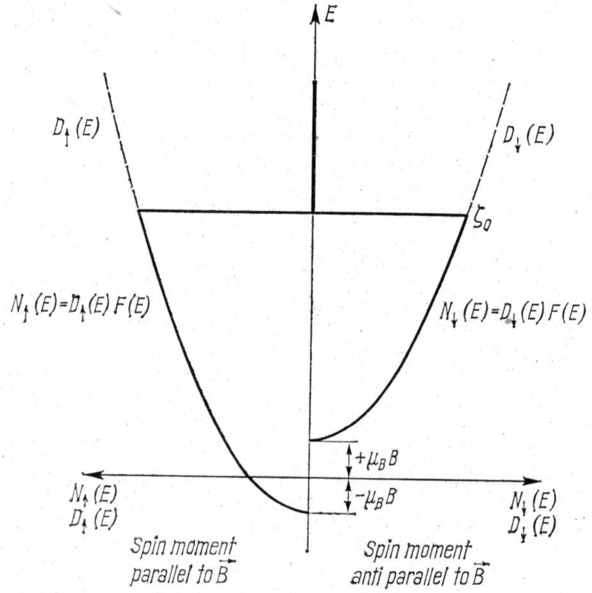

FIG. 182. Distribution function of free electrons in a magnetic field at $T = 0$°K (the orbital motion is not taken into account).

sequence of the quantization of the electron orbits, cf. pp. 254 ff.). The electrons whose spins are parallel to $\boldsymbol{B}$ have the eigenvalue density

$$D_{\uparrow}(E)\, dE = D(E+\mu_B B)\, dE \qquad \text{(K. 252)}$$

and those with spins antiparallel to $\boldsymbol{B}$ (cf. Fig. 182)

$$D_{\downarrow}(E)\, dE = D(E-\mu_B B)\, dE. \qquad \text{(K. 253)}$$

In thermodynamic equilibrium the population of the states in the two systems is governed by the same Fermi function:

$$F(E) = \frac{1}{\exp\left(\dfrac{E-\zeta}{kT}\right)+1}\,. \qquad \text{(K. 254)}$$

Thermodynamic equilibrium is reached when in both systems the states are occupied up to the same maximum energy ζ at $T = 0°\text{K}$. This means that the number of electrons in the partial systems is different: the number $N_{\uparrow}$ with parallel spin moments is larger. Hence it follows that the total electron gas will have a net magnetic moment in the direction of the induction $\boldsymbol{B}$, that is, we observe the Pauli spin paramagnetism.

Before we calculate this moment, we want to show that the Fermi energies $\zeta(\boldsymbol{B} \neq 0)$ and $\zeta_0(\boldsymbol{B} = 0)$ are almost equal. This follows from the condition

$$N = N_{\uparrow}+N_{\downarrow} \qquad \text{(K. 255)}$$

with [cf. eqn. (E. 23)]

$$N = 2\int_0^{\zeta_0} D(E)\, F(E)\, dE, \qquad \text{(K. 256)}$$

$$N_{\uparrow} = 1 \int_{-\mu_B B}^{\zeta} D_{\uparrow}(E)\, F(E)\, dE = \int_{-\mu_B B}^{\zeta} D(E+\mu_B B)\, F(E)\, dE \qquad \text{(K. 257)}$$

and

$$N_{\downarrow} = 1 \int_{+\mu_B B}^{\zeta} D_{\downarrow}(E)\, F(E)\, dE = \int_{+\mu_B B}^{\zeta} D(E-\mu_B B)\, F(E)\, dE. \qquad \text{(K. 258)}$$

According to eqn. (E. 24) the total number of electrons is a function of the Fermi energy. Equation (K. 255) can therefore be written in the form (at $T = 0°\text{K}$ we have $F(E) = 1$)

$$N(\zeta_0) = \tfrac{1}{2}N(\zeta+\mu_B B)+\tfrac{1}{2}N(\zeta-\mu_B B). \qquad \text{(K. 259)}$$

Since even for the highest achievable magnetic inductions (of the order of 10^2 V s/m^2 = 10^6 gauss) the magnetic energy $\mu_B B$ (of the order of 10^{-2} eV) is small compared with the Fermi energy ζ (of the order of 1 eV, cf. Table 16), the functions in eqn. (K. 259) can be expanded at ζ:

$$N(\zeta_0) = \frac{1}{2}\left(N(\zeta)+\mu_B B\frac{dN}{d\zeta}\right)+\frac{1}{2}\left(N(\zeta)-\mu_B B\frac{dN}{d\zeta}\right), \tag{K. 260}$$

that is,

$$N(\zeta_0) = N(\zeta) \tag{K. 261}$$

or

$$\zeta_0 = \zeta. \tag{K. 262}$$

We see that to a first approximation the Fermi energies with and without magnetic field are the same.

The net magnetic moment is given by the difference of electron numbers in the two partial systems

$$M = \mu_B(N_\uparrow - N_\downarrow). \tag{K. 263}$$

At $T = 0°K$ this relation can be immediately computed with the help of (K. 257) and (K. 258). By analogy with (K. 260) we obtain

$$M = \mu_B^2 B\frac{dN}{d\zeta}. \tag{K. 264}$$

Using eqns. (K. 256) and (K. 262) we hence obtain

$$M = 2\mu_B^2 BD(\zeta_0). \tag{K. 265}$$

Thus the paramagnetic susceptibility of the degenerate electron gas at $T = 0°K$ amounts to

$$\varkappa^0_{\text{para}} = 2\mu_0\mu_B^2 D(\zeta_0). \tag{K. 266}$$

A comparison with (K. 243) yields again

$$\varkappa_{\text{para}} = -3\varkappa_{\text{dia}}. \tag{K. 267}$$

We must take into account that eqn. (K. 243) holds in the approximation of $\mu_B B \ll kT$, while eqn. (K. 266) has been derived for $T = 0°K$. A calculation for $T > 0°K$, however, shows that in the range of practically possible temperatures the Pauli spin paramagnetism is a very weak function of the temperature. Relation (K. 267) is therefore physically meaningful.

In order to calculate the Pauli spin paramagnetism at $T > 0°K$ we shall again start from (K. 263). In eqns. (K. 257) and (K. 258) the upper limit of integration must be replaced by ∞ so that the Fermi function is no longer equal to unity in the entire range of integration. The calculation leads us to Fermi integrals (cf. p. 165). In the case of $\zeta \gg kT$ we obtain the result

$$\varkappa_{\text{para}} = 2\mu_0\mu_B^2 D(\zeta_0)\left[1-\frac{\pi^2}{24}\left(\frac{kT}{\zeta_0}\right)^2+\dots\right]. \tag{K. 268}$$

Thus the paramagnetic susceptibility decreases slightly with increasing temperature.

The introduction of a degeneracy temperature T_E [cf. eqn. (E. 29)] stresses the difference between the susceptibilities for non-degenerate and degenerate electron gases. Using (E. 14) and (E. 24) we can rewrite (K. 268) in the form

$$\varkappa_{\text{para}} = \frac{3}{2}\mu_0\mu_B^2\frac{N}{\zeta_0}\left[1-\frac{\pi^2}{24}\left(\frac{kT}{\zeta_0}\right)^2+\dots\right] \qquad \text{(K. 269)}$$

and the substitution of (E. 30) yields

$$\varkappa_{\text{para}} = \frac{3}{5}\mu_0\mu_B^2\frac{N}{kT_E}\left[1-\frac{\pi^2}{150}\left(\frac{T}{T_E}\right)^2+\dots\right]. \qquad \text{(K. 270)}$$

Degeneracy temperatures are of the order of 10^4 °K. A comparison with (K. 248) shows that the paramagnetic susceptibility of the degenerate electron gas is smaller by a factor of 10^2–10^3 than that of the non-degenerate electron gas. This is due to the Pauli principle or the Fermi–Dirac statistics. Only electrons with energies $\zeta \pm kT$ in the region of the Fermi energy can change their energies by orientation of the spin moments and so contribute to the magnetization. The number of these effective electrons is of the order of

$$N_{\text{eff}} \approx N\frac{T}{T_E}. \qquad \text{(K. 271)}$$

When we replace N by N_{eff} in (K. 248), we obtain approximately the paramagnetic susceptibility of the degenerate electron gas, which is smaller by a factor $T/T_E \approx 10^{-2}$–10^{-3} and which is temperature independent in this approxima-

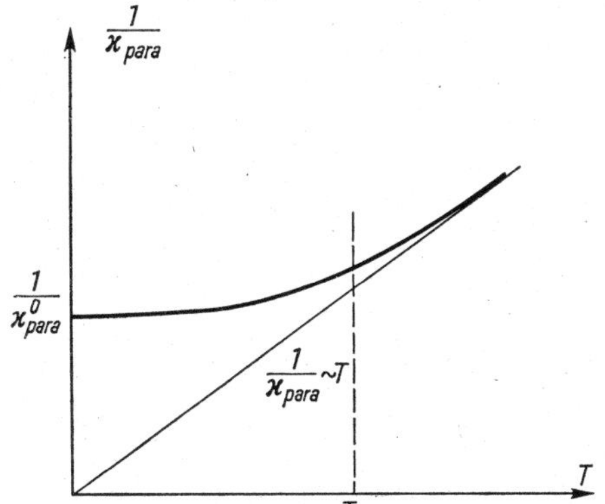

FIG. 183. Spin susceptibility of an electron gas as a function of temperature.

tion. In this connection we should also compare the difference between the specific heats of the non-degenerate and the degenerate electron gas (cf. p. 170).

Figure 183 illustrates the temperature dependence of the paramagnetic susceptibility of the electron gas. According to eqn. (K. 248) the contribution of a single electron to the susceptibility can be taken to be $\mu_0\mu_B^2/kT$, that is, it is inversely proportional to T. The number of effective electrons is at $T \ll T_E$ proportional to T. In this temperature range it is mainly the temperature dependence of the Fermi energy which is decisive (ζ and thus also $D(\zeta)$ decrease slightly with T). At $T > T_E$ the electron gas becomes non-degenerate, the effective number of electrons is then equal to the temperature-independent total number of electrons; the contribution of the individual electrons, however, decreases with T, that is, the Curie law is valid.

5. EXPERIMENTAL RESULTS FOR FREE ELECTRONS

The total susceptibility of the electron gas is obtained as the sum of the diamagnetic and paramagnetic susceptibilities. Because of eqn. (K. 267) we have

$$\varkappa_{\text{tot}} = \tfrac{2}{3}\varkappa_{\text{para}}. \qquad \text{(K. 272)}$$

The electron gas is always paramagnetic. In the particular case of degeneracy, disregarding the temperature correction [cf. eqns. (K. 243) and (K. 266) or (K. 269)], we can write

$$\varkappa_{\text{tot}} \approx \frac{4}{3}\,\mu_0\mu_B^2 D(\zeta_0) = \mu_0\mu_B^2\frac{N}{\zeta_0}. \qquad \text{(K. 273)}$$

When we want to check the theories of the magnetism of the electron gas, it is, apart from considering liquid metals, convenient to work with monovalent metals whose valence electrons behave in a good approximation like a free degenerate electron gas. The alkali metals, for example, have an almost

TABLE 29. Magnetic volume susceptibility of alkali metals, (a) according to Knight shift measurements, (b) according to Pauli (K. 269) (after A. H. Morrish, *The Physical Principles of Magnetism* [Wiley, New York 1965], pp. 211, 220)

	$\varkappa_{\text{spin}}\cdot\frac{10^6}{4\pi}$		$\varkappa_{\text{tot}}\cdot\frac{10^6}{4\pi}$
	experim.[(a)]	theor.[(b)]	experim.
Li	2.08±0.1	1.17	1.89±0.05
Na	0.95±0.1	0.64	0.68±0.03

temperature-independent paramagnetism which is in order-of-magnitude agreement with the theoretical value (K. 273), cf. Table 29. The differences between the measured and the theoretical values are due to the following reasons:

1. The measured value of the susceptibility has a diamagnetic component which is due to the electrons of the atomic core.
2. The valence electrons are in fact not completely free; instead of the rest mass one has to take into account the effective electron mass. Even for quasi-free electrons the given relation between diamagnetism and paramagnetism is no longer valid. We must take into account that in the case of diamagnetism the value of $\mu_B = |e|\hbar/2m$ must be replaced by $\mu_B^* = |e|\hbar/2m^*$; in this theory the quantity μ_B is a mere combination of certain constants. In the case of paramagnetism, however, the decisive spin moment is determined by the Bohr magneton μ_B. The effective mass is here normally contained not in the quantity μ_B but in the eigenvalue density $D(\zeta_0)$. Moreover, more detailed calculations for not completely free electrons lead to other susceptibility contributions, such as, for example, the Van Vleck term (see p. 448), apart from the Landau and Pauli terms [eqns. (K. 243) and (K. 266)].
3. The value of the susceptibility can be influenced by correlation and exchange forces, i.e. electron–electron and electron–ion interactions.

V. Phenomena of Collective Magnetic Ordering

Several elements and numerous compounds have magnetic properties which are in principle different from the diamagnetic or paramagnetic behaviour described above. Such materials are ferromagnetic, antiferromagnetic, or ferrimagnetic. In all these cases there exist strong interactions between the magnetic moments which give rise to a certain ordering with respect to their mutual orientation. It is an essential fact that the establishment of ordering occurs collectively, at a fixed temperature (cf. p. 469). Paramagnetic saturation is also connected with some ordering (cf. p. 444) which, however, is not a collective phenomenon. The type of magnetic ordering can be shown directly by means of neutron diffraction. Although the neutrons are electrically neutral, they possess a spin (spin quantum number $s = \frac{1}{2}$) and thus a magnetic moment which interacts with the magnetic moments of the atoms.

The type of magnetic ordering manifests itself in the macroscopic magnetic properties, such as $\varkappa(T)$, $\boldsymbol{M}(\boldsymbol{B})$. We shall first give a phenomenological survey of ferromagnetism, antiferromagnetism, and ferrimagnetism.

Ferromagnetism. Substances are ferromagnetic if, without an external field, the magnetic moments assume parallel orientations, at least in certain regions (called domains). The elements Fe, Co, Ni, Gd, and numerous alloys, which often contain these elements, are typical ferromagnetic substances. There also exist a few non-metallic ferromagnetic compounds, such as, for instance, $CrBr_3$, EuO, EuS, and $CdCr_2S_4$. The most characteristic property of the ferromagnetic substances is the spontaneous magnetization: below a certain temperature, the so-called ferromagnetic Curie temperature, there exists a net magnetic moment also in the absence of magnetic induction. The spontaneous magnetization can be proved in an impressive experiment with the help of the Mössbauer effect (cf. p. 85), which gives information about the hyperfine splitting of the nuclear levels by the internal magnetic field at the position of the nucleus. This internal field is due to the magnetic polarization of the electron shells

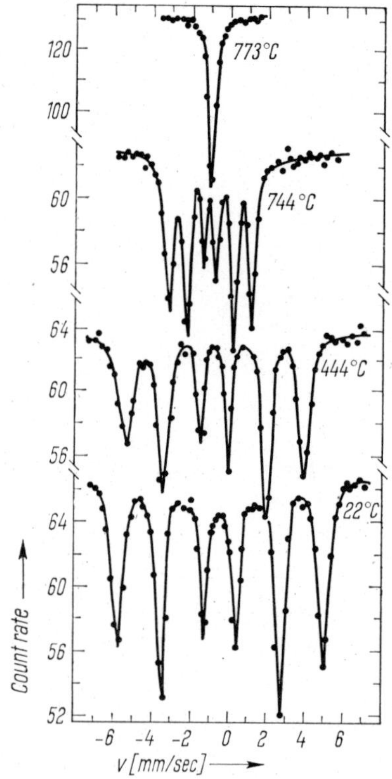

FIG. 184. Mössbauer spectra of metallic iron at various temperatures. The magnetic hyperfine splitting of the nuclear levels disappears above the Curie point $T_C = 770°C$ [after R. S. Preston, S. S. Hanna and J. Heberle, *Phys. Rev.* **128,** 2207 (1962)].

which, in its turn, is caused by the spontaneous magnetic ordering. As shown in Fig. 184, a splitting of the Mössbauer spectrum caused by the hyperfine field can be observed below the Curie point. Immediately above the Curie point the magnetic field at the nucleus vanishes on average in time and so does the hyperfine splitting, and there appears only a single Mössbauer line.

Ferromagnetics have very high values of the susceptibility $\varkappa$. Therefore, owing to (K. 57), the effective magnetic field is not equal to the applied external field. Moreover, there is no single-valued relation between magnetization and induction, as the magnetization curve is a hysteresis loop. The form of the latter depends on many factors, such as temperature, history and treatment of the sample, even for the same material. We shall not discuss these problems, which are mainly of technical interest, here. We only want to mention that the area inside the hysteresis loop is a measure of the magnetization work to be performed for one cycle of magnetization. The work per cycle of magnetization is given by [cf. eqn. (K. 34)]

$$A^{\swarrow} = \oint \delta A = \mu_0 \oint H\, dM. \qquad \text{(K. 274)}$$

In the case of a closed cycle of magnetization the work is independent of the magnetization device (cf. p. 409); we can write

$$\oint \delta A = \oint \delta A_1 = \oint \delta A_1' \qquad \text{(K. 275)}$$

or, using eqns. (K. 30) and (K. 41),

$$A^{\swarrow} = \mu_0 \oint H\, dM = -\mu_0 \oint M\, dH. \qquad \text{(K. 276)}$$

This is easy to verify when we consider the hysteresis loop. The passage through one cycle of magnetization makes it necessary to perform work which grows with the number of cycles and appears as the hysteresis heat. This represents an irreversible cyclic process.

Antiferromagnetism. According to Néel an antiferromagnetic substance has at least two sublattices in which, below a certain critical temperature, the Néel temperature, the magnetic moments assume a parallel orientation which, however, is antiparallel to that of the other sublattice. In the case of equal magnetic moments of the two sublattices, the whole crystal has a zero net moment, that is, the antiferromagnet displays no permanent magnetization. In the curve of the susceptibility versus temperature antiferromagnetism is shown up by a sharp break at the Néel temperature; at this temperature the susceptibility has its highest value.

Moreover, below the Néel temperature the susceptibility is anisotropic (cf. Fig. 185). A field applied perpendicular to the direction of magnetization exerts the same effect on both sublattices and causes only a small rotation of the magnetic moments into the field direction. This rotation is independent of temperature and therefore the susceptibility $\varkappa_\perp$ is temperature independent below

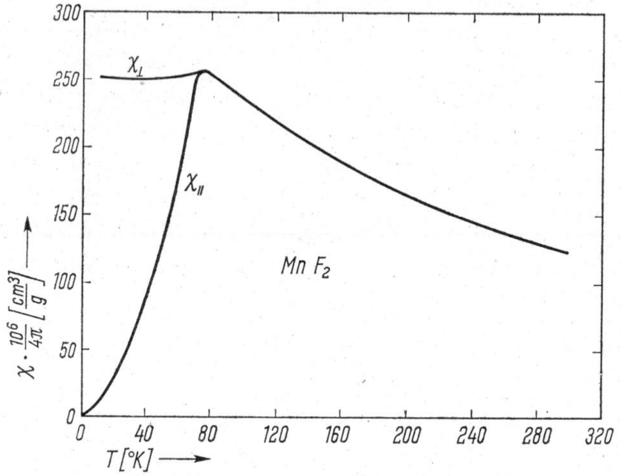

FIG. 185. Temperature dependence of the magnetic susceptibility of an antiferromagnetic material. The susceptibility of a MnF_2 monocrystal is measured for magnetic fields respectively vertical and parallel to the [001] direction of the crystal (after D. H. Martin, *Magnetism in Solids*, M.I.T. Press, Cambridge, Massachusetts 1967).

the Néel point. A field which is parallel or antiparallel to the magnetic moments exerts no torque and thus causes no resulting magnetization. The susceptibility $\varkappa_{||}$ is therefore zero at absolute zero and increases with increasing temperature as soon as the magnetization of one sublattice increases by virtue of thermal motion at the expense of the other.

Several metals are antiferromagnetic, namely those of the transition elements such as Cr and α-Mn and some rare earths, such as Ce and Nd. Moreover, many compounds are antiferromagnetic, mainly the oxides, fluorides and chlorides of transition elements, lanthanides, and actinides.

Ferrimagnetism. Below a certain temperature ferrimagnetic substances display spontaneous magnetization which, in contrast to that of ferromagnetism, is due to an incomplete compensation of the magnetic moments of the sublattices. In the simplest case ferrimagnetism may be considered to be a special form of antiferromagnetism, namely when the magnetic moments of the sublattices are exactly antiparallel but of different magnitude so that the net magnetiza-

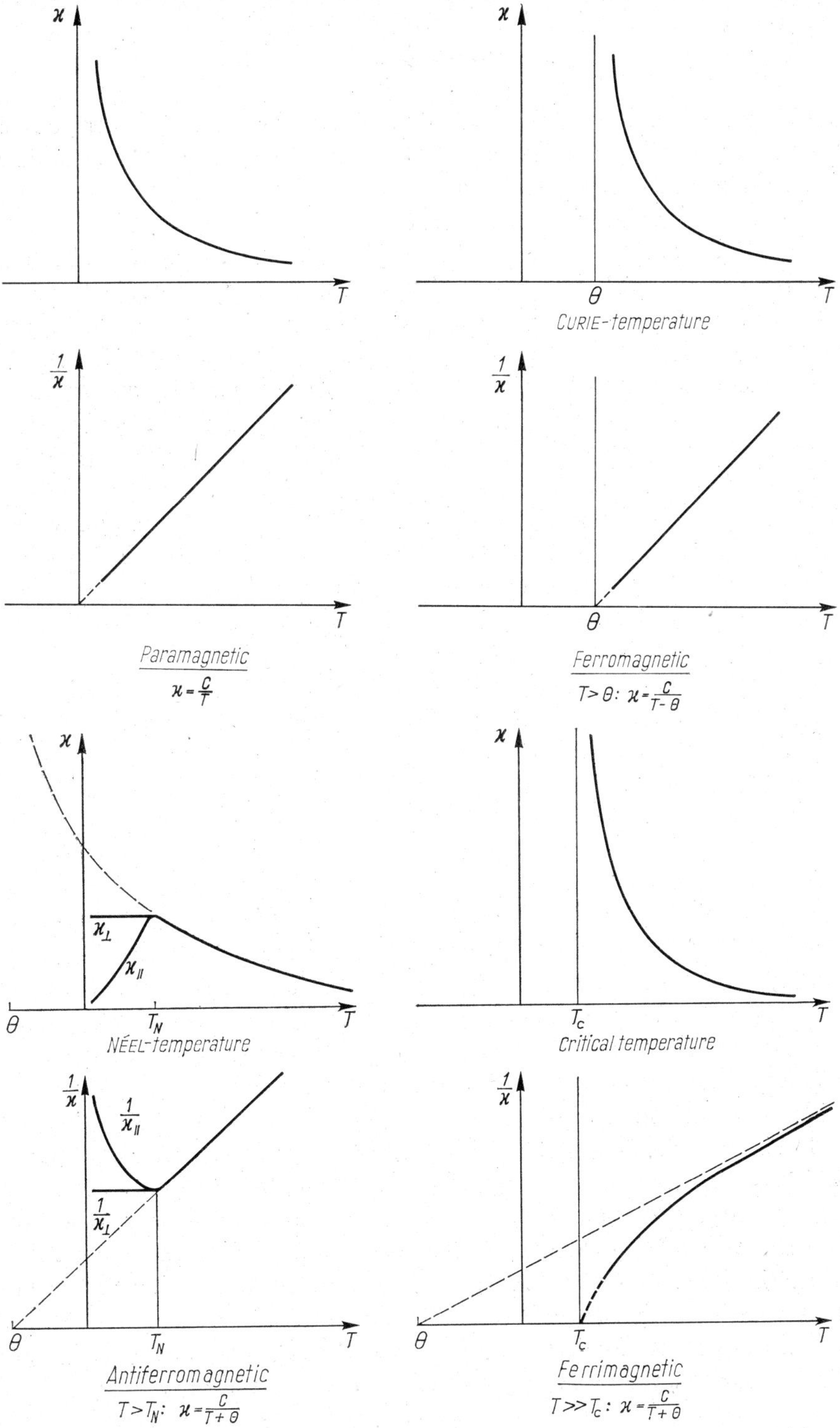

FIG. 186. Temperature dependence of the magnetic susceptibility of substances which are paramagnetic, ferromagnetic, antiferromagnetic, and ferrimagnetic, respectively.

tion is non-zero. So-called Dzyaloshinskii-ferrimagnetism is often called "weak ferromagnetism"; it is due to the fact that the magnetic moments deviate by a small angle from antiparallel orientation.

The best known ferrimagnetic substances, such as magnetite Fe_3O_4 ($= Fe^{2+}Fe_2^{3+}O_4$), are cubic crystals with spinel structure. They have the chemical composition $MFe_2^{3+}O_4$ where M stands for a bivalent metal ion such as Cu, Mg, Zn, Cd, Fe, Mn, Co, or Ni. Ferrimagnets with the spinel structure are also called ferrites. Because of their low electrical conductivity (10^{-2}–10^{-6} ohm^{-1} cm^{-1}) there occur practically no eddy currents in ferrites. They are therefore interesting for many technical applications, up to very high frequencies. They are particularly suitable as transformer cores with small losses for microwave elements or memory elements.

Other ferrimagnets are compounds with garnet ($Mg_3Al_2Si_3O_{12}$) or perovskite ($CaTiO_3$) structure, of which the yttrium–iron–garnet (YIG = $Y_3Fe_5O_{12}$) is the best known one.

The various interactions which give rise to the corresponding collective ordering manifest themselves already above the respective critical temperature in the temperature curve of the susceptibility. In Fig. 186 we show the temperature dependences $\varkappa(T)$ of substances which become ferromagnetic, antiferromagnetic, and ferrimagnetic, respectively.

In the following we shall consider Weiss's phenomenological theory of ferromagnetism.

1. WEISS THEORY OF FERROMAGNETISM

Below the Curie point a ferromagnetic sample need not possess a net magnetization in the absence of an external magnetic field. A very small field, however, will give rise to a magnetization which exceeds by many orders of magnitude the magnetization values of paramagnetic substances. In view of this fact, Weiss made the following assumptions which are basic to his model of ferromagnetism:

1. The material is subdivided into regions (the so-called Weiss domains) which display spontaneous magnetization. The magnetic moments within a domain are parallel to one another. The magnetization curve of a single domain is a single-valued function $\boldsymbol{M}(\boldsymbol{B})$ which depends on the temperature (cf. Fig. 187). The vector sum over the magnetic moments of all domains yields the net magnetization of the sample considered. Certain domain configurations have a zero net magnetization; the application of a weak exciting magnetic field changes the domain configuration and gives rise to strong magnetizations.

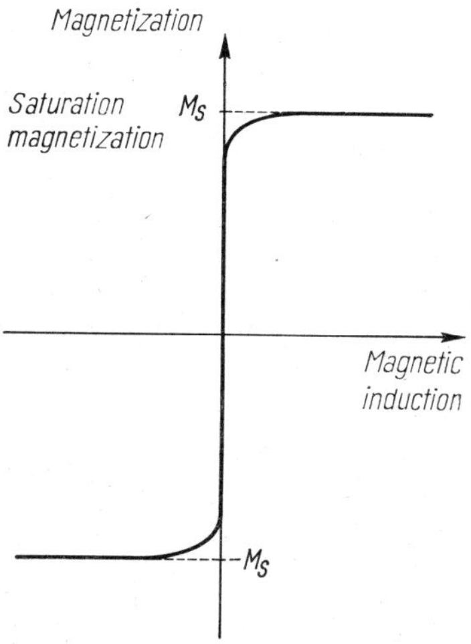

FIG. 187. Magnetization curve of a Weiss domain.

2. The second assumption concerns the property of the spontaneous magnetization of a domain. The individual magnetic moments are affected not only by the external magnetic field but apart from the demagnetizing field [cf. eqn. (K. 57)] there exists also a strong magnetic field H_m (the molecular or Weiss field) which is proportional to the magnetization of the sample. The effective internal magnetic field will then be given by

$$H_i = H + H_m \tag{K. 277}$$

with

$$H_m = WM, \tag{K. 278}$$

where W is the Weiss constant (or molecular-field constant). The Weiss constant is assumed to be independent of H, M, and T. In the framework of the Weiss theory nothing is said about the physical cause of the molecular field which is due to the exchange interaction between the magnetic moments (cf. p. 471).

The form of eqn. (K. 277) is analogous to the Lorentz relation for the internal field which takes into account the classical mutual forces of the magnetic dipoles. According to it the local field inside a cubic crystal is given by

$$H_{\text{loc}} = H + \tfrac{1}{3}M. \tag{K. 279}$$

The following estimate, however, shows that the Lorentz correction is negligible compared with the Weiss field.

Assume that the collective ordering and thus the spontaneous magnetization is caused by the molecular field H_m, that is the molecular induction B_m. The magnetic energy of a dipole will then be of the order of $\mu_B B_m$. Above the Curie point T_C ferromagnetic substances become paramagnetic. This means that a sufficiently high thermal energy kT_C destroys the collective ordering. From the relation

$$\mu_B B_m \approx kT_C \tag{K. 280}$$

we obtain for $T_C \approx 10^3$ °K the order of magnitude of the molecular induction: $B_m \approx 10^3$ Vs/m² $= 10^7$ gauss. Inductions of this order of magnitude cannot be produced in the laboratory, not even with superconducting magnets and pulse methods. From eqn. (K. 277) and the relation

$$B = \mu_0 H_i \tag{K. 281}$$

or, for $H = 0$,

$$B_m = \mu_0 H_m \tag{K. 282}$$

we obtain a value of $H_m \approx 10^9$ A/m for the Weiss field. The spontaneous magnetization of ferromagnetic substances is of an order of magnitude of about $M_s \approx 10^6$ A/m. According to eqn. (K. 278) the Weiss constant is of the order of $W \approx 10^3$, i.e. the Lorentz correction can be neglected against the Weiss field ($W \gg \frac{1}{3}$).

The Weiss theory is based on the theory of the paramagnetism of free atoms (cf. pp. 439 ff.). The collective ordering which causes the ferromagnetism is taken into account by substituting the internal magnetic field H_i for the external magnetic field H. According to eqn. (K. 213) the ratio of magnetization and saturation magnetization is given by the Brillouin function

$$B_J(a) = \frac{M}{M_{\max}}. \tag{K. 283}$$

The Weiss field is taken into account through the argument a. If in eqn. (K. 194) the induction is given in terms of the internal field, we obtain from eqns. (K. 277) and (K. 281)

$$M = \frac{kT}{\mu_0 g J \mu_B W} a - \frac{H}{W} \tag{K. 284}$$

or, after division by $M_{\max}$,

$$\frac{M}{M_{\max}} = \frac{kT}{\mu_0 g J \mu_B M_{\max} W} a - \frac{H}{W M_{\max}}. \tag{K. 285}$$

Thus we have two conditions, (K. 283) and (K. 285), which must be satisfied by the magnetization. The graphical representation of $M/M_{\max}$ versus the parameter a yields two different functions whose point of intersection determines the magnetization of the ferromagnet for a given field H and a given temperature T. The characteristic properties of ferromagnetic substances can be described with the help of this parameter representation (cf. Fig. 188).

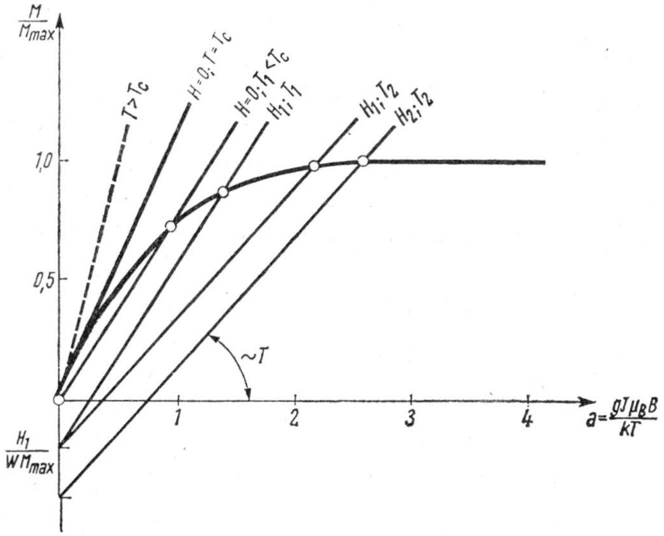

FIG. 188. Graphic method to determine the magnetization M as a function of the magnetic field H and temperature T.

The Brillouin function is universal. Equation (K. 285) yields a straight line whose slope is proportional to the temperature T and whose negative ordinate intercept is proportional to the external magnetic field H. The point of intersection with the Brillouin function lies at magnetizations which are the higher, the stronger the magnetic field and the lower the temperature. In the particular case of $H = 0$ we must distinguish between two cases:

1. At sufficiently high temperatures $T > T_C$ the slope of the straight line exceeds the initial slope of the Brillouin function. The point of intersection of these two functions is then at the origin of coordinates, that is, if $H = 0$ then $M = 0$ (Curie–Weiss paramagnetism, cf. p. 470).
2. When $T < T_C$ the straight line intersects the Brillouin function even if $M_s/M_{\max} \neq 0$, that is, in the absence of an external magnetic field there will exist the spontaneous magnetization M_s.

The Curie point $T = T_C$ is obtained from the condition

$$\frac{kT_C}{\mu_0 g J \mu_B M_{\max} W} = \frac{d}{da} B_J(0). \tag{K. 286}$$

Using eqn. (K. 201) we obtain from it

$$T_C = \frac{\mu_0 g J \mu_B W M_{\max}}{k} \frac{J+1}{3J}. \tag{K. 287}$$

This equation essentially resembles the energy balance equation (K. 280) (since $H \ll H_m$ we have the order-of-magnitude relation $M \approx M_s \approx M_{\max}$).

The temperature dependence $M_s(T)$ of the spontaneous magnetization is graphically obtained from eqns. (K. 283) and (K. 285) for $H = 0$ (cf. Fig. 188). Combining these equations we can write $M_s(T)$ in an implicit form:

$$M_s(T) = M_{\max} B_J \left(\frac{\mu_0 g J \mu_B W M_s(T)}{kT} \right) \tag{K. 288}$$

or, with the help of (K. 287),

$$\frac{M_s(T)}{M_{\max}} = B_J \left(\frac{3J}{J+1} \frac{M_s(T)}{M_{\max}} \frac{T_C}{T} \right). \tag{K. 289}$$

Expressed in terms of the temperature, the spontaneous magnetization has two limits: in the case of low temperatures ($T \to 0$), the Brillouin function tends to unity, that is, the spontaneous magnetization becomes equal to the saturation magnetization,

$$M_s(0) = M_{\max}. \tag{K. 290}$$

The other limit is reached at the Curie point. The curve $M_s(T)$ for $T \to T_C$ ($T < T_C$) is obtained from an expansion of the Brillouin function for small arguments which is broken off after the first non-linear term. Using the series

$$\coth x = \frac{1}{x} + \frac{x}{3} - \frac{x^3}{45} \dots \qquad (|x| < \pi) \tag{K. 291}$$

we obtain an expanded form of (K. 201),

$$B_J(a) \approx \frac{J+1}{J} \frac{a}{3} - \frac{(2J+1)^4 - 1}{(2J)^4} \frac{a^3}{45}. \tag{K. 292}$$

From this we obtain from (K. 289) for the temperature dependence of the spontaneous magnetization near the Curie point

$$\left(\frac{M_s}{M_{\max}}\right)^2_{T\to T_C} = \frac{10}{3}\,\frac{(J+1)^2}{(J+1)^2+J^2}\left(\frac{T}{T_C}\right)^2\left(1-\frac{T}{T_C}\right). \qquad \text{(K. 293)}$$

At the Curie point itself the spontaneous magnetization vanishes:

$$M_s(T_C) = 0. \qquad \text{(K. 294)}$$

From eqn. (K. 293) it follows that the temperature derivative of the spontaneous magnetization becomes infinite at $T = T_C$. This behaviour is characteristic for the appearance of a collective ordering phenomenon.

The shape of the curve $M_s(T)$ between the two limits $M_s(0) = M_{\max}$ and $M_s(T_C) = 0$ depends on the value of the angular momentum quantum number J. When we plot the reduced quantities $M_s/M_{\max}$ versus T/T_C we obtain a universal curve for each value of J (cf. Fig. 189). The experimental values for Fe, Co, and Ni agree best with the theoretical curve for $J = \frac{1}{2}$. This permits

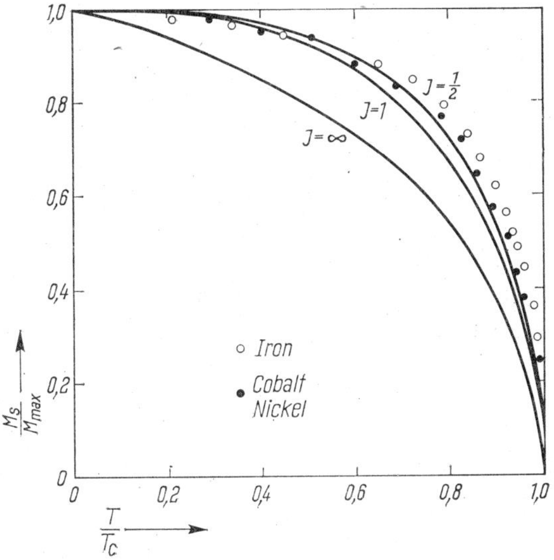

FIG. 189. Temperature dependence of the spontaneous magnetization [after F. Tyler, *Phil. Mag.* **11**, 596 (1931)].

the conclusion that the magnetization is mainly due to the electron spins and that the orbital moments contribute only little to the ferromagnetism of these metals. Independent of the Weiss theory, this result is verified by measurements of the gyromagnetic ratio g (e.g. with the help of the Einstein–de Haas effect) which generally yield $g \approx 2$ for ferromagnetic substances.

2. CURIE–WEISS THEORY OF PARAMAGNETISM

Ferromagnetic substances become paramagnetic above the Curie point. At $T > T_C$ the two curves of $M/M_{\max}$ versus T, eqns. (K. 283) and (K. 285), intersect at parameter values $a \ll 1$. To a first approximation we can therefore replace the Brillouin function by eqn. (K. 201). The substitution of (K. 201) in (K. 285) enables us, using (K. 287), to obtain the Curie–Weiss law

$$\frac{M}{H} = \varkappa = \frac{C}{T-T_C} \tag{K. 295}$$

with the Curie constant (K. 206)

$$C = \frac{T_C}{W} \tag{K. 296}$$

or, using (K. 211) and (K. 287),

$$C = Ng^2J(J+1)\,\mu_0\frac{\mu_B^2}{3k}\,. \tag{K. 297}$$

The Curie–Weiss law is in satisfactory agreement with the observed temperature dependence of the susceptibility. Far above the Curie point the reciprocal susceptibility increases linearly with temperature. The so-called paramagnetic Curie point Θ extrapolated from the linear dependence $1/\varkappa = 0$, however, is higher than the so-called ferromagnetic Curie point T_C determined from the temperature dependence of spontaneous magnetization (cf. p. 469 and Figs. 189 and 190). The ferromagnetic and paramagnetic Curie points as well as the Curie constants are compiled in Table 30 for a few metals. The Curie–Weiss law determined experimentally reads

$$\varkappa_{\text{exp}} = \frac{C}{T-\Theta}\,. \tag{K. 298}$$

Deviations from this linear temperature dependence of the reciprocal susceptibi-

TABLE 30. Ferromagnetic and paramagnetic Curie points T_C and Θ and the Curie constants C of some ferromagnetic metals (after A. H. Morrish, *The Physical Principles of Magnetism* [Wiley, New York 1965], p. 270)

Metal	T_C [°K]	Θ [°K]	$C/4\pi$ [cm³ °K/mole]
Fe	1043	1093	1.26
Co	1394	1428	1.22
Ni	631	650	0.32
Gd	293	302·5	7.8
Dy	85		

lity are observed only above the ferromagnetic Curie point T_C which determines the true point of transition between paramagnetic and ferromagnetic behaviour.

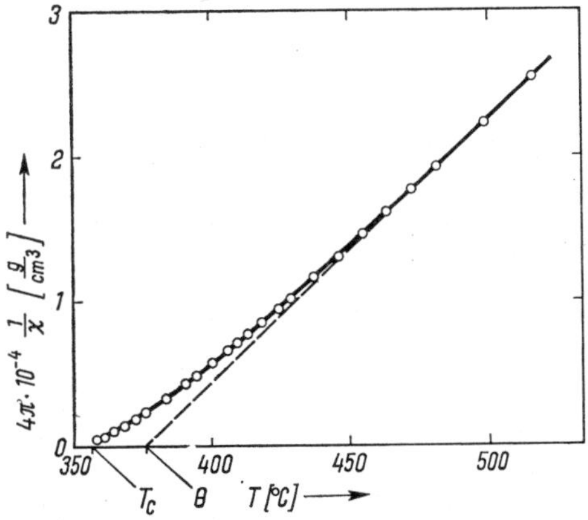

FIG. 190. Temperature dependence of the magnetic susceptibility of nickel above the Curie point [after P. Weiss and R. Forrer, *Ann. Phys., Paris*, **5**, 153 (1926)].

3. THE BASIS OF QUANTUM-MECHANICAL THEORIES OF FERROMAGNETISM

The Weiss theory gives a correct order-of-magnitude description of the ferromagnetic properties. However, the theory makes use of the postulated quantity of the Weiss field whose physical origin remains unexplained. The magnetic dipole interaction alone would be much too small to explain the measured Curie points, which are of the order of 10^3 °K (cf. p. 465).

The decisive interaction, the so-called exchange interaction, which gives rise to the collective magnetic ordering and thus determines the strength of the Weiss field, can only be explained within the framework of quantum mechanics.

According to Dirac, Heisenberg, and Van Vleck, that part of the total energy of a magnetically ordered material which is due to exchange interaction can be determined with the help of the following Hamiltonian (Heisenberg model):

$$\mathcal{H}_{\text{exchange}} = -\sum_j \sum_{i<j} \frac{2}{\hbar^2} A_{ij}(\boldsymbol{S}_i \cdot \boldsymbol{S}_j). \qquad \text{(K. 299)}$$

$\boldsymbol{S}_i$ and $\boldsymbol{S}_j$ are the spin operators of the ith and jth magnetic ion, respectively. The parameter A_{ij} characterizes the strength of the exchange interaction considered. Positive A_{ij} means ferromagnetic, negative A_{ij} antiferromagnetic coupling of the two spins involved. In the general case summation is to be taken over all ion pairs in the crystal. The eigenvalue problem described by eqn. (K. 299) which only accounts for pair interactions of the spins cannot be solved exactly. As regards the various approximations, such as the Ising model, the effective-field method, and the spin-wave approximation, we refer the reader to the specialist literature.

Here we shall only show that the simplest approximation is connected with the molecular field theory dealt with on pp. 464 ff. When we replace in (K. 299) the operator $\boldsymbol{S}_i$ by its expectation value $\langle \boldsymbol{S}_i \rangle$ we obtain the approximation

$$\mathcal{H}_{\text{exchange}} \approx \tfrac{1}{2} \sum_j \mathcal{H}_j \tag{K. 300}$$

where

$$\mathcal{H}_j = -(\boldsymbol{S}_j \cdot \langle \boldsymbol{S}_i \rangle) \sum_{i \neq j} \frac{2}{\hbar^2} A_{ij}. \tag{K. 301}$$

This interaction operator can be expressed in terms of the Weiss molecular field, that is,

$$\mathcal{H}_j = -g\mu_B \left(\frac{\boldsymbol{S}_j}{\hbar} \cdot \boldsymbol{B}_m \right), \tag{K. 302}$$

or, using (K. 278) and (K. 282),

$$\mathcal{H}_j = -\mu_0 g \mu_B \left(\frac{\boldsymbol{S}_j}{\hbar} \cdot W\boldsymbol{M} \right). \tag{K. 303}$$

The magnetization is determined by the mean spin $\langle \boldsymbol{S}_i \rangle$ of the magnetic ions of concentration N:

$$\boldsymbol{M} = g\mu_B \frac{\langle \boldsymbol{S}_i \rangle}{\hbar} N. \tag{K. 304}$$

From eqns. (K. 301) and (K. 302) we get for the molecular field

$$\boldsymbol{B}_m = \frac{1}{g\mu_B} \frac{\langle \boldsymbol{S}_i \rangle}{\hbar} \sum_{i \neq j} 2A_{ij} \tag{K. 305}$$

and from eqns. (K. 301) and (K. 303) for the Weiss factor

$$W = \frac{1}{\mu_0 g^2 \mu_B^2 N} \sum_{i \neq j} 2A_{ij}. \tag{K. 306}$$

The substitution of (K. 306) into (K. 287) and use of eqn. (K. 304) yields a direct relation between the Curie point and the exchange parameter. As the exchange interaction acts only between spins we must replace J by S in eqn. (K. 287), and we get

$$T_C = \frac{2}{3k} S(S+1) \sum_{i \neq j} A_{ij}. \tag{K. 307}$$

In the simplest case, where the exchange interaction is isotropic and involves only the nearest-neighbour ions, eqn. (K. 307) can be rewritten in the form

$$T_C = \frac{2}{3k} S(S+1)\, ZA, \tag{K. 308}$$

where Z is the number of the nearest-neighbour ions and A is the exchange integral.

The range of the exchange interaction is generally much larger than the distance to the nearest neighbour. Moreover, a substance can contain several species of magnetic ions. These are only some of the facts which make it difficult to describe the magnetic behaviour. As to further information on phenomena of magnetic ordering, we refer the reader to the specialist literature.

References

Bates, L. F., *Modern Magnetism* (Cambridge University Press, London 1951).

Becker, R., *Theorie der Elektrizität*, Vol. 1 §§ 49–51, Vol. 2 §§ 29–31 (Teubner, Leipzig 1949).

Becker, R., and Döring, W., *Ferromagnetismus* (Springer, Berlin 1939).

Belov, K. P., *Magnetic Transitions* (Consultants Bureau, New York 1961).

Bozorth, R. M., *Ferromagnetism* (Van Nostrand, Princeton 1951).

Chikazumi, S., and Charap, S. H., *Physics of Magnetism* (Wiley, New York 1964).

Dorfman, J. G., *Diamagnetismus und chemische Bindung* (Deutsch, Frankfurt/M. 1964).

Guggenheim, E. A., *Thermodynamics*, Ch. 13 (North-Holland, Amsterdam 1949).

Herpin, A., *Magnétisme*, in *Low Temperature Physics*, edited by C. de Witt *et al.* (Gordon and Breach, New York 1962).

Herpin A., *Théorie du Magnétisme* (Presses Universitaires de France, Paris 1968).

Kneller, E., *Ferromagnetismus* (Springer, Berlin 1962).

Krupička, S., and Sternberk, J. (Editors), *Elements of Theoretical Magnetism* (Iliffe, New York 1968).

Marshall, W. (Editor), *Theory of Magnetism in Transition Metals* (Academic Press, New York 1967).

Martin, D. H., *Magnetism in Solids* (M.I.T. Press, Cambridge, Mass. 1967).

Morrish, A. H., *The Physical Principles of Magnetism* (Wiley, New York 1965).

Pryce, M. H. L., *et al.*, *Varenna Lectures on Magnetism*, in *Nuovo Cimento*, Suppl. 6, pp. 895 ff. (1957).

Rado, G. and Suhl, H. (Editors), *Magnetism*, Vols. 1–4 (Academic Press, New York 1962–1966).

Schieber, M. M., *Experimental Magnetochemistry* (North-Holland, Amsterdam 1967).

Selwood, P. W., *Magnetochemistry* (Interscience, New York 1956).

Smart, J. S., *Effective Field Theories of Magnetism* (Saunders, Philadelphia 1966).

Stoner, E. C., *Magnetism and Matter* (Methuen, London 1934).

Van Vleck, J. H., *The Theory of Electric and Magnetic Susceptibilities* (Oxford University Press, London 1932).

Vogt, E., *Physikalische Eigenschaften der Metalle*, Vol. 1, Ch. 4, 5 (Akademische Verlagsgesellschaft, Leipzig 1958).

Wijn, H. P. J. (Editor), *Magnetismus*, in *Handbuch der Physik*, Vol. 18/1 (Springer, Berlin 1968).

Wijn, H. P. J. (Editor), *Ferromagnetismus*, in *Handbuch der Physik*, Vol. 18/2 (Springer, Berlin 1966).

Wonsowski, S. W., *Moderne Lehre vom Magnetismus* (Deutscher Verlag der Wissenschaften, Berlin 1956).

Physical Constants and Conversion Factors

I. Physical Constants†

Speed of light *in vacuo*	$c = 2.998\times 10^{8}$ m/s
Magnetic permeability of the vacuum	$\mu_0 = 4\pi\times 10^{-7}$ V s/A m
Electric permeability of the vacuum	$\varepsilon_0 = \dfrac{1}{\mu_0 c^2} = 8.854\times 10^{-12}$ A sec/ V m
Avogadro number	$L = 6.022\times 10^{26}$ kmole^{-1}
Universal gas constant	$R_0 = Lk_B = 8.314\times 10^{3}$ J/°K kmole
	$= 1.986$ kcal/°K kmole
Boltzmann's constant	$k_B = 1.381\times 10^{-23}$ J/°K kmole
Planck's quantum of action	$h = 6.626\times 10^{-34}$ J s$= 4.136\times 10^{-15}$ eV s
	$\hbar = \dfrac{h}{2\pi} = 1.055\times 10^{-34}$ J s
	$= 6.582\times 10^{-16}$ eV s
Electric elementary charge	$e = 1.602\times 10^{-19}$ A s
	$= 4.803\times 10^{-10}$ e.s.u.
Quantum of magnetic flux	$\Phi_0 = \left[\dfrac{1}{c}\right]\left(\dfrac{hc}{2e}\right) = 2.068\times 10^{-15}$ V s
	$= 2.068\times 10^{-7}$ gauss cm^2
Rest mass of the electron	$m_e = 0.9110\times 10^{-30}$ kg
Specific charge of the electron	$\dfrac{e}{m_e} = 1.759\times 10^{11}$ A sec/kg
Classical radius of the electron	$r_e = \left[\dfrac{1}{4\pi\varepsilon_0}\right]\left(\dfrac{e^2}{m_e c^2}\right) = \alpha^2 a_0 = 2.818\times 10^{-15}$ m
Compton wavelength of the electron	$\lambda_e = \dfrac{h}{m_e c} = 2.426\times 10^{-12}$ m
Rest mass of the neutron	$m_N = 1.6749\times 10^{-27}$ kg
Rest mass of the proton	$m_P = 1.6726\times 10^{-27}$ kg $= 1836 m_e$

† From B. N. Taylor, W. H. Parker, and D. N. Langenberg, *Rev. Mod. Phys.* **41**, 375 (1969) and E. R. Cohen and J. W. M. Dumond, *Rev. Mod. Phys.* **37**, 537 (1965).

Atomic mass unit (12 units for the ^{12}C atom) $m_u = 1.6605 \times 10^{-27}$ kg $= 1$ u

Rydberg's constant (for infinite nuclear mass) $R_\infty = \left[\frac{1}{4\pi\varepsilon_0}\right]^2 \left(\frac{m_e e^4}{4\pi\hbar^3 c}\right) = 1.0974 \times 10^7 \text{ m}^{-1}$

$$\nu_{Ry} = R_\infty c = 3.290 \times 10^{15} \text{ s}^{-1}$$

Rydberg energy (for infinite nuclear mass) $E_{Ry} = h\nu_{Ry} = 13.606$ eV

Bohr radius (for infinite nuclear mass) $a_0 = [4\pi\varepsilon_0]\left(\frac{\hbar^2}{m_e e^2}\right) = \frac{\alpha}{4\pi R_\infty} = 5.292 \times 10^{-11}$ m

Sommerfeld's fine structure constant $\alpha = \left[\frac{1}{4\pi\varepsilon_0}\right]\left(\frac{e^2}{\hbar c}\right) = \frac{\lambda_c}{2\pi a_0} = \frac{1}{137.04}$

Bohr magneton

in the Giorgi–Mie system of units $\{B = \mu_0(H+M)\}$ $\mu_B^{Mie} = \frac{e\hbar}{2m_e} = 0.9274 \times 10^{-23}$ A m^2

in the Giorgi–Pohl system of units $\{B = \mu_0 H + M\}$ $\mu_B^{Pohl} = \mu_0 \frac{e\hbar}{2m_e} = 1.165 \times 10^{-29}$ V s/m

in the Gaussian system of units $\{B = H + 4\pi M\}$ $\mu_B^{Gauss} = \frac{e\hbar}{2m_e c} = 0.9274 \times 10^{-20}$ erg/oe

II. Energy Conversion

1 J = 1 V A s = 1 W s = 10^7 erg = $4\pi \times 10^7$ cm^3 gauss oe
(1 A/m = $4\pi \times 10^{-3}$ oe, 1 V s/m^2 = 10^4 gauss; the gauss is considered here as the unit of the magnetic induction B. In the literature very often no difference is made between gauss and oersted. Sometimes the unit for the magnetization M is given in gauss for magnetization = 4π gauss for B).

1 eV = 0.07350 Rydberg = 1.602×10^{-19} J = 3.827×10^{-20} cal = k_B (11605°K)

$= 2\mu_B^{Mie}$ (0.8638×10^4 V s/m^2) = $2\mu_B^{Pohl}$ (0.6874×10^{10} A/m) = $2\mu_B^{Gauss}$ (0.8638×10^8 oe)

k_B (1°K) = 0.8617×10^{-4} eV = $2\mu_B^{Mie}$ (0.7443 V s/m^2)

$$\exp\left(\frac{E}{k_B T}\right) = \exp\left(11605 \frac{E\,[\text{eV}]}{T\,[°\text{K}]}\right) = 10^{5039\,E(\text{eV})/T(°\text{K})}$$

1 cal = 4.187 J; 1 J = 0.2388 cal

1 eV per molecule corresponds to 23.05 kcal/mole

(The value given here for 1 cal corresponds to the international definition of vapour pressure tables.)

III. Particles and Waves

1. *Energy E and momentum p in relation to frequency ν and wavelength λ*

$$E = h\nu = \hbar\omega, \quad p = \frac{h}{\lambda} = \hbar k$$

Dispersion relation: $E = E(p) = E(k)$ or $\omega = \omega(k)$ or $\nu = \nu(\lambda)$

Phase velocity: $v_p = \frac{E}{p} = \frac{\omega}{k} = \nu\lambda$

Group velocity: $v_g = \frac{\partial E}{\partial p} = \frac{\partial \omega}{\partial k}$

2. *Particles or quasiparticles without dispersion*

$$\frac{E}{p} = \frac{\partial E}{\partial p} = \text{const.}$$

Example: photons *in vacuo*

Speed of light $c = v_g = v_p = 2.998\times 10^{-10}$ cm/s

Energy and wavelength: $E = \frac{hc}{\lambda}$ or $E\,(\text{eV}) = \frac{1.2399}{\lambda\,[\mu\text{m}]}$

For $E = 1$ eV the wavelength is given by $\lambda = 12399$ Å $\approx 1.240\ \mu$m,

the wave number is then $\frac{1}{\lambda} = \frac{k}{2\pi} = 8065\ \text{cm}^{-1}$,

the frequency is $\nu = \frac{E}{h} = 2.418\times 10^{14}$ Hz.

3. *Free particles with dispersion*

$$E = mc^2 = \frac{m_0c^2}{\left(1-\frac{v_g^2}{c^2}\right)^{1/2}} = m_0c^2 + E_{\text{kin}}\,, \quad p = mv_g = \frac{m_0 v_g}{\left(1-\frac{v_g^2}{c^2}\right)^{1/2}}$$

In the non-relativistic case we have:

$$E_{\text{kin}} = \frac{p^2}{2m_0}, \quad p = m_0 v$$

Example: free electrons

Rest mass $m_0 = m_e = 0.9110\times 10^{-30}$ kg

Rest energy $m_ec^2 = 5.110\times 10^5$ eV

For $E_{\text{kin}} = 1$ eV we have:

Velocity $v_g = 5.931\times 10^7$ cm/s,

Wavelength $\lambda = \frac{h}{p} = \left(\frac{h^2}{2m_eE_{\text{kin}}}\right)^{1/2} = \left(\frac{150.4}{E_{\text{kin}}\,[\text{eV}]}\right)^{1/2}$ Å $= 12.26$ Å,

Wave number $\frac{1}{\lambda} = \frac{k}{2\pi} = 8.154 \times 10^{6}\ \text{cm}^{-1}$,

Frequency $\nu = \frac{E}{h} = \frac{m_e c^2}{h} = 1.236 \times 10^{20}\ \text{Hz}.$

For high kinetic energies ($E_{\text{kin}} \gtrsim 10^5$ eV) or high velocities ($v_g > 10^9$ cm/s) the relativistic correction has to be taken into account. Then we have

$$\lambda = \frac{h}{p} = \frac{h}{m_0 v_g}\left(1 - \frac{v_g^2}{c^2}\right)^{1/2} = \left(\frac{h^2}{2m_0 E_{\text{kin}}\left(1 + \frac{E_{\text{kin}}}{2m_0 c^2}\right)}\right)^{1/2},$$

$$\nu = \frac{E}{h} = \frac{m_0 c^2}{h\left(1 - \frac{v_g^2}{c^2}\right)^{1/2}}.$$

Data of the "Standard" Metal

THE data of the "standard" metal are taken from A. B. Pippard [*Rep. Prog. Phys.* **23,** 176 (1960)] in order to compare with the data of real metals. Pippard has given data of an idealized metal on the basis of the free electron model. In this model the electron concentration is assumed to be $n = 6{\cdot}0 \times 10^{22}\ \mathrm{cm}^{-3}$ (this corresponds to an atomic volume A/ϱ of approximately 10 cm^3 for a monovalent metal). For $T = 0°\mathrm{K}$ we have the following values:

Fermi energy
$$\zeta_0 = \frac{\hbar^2}{2m_e}(3\pi^2 n)^{2/3} = 8.95 \times 10^{-12}\ \mathrm{erg} = 5.59\ \mathrm{eV}$$

Temperature of degeneracy
$$T_E = \frac{2}{5}\frac{\zeta_0}{k_B} = 25900°\mathrm{K}$$

Radius of the Fermi sphere
$$k(\zeta_0) = (3\pi^2 n)^{1/3} = 1.21 \times 10^8\ \mathrm{cm}^{-1}$$

Eigenvalue density at the Fermi per unit volume (spin degeneracy not taken into account)
$$D(\zeta_0) = \frac{1}{4\pi^2}\left(\frac{2m_e}{\hbar^2}\right)^{3/2}\zeta_0^{1/2} = \frac{m_e}{2\pi^2\hbar^2}k(\zeta_0) =$$
$$= 5.03 \times 10^{33}\ \mathrm{erg}^{-1}\ \mathrm{cm}^{-3}$$
$$= 0.805 \times 10^{22}\ \mathrm{eV}^{-1}\ \mathrm{cm}^{-3}$$

Specific heat of electrons per unit volume
$$c_V^{El} = \frac{2}{3}\pi^2 k_B^2 D(\zeta_0)T = \gamma T = 630\,T\ \mathrm{erg/cm^3\,°K}$$

Fermi velocity
$$v_F = \frac{\hbar k(\zeta_0)}{m_e} = 1.40 \times 10^8\ \mathrm{cm/s}$$

Surface of the Fermi sphere
$$S = 4\pi k^2(\zeta_0) = 1.84 \times 10^{17}\ \mathrm{cm}^{-2}$$

Maximum cross-section of the Fermi sphere
$$A_{\max} = \frac{S}{4} = 4.61 \times 10^{16}\ \mathrm{cm}^{-2}$$

Cyclotron frequency of electrons in a magnetic field H
$$\omega_c = \frac{e}{m_e}B \approx \frac{2\mu_B^{Mie}}{\hbar}\mu_0 H =$$
$$= 2.21 \times 10^7\left(H\left[\frac{\mathrm{A}}{\mathrm{cm}}\right]\right)\ \mathrm{s}^{-1}$$

Classical cyclotron radius for electrons at the Fermi energy
$$R_c = \frac{v_F}{\omega_c} = \frac{6.33}{H\,[\mathrm{A/cm}]}\ \mathrm{cm}$$

Electrical conductivity of electrons having the Fermi energy and a mean free path Λ
$$\sigma = \frac{ne^2}{m_e v_F}\Lambda = 1.21 \times 10^{11}\ (\Lambda\ [\mathrm{cm}])\ \Omega^{-1}\ \mathrm{cm}^{-1}$$

Periodic System and Periodic Tables of the Physical Properties of the Elements

THE periodic Tables P1–5 contain the following physical data:

Tables P1–2 P4–5	**A**	Atomic number
Tables P1–5	**B**	Symbol
Table P1	**C**	Name (*CRC Handbook Chem. Phys.* 1970)
	D	Atomic weight in g/g-atom (*CRC Handbook Chem. Phys.* 1970)
	E	Density in g/cm^3 (TAYLOR and KAGLE 1963; LANDOLT-BÖRNSTEIN IV/4a, 1967)
	F	Atomic volume in cm^3/g-atom (Periodic Tables, SARGENT 1964; GSCHNEIDNER 1964)
Table P2	**G**	Normal electron configuration of the atom (MOORE 1949; WYBOURNE 1965; SAMSONOV 1968)
	H	Ground state of the atom (MOORE 1949–1958; WYBOURNE 1965)
	I	Ground state of the singly, doubly or three-fold ionized atom (see under **H**)
	J	First, second and third energy of ionization in eV (MOORE 1958; SAMSONOV 1968)
Table P3	**K**	Electronegativity (RICH 1965)
	L	Oxidation states (JØRGENSEN 1969)
	M	Crystal structure (LANDOLT-BÖRNSTEIN I/4, 1955; PEARSON 1967)
	N	Atomic radius in Å (SLATER 1965)
	O	Ionic radius in Å (*CRC Handbook Chem. Phys.* 1970)
Table P4	**P**	Melting temperature in °K (GSCHNEIDNER 1964; LANDOLT-BÖRNSTEIN IV/4a, 1967)
	Q	Boiling temperature in °K (see under **P**)
	R	Enthalpy of melting in kcal/g-atom (see under **P**)
	S	Enthalpy of evaporation in kcal/g-atom (Periodic Tables, SARGENT 1964; SANDERSON 1967; *CRC Handbook Chem. Phys.* 1970)
	T	Enthalpy of atomization in kcal/g-atom (GSCHNEIDNER 1964; RICH 1965; SANDERSON 1967)
Table P5	**U**	Magnetic transition temperature in °K (KEFFER 1966; TEBBLE and CRAIK 1969; CONNOLLY and COPENHAVER 1970)
	V	Superconducting transition temperature in °K (MÜLLER 1968; GLADSTONE *et al.* 1969)
	W	Work function in eV (FOMENKO and SAMSONOV 1966; EASTMAN 1970)

TABLE P1
Periodic system of the elements

Period									
1	1 1.00797 [liqf.] 0.071 H 14.2 Hydrogen								2 4.0026 [liqf.] 0.126 He 31.8 Helium
2	3 6.939 0.53 Li 13.0 Lithium	4 9.0122 1.84 Be 4.89 Beryllium		5 10.811 [α] 2.46 B 4.40 Boron	6 12.01115 [G] 2.26 C 5.30 Carbon	7 14.0067 [liqf.] 0.81 N 17.3 Nitrogen	8 15.9994 [liqf.] 1.14 O 14.0 Oxygen	9 18.9984 [liqf.] 1.50 F 12.6 Fluorine	10 20.183 [liqf.] 1.21 Ne 16.7 Neon
3	11 22.9898 0.97 Na 23.8 Sodium	12 24.312 1.74 Mg 14.0 Magnesium		13 26.9815 2.70 Al 10.0 Aluminum	14 28.086 2.33 Si 12.1 Silicon	15 30.9738 [w] 1.80 P 17.2 Phosphorus	16 32.064 [α] 2.08 S 15.4 Sulfur	17 35.453 [liqf.] 1.56 Cl 22.7 Chlorine	18 39.948 [liqf.] 1.41 Ar 28.4 Argon
4	19 39.102 0.86 K 45.5 Potassium	20 40.08 1.53 Ca 26.2 Calcium	30 65.37 7.13 Zn 9.17 Zinc	31 69.72 5.91 Ga 11.8 Gallium	32 72.59 5.33 Ge 13.6 Germanium	33 74.9216 5.78 As 13.0 Arsenic	34 78.96 [hex] 4.81 Se 16.4 Selenium	35 79.904 [liq.] 3.12 Br 25.6 Bromine	36 83.80 [liqf.] 2.40 Kr 35 Krypton
5	37 85.47 1.53 Rb 55.8 Rubidium	38 87.62 2.58 Sr 33.9 Strontium	48 112.40 8.64 Cd 13.0 Cadmium	49 114.82 7.29 In 15.7 Indium	50 118.69 [β] 7.29 Sn 16.3 Tin	51 121.75 6.69 Sb 18.2 Antimony	52 127.60 6.24 Te 20.5 Tellurium	53 126.9044 4.93 J 25.7 Iodine	54 131.30 [liqf.] 3.06 Xe 43 Xenon
6	55 132.905 1.90 Cs 70.0 Cesium	56 137.34 3.6 Ba 38.1 Barium	80 200.59 [liq.] 13.6 Hg 14.8 Mercury	81 204.37 11.9 Tl 17.2 Thallium	82 207.19 11.3 Pb 18.3 Lead	83 208.980 9.81 Bi 21.3 Bismuth	84 [209] [α] 9.32 Po 22.5 Polonium	85 [210] – At – Astatine	86 [222] [liqf.] *4.5* Rn *50* Radon
7	87 [223] – Fr *73* Francium	88 [226] 6.0 Ra *38* Radium							

Key (example cell):

A → 1	1.00797	← D
[liqf.]		
E → 0.071	H	← B
F → 14.2		
	Hydrogen	← C

A Atomic number
B Symbol [a])
C Name [a])
D Atomic weight A [a])

E Density ϱ in $\frac{\text{g}}{\text{cm}^3}$ [b])

F Atomic volume $V_A = A/\varrho$ in $\frac{\text{cm}^3}{\text{g-atom}}$ [b])

Re E and F:
(liqf.): liquefied at the boiling point
(liq.): liquid at room temperature
(α, β, γ, w, G, hex): Designation of a crystal modification at room temperature specified in more detail in Table P3
(^{99}Tc, ^{147}Pm): refers to the isotopes ^{99}Tc and ^{147}Pm respectively

Data printed in italics are either uncertain or else were obtained by extrapolation.

Remarks [a]) and [b]) follow the tables (p. 490).

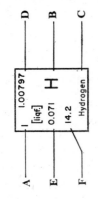

Elements of the main groups

4		21 44.956 Sc 2.99 15.0 Scandium		22 47.90 Ti 4.51 10.6 Titanium	23 50.942 V 6.09 8.37 Vanadium	24 51.996 Cr 7.19 7.23 Chromium	25 54.9380 [α] Mn 10.2 9.39 Manganese	26 55.847 Fe 7.87 7.10 Iron	27 58.9332 Co 8.8 6.7 Cobalt	28 58.71 Ni 8.9 6.6 Nickel	29 63.546 Cu 8.93 7.11 Copper
5		39 88.905 Y 4.47 19.9 Yttrium		40 91.22 Zr 6.56 14.0 Zirconium	41 92.906 Nb 8.58 10.8 Niobium	42 95.94 Mo 10.2 9.39 Molybdenum	43 [97] [^{99}Tc] Tc 11.5 8.64 Technetium	44 101.07 Ru 12.4 8.18 Ruthenium	45 102.905 Rh 12.4 8.29 Rhodium	46 106.4 Pd 12.0 8.88 Palladium	47 107.868 Ag 10.5 10.3 Silver
6	57 138.91 La 6.17 22.5 Lanthanum		71 174.97 Lu 9.74 17.9 Lutetium	72 178.49 Hf 13.2 13.5 Hafnium	73 180.948 Ta 16.6 10.8 Tantalum	74 183.85 W 19.3 9.53 Tungsten	75 186.2 Re 21.0 8.87 Rhenium	76 190.2 Os 22.6 8.42 Osmium	77 192.2 Ir 22.7 8.48 Iridium	78 195.09 Pt 21.5 9.08 Platinum	79 196.967 Au 19.3 10.2 Gold
7	89 [227] Ac 10.1 22.6 Actinium		103 [257] (Lr) – – (Lawrencium)	104 [257] (Rf) – – (Rutherfordium)	105 [260] (Ha) – – (Hahnium)						

d-elements and precious metals

f-elements (lanthanides and actinides)

6	58 140.12 [γ] Ce 6.77 20.7 Cerium	59 140.907 Pr 6.77 20.8 Praseodymium	60 144.24 Nd 7.00 20.6 Neodymium	61 [145] [^{147}Pm] Pm *7.26* *20.3* Prometheum	62 150.35 Sm 7.54 20.0 Samarium	63 151.96 Eu 5.25 29.0 Europium	64 157.25 Gd 7.87 19.9 Gadolinium	65 158.924 Tb 8.27 19.2 Terbium	66 162.50 Dy 8.53 19.0 Dysprosium	67 164.930 Ho 8.80 18.8 Holmium	68 167.26 Er 9.04 18.5 Erbium	69 168.934 Tm 9.33 18.1 Thulium	70 173.04 Yb 6.97 24.8 Ytterbium
7	90 232.038 Th 11.7 19.8 Thorium	91 [231] Pa 15.4 15.0 Protactinium	92 238.03 [α] U 19.0 12.5 Uranium	93 [237] Np 20.5 11.6 Neptunium	94 [244] Pu 19.8 12.1 Plutonium	95 [243] Am 13.7 17.7 Americium	96 [247] Cm – – Curium	97 [247] Bk – – Berkelium	98 [251] Cf – – Californium	99 [254] Es – – Einsteinium	100 [257] Fm – – Fermium	101 [256] Md – – Mendelevium	102 [254] No – – Nobelium

1	H 13.6 1 1s $^2S_{1/2}$								He 24.6 54.4 $^2S_{1/2}$ 2 $1s^2$ $^1S_0$
2	Li 5.39 75.6 122 $^1S_0$ $^2S_{1/2}$ 3 $1s^2 2s$ $^2S_{1/2}$	Be 9.32 18.2 154 $^2S_{1/2}$ $^1S_0$ $^2S_{1/2}$ 4 $1s^2 2s^2$ $^1S_0$		B 8.30 25.1 37.9 $^1S_0$ $^2S_{1/2}$ $^1S_0$ 5 $2s^2 2p$ $^2P_{1/2}$	C 11.3 24.4 47.9 $^2P_{1/2}$ $^1S_0$ $^2S_{1/2}$ 6 $2s^2 2p^2$ $^3P_0$	N 14.5 29.6 47.4 $^3P_0$ $^2P_{1/2}$ $^1S_0$ 7 $2s^2 2p^3$ $^4S_{3/2}$	O 13.6 35.1 54.9 $^4S_{3/2}$ $^3P_0$ $^2P_{1/2}$ 8 $2s^2 2p^4$ $^3P_2$	F 17.4 35.0 62.6 $^3P_2$ $^4S_{3/2}$ $^3P_0$ 9 $2s^2 2p^5$ $^2P_{3/2}$	Ne 21.6 41.1 63.5 $^2P_{3/2}$ $^3P_2$ $^4S_{3/2}$ 10 $2s^2 2p^6$ $^1S_0$
3	Na 5.14 47.3 71.8 $^1S_0$ $^2P_{3/2}$ $^3P_2$ 11 $2p^6 3s$ $^2S_{1/2}$	Mg 7.64 15.0 *78.1* $^2S_{1/2}$ $^1S_0$ $^2P_{3/2}$ 12 $2p^6 3s^2$ $^1S_0$		Al 5.98 18.8 28.4 $^1S_0$ $^2S_{1/2}$ $^1S_0$ 13 $3s^2 3p$ $^2P_{1/2}$	Si 8.15 16.3 33.5 $^2P_{1/2}$ $^1S_0$ $^2S_{1/2}$ 14 $3s^2 3p^2$ $^3P_0$	P 10.5 19.7 30.2 $^3P_0$ $^2P_{1/2}$ $^1S_0$ 15 $3s^2 3p^3$ $^4S_{3/2}$	S 10.4 23.4 34.8 $^4S_{3/2}$ $^3P_0$ $^2P_{1/2}$ 16 $3s^2 3p^4$ $^3P_2$	Cl 13.0 23.8 39.9 $^3P_2$ $^4S_{3/2}$ $^3P_0$ 17 $3s^2 3p^5$ $^2P_{3/2}$	Ar 15.8 27.6 40.9 $^2P_{3/2}$ $^3P_2$ $^4S_{3/2}$ 18 $3s^2 3p^6$ $^1S_0$
4	K 4.34 31.8 45.9 $^1S_0$ $^2P_{3/2}$ $^3P_2$ 19 $3p^6 4s$ $^2S_{1/2}$	Ca 6.11 11.9 51.2 $^2S_{1/2}$ $^1S_0$ $^2P_{3/2}$ 20 $3p^6 4s^2$ $^1S_0$	Zn 9.39 18.0 39.7 $^2S_{1/2}$ $^1S_0$ *$^2D_{5/2}$* 30 $3d^{10} 4s^2$ $^1S_0$	Ga 6.00 20.5 30.7 $^1S_0$ $^2S_{1/2}$ $^1S_0$ 31 $4s^2 4p$ $^2P_{1/2}$	Ge 7.88 15.9 34.2 $^2P_{1/2}$ $^1S_0$ $^2S_{1/2}$ 32 $4s^2 4p^2$ $^3P_0$	As 9.81 18.7 28.3 $^3P_0$ $^2P_{1/2}$ $^1S_0$ 33 $4s^2 4p^3$ $^4S_{3/2}$	Se 9.75 21.5 32.0 $^4S_{3/2}$ $^3P_0$ $^2P_{1/2}$ 34 $4s^2 4p^4$ $^3P_2$	Br 11.8 21.6 35.9 $^3P_2$ $^4S_{3/2}$ $^3P_0$ 35 $4s^2 4p^5$ $^2P_{3/2}$	Kr 14.0 24.6 36.9 $^2P_{3/2}$ $^3P_2$ $^4S_{3/2}$ 36 $4s^2 4p^6$ $^1S_0$
5	Rb 4.18 27.6 40 $^1S_0$ $^2P_{3/2}$ – 37 $4p^6 5s$ $^2S_{1/2}$	Sr 5.69 11.0 43.6 $^2S_{1/2}$ – $^2P_{3/2}$ 38 $4p^6 5s^2$ $^1S_0$	Cd 8.99 16.9 *44* $^2S_{1/2}$ $^1S_0$ $^2D_{5/2}$ 48 $4d^{10} 5s^2$ $^1S_0$	In 5.79 18.9 28.0 $^1S_0$ $^2S_{1/2}$ $^1S_0$ 49 $5s^2 5p$ $^2P_{1/2}$	Sn 7.33 14.6 30.7 $^2P_{1/2}$ $^1S_0$ $^2S_{1/2}$ 50 $5s^2 5p^2$ $^3P_0$	Sb 8.64 16.7 24.8 $^3P_0$ $^2P_{1/2}$ $^1S_0$ 51 $5s^2 5p^3$ $^4S_{3/2}$	Te 9.01 18.8 31.0 $^4S_{3/2}$ $^3P_0$ $^2P_{1/2}$ 52 $5s^2 5p^4$ $^3P_2$	J 10.4 19.0 33 $^3P_2$ $^4S_{3/2}$ $^3P_0$ 53 $5s^2 5p^5$ $^2P_{3/2}$	Xe 12.1 21.2 32.1 $^2P_{3/2}$ $^3P_2$ $^4S_{3/2}$ 54 $5s^2 5p^6$ $^1S_0$
6	Cs 3.89 25.1 34.6 $^1S_0$ $^2P_{3/2}$ $^3P_2$ 55 $5p^6 6s$ $^2S_{1/2}$	Ba 5.21 10.0 *37* $^2S_{1/2}$ – $^2P_{3/2}$ 56 $5p^6 6s^2$ $^1S_0$	Hg 10.4 18.8 34.2 $^2S_{1/2}$ $^3D_3$ $^4F_{9/2}$ 80 $5d^{10} 6s^2$ $^1S_0$	Tl 6.11 20.4 29.8 $^1S_0$ $^2S_{1/2}$ $^3D_3$ 81 $6s^2 6p$ $^2P_{1/2}$	Pb 7.42 15.0 31.9 $^2P_{1/2}$ $^1S_0$ $^2S_{1/2}$ 82 $6s^2 6p^2$ $^3P_0$	Bi 7.28 *16.7* 25.6 $^3P_0$ $^2P_{1/2}$ $^1S_0$ 83 $6s^2 6p^3$ $^4S_{3/2}$	Po 8.2 19.4 27 – – – 84 $6s^2 6p^4$ $^3P_2$	At 9.2 20.1 29 – – – 85 $6s^2 6p^5$ $^2P_{3/2}$	Rn 10.7 21.4 29 $^2P_{3/2}$ – – 86 $6s^2 6p^6$ $^1S_0$
7	Fr 3.98 22.5 33 – – – 87 $6p^6 7s$ $^2S_{1/2}$	Ra 5.28 10.1 *34* $^2S_{1/2}$ – – 88 $6p^6 7s^2$ $^1S_0$							

TABLE P2
Atomic data

B —	Si	8.15 16.3 33.5		— J
I —	$^2P_{1/2}$	$^1S_0$	$^2S_{1/2}$	
A —	14	$3s^2 3p^2$ (— G)	$^3P_0$	— H

- **A** Atomic number
- **B** Symbol
- **G** Normal electron configuration of the atom [c])
- **H** Ground state of the atom [d])
- **I** From left to right: ground state of the singly, doubly and threefold ionized atom [d])
- **J** From top to bottom: first, second and third ionization energy in eV

Data given in italics are either uncertain or were obtained by extrapolation.

Period										
4	Sc 6.56 12.9 24.8 $^3D_1$ $^2D_{3/2}$ $^1S_0$ 21 $3d\,4s^2$ $^2D_{3/2}$		Ti 6.83 13.6 28.1 $^4F_{3/2}$ $^3F_2$ $^2D_{3/2}$ 22 $3d^2 4s^2$ $^3F_2$	V 6.74 14.2 29.7 $^5D_0$ $^4F_{3/2}$ $^3F_2$ 23 $3d^3 4s^2$ $^4F_{3/2}$	Cr 6.76 16.5 31 $^6S_{5/2}$ $^5D_0$ $^4F_{3/2}$ 24 $3d^5 4s$ $^7S_3$	Mn 7.43 15.6 33.7 $^7S_3$ $^6S_{5/2}$ $^5D_0$ 25 $3d^5 4s^2$ $^6S_{5/2}$	Fe 7.9 16.2 30.6 $^6D_{9/2}$ $^5D_4$ $^6S_{5/2}$ 26 $3d^6 4s^2$ $^5D_4$	Co 7.86 17.1 33.5 $^3F_4$ $^4F_{9/2}$ $^5D_4$ 27 $3d^7 4s^2$ $^4F_{9/2}$	Ni 7.63 18.2 35.2 $^2D_{5/2}$ $^3F_4$ — 28 $3d^8 4s^2$ $^3F_4$	Cu 7.72 20.3 36.8 $^1S_0$ $^2D_{5/2}$ — 29 $3d^{10} 4s$ $^2S_{1/2}$
5	Y 6.38 12.2 20.5 $^1S_0$ $^2D_{3/2}$ — 39 $4d\,5s^2$ $^2D_{3/2}$		Zr 6.84 12.9 *24* $^4F_{3/2}$ $^3F_2$ $^2D_{3/2}$ 40 $4d^2 5s^2$ $^3F_2$	Nb 6.88 13.9 *28* $^5D_0$ $^4F_{3/2}$ $^3F_2$ 41 $4d^4 5s$ $^6D_{1/2}$	Mo 7.13 15.7 *29* $^6S_{5/2}$ $^5D_0$ $^4F_{3/2}$ 42 $4d^5 5s$ $^7S_3$	Tc 7.23 14.9 *31* $^7S_3$ — — 43 $4d^5 5s^2$ $^6S_{5/2}$	Ru 7.36 16.6 *30* $^4F_{9/2}$ $^5D_4$ — 44 $4d^7 5s$ $^5F_5$	Rh 7.46 *18.1* *31* $^3F_4$ $^4F_{9/2}$ — 45 $4d^8 5s$ $^4F_{9/2}$	Pd 8.33 19.4 *33* $^2D_{5/2}$ $^3F_4$ — 46 $4d^{10}$ $^1S_0$	Ag 7.57 21.5 *36* $^1S_0$ $^2D_{5/2}$ — 47 $4d^{10} 5s$ $^2S_{1/2}$
6	La 5.61 11.4 19.2 $^3F_2$ $^2D_{3/2}$ $^1S_0$ 57 $5d\,6s^2$ $^2D_{3/2}$	Lu 6.15 14.7 *22* $^1S_0$ $^2S_{1/2}$ $^1S_0$ 71 $4f^{14} 5d6s^2$ $^2D_{3/2}$	Hf *7* 14.9 *23* $^2D_{3/2}$ — — 72 $5d^2 6s^2$ $^3F_2$	Ta 7.7 16.2 *22* $^5F_1$ — — 73 $5d^3 6s^2$ $^4F_{3/2}$	W 7.98 17.7 *24* $^6D_{1/2}$ — — 74 $5d^4 6s^2$ $^5D_0$	Re 7.87 16.6 *26* $^7S_3$ — — 75 $5d^5 6s^2$ $^6S_{5/2}$	Os 8.7 17 *25* $^6D_{9/2}$ — — 76 $5d^6 6s^2$ $^5D_4$	Ir 9.2 17.0 *27* — — — 77 $5d^7 6s^2$ $^4F_{9/2}$	Pt 8.96 18.5 *29* $^2D_{5/2}$ — — 78 $5d^9 6s$ $^3D_3$	Au 9.22 20.5 *30* $^1S_0$ — — 79 $5d^{10} 6s$ $^2S_{1/2}$
7	Ac 6.89 11.5 *20* $^1S_0$ $^2S_{1/2}$ — 89 $6d\,7s^2$ $^2D_{3/2}$	(Lr) — — — $^1S_0$ $^2S_{1/2}$ $^1S_0$ 103 $5f^{14} 6d7s^2$ $^2D_{3/2}$	(Rf) — — — — — — 104 $6d^2 7s^2$ —	(Ha) — — — — — — 105 — —						

Period							
6	Ce 6.54 12.3 19.9 $^2G_{7/2}$ $^3H_4$ $^2F_{5/2}$ 58 $4f5d6s^2$ $^1G_4$	Pr *5.76* *11.5* *21* $^5I_4$ $^4I_{9/2}$ $^3H_4$ 59 $4f^3 6s^2$ $^4I_{9/2}$	Nd *6.31* *12.1* *20.5* $^6I_{7/2}$ $^5I_4$ $^4I_{9/2}$ 60 $4f^4 6s^2$ $^5I_4$	Pm *5.9* *11.7* *22* $^7H_2$ $^6H_{5/2}$ $^5I_4$ 61 $4f^5 6s^2$ $^6H_{5/2}$	Sm 5.56 11.4 *24* $^8F_{1/2}$ $^7F_0$ $^6H_{5/2}$ 62 $4f^6 6s^2$ $^7F_0$	Eu 5.61 11.2 *24.5* $^9S_4$ $^8S_{7/2}$ $^7F_0$ 63 $4f^7 6s^2$ $^8S_{7/2}$	Gd 5.98 *12* *23* $^{10}D_{5/2}$ $^9D_2$ $^8S_{7/2}$ 64 $4f^7 5d6s^2$ $^9D_2$
7	Th 6.95 11.5 20 $^2D_{3/2}$ $^3F_2$ $^2F_{5/2}$ 90 $6d^2 7s^2$ $^3F_2$	Pa — — — $^3H_4$ $^4K_{11/2}$ $^3H_4$ 91 $5f^2 6d7s^2$ $^4K_{11/2}$	U 5.65 14.6 25 $^4I_{9/2}$ $^5I_4$ $^4I_{9/2}$ 92 $5f^3 6d7s^2$ $^5L_6$	Np — — — $^5I_4$ $^6H_{5/2}$ $^5I_4$ 93 $5f^4 6d7s^2$ $^6L_{11/2}$	Pu — — — $^8F_{1/2}$ $^7F_0$ $^6H_{5/2}$ 94 $5f^6 7s^2$ $^7F_0$	Am — — — $^9S_4$ $^8S_{7/2}$ $^7F_0$ 95 $5f^7 7s^2$ $^8S_{7/2}$	Cm — — — $^{10}D_{5/2}$ $^9D_2$ $^8S_{7/2}$ 96 $5f^7 6d7s^2$ $^9D_2$

Period						
6	Tb *6.74* *12.5* *22* $^7H_8$ $^6H_{15/2}$ $^7F_6$ 65 $4f^9 6s^2$ $^6H_{15/2}$	Dy 5.80 *12.6* *22* $^6I_{17/2}$ $^5I_8$ $^6H_{15/2}$ 66 $4f^{10} 6s^2$ $^5I_8$	Ho 5.85 *12.7* *22* $^5I_8$ $^4I_{15/2}$ $^5I_8$ 67 $4f^{11} 6s^2$ $^4I_{15/2}$	Er 6.11 *12.5* *23.0* $^4H_{13/2}$ $^3H_6$ $^4I_{15/2}$ 68 $4f^{12} 6s^2$ $^3H_6$	Tm 5.85 *12.4* *24* $^3F_4$ $^2F_{7/2}$ $^3H_6$ 69 $4f^{13} 6s^2$ $^2F_{7/2}$	Yb 5.90 12.1 *25.5* $^2S_{1/2}$ $^1S_0$ $^2F_{7/2}$ 70 $4f^{14} 6s^2$ $^1S_0$
7	Bk — — — $^9G_8$ $^8G_{15/2}$ $^7F_6$ 97 $5f^8 6d7s^2$ $^8G_{15/2}$	Cf — — — $^6I_{17/2}$ $^5I_8$ $^6H_{15/2}$ 98 $5f^{10} 7s^2$ $^5I_8$	Es — — — $^5I_8$ $^4I_{15/2}$ $^5I_8$ 99 $5f^{11} 7s^2$ $^4I_{15/2}$	Fm — — — $^4H_{13/2}$ $^3H_6$ $^4I_{15/2}$ 100 $5f^{12} 7s^2$ $^3H_6$	Md — — — $^3F_4$ $^2F_{7/2}$ $^3H_6$ 101 $5f^{13} 7s^2$ $^2F_{7/2}$	No — — — $^2S_{1/2}$ $^1S_0$ $^2F_{7/2}$ 102 $5f^{14} 7s^2$ $^1S_0$

TABLE P3
Crystal-chemical data

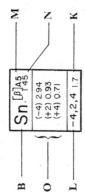

1	H [n]"A3" 0.25 (−1) *1.54* −1,1 2.1								He [n] A3 — (0) 1.28 0
2	Li [n]A2 1.45 (1) 0.68 1 1.0	Be [α]A3 1.05 (2) 0.35 2 1.5		B [α]∗ 0.85 (3) 0.23 3 2.0	C [G]A9 0.70 (−4) 2.60 (+4) 0.16 −4,2,4 2.5	N [β]"A3" 0.65 (−3) 1.71 (+3) 0.16 (+5) 0.13 −3,2,3,4,5 3.05	O [γ]"A1" 0.60 (−2) 1.40 −2 3.5	F [β]∗ 0.50 (−1) 1.36 −1 4.1	Ne A1 — (0) 1.39 0
3	Na [n]A2 1.80 (1) 0.98 1 1.0	Mg A3 1.50 (2) 0.66 2 1.25		Al A1 1.25 (3) 0.51 3 1.45	Si A4 1.10 (−4) 2.71 (+4) 0.42 −4,2,4 1.75	P [w]∗ 1.00 (−3) 2.12 (+3) 0.44 (+5) 0.35 −3,3,5 2.05	S [α]A16 1.00 (−2) 1.84 (+4) 0.37 (+6) 0.30 −2,2,4,6 2.45	Cl [n]A14 1.00 (−1) 1.81 (+5) 0.34 (+7) 0.27 −1,1,5,7 2.9	Ar A1 — (0) 1.71 0
4	K A2 2.20 (1) 1.33 1 0.9	Ca [α]A1 1.80 (2) 0.99 2 1.05	Zn A3 1.35 (2) 0.74 2 1.65	Ga [n]A11 1.30 (3) 0.62 3 1.8	Ge A4 1.25 (−4) 2.72 (+2) 0.73 (+4) 0.53 −4,2,4 2.0	As [n]A7 1.15 (−3) 2.22 (+3) 0.58 (+5) 0.46 −3,3,5 2.2	Se [n]A8 1.15 (−2) 1.91 (+4) 0.50 (+6) 0.42 −2,2,4,6 2.5	Br A14 1.15 (−1) 1.95 (+5) 0.47 (+7) 0.39 −1,1,5,7 2.75	Kr A1 — (0) 1.80 0,2
5	Rb A2 2.35 (1) 1.47 1 0.9	Sr [α]A1 2.00 (2) 1.12 2 1.0	Cd A3 1.55 (2) 0.97 2 1.45	In A6 1.55 (3) 0.81 1,3 1.5	Sn [β]A5 1.45 (−4) 2.94 (+2) 0.93 (+4) 0.71 −4,2,4 1.7	Sb [n]A7 1.45 (−3) 2.45 (+3) 0.76 (+5) 0.62 −3,3,5 1.8	Te A8 1.40 (−2) 2.21 (+4) 0.70 (+6) 0.56 −2,2,4,6 2.0	J A14 1.40 (−1) 2.16 (+5) 0.62 (+7) 0.50 −1,1,3,5,7 2.2	Xe A1 — (0) 2.0 0,2,6,8
6	Cs A2 2.60 (1) 1.67 1 0.85	Ba [n]A2 2.15 (2) 1.34 2 0.95	Hg [α]A10 *1.5* (2) 1.10 2 1.45	Tl [α]A3 *1.9* (1) 1.47 (3) 0.95 1,3 1.45	Pb A1 *1.8* (2) 1.20 (4) 0.84 −4,2,4 1.55	Bi [n]A7 *1.6* (3) 0.96 (5) 0.74 −3,3,5 1.65	Po [α]∗ *1.6* (−2) 2.3 −2,2,4 1.75	At — — (−1) 2.25 — 1.9	Rn — — (0) 2.2 0
7	Fr — — (1) 1.80 1 0.85	Ra A2 2.15 (2) 1.43 2 0.95							

Key (example): B — Sn; M — [β]A5; N — 1.45; O — (−4) 2.94, (+2) 0.93, (+4) 0.71; L — −4,2,4; K — 1.7

B Symbol
K Electronegativity e)
L Oxidation states
M Crystal structure f)
N Atomic radius in Å g)
O Ionic radius or van der Waals radius in Å h)

Re L:
Values in small print show oxidation states occurring comparatively rarely

Re M
[α, β, γ, G, w, n]: Crystal modification f)
A1, A2...A20: Type of structure according to *Strukturbericht*, i.e.:
A1: Cu, cubic face-centred, densest packing; A2: α-W, cubic body-centred; A3: Mg, hexagonal, densest packing; A4: diamond, cubic face-centred; A5: β-Sn, tetragonal, body-centred, A6: In, tetragonal, body-centred; A7: As, rhombohedral; A8: Se, hexagonal; A9: graphite, hexagonal; A10: Hg, rhombohedral; A11: Ga, orthorhombic, base-face-centred; A12: α-Mn, cubic, body-centred; A13: β-Mn, cubic primitive; A14: I, orthorhombic, base-face-centred; A16: α-S, orthorhombic, face-centred; A17: black phosphorus, orthorhombic, base-face-centred; A18: Cl, tetragonal, prim.; A20: α-U, orthorhombic, base-face-centred.
"La": **α-La, α-Pr, α-Nd, Pm, α-Am, α-Cm,** stacking variant of the A3 structure having a unit cell of double height.
"A3": H, β-N, A3 structure for the centres of masses of the H_2 and N_2 molecules, respectively.

Period										
4	Sc [α]A3 1.60 (3)0.81 3 1.2		Ti [α]A3 1.40 (2)0.94 (3)0.76 (4)0.68 2,3,4 1.3	V A2 1.35 (2)0.88 (3)0.74 (4)0.63 (5)0.59 2,3,4,5 1.45	Cr [α]A2 1.40 (2)0.84 (3)0.63 (6)0.52 2,3,5,6 1.55	Mn [α]A12 1.40 (2)0.80 (3)0.66 (4)0.60 (7)0.46 2,3,4,6,7 1.6	Fe [α]A2 1.40 (2)0.74 (3)0.64 2,3,4,6 1.65	Co [α]A3 1.35 (2)0.72 (3)0.63 2,3,4 1.7	Ni A1 1.35 (2)0.69 (3)0.62 2,3,4 1.75	Cu A1 1.35 (1)0.96 (2)0.72 1,2,3 1.75
5	Y A3 1.80 (3)0.89 3 1.1		Zr [α]A3 1.55 (4)0.79 3,4 1.2	Nb A2 1.45 (5)0.69 3,5 1.25	Mo A2 1.45 (4)0.70 (6)0.62 2,3,4,5,6 1.3	Tc A3 1.35 (7)0.57 4,6,7 1.35	Ru A3 1.30 (4)0.67 2,3,4,6,8 1.4	Rh A1 1.35 (3)0.68 2,3,4,6 1.45	Pd A1 1.40 (2)0.80 (4)0.65 2,4 1.35	Ag A1 *1.6* (1)1.26 (2)0.89 1,2,3 1.4
6	La [α]"La" 1.95 (3)1.02 3 1.05	Lu [α]A3 *1.7* (3)0.85 3 1.15	Hf [α]A3 1.55 (4)0.78 4 1.25	Ta A2 1.45 (5)0.68 5 1.35	W [α]A2 1.35 (4)0.72 (6)0.56 4,5,6 1.4	Re A3 1.35 (4)0.72 (7)0.56 4,5,6,7 1.45	Os A3 1.30 (4)0.69 2,3,4,6,8 1.5	Ir A1 1.35 (4)0.68 3,4,6 1.55	Pt A1 1.35 (2)0.80 (4)0.65 2,4,6 1.45	Au A1 *1.4* (1)1.37 (3)0.85 1,3 1.4
7	Ac A1 1.95 (3)1.18 3 1.0	(Lr) — — — —	(Rf) — — — —	(Ha) — — — —						

Period														
6	Ce [γ]A1 1.85 (3)1.03 (4)0.92 3,4 1.05	Pr [α]"La" 1.85 (3)1.01 (4)0.90 3,4 1.05	Nd [α]"La" 1.85 (3)1.00 2,3,4 1.05	Pm "La" 1.85 (3)0.98 3 1.05	Sm [α]* 1.85 (3)0.97 2,3 1.05	Eu A2 1.85 (2)1.09 (3)0.95 2,3 1.0	Gd [α]A3 1.80 (3)0.94 3 1.1	Tb [α]A3 1.75 (3)0.92 (4)0.84 3,4 1.1	Dy [α]A3 1.75 (3)0.91 2,3,4 1.1	Ho [α]A3 1.75 (3)0.89 3 1.1	Er [α]A3 1.75 (3)0.88 3 1.1	Tm [α]A3 1.75 (3)0.87 2,3 1.1	Yb [α]A1 1.75 (2)0.93 (3)0.86 2,3 1.05
7	Th [α]A1 1.80 (4)1.02 4 1.1	Pa * 1.80 (4)0.98 (5)0.89 4,5 1.15	U [α]A20 1.75 (4)0.97 (6)0.80 3,4,5,6 1.2	Np [α]* 1.75 (3)1.10 (4)0.95 (7)0.71 3,4,5,6,7 1.2	Pu [α]* 1.75 (3)1.08 (4)0.93 3,4,5,6,7 1.2	Am [α]"La" 1.75 (3)1.07 (4)0.92 3,4,5,6 1.2	Cm [α]"La" — — 3,4 *1.2*	Bk [n]A1 — — 3,4 —	Cf — — — 3 —	Es — — — 3 —	Fm — — — 3 —	Md — — — 2,3 —	No — — — — —

the centres of the double molecules $2.O_2$ form an A1 type.

*: α-B, rhombohedral; β-F, cubic primitive; **white phosphorus**, cubic; **α-Sm**, rhombohedral; **α-Po**, simple cubic; **Pa**, tetragonal, body-centred; **α-Np**, orthorhombic, primitive; **α-Pu**, monoclinic primitive.

Re O:
(−4, ...−1, 1, ...7): Oxidation state pertaining to the given ionic radius
(0): Interpolated VAN DER WAALS radius of the noble gases h).

Data printed in italics are either uncertain or have been obtained by extrapolation.

Remarks e), f), g) and h) follow the tables (pp. 491, 492).

Period	A	B	T	Q	S	P	R
1	1	H	52	20.38	0.109	14.0	0.014
1	2	He	0	4.22	0.020	*	
2	3	Li	38	1600	32	454	0.72
2	4	Be	78	3140	70	1557	3.52
2	5	B	130	4050	*120*	2500	*5.7*
2	6	C	[G] 170	4000 (S)	170	*4100* *	*25*
2	7	N	113	77.35	0.67	63.2	0.086
2	8	O	[O_2] 60	90.17	0.81	54.4	0.053
2	9	F	19	85	0.76	53.5	0.061
2	10	Ne	0	27.09	0.42	24.6	0.080
3	11	Na	26	1154	23	370.8	0.62
3	12	Mg	36	1385	32	923	2.14
3	13	Al	77	2333	68	933.2	2.56
3	14	Si	105	2753	*100*	1685	12.0
3	15	P	[w] 75	553	3.0	317.2	0.15
3	16	S	66	717.75	*2.5*	392	0.335
3	17	Cl	29	239.1	2.44	172.2	0.77
3	18	Ar	0	87.29	1.57	84.0	0.28
4	19	K	21	1027	19	336.6	0.56
4	20	Ca	42	1765	36	1112	2.07
4	30	Zn	31	1175	27	692.655	1.77
4	31	Ga	65	2510	*60*	302.8	1.34
4	32	Ge	90	3100	*70*	1209	7.6
4	33	As	*72*	886 (S)	7.75	1090 *	*6.6*
4	34	Se	49	958	*6*	490	1.30
4	35	Br	27	331.4	3.58	266	1.26
4	36	Kr	0	119.79	2.16	116.2	0.39
5	37	Rb	20	959	18	311.8	0.56
5	38	Sr	39	1645	34	1045	*2.2*
5	48	Cd	27	1038	24	594.18	1.48
5	49	In	57	2280	54	429.76	0.78
5	50	Sn	[β] 72	2765	70	505.06	1.71
5	51	Sb	63	1905	*46*	903.6	4.74
5	52	Te	46	1163	12	722.8	4.18
5	53	J	26	456	5.2	386.8	1.87
5	54	Xe	0	165.02	3.0	161.3	0.55
6	55	Cs	19	939	16	301.8	0.51
6	56	Ba	42	1910	36	998	1.83
6	80	Hg	15	629.73	14	234.28	0.549
6	81	Tl	43	1939	39	576	1.02
6	82	Pb	47	2020	42	600.576	1.14
6	83	Bi	50	1825	40	544.525	2.60
6	84	Po	35	1235	*25*	519	*0.9*
6	85	At	*22*	607	8	575	—
6	86	Rn	0	211.2	4.4	202.2	0.69
7	87	Fr	*18*	*1020*	—	*297*	*0.5*
7	88	Ra	*42*	*1900*	27	973	*1.7*

TABLE P4
Thermochemical data

Key (example cell):

A, B	T
80 Hg	15
Q: 629.73	S: 14
P: 234.28	R: 0.549

A Atomic number
B Symbol
P Melting temperature T_m in °K [i])
Q Boiling temperature T_v in °K for $p = 1$ atm [i])
R Enthalpy of melting $\Delta H_m(T_m)$ in $\frac{\text{kcal}}{\text{g-atom}}$ at T_m, for $p = 1$ atm
S Enthalpy of evaporation $\Delta H_v(T_v)$ in $\frac{\text{kcal}}{\text{g-atom}}$ at T_v for $p = 1$ atm [j])
T Enthalpy of atomization ΔH_A (300°K) in $\frac{\text{kcal}}{\text{g-atom}}$, at room temperature and $p = 1$ atm [j]), [k])

Re P and Q:
0°C = 273.15°K
*: He crystallizes only for $p \geqq 25$ atm even when $T \to 0$°K.
*, (s): As sublimes at normal pressure; T_m denotes the melting temperature at $p = 36$ atm., T_v the sublimation temperature at $p = 1$ atm.
For C (graphite) T_m is extrapolated, T_v is the sublimation temperature at $p = 1$ atm.

Re T:
[β, γ, G, w]: Crystal modification (cf. Table P3).
[O_2]: In the case of oxygen data refer to O_2 rather than to O_3.

Data printed in italics are either uncertain or [illegible]

4	21 Sc 80 3540 75 1812 3.70		22 Ti 112 3586 105 1940 3.4	23 V 123 3580 110 2180 3.8	24 Cr 95 2920 75 2150 3.47	25 Mn 67 2368 53 1517 3.50	26 Fe 100 3160 84 1808 3.67	27 Co 102 3229 92 1765 3.70	28 Ni 102 3055 90 1726 4.21	29 Cu 81 2810 73 1356 3.12
5	39 Y 98 3670 93 1775 2.73		40 Zr 146 4650 130 2123 3.7	41 Nb 175 4813 165 2740 4.8	42 Mo 157 5790 130 2888 6.66	43 Tc 152 5300 130 2440 5.4	44 Ru 154 4325 140 2550 5.7	45 Rh 133 3960 120 2233 5.0	46 Pd 90 3200 90 1825 4.10	47 Ag 68 2470 61 1234.0 2.78
6	57 La 102 3710 96 1193 1.48	71 Lu 99 4140 90 1925 2.9	72 Hf 150 4570 150 2495 4.4	73 Ta 187 5760 180 3270 5.8	74 W 200 6000 185 3653 8.4	75 Re 186 6030 160 3430 7.9	76 Os 187 5300 155 3300 7.6	77 Ir 159 4820 140 2716 6.2	78 Pt 135 4100 122 2042 4.7	79 Au 88 3240 80 1336.2 2.96
7	89 Ac 104 3200 – 1320 3.0	103 (Lr) – – – – –	104 (Rf) – – – – –	105 (Ha) – – – – –						

6	58 Ce [γ] 98 3972 95 1070 1.24	59 Pr 86 3616 80 1208 1.65	60 Nd 76 2956 70 1297 1.71	61 Pm 64 2730 – 1441 1.9	62 Sm 50 2140 46 1345 2.06	63 Eu 43 1971 42 1100 2.20	64 Gd 82 3540 72 1585 2.44	65 Tb 90 3810 70 1630 2.46	66 Dy 67 3011 65 1680 2.5	67 Ho 70 3228 65 1735 3.38	68 Er 70 3000 65 1770 2.6	69 Tm 58 2266 58 1818 4.22	70 Yb 40 1970 38 1097 1.83
7	90 Th 137 4500 130 2024 3.6	91 Pa 132 4680 130 1698 3.0	92 U 125 3950 110 1404 2.5	93 Np 113 4150 – 910 1.6	94 Pu 92 3727 – 913 0.68	95 Am 65 – 52 – –	96 Cm – – – – –	97 Bk – – – – –	98 Cf – – – – –	99 Es – – – – –	100 Fm – – – – –	101 Md – – – – –	102 No – – – – –

Remarks [i]), [j]) and [k]) follow the tables (p. 492).

Period									
1	1 H –								2 He –
2	3 Li 2.4	4 Be 3.9		5 B 4.5	6 C 4.7	7 N –	8 O –	9 F –	10 Ne –
							[α] (N) 24		
		0.026							
3	11 Na 2.4	12 Mg 3.7		13 Al 4.3	14 Si 4.8	15 P –	16 S –	17 Cl –	18 Ar –
				1.19					
4	19 K 2.2	20 Ca 2.8	30 Zn *4.2*	31 Ga 4.0	32 Ge 4.8	33 As 5.1	34 Se 4.7	35 Br –	36 Kr –
			0.88	1.09					
5	37 Rb 2.2	38 Sr 2.4	48 Cd 4.1	49 In 3.8	50 Sn 4.4	51 Sb *4.1*	52 Te 4.7	53 J –	54 Xe –
			0.56	3.40	[β] 3.72				
6	55 Cs 1.8	56 Ba 2.5	80 Hg 4.5	81 Tl 3.7	82 Pb 4.0	83 Bi 4.4	84 Po *4.8*	85 At –	86 Rn –
			[α] 4.15	2.39	7.19				
7	87 Fr *1.8*	88 Ra *3.2*							

TABLE P5

Transition temperatures, work functions

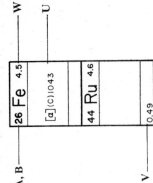

A, B —	26 Fe 4.5	— W
	[α] (C) 1043	— U
	44 Ru 4.6	
V —	0.49	

A Atomic number

B Symbol

U Magnetic transition temperature in °K, at $p = 1$ atm. [l])

V Superconducting transition temperature in °K, at $p=1$ atm. [m])

W Work function in eV [n])

Re U:

[α, β, γ]: Crystal modification. α-**O** is monoclinic, base-face-centred. γ-**Co**: cubic face-centred. β-**Ce**: hexagonal (α-La type).

(C): ferromagnetic Curie temperature.[l])

(FA): Temperature at which the transition of a ferromagnetic or ferrimagnetic phase to an antiferromagnetic phase occurs at rising temperature.[l])

(AA): Temperature at which transition from one antiferromagnetic phase to another one takes place.[l])

*: The cubic face-centred phases of **Pr** and **Nd** are ferromagnetic.[l])

Data printed in italics are either uncertain or else were obtained by extrapolation.

Remarks [l]), [m]) and [n]) follow the tables (pp. 492–493).

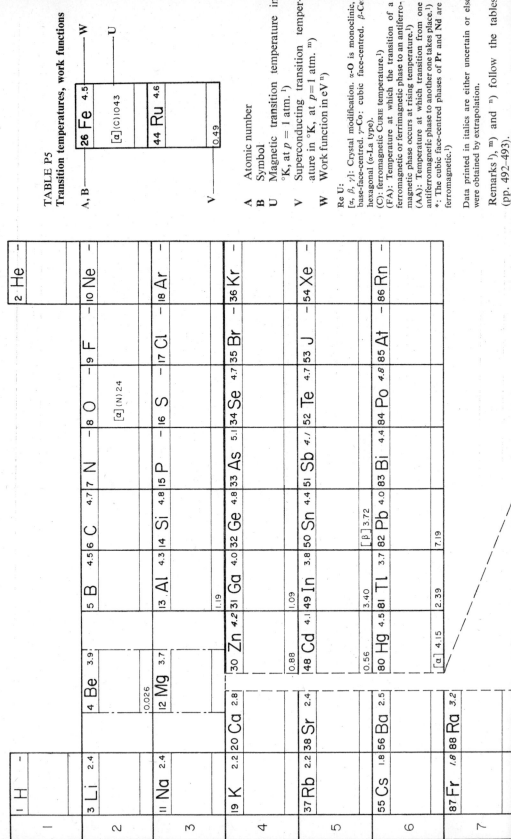

4		21 Sc 3.5	22 Ti 4.3 0.39	23 V 4.3 5.30	24 Cr 4.5 (N) 311 (AA) 120	25 Mn 4.1 [α](N)95	26 Fe 4.5 [α](C)1043	27 Co 5.0 [γ](C)1394	28 Ni 5.1 (C)631	29 Cu 4.6
5		39 Y 3.1	40 Zr 4.0 0.59	41 Nb 4.3 9.5	42 Mo 4.6 0.92	43 Tc – 8.82	44 Ru 4.6 0.49	45 Rh 4.8	46 Pd 5.5	47 Ag 4.0
6	57 La 3.5 [α] 4.90	71 Lu 3.1	72 Hf 3.9 0.09	73 Ta 4.1 4.48	74 W 4.5 0.01	75 Re 5.0 1.70	76 Os 4.7 0.66	77 Ir 5.3 0.105	78 Pt 5.6	79 Au 5.1
7	89 Ac 2.7	103(Lr) –	104(Rf) –	105(Ha) –						

6	58 Ce 2.9 [β](N) 14	59 Pr 2.8 * [α](N) 23	60 Nd 3.2 * [α](N) 19.5 (AA) 7.0	61 Pm 3.1 –	62 Sm 2.7 (N) 15	63 Eu 2.5 (N) 90	64 Gd 3.1 (C) 293	65 Tb 3.1 (N) 228 (FA) 220	66 Dy 3.1 (N) 178.5 (FA) 85	67 Ho 3.1 (N) 132 (FA) 20	68 Er 3.1 (N) 85 (AA) 53 (FA) 20	69 Tm 3.1 (N) 55 (AA) 40 (FA) 22	70 Yb 2.6
7	90 Th 3.3 1.37	91 Pa –	92 U 3.3	93 Np –	94 Pu –	95 Am –	96 Cm –	97 Bk –	98 Cf –	99 Es –	100 Fm –	101 Md –	102 No –

Remarks relating to Tables P1–P5

The main literature sources are given in brackets (cf. p. 479 and pp. 493–495). These are mostly critical collections of data taken from numerous original papers. In many cases data from different sources had to be taken for a given physical property. In some of these cases the choice had to be made arbitrarily. It is, therefore, advisable to refer to more recent original papers in critical cases.

Many numerical values given here have been rounded off. Thus, for the melting temperature of Rhenium the literature gives 3433±20°K; Table P4 shows T_m(Re) = 3430°K instead. Uncertain data or values obtained by extrapolation are printed in italics, e.g. T_m(Fr) = *297*°K.

Periodic tables may be laid out in various ways. The method used here, in which the transition elements (*d*- and *f*-elements) are shown separately, was devised in essence by SANDERSON (1967). The two heavy horizontal lines between Al, Si, ... and Zn, Ga, ..., and between Zr, Nb, ... and Lu, Hf, ... are the boundary lines below which the 3*d* and the 4*f* shells of the elements are completely filled.

The selected physical quantities are given mainly with the special interests of solid-state physicists in mind.

The following remarks supplement the explanations given in the individual tables:

[a]) **B, C, D** (Table P1)

Thinly printed symbols relate to elements which can only be prepared artificially. The names and symbols of the elements 103, 104 and 105 have not been designated officially as yet. The symbol Lw is frequently employed instead of Lr and for element 104 the names Kurtschatovium (Ku) and Dubnium (Du) have been proposed in addition to Rutherfordium (Rf). The symbols Lr, Rf and Ha employed here were proposed by GHIORSO *et al.* (1970).

The atomic weights are the international values of 1966, referred to $^{12}C \triangleq 12.000\,00$. Values in square brackets show the most stable isotope of the element in question known up to now.

[b]) **E, F** (Table P1)

The density and the atomic volume depend on the temperature and the state of matter. In the case of elements in the crystalline state, the X-ray density ϱ_X is generally given (cf. p. 16). The difference between the X-ray density and the density determined macroscopically is in most cases only slight and rarely exceeds a value of 1% in the case of the elements.

For solid and liquid elements the data refer to room temperature. In the case of elements having several crystal modifications, the table considers only one form in each case: e.g. for C, ϱ (graphite) = 2.26 instead of ϱ (diamond) = 3.52; for Sn, ϱ (β, white) = 7.29, instead of ϱ (α, grey) = 5.77; for P, ϱ (white) = 1.80 instead of ϱ (red) = 2.35 and ϱ (black) = = 2.62; for S, ϱ (α, orthorhombic) = 2.09 instead of ϱ (monoclinic) = 1.94 and ϱ (rhombohedral) = 2.21 (ϱ given in g/cm^3).

In the case of elements gaseous under standard conditions the data refer to the liquid state at the boiling point (cf. Table P4): e.g. for Ne, ϱ (liquefied; 27.09°K) = 1.21 as compared with ϱ (solid; 4°K) = 1.51 and ϱ (solid; triple point 24.56°K) = 1.44 (cf. POLLACK 1964 and *CIG Data Book* 1970).

[c]) **G** (Table P2)

The ground state electron configurations of the atoms are mostly given in an abbreviated form, i.e. the inner completed shells are not shown. For Tc, Ce, Tb and Bk varying data are given in the literature. In this publication the data of MOORE (1958) for Tc and those of WYBOURNE (1965) for Ce, Tb and Bk are employed.

[d]) **H, I** (Table P2)

Data relating to the ground states of the atoms and ions are clearly systematic, especially in the case of the main groups. This is due to the fact that atoms and ions containing the same number of electrons (isoelectronic sequence) or the same number of outermost electrons frequently possess the same ground state (Examples: He, Li^{+}, Be^{2+}, B^{3+}, C^{4+} or Li, Na, K, Rb, Cs). Deviations from this systematic behaviour occur with certain transition elements in which electrons in different outer shells have comparable energies. For the transuranic elements many data represent theoretical predictions still requiring experimental confirmation.

[e]) **K** (Table P3)

The term "electronegativity" was introduced by PAULING and it may be defined in various ways [cf. e.g. SANDERSON (1967) and PHILLIPS (1969)]. The values used here are of the ALLRED-ROCHOW type [according to RICH (1965)]. They deviate considerably from the Pauling values for some of the 4*d* and 5*d* elements.

In the cases of compounds formed from two elements, the more electronegative element always forms the anion and the more electropositive element the cation. The magnitude of the difference in the electronegativities of the two elements is a measure of the ionic contribution to the chemical bond.

[f]) **M** (Table P3)

Many elements possess several modifications (polymorphism or polytypism) (cf. also p. 11). Examples are B, tetragonal and rhombohedral; C, diamond (A4) and graphite (A9); Sn, α-Sn (grey tin, A4) and β-Sn (white tin, A5); P, white (cubic), red (cubic) and black (A17); S, orthorhombic (α-S, A16), monoclinic and rhombohedral. Detailed information may be found *inter alia* in PEARSON (1967, Vol. 2, p. 79) and SAMSONOV (1968, p. 110), where various high-pressure modifications are also listed.

The table lists only one modification in each case. In the case of elements which are solid under standard conditions, this is the room temperature modification (e.g. α-Fe; with increasing temperature α-Fe is transformed to the β-, γ- and δ-phases; melting data obtained for Fe (cf. Table P4) relate, therefore, to δ-Fe). The conditions under which the crystalline modifications of the other elements exist cannot be stated explicitly here on account of lack of space [cf. e.g. ADDISON (1964) and SAMSONOV (1968)].

Where a certain modification has no specific designation it is denoted by [n] in the table (e.g. Na: [n] A2).

[g]) **N** (Table P3)

Atoms present in crystals have different types of radii, though their definitions are not always identical. Three of these radii, namely the covalent radius, the atomic radius and the metallic radius, have comparable values. For example, for Li their values are 1.34 Å, 1.45 Å and 1.55 Å, respectively.

This table lists the values of the atomic radii according to SLATER. They show satisfactory agreement with the theoretical values [cf. SLATER (1965, Table 4-3), where also the absence of corresponding data for the noble gases is explained (§ 4-2)]. Moreover, these values hold as a first approximation not only for covalent and metallic bonds but also for ionic bonds. Thus making use of the atomic radii of Na and Cl (r_{Na} = 1.80 Å and r_{Cl} = 1.00 Å) we obtain the following approximate values for the distances between next neighbours in the ionic crystal NaCl, the metal Na, and the molecular crystal Cl_2: d_{Na-Cl} = 2.80 Å, d_{Na-Na} = 3.60 Å, d_{Cl-Cl} = 2.00 Å. These values show satisfactory agreement with the corresponding experimental data of 2.82, 3.72 and 2.02 Å respectively.

[h]) **O** (Table P3)

The literature gives different types of ionic radii [cf. e.g. SAMSONOV (1968, p. 97)]. Data given here are mainly taken from PAULING.

The values for the van der Waals radii of the noble gases were calculated by interpolation of the ionic radii in isoelectronic sequences, e.g. using the series O^{2-}, F^-, Ne^0, Na^+, Mg^{2+} [cf. PAULING (1960, p. 514) and RICH (1965, p. 34)].

[i]) **P, Q** (Table P4)

The boiling, melting and triple points of several elements are used as fixed points of the international practical temperature scale, e.g. the boiling points of He, H_2, Ne, N_2, Ar, O_2, Kr, Xe, Hg and S [cf. LANDOLT-BÖRNSTEIN (1967, Vol. IV/4a, p. 172)] and the melting points of Hg, In, Sn, Cd, Pb, Zn, Al, Ag, Au, Cu, Ni, Co, Pd, Pt, Rh, Ir and W [cf. MCLAREN (1962, p. 185), STIMSON (1962, p. 59), GSCHNEIDNER (1964, p. 326)]. The underlined elements give primary fixed points.

[j]) **R, S, T** (Table P4)

The enthalpy values are given in units of kcal/g-atom rather than kcal/mole (e.g. for H_2 the value given for the enthalpy of evaporation refers only to 1/2 mole and not to 1 mole).

[k]) **T** (Table P4)

The enthalpy of atomization is the enthalpy required to convert a substance to the monoatomic gaseous state at a given temperature and pressure [cf. e.g. SANDERSON (1967, p. 63)]. Thus, in the case of polyatomic gaseous substances it represents the enthalpy of dissociation whilst in the case of solid substances the enthalpy of sublimation, provided a monoatomic gas is formed after sublimation (this does not apply, for instance, to As, since As_4 vapour is formed during sublimation).

The values of ΔH_A (300°K) should, strictly speaking, be related to the values of $\Delta H_m(T_m)$ and $\Delta H_v(T_v)$. This is, however, not the case in general in this table, since the values had to be taken from different sources.

[l]) **U** (Table P5)

In the absence of a magnetic field, magnetically ordered materials exhibit a finite magnetization within each domain (as in the case of ferro- or ferrimagnetism) or zero magnetization when the magnetic moments cancel one another [as in the case of antiferromagnetism or in special cases of helimagnetism (helical array of magnetic moments)]. The temperature at which the transition from the magnetically ordered to the paramagnetic phase takes place is frequently referred to as the ferromagnetic Curie temperature in the former case, but as the Néel temperature in the latter.

Furthermore, a substance may have two or several magnetically ordered phases, depending on the temperature and the magnetic field. The corresponding transition temperatures were appropriately denoted by the symbols T_{FA}, T_{AF}, T_{FF} and T_{AA}, the index F referring to ferromagnetism or ferrimagnetism and A to antiferromagnetism [cf. e.g. VOGT (1969, p. 296)].

Different crystallographic phases as a rule display different magnetic properties. For instance, the hexagonal phases of Pr and Nd show antiferromagnetism whilst their cubic face-centred phases which are metastable at low temperatures are ferromagnetic and possess Curie temperatures of 8.7 and 29°K respectively [according to BUCHER *et al.* (1969)].

[m]) **V** (Table P5)

The superconducting transition temperatures apply to substances as pure as possible and under normal pressure. Many elements become superconducting under high pressure, e.g. Cs, Ba, Y, Ce, U, Si, Ge, P, Sb, Bi, Se and Te [cf. BRANDT and GINZBURG (1969), BOUGHTON *et al.* (1970) and WITTIG (1970)].

[n]) **W** (Table P5)

The work function can be determined by various methods, e.g. using measurements of the thermionic, field or photo emission. Data obtained from different types of measurements carried out on a given substance are, as a rule, comparable. It must, however, be mentioned that the purity and the state of the surface of the substance exert a considerable influence on the results of the measurements (cf. p. 181). This table gives the results obtained by EASTMAN (1970) from the photo emission of films of high purity for Sc, Ti, V, Cr, Mn, Fe, Co, Ni, Cu, Y, Zr, Nb, Mo, Pd, Ag, La, Ce, Nd, Sm, Eu, Gd, Hf, Pt and Au. The values given for the remaining elements are those compiled by FOMENKO and SAMSONOV (1966) on the basis of various experiments. A comparison of these data given in the two publications shows that the values of EASTMAN are, in general, somewhat higher than those of FOMENKO and SAMSONOV.

The References listed below contain not only the main sources but also additional ones which were used in compiling the tables.

References

ADDISON, W. E., *The Allotropy of the Elements* (Oldbourne Press, London 1964) [on M].

BOUGHTON, R. I., OLSEN, J. L., and PALMY, C., *Pressure Effect in Superconductors*, in *Progress in Low Temperature Physics*, C. J. GORTER (Editor), Vol. 6, p. 163 (North-Holland, Amsterdam 1970).

BRANDT, N. B., and GINZBURG, N. I., *Superconductivity at High Pressures, Contemp. Phys.* **10,** 355 (1969).

BUCHER, E., CHU, C. W., MAITA, J. P., ANDRES, K., COOPER, A. S., BUEHLER, E., and NASSAU, K., *Electronic Properties of Two New Elemental Ferromagnets: fcc Pr and Nd, Phys. Rev. Letters* **22,** 1260 (1969).

CIG Data Book and Buyers Guide, Cryogenics and Industrial Gases **5,** No. 3 (Business Communication Inc., Cleveland, Ohio 1970) [pp. 51 ff. on **E, P, Q,** and **S**].

CONNOLLY, T. F., and COPENHAVER, E. D., *Bibliography of Magnetic Materials and Tabulation of Magnetic Transition Temperatures* (Oak Ridge National Laboratories, Oak Ridge, Tennessee, 1970) [on **U**].

COOPER, B. R., *Magnetic Properties of Rare Earth Metals*, in *Solid State Physics*, Vol. 21 (Academic Press, New York 1968) [p. 393 on **U**].

CRC Handbook of Chemistry and Physics, R. C. WEAST (Editor), 50th ed. (Chemical Rubber Company, Cleveland, 1970) [pp. B-8 ff. on **C,** p. B-256 on **D,** p. B-253 on **F,** p. F-152 on **O,** and p. D-56 on **S**].

D'ANS-LAX, *Taschenbuch für Chemiker und Physiker*, 3rd ed., Vol. 3 (Springer, Berlin 1970) [p. 89, p. 346 on **G, H,** and **J**].

EASTMAN, D. E., *Photoelectric Work Functions of Transition, Rare Earth and Noble Metals, Phys. Rev.* **B2,** 1 (1970) [on **W**].

EBERT, H., *Physikalisches Taschenbuch*, 4th ed. (Vieweg, Braunschweig 1967).

FALGE, Jr., R. L., *Superconductivity of Hexagonal Beryllium, Physics Letters* **24A,** 579 (1967).

FOMENKO, V. S., and SAMSONOV, G. W., *Handbook of Thermionic Properties* (Plenum, New York 1966) [on **W**].

GAYDON, A. G., *Dissociation Energies and Spectra of Diatomic Molecules*, 3rd ed. (Chapman & Hall, London 1968) [pp. 260 ff. on **T**].

GHIORSO, A., NURMIA, M., ESKOLA, K., HARRIS, J., and ESKOLA, P., *New Element Hahnium, Atomic Number 105, Phys. Rev. Letters* **24,** 1498 (1970).

GLADSTONE, G., JENSEN, M. A., and SCHRIEFFER, J. R., *Superconductivity in the Transition*

Metals: Theory and Experiment, in *Superconductivity*, R. D. PARKS (Editor), Vol. 2 (Dekker, New York 1969) [p. 772 on **V**].

GSCHNEIDNER, Jr., K. A., *Physical Properties and Interrelationships of Metallic and Semimetallic Elements*, in *Solid State Physics*, Vol. 16 (Academic Press, New York 1964) [p. 275 on **F, P, Q, R,** and **T**].

JØRGENSEN, C. K., *Oxidation Number and Oxidation States* (Springer, Berlin 1969) [pp. 259 ff. on **L**].

KEFFER, F., *Spin Waves*, in *Handbuch der Physik*, S. FLÜGGE (Editor), Vol. 18/2, *Ferromagnetismus* (Springer, Berlin 1966) [pp. 4, 151, 153 on **U**].

KOEHLER, W. C., *Magnetic Properties of Rare Earth Metals and Alloys*, *J. Appl. Phys.* **36,** 1078 (1965) [on **U**].

LANDOLT-BÖRNSTEIN, *Zahlenwerte und Funktionen*, Vol. I/4 (Springer, Berlin 1955) [p. 81 on **M**]; Vol. IV/4a (1967) [pp. 172 ff. on **E, P, Q, S,** and **T**].

LEVERENZ, H. W., *Periodic Chart of the Elements* (RCA Laboratories, Princeton, New Jersey 1953).

MCLAREN, E. H., *The Freezing Points of High Purity Metals as Precision Temperature Standards*, in *Temperature, Its Measurement and Control in Science and Industry*, C. M. HERZFELD (Editor), Vol. 3/1 (Reinhold, New York 1962) [p. 185 on **P**].

MENDELSOHN, K., *Geschichte der Supraleitung*, in *Vorträge über Supraleitung*, Physik. Gesellschaft Zürich (Editor) (Birkhäuser, Basel 1968) [p. 18 on **V**].

MOORE, C. E., *Atomic Energy Levels*, 3 vols., Circular of the National Bureau of Standards Nr. 467 (Washington, D.C. 1949, 1952, 1958) [Vol. 1, p. XI on **G, H,** and **I**; Vol. 3, p. XXXIV on **J**].

MÜLLER, J., *Supraleitende Materialien*, in *Vorträge über Supraleitung*, Physik. Gesellschaft Zürich (Editor) (Birkhäuser, Basel 1968) [p. 96 on **V**].

NESMEYANOV, A. N., *Vapour Pressure of the Elements* (Infosearch, London 1963) [p. 448 on **P, Q, R,** and **T**].

PALLMER, P. G., and CHIKALLA, T. D., *The Crystal Structure of Promethium*, *J. Less-Common Metals* **24,** 233 (1971).

PAULING, L., *The Nature of the Chemical Bond* (Cornell University Press, Ithaca 1960) [pp. 88 ff. on **K,** and pp. 505 ff. on **O**].

PEARSON, W. B., *A Handbook of Lattice Spacings and Structures of Metals and Alloys*, Vol. 2 (Pergamon, Oxford 1967) [p. 79 on **E** and **M**].

Periodic Table of the Elements and *Table of Periodic Properties of the Elements* (Sargent, Chicago 1964).

PHILLIPS, J. C., *Covalent Bonding in Crystals, Molecules and Polymers* (University of Chicago Press, Chicago 1969) [on **K**].

POLLACK, G. L., *The Solid State of Rare Gases*, *Rev. Mod. Phys.* **36,** 748 (1964) [p. 760 on **Q**].

RICH, R., *Periodic Correlations* (Benjamin, New York 1965) [p. 50 on **K**; p. 36 on **O**; p. 80 on **T**].

SAMSONOV, G. V. (Editor), *Handbook of the Physicochemical Properties of the Elements* (Plenum, New York 1968) [pp. 16, 24 on **G, H, I,** and **J**; pp. 110, 124 on **M**; p. 97 on **O**; pp. 248 ff. on **P, Q, R,** and **T**].

SANDERSON, R. T., *Inorganic Chemistry* (Reinhold, New York 1967) [p. 78 on **K**; pp. 64 ff. on **S** and **T**].

SCHIRBER, J. E., and SWENSON, C. A., *Superconductivity of α- and β-Mercury*, *Phys. Rev.* **123,** 1115 (1961).

SLATER, J. C., *Quantum Theory of Molecules and Solids*, Vol. 2 (McGraw-Hill, New York 1965) [p. 55 on **N**].

STIMSON, H. F., *The Text Revision of the International Temperature Scale of 1948*, in *Temperature, Its Measurement and Control in Science and Industry*, C. M. HERZFELD (Editor), Vol. 3/1 (Reinhold, New York 1962) [p. 59 on **P**].

TAYLOR, A., and KAGLE, B. J., *Crystallographic Data on Metal and Alloy Structures* (Dover, New York 1963) [pp. 245 ff. on **E** and **M**].
TEBBLE, R. S., and CRAIK, D. J., *Magnetic Materials* (Wiley–Interscience, New York 1969) [pp. 60, 182 on **U**].
TOULOUKIAN, Y. S. (Editor), *Thermophysical Properties of High Temperature Solid Materials*, 6 vols. (Collier–Macmillan, New York 1967) [Vol. 1 on **P, R, S,** and **T**].
VOGT, E., *Magnetic Moments and Transition Temperatures*, in *Magnetism and Metallurgy*, A. E. BERKOWITZ and E. KNELLER (Editors), Vol. 1 (Academic Press, New York 1969) [pp. 249 ff., 262, 296 on **U**].
WILKS, J., *The Properties of Liquid and Solid Helium* (Oxford University Press, London 1967).
WITTIG, J., *Pressure-Induced Superconductivity in Cesium and Yttrium, Phys. Rev. Letters*, **24,** 812 (1970).
WYBOURNE, B. G., *Spectroscopic Properties of Rare Earths* (Wiley–Interscience, New York 1965) [pp. 3 ff. on **G, H,** and **I**].
WYCKOFF, R. W. G., *Crystal Structures*, Vol. 1 (Wiley–Interscience, New York 1965) [p: 7–83 on **M**].
YAMADA, T., and TAZAWA, S., *The Magnetic Symmetry of α-Mn and the Parasitic Ferromagnetism from Magnetic Torque Measurements, J. Phys. Soc. Japan* **28,** 609 (1970).

Problems

B. Interference Effects in Crystals

B-1. CUBIC BRAVAIS LATTICES

Determine the following characteristic quantities for the three cubic Bravais lattices with lattice constant a: number of lattice points per cubic unit cell, distance between nearest neighbours, number of nearest neighbours (coordination number), distance and number of the second-nearest neighbours, volume ratio of the cubic and primitive unit cells.

B-2. UNIT CELL AND X-RAY DENSITY

(a) The unit cell of NaCl of volume a^3 contains 4 Na and 4 Cl atoms. Calculate the lattice constant a from the macroscopic density of NaCl at 291°K, $\varrho = 2.163$ g/cm³.

(b) The unit cell of Si of volume a^3 contains 8 atoms. Calculate the X-ray density ϱ_X from the lattice constant which at 298°K has been determined to be $a = 5.43$ Å.

B-3. PEROVSKITE STRUCTURE

The unit cell of the perovskite structure ABX_3 (perovskite $CaTiO_3$) is cubic. The corners of the cube are occupied by the large cations A, the centre of the cube is occupied by the smaller cation B, and the centres of the cube faces {100} are occupied by the anions X. The basis coordinates are: 000 for A, $\frac{1}{2}\frac{1}{2}\frac{1}{2}$ for B, and $\frac{1}{2}\frac{1}{2}0$, $0\frac{1}{2}\frac{1}{2}$, $\frac{1}{2}0\frac{1}{2}$ for X. A and X form a cubic close-packed structure. This structure can exist under the following condition for the ionic radii (the ions are assumed to be rigid spheres):

$$r_A + r_X = t\times\sqrt{2}\,(r_B + r_X)$$

with $0.88 < t < 1.0$, where t is a tolerance factor (for values of t outside of the given limits the structure will become distorted or hexagonal). Derive the given relation for the special case $r_A = r_X$. Calculate the filling factor, i.e., the volume of the spheres per unit cell divided by the volume of the unit cell.

B-4. DIRECT AND RECIPROCAL LATTICE

(a) Show that the simple primitive unit cell of the body-centred cubic (bcc) and of the face-centred cubic (fcc) lattice is defined by the following translation vectors, respectively:

bcc lattice	fcc lattice
$\boldsymbol{a}_{1b} = d(\boldsymbol{i}+\boldsymbol{j}-\boldsymbol{k})/2$	$\boldsymbol{a}_{1f} = d(\boldsymbol{i}+\boldsymbol{j})/2$
$\boldsymbol{a}_{2b} = d(-\boldsymbol{i}+\boldsymbol{j}+\boldsymbol{k})/2$	$\boldsymbol{a}_{2f} = d(\boldsymbol{j}+\boldsymbol{k})/2$
$\boldsymbol{a}_{3b} = d(\boldsymbol{i}-\boldsymbol{j}+\boldsymbol{k})/2$	$\boldsymbol{a}_{3f} = d(\boldsymbol{k}+\boldsymbol{i})/2$

$\boldsymbol{i}, \boldsymbol{j}, \boldsymbol{k}$ are the unit vectors in the x, y, z direction, respectively, and d is the length of the cubic unit cell.

(b) Show that the reciprocal lattice of a body-centred cubic lattice is face-centred cubic, and vice versa.

B-5. WAVELENGTH AND ENERGY

(a) X-ray diffraction: Estimate the necessary voltage between the cathode and the Cu-anticathode of an X-ray tube to generate CuKα_1 radiation ($\lambda = 1.54$ Å).

(b) Electron diffraction: Calculate the voltage between cathode and anode of an electron microscope whose electron beam has a wavelength of 1.54 Å.

(c) Neutron diffraction: Show that the kinetic energy of free neutrons corresponding to a wavelength of 1.54 Å is of the same order of magnitude as the energy of thermal neutrons.

B-6. ATOMIC SCATTERING FACTOR

The atomic scattering factor ψ may be defined as the ratio between the radiation amplitude scattered by an atom and that scattered by a single point-like electron. Show that to a first approximation for small scattering angles θ the atomic scattering factor equals the number, Z, of electrons of the atom.

Remark: The total amplitude scattered by the atom follows from an integration over the differential scattering amplitudes of all charge elements of the whole electron shell if the phase differences of radiation scattered from different volume elements are taken into account.

B-7. STRUCTURE FACTOR AND SCATTERING AMPLITUDE

(a) Calculate the structure factors S for the body-centred cubic structure (W-type) and the CsCl-structure. The basis coordinates $u_j v_j w_j$ are for the bcc structure: W at 000 and $\frac{1}{2}\frac{1}{2}\frac{1}{2}$, and for the CsCl-structure: Cs at 000, Cl at $\frac{1}{2}\frac{1}{2}\frac{1}{2}$. Compare the conditions for extinction of reflections $h_1h_2h_3$ in both structures. How do the intensities of the reflections vary from CsCl to CsBr to CsI, which all crystallize in the CsCl-structure?

(b) Calculate the absolute value of the structure factor of Si for the following reflections: 200, 210, and 111. Si crystallizes in the diamond structure, which is composed of two identical face-centred cubic lattices displaced from each other by a vector $\frac{1}{4}\frac{1}{4}\frac{1}{4}$.

B-8. CRYSTAL PLANES

The lattice vector $\boldsymbol{h}(h_1, h_2, h_3)$ of the reciprocal lattice has the following general properties:

(a) $\boldsymbol{h}$ is perpendicular to the crystal planes defined by the Miller indices $h_1h_2h_3$.

(b) The length $|\boldsymbol{h}|$ is inversely proportional to the distance d between two adjacent planes $(h_1h_2h_3)$. Prove these two properties graphically for the special case of the (120)- and $(1\bar{2}0)$-planes of a simple orthorhombic lattice with the lattice constants $a_1 = 2$ Å, $a_2 = 4$ Å, $a_3 = 3$ Å.

B-9. VON LAUE DIFFRACTION PATTERN

Calculate the maximum possible number of Laue spots for a simple cubic crystal with lattice constant 2 Å, if the voltage applied to the X-ray tube is 60 kV.

B-10. TWO-DIMENSIONAL BRILLOUIN ZONE

A two-dimensional reciprocal lattice is defined by the vectors $\boldsymbol{b}_1(1,0)$ Å^{-1} and $\boldsymbol{b}_2(\frac{1}{2}, \frac{1}{2}\sqrt{3})$Å^{-1}.
(a) Construct the first Brillouin zone.
(b) Determine the maximum wavelength λ_{max} [Å] of X-rays for which coherent interference can be obtained.

C. Lattice Dynamics

C-1. MADELUNG CONSTANT AND COULOMB ENERGY

(a) Calculate the Madelung constant for an infinite one-dimensional ionic crystal. Using this result, calculate the Coulomb energy (in eV) of an ion in a one-dimensional NaCl chain. The distance between next-neighbour ions is $r_1 = 2.8$ Å.
(b) Estimate the Madelung constant for a two-dimensional NaCl crystal.

Remark: Since a direct summation yields only very slow convergence, a different method of summation, due to Evjen, is applied. One adds the contributions of ions in areas F_1, $F_2 - F_1$, etc., where $F_1 = a^2$, $F_2 = (2a)^2$, etc., are squares around a reference ion, and a is the lattice constant. Calculate the contributions from the first two squares. Explain why this summation converges more rapidly.

C-2. VAN DER WAALS CRYSTALS

All noble gases except He form face-centred cubic crystals as a consequence of van der Waals forces. The attractive potential ϕ_{ij} between atoms i and j is given by

$$\phi_{ij} = -\frac{C}{r_{ij}^6},$$

where C is the van der Waals constant, and r_{ij} is the distance between atoms i and j.
(a) Estimate the heat of sublimation (in kcal/mole) for Ne. *Remark:* The "effective" number Z of next neighbours can be defined as a dimensionless lattice sum

$$Z = \sum_{j \neq 0} \left(\frac{r_1}{r_{0j}}\right)^6 = 14.46,$$

where $r_1 = 3.2$ Å is the distance between next neighbours. The van der Waals constant for Ne is $C_{Ne} = 7.97 \times 10^{-60}$ erg cm^6.
(b) Take into account a repulsive potential of the form

$$\psi_{ij} = \frac{D}{r_{ij}^n}.$$

Determine the exponent n using the experimental value of the heat of sublimation, $Q_{Ne} =$ $= 0.52$ kcal/mole.

C-3. FRACTURE STRESS AND HEAT OF FUSION OF AN IDEAL IONIC CRYSTAL

(a) Estimate the fracture stress of an ideal NaCl crystal under uniform dilation. *Suggestion:* Use the Mie potential

$$\Phi(r) = -\alpha_M \frac{e^2}{4\pi\varepsilon_0} \frac{1}{r}\left[1-\frac{1}{9}\left(\frac{r_0}{r}\right)^8\right]$$

to calculate the maximum force F_{max}. For uniform dilation, the fracture stress is given by $\sigma_f = F_{max}/r_f^2$, where r_f is the ion distance at F_{max}. The equilibrium ion distance in NaCl is $r_0 = 2.8$ Å; the Madelung constant is $\alpha_M = 1.75$.

Since the crystal is assumed to be ideal, the calculated fracture stress will be about 10^3 times larger than the experimental values.

(b) Estimate the molar heat of fusion for NaCl assuming the lattice energy to be given by

$$U_l(r) = -\alpha_M N_P \frac{e^2}{4\pi\varepsilon_0} \frac{1}{r}\left[1-\frac{1}{9}\left(\frac{r_0}{r}\right)^8\right].$$

Suggestion: Calculate the ion distance r_f for which the attractive force reaches a maximum. The heat of fusion is approximated by the energy difference which is required to increase the ion distance from r_0 to r_f.

C-4. LATTICE ENERGY AND COMPRESSIBILITY

Assuming the lattice energy to be given by a Mie potential, derive the following relation between the lattice energy U_l and the compressibility $\varkappa$,

$$U_l = -\frac{gV_{mol}}{mn}\frac{1}{\varkappa},$$

where m and n are the exponents of the attractive and repulsive contributions to the lattice energy, and V_{mol} is the molar volume. Calculate the lattice energy of NaI using $n = 9$ and $\varkappa = 6.7\times10^{-11}$ m²/N. NaI crystallizes in the NaCl-structure with a lattice constant of $a = 6.5$ Å.

C-5. MEAN ENERGY OF THE LINEAR OSCILLATOR AND EQUIPARTITION VALUE

Calculate above which temperature the difference between the mean energy $\bar{u}(\omega, T)$ of a linear oscillator and the equipartition value kT is less than 1%. The oscillator frequency is assumed to be $\omega = 2\pi\times10^{13}$ sec⁻¹.

C-6. DEBYE TEMPERATURE FROM ELASTIC DATA

The equations of motion for elastic waves in a cubic crystal are

$$\varrho \frac{\partial^2 \xi}{\partial t^2} = c_{44} \frac{\partial^2 \xi}{\partial x^2} + c_{44} \left(\frac{\partial^2 \xi}{\partial y^2} + \frac{\partial^2 \xi}{\partial z^2} \right) + (c_{12}+c_{44}) \left(\frac{\partial^2 \eta}{\partial x\, \partial y} + \frac{\partial^2 \zeta}{\partial x\, \partial z} \right)$$

$$\varrho \frac{\partial^2 \eta}{\partial t^2} = c_{11} \frac{\partial^2 \eta}{\partial y^2} + c_{44} \left(\frac{\partial^2 \eta}{\partial z^2} + \frac{\partial^2 \eta}{\partial x^2} \right) + (c_{12}+c_{44}) \left(\frac{\partial^2 \zeta}{\partial y\, \partial z} + \frac{\partial^2 \xi}{\partial y\, \partial x} \right)$$

$$\varrho \frac{\partial^2 \zeta}{\partial t^2} = c_{11} \frac{\partial^2 \zeta}{\partial z^2} + c_{44} \left(\frac{\partial^2 \zeta}{\partial x^2} + \frac{\partial^2 \zeta}{\partial y^2} \right) + (c_{12}+c_{44}) \left(\frac{\partial^2 \xi}{\partial z\, \partial x} + \frac{\partial^2 \eta}{\partial z\, \partial y} \right),$$

where ϱ is the density, ξ, η, ζ are the displacements of a volume element in the x, y, z direction, respectively, and c_{11}, c_{12}, c_{44} are the three non-vanishing components of the elastic stiffness tensor for cubic crystals. For, isotropic media $c_{44} = \frac{1}{2}(c_{11} - c_{12})$, and thus only two components of the stiffness tensor remain independent.

(a) Show that the longitudinal and transverse sound velocities v_l and v_t of an isotropic medium are given by

$$v_l = (c_{11}/\varrho)^{1/2} \quad \text{and} \quad v_t = (c_{44}/\varrho)^{1/2}.$$

Remark: For isotropic media v_l and v_t are independent of the direction of the wave vector $\mathbf{k}$. It is sufficient, therefore, to calculate v_l and v_t for wave propagation in the x direction.

(b) Calculate the Debye temperature of Pb from its elastic stiffness components $c_{11} = 0.55 \times 10^{12}$ dyne/cm² and $c_{44} = 0.19 \times 10^{12}$ dyne/cm² and its density $\varrho = 11.6$ g/cm³.

C-7. LINDEMANN RULE

(a) According to Lindemann, the melting temperature T_m is related to the Debye temperature θ by

$$T_m = CMV_M^{2/3}\theta^2,$$

where $C = 5.3 \times 10^{-5}$ g^{-1} cm^{-2} deg^{-2} is an empirical factor, M is the atomic weight expressed in g, and $V_M = M/\varrho$ is the molar volume expressed in cm³. Check the formula for Pb ($M = 207.2$ g, $\varrho = 11.3$ g/cm³). The Debye temperature from specific heat data is $\theta = 88$°K.

(b) Derive the Lindemann formula considering the following model: The crystal melts as soon as its thermal energy ($3kT$ per atom according to Dulong and Petit) becomes so large that the amplitudes ξ_0 of the vibrating atoms reach a certain fraction $1/n$ of the average distance d of the atoms, typically $\xi_0 = d/n \approx d/10$. The vibrations of the atoms are assumed to be harmonic, i.e., $\xi = \xi_0 \sin \omega t$. With the Einstein model, $\omega = \omega_E$, the maximum velocity of the atoms is $v_{max} = \xi_0 \omega_E$. The melting temperature T_m follows from the equation $3kT_m = \frac{1}{2} m v_{max}^2 = \frac{1}{2} m\omega_E^2 (d/n)^2$, where m is the mass of the vibrating atom.

C-8. DEBYE–WALLER FACTOR

Calculate the Debye–Waller factor for the (111) and (511) X-ray reflections from Pb at 4°K and 300°K. Pb is face-centred cubic with lattice constant $a = 4.95$ Å. The Debye temperature is $\theta = 88$°K. Note that for a cubic lattice $h^2 = (2\pi/a)^2 (h_1^2 + h_2^2 + h_3^2)$.

C-9. RESONANCE FLUORESCENCE

The effect of resonance fluorescence in gases is much more probable for optical transitions between atomic levels than for γ transitions between nuclear levels. Prove this statement by considering the two following cases:

(a) Emission and absorption of the yellow line of Na atoms; the mean wavelength of the D doublet is $\lambda_D = 5893$ Å.

(b) Emission and absorption of γ radiation by ^{57}Fe nuclei; the energy of the γ quanta is $E_0 = E_{exc} - E_g = 14.4$ keV.

Suggestion: Calculate the recoil energy E_r for both cases and show the resulting spectra of emission and absorption graphically. In both cases, the line widths are of the same order of magnitude, namely $\Gamma \approx 10^{-8}$ eV. The Doppler effect may be neglected.

C-10. DEBYE TEMPERATURE OF IONIC CRYSTALS

The Debye temperature of a crystal is proportional to an upper cut-off frequency ω_{max} of the lattice vibrations, and ω_{max} is proportional to the square root of the ratio between the force constant f and the mass M of the lattice particles, i.e.,

$$\theta \propto \omega_{max} \propto \left(\frac{f}{M}\right)^{1/2}.$$

Using a Mie potential, prove the following relation between the Debye temperatures, lattice constants, and particle masses of two ionic crystals 1 and 2 with the same structure ($\alpha_1 = \alpha_2$) and with similar repulsive exponents ($n_1 \approx n_2$):

$$\frac{\theta_1}{\theta_2} = \left(\frac{a_2^3 M_2}{a_1^3 M_1}\right)^{1/2}.$$

Calculate the Debye temperature of KCl from that of NaCl, $\theta_{NaCl} = 315$°K. For M, use the reduced masses ($M_{Na} = 22.99$, $M_K = 39.10$, $M_{Cl} = 35.45$). The lattice constants of NaCl and KCl are $a_{NaCl} = 5.62$ Å and $a_{KCl} = 6.28$ Å.

C-11. LINEAR DIATOMIC CRYSTAL MODEL

A linear model of NaCl is characterized by the particle masses $M_{Na} = 23.0$ and $M_{Cl} = 35.5$, the distance between Na and Cl, $d = 2.8$ Å, and the force constant $f = 1.5 \times 10^4$ dyne/cm.

(a) Show that no dispersion exists around $q = 0$ for acoustical phonons.

(b) Calculate the phase and group velocity of optical phonons for $q = 0$ and $q = \pi/2d$.

(c) Calculate the widths (in eV) of the two allowed energy bands of the phonons and that of the forbidden energy range.

(d) Compare the relative widths and distances of the energy bands (relative to $\hbar\omega_+$ at $q = \pi/2d$) for linear models of KCl and PuN. The particle masses are given by the atomic weights: $M_N = 14$, $M_{Cl} = 35.5$, $M_K = 39$, $M_{Pu} = 242$. Discuss the limiting cases $m = M$, and $m \rightarrow 0$.

C-12. OPTICAL ABSORPTION IN THE SPECTRAL RANGE OF "RESTSTRAHLEN"

A linear chain of NaCl is polarized by irradiation with light of the far infrared. Estimate the wavelength of the absorbed light, and calculate the polarizability α of the chain as a function of the frequency of the electromagnetic radiation. The particle masses are given by the atomic weights: $M_{Na} = 23$, $M_{Cl} = 35.5$. The lattice constant is $2d = 5.6$ Å, and the force constant is $f = 1.5 \times 10^4$ dyne/cm.

Remark: The wave number q of the optically excited phonons is $q \approx 0$. Prove this subsequently from the condition of energy and momentum conservation. In order to calculate the polarizability, consider the lattice vibrations due to the electric field of the radiation. The damping is assumed to be negligible except for the resonance frequency.

C-13. INELASTIC PHOTON–PHONON SCATTERING

The inelastic scattering of light in a crystal can give rise to the emission or absorption of phonons. The scattering of photons by acoustical phonons is known as Brillouin scattering, and the scattering of photons by optical phonons is called Raman scattering.

A crystal with refractive index $n = 1.5$ is irradiated by light whose wavelength *in vacuo* is $\lambda = 5000$ Å. The velocity of the acoustical phonons (sound velocity) is $c_s = 2 \times 10^5$ cm/sec. Derive the relation $\Omega(\omega)$ between the frequency ω of the incident photon and the frequency Ω of the emitted or absorbed acoustical phonon. Determine the maximum frequency of the phonon.

Suggestion: Show the conservation of momentum graphically under the assumption that the magnitude of the photon wave vector remains to a first approximation unchanged in the scattering event, $|\boldsymbol{k}| \approx |\boldsymbol{k}'|$. Justify this assumption subsequently.

C-14. INELASTIC NEUTRON–PHONON SCATTERING

The scattering of a neutron of wavelength $\lambda_n = 4$ Å results in the absorption of a phonon of wavelength $\lambda_p = 10$ Å. Calculate the wavelength λ_n' and the relative energy increase $(E_n' - E_n)/E_n$ of the scattered neutron assuming the phonon velocity to be $c_s = 10^5$ cm/sec. Determine the scattering angle of the neutron graphically.

D. Imperfections

D-1. LOW-ANGLE GRAIN BOUNDARY

A bicrystal comprises two single crystals whose crystallographic axes are tilted relative to each other by a certain angle. The boundary between the two single crystals is called a grain boundary. Consider a simple cubic bicrystal, in which the single crystal portions are tilted around the [001] axis and thus the grain boundary is formed in the (010) plane. Show graphically that for small angles φ of tilt the grain boundary consists of a regular array of edge dislocations. Calculate the number of dislocations per unit length in the [100] direction as a function of the angle φ (use as an example $\varphi = 0.5°$, and the lattice constant $a = 5$ Å).

D-2. SCHOTTKY AND FRENKEL DEFECTS IN ALUMINIUM

In Al the activation energy for Schottky defects is $W_S = 0.75$ eV, and that for Frenkel defects is $W_F \approx 3$ eV. What is the ratio of the numbers n_S of Schottky defects and n_F of Frenkel defects at 300°K and 900°K?

Remark: Al is face-centred cubic. Each close-packed structure contains one octahedral and two tetrahedral interstices per atom. In the case of Al, the tetrahedral interstices are too small for interstitials.

D-3. ORDER–DISORDER TRANSITION

The order–disorder transition can be theoretically treated in a rather simple way, if only bonds between next neighbours are taken into account and if the dependence of the lattice constant on the degree of disorder is neglected. We consider a metallic alloy, consisting of N units AB, with the CsCl structure (coordination number 8), such as β-CuZn, AgMg, or CoFe. In the ordered state, there are only AB pairs with a bond energy U_{AB}. The existence of AA and BB pairs represents disorder. The bonds between these pairs are weaker, and correspondingly $U_{AA}+U_{BB} > 2U_{AB}$. The degree of order is given by an order parameter w, where $0 \leqslant \leqslant |w| \leqslant 1$, defined as follows: there are $N(1+w)/2$ "right" atoms A on sites a and $N(1-w)/2$ "wrong" atoms A on sites b, and $N(1+w)/2$ "right" atoms B on sites b and $N(1-w)/2$ "wrong" atoms B on sites a. Derive the temperature dependence of the order parameter $w(T)$ from the condition that the free energy be a minimum with respect to w. Determine the transition temperature in the limit as $w \to 0$.

D-4. SPECIFIC HEAT ANOMALY

The energy required to form lattice defects causes a contribution to the specific heat of crystals. This specific heat anomaly amounts to $\Delta c_V = 20$ cal/mole deg for AgBr close to its melting point at $T_m = 690$°K.

(a) What percentage of the total specific heat of AgBr does the specific heat anomaly represent at this temperature? The Debye temperature of AgBr is $\theta = 140$°K.

(b) The lattice defects in AgBr are Frenkel defects (Ag^+ ions) with an activation energy $W_F = 29.4$ kcal/mole. Calculate the concentration of Frenkel defects at 600°K from the specific heat anomaly given above.

D-5. EINSTEIN RELATION

Derive the Einstein relation $qD = bkT$ which links the diffusion coefficient D with the mobility b of a current carrier species of charge q.

Suggestion: Consider a medium in which both an electric field $F = -dV/dx$ and a position-dependent carrier concentration $n(x)$ exist. In thermal equilibrium, the field current i_F due to the electric field and the diffusion current i_D due to the concentration gradient cancel each other ($i_F+i_D = 0$), and the carrier concentration assumes a Boltzmann distribution $n(x) = n_0 \exp(-qV(x)/kT)$.

D-6. MOBILITY OF CURRENT CARRIERS

(a) For an ionic crystal with NaCl structure, estimate the ionic mobility in the [110] direction at 300, 600, and 900°K. The crystal is characterized by a lattice constant $a = 5$ Å, a jump frequency $\nu = 5 \times 10^{12}\ \text{sec}^{-1}$, and a threshold energy $w = 0.01$ eV.

(b) Calculate the diffusion coefficient D_n of electrons in Ge from the electron mobility b_n = 3[illegible]00 $\text{cm}^2/\text{V sec}$ at $T = 300$°K.

D-7. SUBSTITUTIONAL OR INTERSTITIAL IMPURITY INCORPORATION

The doping of a Si single crystal with 0.314 at.% of B ($M_B = 10.811$) results in an increase of the density by 0.00154 g/cm^3 and a decrease of the lattice constant by 0.0046 Å. Originally, the density of the Si ($M_{Si} = 28.086$) was $\varrho = 2.3306\ \text{g/cm}^3$, and the lattice constant $a = 5.4295$ Å. The measurements before and after doping were performed at a temperature $T = 298.3$°K. Determine on the basis of these data whether B is incorporated into the Si crystal substitutionally or interstitially.

Suggestion: Consider the density for a certain number of unit cells.

D-8. ELECTRONIC AND IONIC POLARIZABILITY

(a) Show that in the classical approximation the static electronic and ionic polarizabilities are given by:

$$\alpha_{el} = \frac{e^2}{\omega_0^2}\,\frac{1}{m}, \quad \text{and} \quad \alpha_{ion} = \frac{e^2}{\omega_R^2}\left(\frac{1}{M_+} + \frac{1}{M_-}\right),$$

where m, M_+, M_- are the masses of the electron, and the positive or negative ion, respectively $\hbar\omega_0$ is the binding energy between electron and nucleus, and ω_R is the "Reststrahlen" frequency.

(b) Estimate α_{el} for the H atom, and α_{ion} for an ion pair NaCl.

E. Foundations of the Electron Theory of Metals

E-1. THE SPECIFIC CHARGE E/M OF ELECTRONS IN METALS

Tolman and Stewart were the first to prove (1916) that the current carriers in metals are electrons. The specific electron charge e/m was measured by the following experiment: a wire of length L and electrical resistance R is wound on a cylinder of radius r. The cylinder is rotated with a large angular velocity ω about its axis and then suddenly stopped. Thereby, a voltage pulse is generated between the ends of the wire, causing a current pulse in the measuring circuit of the resistance R_e. Assuming $R_e \gg R$, calculate e/m as a function of the total charge Q which has passed through the circuit.

E-2. FERMI–DIRAC STATISTICS

Derive the Fermi–Dirac function $f(E)$ from the free energy of the conduction electrons.
Remark. The free energy of a system of N electrons is

$$F = U - TS = \sum_i (N_i E_i - kT \ln W_i),$$

where U is the internal energy, S is the entropy, and W_i is the number of possibilities for arranging N_i electrons of energy E_i in $D(E_i)$ available states. By applying the method of Lagrangian multipliers, formulate the condition of thermodynamic equilibrium, subject to the additional conditions that the total number of electrons and their total energy are constant. The chemical potential of the electrons is $\partial F/\partial N_i = \zeta$, which is equal to the Fermi energy of a free electron gas.

E-3. FERMI–DIRAC FUNCTION

(a) Calculate the energy range (in units of kT) in which the Fermi function drops from 0.9 to 0.1.

(b) Above which energies measured from the Fermi level can the Fermi function be approximated by a Boltzmann factor with an error of less than 1% and 10%, respectively, at temperatures of 300°K and 1000°K?

E-4. DEGENERACY OF THE ELECTRON GAS

From Heisenberg's uncertainty relation $\Delta p \Delta q \geq \hbar$, estimate the degeneracy temperature above which the electron gas of a metal can be considered to be a classical gas and thus be described by Maxwell–Boltzmann statistics.

Remark: For a classical gas, the uncertainties of particle position and momentum are $\Delta q \ll n^{-1/3}$ and $\Delta p \ll \bar{p}$, where n is the concentration and $\bar{p}$ the average momentum of the particles.

E-5. FERMI ENERGY OF Cu

(a) Calculate the Fermi energy of Cu at temperatures $T = 0$°K, 300°K, 1350°K, assuming that the 4s electrons of the Cu atoms behave like free conduction electrons. Cu is face-centred cubic with a lattice constant of $a = 3.60$ Å.

(b) Physical measurements are normally performed at constant pressure rather than at constant volume. The lattice constant of Cu varies with temperature as follows: $a(303°\text{K}) = 3.6147$ Å, $a(503°\text{K}) = 3.6276$ Å, $a(803°\text{K}) = 3.6486$ Å. How does this variation affect the Fermi energy?

E-6. SPECIFIC HEAT OF THE DEGENERATE ELECTRON GAS

Show that the specific heat of a degenerate electron gas is given by the general relation

$$c_V^{\text{el}} = \tfrac{2}{3}\pi^2 k^2 D(\zeta)\, T,$$

which holds for any dispersion relation and thus for any form of the eigenvalue density of electrons.

Suggestion: Calculate the energy $U(T)$ of the electron gas by applying Sommerfeld's expansion for $\zeta/kT \gg 1$. The temperature dependence of the Fermi energy is given by the general relation:

$$\zeta = \zeta_0 - \frac{\pi^2}{6}(kT)^2 \left[\frac{\partial \ln D(E)}{\partial E}\right]_{E=\zeta_0} + \ldots,$$

which can also be dervied by using Sommerfeld's expansion.

E-7. FERMI VELOCITY AND DRIFT VELOCITY

Based on the model of free electrons, calculate the Fermi velocity v_F of the conduction electrons in Cu, and compare this value with the drift velocity v_d at a current density of 5 A/cm^2. The concentration of conduction electrons in Cu is $n \approx 10^{23}$ cm^{-3}, and the conductivity is $\sigma \approx 5\times 10^5\ \Omega^{-1}$ cm^{-1}.

E-8. RADIUS OF THE FERMI SPHERE

Many properties of the free electron gas depend on a certain power m of the radius k_F of the Fermi sphere. Compile these k_F^m dependences for the following quantities: Fermi energy ζ_0 at $T = 0°$K, electron concentration n, total energy U_0 of the electron gas at $T = 0°$K, eigenvalue density $D(\zeta_0)$ at the Fermi level ζ_0, and specific heat c_V^{el} of the electrons.

E-9. RICHARDSON EQUATION

The measurement of thermionic emission from a metal usually extends over a relatively small temperature range. Since the temperature dependence of the work function $W(T)$ is small, the experimental results, plotted as $\ln(I/T^2)$ vs. $1/T$, lie on a straight line. This experimental line is defined by $\ln(I/T^2) = \ln A^* - W^*/kT$, where A^* and W^* are constants. It represents a tangent to the true Richardson equation $\ln(I/T^2) = \ln(A\delta) - W(T)/kT$ at an intermediate temperature T_i. Calculate A^* and W^* as a function of A, δ, and $W(T)$. The true values of $A\delta$ and $W(T)$ can be obtained from A^* and W^*, if $dW(T_i)/dT$ is known.

E-10. SCHOTTKY EFFECT

A positive electric field of $F = 10^4$ V/cm is applied perpendicularly to the emitting surface of a thermionic cathode. Calculate the decrease ΔW in the work function due to the image force. How far from the surface is the maximum of the potential energy? By how many percent does the emission current increase at a cathode temperature $T = 2000°$K due to an applied field F?

E-11. THERMIONIC CONVERTER

A thermionic converter consists of two metals 1 and 2 in vacuum which form the plates of a capacitor and are connected by an external resistor R. The metals have work functions W_1 and W_2, and are kept at temperatures T_1 and T_2. It is assumed that $W_1 > W_2$ and $T_1 > T_2$.

(a) Calculate R for maximum power delivery.

(b) Plot the current–voltage relation for the load resistor, neglecting the thermoelectric voltage, the space charge, and the temperature dependences of W_1 and W_2.

E-12. FIELD EMISSION AND THERMIONIC EMISSION

Estimate the cathode temperature at which for zero field the same emission current density is drawn as from a cold cathode at an applied field of 2.2×10^7 V/cm. The work function of the cathode material is $W = 4.5$ eV. The temperature dependence of the work function and the image force are neglected. The transmission factor is assumed to be $\bar{\delta} = 1$.

E-13. OPTICAL PROPERTIES OF FREE ELECTRONS (DRUDE'S THEORY)

The equation of motion of a free electron in the electric field of light is

$$m\frac{d^2x}{dt^2}+\frac{m}{\tau}\frac{dx}{dt} = eE = eE_0 \exp(i\omega t),$$

where τ is the relaxation time of the electrons. The polarization of N electrons per cm^3, $P = Nex$, is related to a complex dielectric constant by

$$\varepsilon^* \equiv (n-i\varkappa)^2 = 1+\frac{1}{\varepsilon_0}\frac{P}{E},$$

where n is the refractive index, and $\varkappa$ is the extinction coefficient. The imaginary part of ε^*, $2n\varkappa$, determines the absorption of light; the absorption constant is given by $\alpha = 2\varkappa\omega/c$, where c is the light velocity. Calculate the absorption constant α of free electrons for the limiting cases $\omega\tau \gg 1$, and $\omega\tau \ll 1$.

E-14. FARADAY ROTATION DUE TO FREE CURRENT CARRIERS

The Faraday effect is a magneto-optical phenomenon in which the plane of linearly polarized light under the influence of a magnetic field parallel to the wave vector is turned over an angle θ per distance d of light propagation. This rotation is caused by different propagation velocities for right- and left-circularly polarized light, and is generally given by

$$\frac{\theta}{d} = \frac{1}{2}\frac{\omega}{c}(n_- - n_+),$$

where ω is the frequency of the light, c is the light velocity in vacuum, and n_- and n_+ are the refractive indices for left- and right-circular polarization, respectively.

Derive the following relation for Faraday rotation due to free current carriers,

$$\frac{\theta}{d} = \frac{1}{2}\frac{\omega}{c}\left(\frac{\omega_p}{\omega}\right)^2 \frac{\frac{\omega_e}{\omega}}{1-\left(\frac{\omega_e}{\omega}\right)^2},$$

where $\omega_e = eB/m$ is the cyclotron frequency, and $\omega_p = (Ne^2/\varepsilon_0 m)^{1/2}$ is the plasma frequency of the current carriers.

Suggestion: Assume weak absorption and thus $n_\pm^2 = \varepsilon_\pm$. In order to find ε, solve the equation of motion for an electron under the influence of the electric field of light and a static magnetic field perpendicular to it.

F. Electrons in a Periodic Potential

F-1. KRONIG–PENNEY MODEL

In 1931 Kronig and Penney studied the exact solutions of the Schrödinger equation for a one-dimensional periodic series of rectangular potentials and thus explained many properties of the energy band structure of crystalline solids. The periodic rectangular potential can be further simplified by a series of δ-functions, i.e.,

$$V(x) = V_p a \sum_{m=-\infty}^{+\infty} \delta(x - ma).$$

For this periodic potential, the energy eigenvalues E as a function of the wave number K are given by the relation

$$F(\xi a) \equiv \cos \xi a + \frac{ma^2 V_p}{\hbar^2} \frac{\sin \xi a}{\xi a} = \cos Ka$$

with $\xi \equiv (2mE/\hbar^2)^{1/2}$. Hence real K values exist for $|F(\xi a)| \leq 1$.

(a) Plot the function $F(\xi a)$ and mark the allowed and forbidden ξa-ranges. Assume $V_p = 4\pi\hbar^2/2ma^2$.

(b) Determine the energies and wave numbers of the upper band edges.

(c) Discuss the widths of the allowed and forbidden energy ranges for the limits $V_p \to \infty$ and $V_p \to 0$.

(d) Verify that for states close to the lower edge of the first band (i.e., for small values of K) the following approximation holds:

$$E = E_0 + \frac{\hbar^2}{2m^*} K^2 \quad \text{with} \quad \frac{m^*}{m} = -\frac{F_0'}{\xi_0 a},$$

where $F' = dF/d(\xi a)$. The subscript 0 refers to the values at $K = 0$.

F-2. APPROXIMATION FOR WEAKLY BOUND ELECTRONS

For a weak one-dimensional periodic potential, the electron energy close to the band edges is given by

$$E(K) = E_n + \varepsilon_\pm, \quad \text{with} \quad K = \pm\frac{\pi}{a} n \pm \varkappa, \quad E_n = \frac{\hbar^2}{2m}\left(\frac{\pi}{a} n\right)^2,$$

and

$$\varepsilon_\pm = \frac{\hbar^2}{2m} \varkappa^2 \pm \left[|V_n|^2 + \left(\frac{\hbar^2}{m} \frac{\pi}{a} n\varkappa\right)^2 \right]^{1/2}.$$

(a) Show that for small $\varkappa$ values the $E(K)$ relation can be approximated by

$$E(K) \approx E_\pm + \frac{\hbar^2}{2m_\pm^*} \varkappa^2,$$

where $m_\pm^*$ are the effective masses which can assume either sign. Verify the relation

$$\frac{m}{m_\pm^*} = 1 \pm \frac{2E_n}{|V_n|}.$$

(b) Calculate $m/m_\pm^*$ for the case $n = 2$, $|V_2| = 1$ eV, and $a = 4$ Å.

F-3. TIGHT BINDING APPROXIMATION

The energy eigenvalues of strongly bound electrons are given by

$$E(\boldsymbol{K}) \approx E_{at} + C + \sum_{\boldsymbol{R}_j \neq 0} A_j \exp \{i(\boldsymbol{K} \cdot \boldsymbol{R}_j)\},$$

where E_{at} is the eigenvalue for a free atom, C and A_j are constant integral expressions, and the $\boldsymbol{R}_j$ are the lattice vectors to the other atoms. Verify that in a one-dimensional crystal model the $E(K)$ relation goes over into

$$E(K) = E_{at} + C + 2A_1 \cos (Ka),$$

if only next-neighbour atoms are taken into account. Show that the $E(K)$ relation is quadratic close to the band edges.

F-4. VELOCITY AND EFFECTIVE MASS FOR ELECTRONS IN A STRONG POTENTIAL

The band structure of electrons in a one-dimensional strong potential is given by

$$E_n(k) = E_{a,n} + S_n \cos (ka).$$

Calculate the group velocity v and the effective mass m^* as a function of S_n and k. Determine the values of v and m/m^* for $S = 0.2$ eV and $k = 0.01/a$ with $a = 4$ Å. Verify that for this value of k the quadratic approximation $E(k) \approx E_0 + \hbar^2 k^2/2m^*$ is justified.

F-5. EFFECTIVE MASS OF THE CONDUCTION ELECTRONS IN Bi

The reciprocal effective mass tensor for electrons at the lower edge of the conduction band of Bi is usually written in non-diagonalized form,

$$\frac{m}{m^*} = \begin{pmatrix} \alpha_{11} & 0 & 0 \\ 0 & \alpha_{22} & \alpha_{23} \\ 0 & \alpha_{32} & \alpha_{33} \end{pmatrix}.$$

(a) Derive the dispersion relation $E(\boldsymbol{k})$ for electron energies close to the band edge E_R. What is the shape of the corresponding surfaces of constant energy?

Suggestion: Set $\boldsymbol{\varkappa} = \boldsymbol{k} - \boldsymbol{k}_R$, where $\boldsymbol{k}_R$ is the wave vector at the band edge E_R, and approximate E by a Taylor series around $\boldsymbol{k}_R$ for small values of $\boldsymbol{\varkappa}$.

(b) Calculate the components $(m^*/m)_{ij}$ of the effective mass tensor.

Remark: The components of the tensor reciprocal to (α) are given by $(\alpha^{-1})_{ij} = A_{ji}/\det \alpha$, where A_{ji} is the co-factor of α_{ij}.

F-6. BRILLOUIN ZONE OF THE SEMICONDUCTING COMPOUND Mg_2Sn

Mg_2Sn crystallizes in the cubic CaF_2 structure. The unit cell contains 12 atoms at the following positions: Sn at 000, $0\frac{1}{2}\frac{1}{2}$, $\frac{1}{2}0\frac{1}{2}$, $\frac{1}{2}\frac{1}{2}0$, and Mg at $\pm\frac{1}{4}\frac{1}{4}\frac{1}{4}$, $\pm\frac{1}{4}\frac{3}{4}\frac{3}{4}$, $\pm\frac{3}{4}\frac{1}{4}\frac{3}{4}$, $\pm\frac{3}{4}\frac{3}{4}\frac{1}{4}$. Show that Mg_2Sn is a semiconductor.

Remark: The Brillouin zone bounded by the {220} planes is just completely filled. From its volume calculate the number of electron states contained in the atomic volume. Show that this number equals the average number of valence electrons per atom in Mg_2Sn.

F-7. ELECTROSTATIC SCREENING

The introduction of a point charge at rest into an electron gas causes a perturbing position-dependent potential. The range over which this potential extends is effectively limited by a local rearrangement of the electron concentration. The electron gas thus "screens" the point charge. In order to calculate this screening effect, we assume a perturbing potential $\varphi(\boldsymbol{r})$ at the point $\boldsymbol{r}$. The electron energy is changed by $e\varphi(\boldsymbol{r})$. Since in thermal equilibrium the Fermi energy must remain constant, the electron concentration changes locally by $-\delta n(\boldsymbol{r}) = 2D(\zeta_0)e\varphi(\boldsymbol{r})$. A position-dependent electron concentration corresponds to a space charge which must satisfy the Poisson equation

$$\nabla^2\varphi(\boldsymbol{r}) = -\frac{e}{\varepsilon_0}\,\delta n(\boldsymbol{r}) = \frac{1}{\lambda^2}\,\varphi(\boldsymbol{r}),$$

where the screening radius λ is given by $\lambda^2 = \varepsilon_0/2e^2D(\zeta_0)$. For a point charge q the solution is

$$\varphi(r) = \frac{q}{4\pi\varepsilon_0 r}\exp\,(-r/\lambda)$$

where r is the distance from the point charge. Calculate the screening radius λ for a Zn^{2+}-ion in Cu. The electron concentration in Cu is $n = 8.5\times10^{22}$ cm^{-3}.

F-8. CYCLOTRON FREQUENCY

Derive the expression for the cyclotron frequency ω_c of free electrons classically. Calculate the radius of the cyclotron orbit of an electron under the influence of the magnetic induction $B = 1$ V sec/m^2. Assume free electrons with a Fermi velocity $v_F = 1.6\times10^6$ m/sec, as in Cu, for instance.

F-9. QUANTIZATION OF CYCLOTRON ORBITS IN REAL SPACE

Free electrons under the influence of a magnetic induction $\boldsymbol{B}(0, 0, B_z)$ move on certain allowed orbits in real space. Show that the cross-section C_l of the lth orbit projected onto a plane perpendicular to $\boldsymbol{B}$ is given by

$$C_l = \pi r_l^2 = \frac{e}{h}\left(l+\frac{1}{2}\right)\frac{1}{B_z},$$

where r_l is the radius of the lth orbit in real space, and l is the Landau quantum number. Calculate the change $\Delta\Phi = \Phi_{l+1} - \Phi_l$ of the magnetic flux for an electron transition from the lth to the $(l+1)$th cyclotron orbit.

F-10. CYCLOTRON MOTION OF CRYSTAL ELECTRONS

Under the influence of a static magnetic induction $\boldsymbol{B}(0, 0, B_z)$, a crystal electron moves on a cyclotron orbit in real space ($\boldsymbol{r}$-space). Correspondingly, the wave vector $\boldsymbol{K}(K_x, K_y, K_z)$ of the electron changes in such a way that the end-point of $\boldsymbol{K}$ moves in the reciprocal space ($\boldsymbol{K}$-space) on a curve defined by the intersection of a surface of constant energy $E =$ const. and a plane $K_z =$ const. perpendicular to $\boldsymbol{B}$. For $K_z = 0$, the $E(\boldsymbol{K})$ relation is given by

$$E = E_0 + \frac{\hbar^2}{2}\left(\frac{K_x^2}{m_x} + \frac{K_y^2}{m_y}\right),$$

where $m_x = m$, and $m_y = 0.25\,m$. Sketch the cyclotron orbits in $\boldsymbol{r}$- and $\boldsymbol{K}$-space. Note that the electron velocity v is always perpendicular to the surface of constant energy in $\boldsymbol{K}$-space. Calculate the ratio of the velocities at $x = 0$ and $y = 0$.

F-11. DISPLACEMENT OF THE OPTICAL ABSORPTION EDGE DUE TO A MAGNETIC FIELD

In the presence of a magnetic induction B, the energy gap $\Delta E(B)$ between valence and conduction band is given by

$$\Delta E(B) = \Delta E(0) + \left(l + \frac{1}{2}\right)\hbar e\left(\frac{1}{m_n} + \frac{1}{m_p}\right)B,$$

where $\Delta E(0)$ is the width of the energy gap for $B = 0$, and m_n and m_p are the cyclotron masses for electrons and holes, respectively. For $l = 0$ and $B = 1$ V sec/m^2, calculate the displacement of the optical absorption edge in InSb ($m_n = 0.014m$, $m_p = 0.18m$).

G. Semiconductors

G-1. DEGENERACY TEMPERATURE OF THE CONDUCTION ELECTRONS IN AN INTRINSIC SEMICONDUCTOR

An intrinsic semiconductor is assumed to be characterized by an energy gap $\Delta E_i = 0.3$ eV and electron and hole masses $m_n = 0.027m$ and $m_p = 0.20m$. Calculate the degeneracy temperature T_E at which the Fermi level reaches the lower edge E_{0n} of the conduction band:

(a) Approximate T_E by extrapolating the temperature dependence of the Fermi level $\zeta(T)$ which holds for non-degenerate intrinsic semiconductors.

(b) Use an exact formulation of the neutrality condition.

G-2. SPECIFIC HEAT OF THE CURRENT CARRIERS OF A SEMICONDUCTOR

Calculate the specific heat of the current carriers in intrinsic Ge at the melting point $T_m = 1209$°K. Calculate also the ratio of the specific heat contributions by the carriers and the lattice vibrations.

Suggestion: Verify that the carrier concentrations of Ge at the melting point are non-degenerate. Hence, the equipartition principle holds for the carrier gases. Note that in addition to the kinetic energy of the carriers also the creation energy of electron–hole pairs contributes to the specific heat. The energy gap of Ge is temperature dependent, namely $\Delta E = \Delta E_0 - \beta T$, where $\Delta E_0 = 0.76$ eV and $\beta = 3.9 \times 10^{-4}$ eV/deg. The Debye temperature of Ge is $\theta = 375$°K.

G-3. MAXIMUM RESISTIVITY OF InSb

For InSb at room temperature, calculate the maximum resistivity which is obtainable by doping with acceptors. The intrinsic conductivity at room temperature is $\sigma_i = 200\ \Omega^{-1}\ \text{cm}^{-1}$, and the electron and hole mobilities are $b_n = 77{,}000\ \text{cm}^2/\text{V sec}$ and $b_p = 700\ \text{cm}^2/\text{V sec}$, respectively.

G-4. EXTRINSIC AND INTRINSIC CONDUCTION

A Si crystal has a purity of 99.999%, or 99.9999%, respectively. The residual impurity is Sb, having an activation energy $\Delta E_D = 0.039$ eV. For the two given purities, determine graphically at which temperature the conduction becomes intrinsic. Assume that at this temperature, the impurities are fully ionized. Check this assumption, and determine the temperature above which all impurities are ionized.

G-5. IMPURITY SEMICONDUCTORS AND IMPURITY BAND CONDUCTION

The wave functions of hydrogen-like impurity atoms extend over several lattice constants. The model of localized impurities becomes invalid, if the Bohr radii of adjacent impurities overlap. The interaction between the donor electrons causes the impurity level to split and thus to form an impurity band which gives rise to impurity band conduction. Estimate the following quantities for Te-doped InSb ($\varepsilon = 17$, $m_n = 0.014m$):

(a) the ionization energy ΔE_D for Te donors,

(b) the first Bohr radius,

(c) the minimum Te concentration N_D above which impurity band conduction occurs. In which temperature range is impurity band conduction detectable?

G-6. IMPURITY CONDUCTION IN SEMICONDUCTORS

(a) Calculate the electron concentration of an n-type semiconductor with donor concentration N_D, when the Fermi level ζ coincides with the donor level E_D.

(b) Show that in the range of impurity conduction the degree of degeneracy, $\alpha = (\zeta - E_{0n})/kT$, has a maximum α_M at a temperature T_M. Calculate T_M for two cases of degeneracy, $\alpha_M = 0$, and $\alpha_M = 2$.

Suggestion: Verify that for $d\alpha/dT = 0$ the neutrality condition yields

$$\delta_M \equiv \frac{\Delta E_D}{kT_M} = \frac{3}{2} + \frac{3}{4} \exp\left[-(\alpha_M + \delta_M)\right].$$

G-7. PARTLY COMPENSATED SEMICONDUCTOR

An n-type semiconductor usually contains a certain concentration N_A of acceptor impurities in addition to the intentional donor concentration N_D. Show that for non-degeneracy the electron concentration n is given by

$$n \propto \exp(-\delta) \quad \text{for} \quad n \ll N_A \ll N_D,$$

and by

$$n \propto \exp(-\delta/2) \quad \text{for} \quad N_A \ll n \ll N_D,$$

where $\delta \equiv \Delta E_D/kT$.

Suggestion: Use the neutrality condition, assuming sufficiently low temperatures, and the acceptors to be completely ionized, $n_A^{\text{ion}} = N_A$.

G-8. VALENCE BANDS OF GERMANIUM

The energy E_{0p} at $\boldsymbol{k} = 0$ is the common upper band edge of the two highest valence bands of Ge. The holes in these bands have nearly isotropic effective masses, and to a first approximation, the surfaces of constant energy are spheres. The hole masses for the heavy and light holes are $m_{p1} = 0.28m$, and $m_{p2} = 0.044m$, respectively. The eigenvalue density resulting from the two valence bands is

$$D_p(E)\,dE = \frac{V}{4\pi^2}\left(\frac{2m_p}{\hbar^2}\right)^{3/2}(E_{0p}-E)^{1/2}\,dE,$$

where V is the crystal volume. Calculate m_p using the values of m_{p1} and m_{p2}.

G-9. HEAVY AND LIGHT HOLES

Consider an intrinsic non-degenerate semiconductor which is characterized by a parabolic conduction band with its lower band edge at the energy E_{0n} and by two parabolic valence bands with a common upper band edge at E_{0p}. The effective mass of the electrons is m_n, and that of the heavy and light holes is m_{p1} and m_{p2}, respectively. From the neutrality condition, calculate the concentrations p_1 and p_2 of the heavy and light holes.

G-10. EFFECTIVE MASS OF THE CONDUCTION ELECTRONS IN InSb

According to Kane, the conduction band of the semiconducting intermetallic compound InSb can be approximated close to the energy minimum by the following $E(k)$ relation, which holds for $kP \ll 0.9$ eV,

$$E = E_{0p} + \frac{E_G}{2} + \frac{\hbar^2}{2m}k^2 + \left(\frac{E_G^2}{4} + \frac{2}{3}P^2k^2\right)^{1/2},$$

where E_{0p} is the energy of the valence band maximum, E_G is the width of the energy gap, and $P^2 \approx (20\text{ eV})\hbar^2/2m$.

(a) Using this relation, calculate the effective mass of a conduction electron as a function of k.

(b) Show that for $k^2 \ll 3E_G^2/8P^2$ the $E(k)$ relation can be further simplified to

$$E \approx E_{0n} + \frac{\hbar^2}{2m^*}k^2,$$

where E_{0n} is the energy of the conduction band minimum. Calculate the effective mass m^* and compare it with its experimental value $m^*_{\text{exp}} = 0.014m$.

G-11. CYCLOTRON RESONANCE OF THE CONDUCTION ELECTRONS IN SILICON

The conduction band of Si has six energy minima which are symmetrically located in the $\langle 100 \rangle$ directions of $\boldsymbol{k}$-space. Consider the [010] minimum around which the energy is determined by

$$E(\boldsymbol{\varkappa}) = E_{010} + \frac{\hbar^2}{2}\left(\frac{\varkappa_y^2}{m_l} + \frac{\varkappa_x^2 + \varkappa_z^2}{m_t}\right),$$

where $\boldsymbol{\varkappa}$ is the wave vector originating at the energy minimum, E_{010} is the energy value of the minimum, $m_l = 0.98m$ and $m_t = 0.19m$ are the longitudinal and transverse effective masses, respectively. Calculate the cyclotron frequency ν_c of the electrons for two directions of the magnetic induction, namely $\boldsymbol{B}(0, 0, B_z)$ and $\boldsymbol{B}(B_x, B_y, B_z)$ with $B_x = B_y = B_z$. The magnitude of the magnetic induction is assumed to be $|\boldsymbol{B}| = 0.1$ V sec/m^2.

G-12. OVERLAPPING ENERGY BANDS

The overlapping bands of a semimetal are given by

$$E_1(k) = E_1(0) - \frac{\hbar^2}{2m_1}, \quad \text{with} \quad m_1 = 0.18m$$

and

$$E_2(k) = E_2(k_0) + \frac{\hbar^2}{2m_2}(k-k_0)^2, \quad \text{with} \quad m_2 = 0.06m,$$

where $E_1(0)$ is the energy of the upper edge of band 1, and $E_2(k_0)$ is the energy of the lower edge of band 2, and $E_1(0) - E_2(k_0) = 0.1$ eV is the overlapping energy range. Due to the overlapping bands, electrons pass from band 1 to band 2, thus creating holes in band 1. Calculate the Fermi energy at $T = 0$°K.

Suggestion: Sketch the eigenvalue densities $D(E)$ for both bands, and count the Fermi energy ζ_0 from $E_2(k_0)$.

H. Contact Effects

H-1. CONTACT POTENTIAL DIFFERENCE

The contact potential difference between two metals is assumed to be 1 eV. When these two metals are brought into contact, electrons are exchanged between them. The contact is assumed to be established, if the metal surfaces are within a distance of the order of a lattice constant, namely 3×10^{-8} cm. How many electrons are exchanged in establishing the contact?

H-2. SCHOTTKY BOUNDARY LAYER

The contact potential difference between a metal and n-type Ge is assumed to be $W_m - W_n = 0.5$ eV. The static dielectric constant of Ge is $\varepsilon = 16$, and the donor concentration in Ge is assumed to be $N_D = 5 \times 10^{17}$ cm^{-3}. Calculate (a) the thickness d of the Schottky boundary layer; (b) the electric field $F(0)$ at the metal–semiconductor interface; (c) the surface charge density Q; (d) the capacitance per unit area of the boundary layer.

H-3. CARRIER INJECTION

Metal–semiconductor contacts may be used to inject excess carriers into semiconductors. The thermal equilibrium which is thus locally disturbed can be restored by recombination of the excess carriers. How far do excess holes diffuse in Ge at room temperature before they recombine with the electrons? The hole diffusion constant and lifetime are assumed to be $D_p = 44$ cm^2/sec and $\tau_p = 10^{-3}$ sec, respectively.

Suggestion: Solve the steady-state equation for excess holes,

$$D_p \frac{d^2p}{dx^2} - \frac{p}{\tau_p} = 0$$

by setting

$$p = p_0 \exp\left(\frac{-x}{L_p}\right),$$

where L_p is the hole diffusion length.

H-4. METAL–SEMICONDUCTOR CONTACT

Metal–semiconductor contacts act as rectifiers provided that the frequency of the applied voltage U is sufficiently low. The current–voltage characteristic is given by

$$j = j_s \left[\exp\left(\frac{eU}{kT}\right) - 1\right]$$

with

$$j_s = A_0 \frac{m_n}{m} T^2 \exp\left(\frac{-W_{ms}}{kT}\right)$$

where j_s is the saturation current density, m_n is an electron effective mass, W_{ms} is the height of the potential barrier between metal and semiconductor, and $A_0 = 120$ A/cm^2 deg^2 is the universal thermionic constant. Calculate the frequency limit for rectification.

Suggestion: Calculate the d.c. resistance of the contact in the limit of small voltages ($eU \ll kT$) and compare it with the reactance. Assume an n-type semiconductor with donor concentration $N_D = 5 \times 10^{18}$ cm^{-3}, donor activation energy $\Delta E_D = 0.05$ eV, and dielectric constant $\varepsilon = 11$. It is further assumed that $W_{ms} = 0.5$ eV, $m_n = 0.2m$, and $T = 300$°K.

H-5 COMPARISON OF SPACE CHARGE LIMITED CURRENTS IN VACUUM DIODES AND IN SOLIDS

According to Langmuir and Schottky, the space-charge limited current density between plane-parallel electrodes in a vacuum diode is given by

$$j = \frac{4\pi\varepsilon_0}{9} \left(\frac{2e}{m}\right)^{1/2} \frac{V^{3/2}}{l^2}$$

where V is the anode voltage and l is the distance between the electrodes.

(a) Derive a similar relation for the case of a rod-shaped crystal where at one end ($x = 0$) a space charge density $en(0)$ is maintained by carrier injection. The length of the crystal is l, and the potential difference between the two ends is V.

Suggestion: Express the current density due to carrier injection as a function of the electric field and the diffusion constant. Use the Poisson equation and the Einstein relation.

(b) Compare the magnitudes of the space-charge limited currents in vacuum and in solids. Explain the difference by comparing the drift velocities in both cases.

H-6. SOLAR CELLS

The solar cell is basically a *pn* junction which can convert solar radiation directly into electrical energy.

(a) Estimate the maximum power output of a Si solar cell from its short-circuit current $j_{h\nu}$ and its open-circuit voltage U_{0c}. For Si the number of photons in the solar spectrum which is available for electron–hole pair creation is $n_{ph} = 4 \times 10^{17} \text{cm}^{-2} \text{sec}^{-1}$. The quantum efficiency of the solar cell is defined as the ratio between the short-circuit current and the absorbed photon flux, i.e., $j_{h\nu}/en_{ph}$, which for well-designed cells is about 90%.

(b) Calculate the maximum power output (maximum power rectangle) $P_{mp} = j_{mp}U_{mp}$ from the current–voltage characteristic of the solar cell, where j_{mp} and U_{mp} are the values of the current density and the voltage, respectively, at maximum power output. Assume the saturation current density of the *pn* junction as $j_s = 10^{-13}$ A/cm². Neglecting losses due to reflection and conductance leakage, and the effect of series resistance, estimate the power efficiency of a Si solar cell. The solar power density above the atmosphere is 135 mW/cm².

H-7. INJECTION LASERS

Estimate the current density required for laser action in a GaAs laser diode.

Remark: Assume a carrier lifetime of $\tau_{n,p} \approx 10^{-9}$ sec, and a carrier diffusion length of $L_{n,p} \approx 1\,\mu$m. The condition of inverted population is achieved if the injected carrier concentrations are degenerate. Show that 10^{19} cm^{-3} represents degeneracy for electrons as well as holes in GaAs at room temperature (the effective masses for electrons and holes are $m_n = 0.07m$ and $m_p = 0.5m$, respectively).

J. Transport Phenomena

J-1. PERTURBATION OF THE DISTRIBUTION FUNCTION

Assuming the eigenvalue density to remain unchanged by external forces, the perturbation of the distribution function is solely due to a change of the occupation probability $F(\boldsymbol{k}, T)$.

(a) Show that the occupation probability perturbed by an electrostatic field $\boldsymbol{E}$ can be expressed as

$$F(\boldsymbol{k}) = F_0\left(\boldsymbol{k} - \frac{e}{\hbar}\tau\boldsymbol{E}\right),$$

where F_0 is the unperturbed occupation probability (Fermi–Dirac function) and τ is the relaxation time. Sketch the $F(\boldsymbol{k})$ relation and the corresponding Fermi surface.

(b) Show that the occupation probability perturbed by a temperature gradient $\text{grad}_{\boldsymbol{r}}\, T$ can be expressed to a first approximation as

$$F(\boldsymbol{k}, T) = F_0(\boldsymbol{k}, T - \tau(\boldsymbol{v}_k \cdot \text{grad}_{\boldsymbol{r}}\, T)).$$

What approximation does this relation imply? Sketch the $F(\boldsymbol{k}, T)$ relation.

J-2. MEAN FREE PATH

(a) From the electrical conductivity σ of an isotropic metal, calculate the mean free path Λ of the conduction electrons. The electron mobility b is related to the relaxation time τ by $b = e\tau/m$, where τ can be approximated by the mean free time between two electron collisions. Choose as an example Cu, for which $\sigma = 6\times10^5\,\Omega^{-1}\,\text{cm}^{-1}$ at 300°K, the electron concentration $n = 8.5\times10^{22}\,\text{cm}^{-3}$, and the Fermi velocity $v_F = 1.56\times10^8$ cm/sec.

(b) Show that the calculated value of Λ cannot be explained by electron scattering at the individual Cu^+ ions.

Suggestion: Estimate the classical mean free path Λ_{cl} by assuming an ionic cross-section for the scattering process; the ionic radius of Cu^+ is 0.96 Å. Show that $\Lambda \gg \Lambda_{cl}$.

J-3. JOULE HEAT

Calculate the Joule heat generated in an electronic conductor. Use a simple kinetic consideration and assume that with each collision the electrons lose the energy they have gained in the electric field.

J-4. HOT ELECTRONS

Ohm's law holds under the condition that the perturbation of the distribution function is small. This condition is violated at high electric fields, when the drift velocity of the electrons approaches their thermal velocity. Estimate the required electric field for (a) a typical metal ($m^* = m$, $b = 10$ cm²/V sec), and for (b) n-type Ge ($m^* = 0.2m$, $b = 3600$ cm²/V sec), both at room temperature. Discuss the temperatures and kind of materials for which hot electron effects can be most easily observed.

J-5. ELECTRON AND PHONON CONTRIBUTIONS TO THE THERMAL CONDUCTIVITY OF METALS

In elementary kinetic theory, the thermal conductivity is given by

$$\varkappa = \tfrac{1}{3}Cv\Lambda,$$

where C is the specific heat, v is the velocity, and Λ is the mean free path of the electrons and phonons, respectively. The mean free path for electrons and phonons is related to the relaxation time τ_{ep} for electron–phonon collisions and τ_{pe} for phonon–electron collisions, respectively. Estimate τ_{pe} at room temperature and the ratio $\varkappa_e/\varkappa_p$. Assume that the number of electron–phonon collisions per unit time equals that of phonon–electron collisions; hence $n_e/\tau_{ep} = n_p/\tau_{pe}$, where n_e and n_p are the concentrations of electrons and phonons, respectively, which contribute to the thermal conductivity. Close to the Debye temperature (about room temperature for metals), n_p is about equal to the concentration of atoms in the metal. Furthermore, use the following data: relaxation time for electron–phonon collisions at room temperature, $\tau_{ep} = 10^{-13}$ sec, degeneracy temperature of the electron gas, $T_E = 3\times10^4$ °K, electron velocity $v_e = 1\times10^8$ cm/sec, phonon velocity $v_p = 3\times10^5$ cm/sec, specific heat of the electrons and phonons, $C_e = 0.1R$ and $C_p = 3R$, respectively, where R is the universal gas constant.

J-6. ELECTRON–PHONON INTERACTIONS

Show that in an electronic conductor the Fermi distribution of the electrons is in equilibrium with the Bose distribution of the phonons.

Remark: In thermodynamic equilibrium the number of electron transitions $\boldsymbol{k} \to \boldsymbol{k}'$ must be compensated by that of the electron transitions $\boldsymbol{k}' \to \boldsymbol{k}$. For inelastic interactions, the energy conservation requires $E_{\boldsymbol{k}'} = E_{\boldsymbol{k}} + \hbar\omega$.

J-7. THERMOCOUPLE AND FIGURE OF MERIT

The maximum efficiency of a thermocouple of two electronic conductors 1 and 2 is given by

$$\eta_{\max} = \eta_{\text{Carnot}} \frac{\xi - 1}{\xi + \dfrac{T_2}{T_1}}$$

with

$$\eta_{\text{Carnot}} = \frac{T_1 - T_2}{T_1}, \quad \text{and} \quad \xi = (1 + Z\bar{T})^{1/2},$$

where the figure of merit is

$$Z = \frac{(S_1 - S_2)^2}{\left[\left(\dfrac{\varkappa_1}{\sigma_1}\right)^{1/2} + \left(\dfrac{\varkappa_2}{\sigma_2}\right)^{1/2}\right]^2},$$

and $\bar{T}$ is the arithmetic mean of T_1 and T_2, $S_{1,2}$ are the thermoelectric powers, $\varkappa_{1,2}$ are the thermal conductivities, and $\sigma_{1,2}$ are the electrical conductivities of the conductors 1 and 2. Under the assumptions $S_1 = -S_2 = S$, and $\varkappa_1/\sigma_1 = \varkappa_2/\sigma_2 = \varkappa/\sigma$, one obtains

$$Z = S^2\sigma/\varkappa.$$

In this form, the quantity Z is also used to characterize the thermoelectric quality of a given conductor material.

Explain the following method of determining Z: one measures the voltage differences ΔV_{dc} and ΔV_{ac} which are obtained by passing a dc current I_{dc} and an equivalent ac current I_{ac}, respectively, through a thermally insulated sample. The distance between the voltage probes is Δx. Z follows from the relation

$$Z = \frac{1}{T} \frac{\Delta V_{\text{dc}} - |\Delta V_{\text{ac}}|}{|\Delta V_{\text{ac}}|}.$$

Remark: The dc current gives rise to a temperature difference between the voltage probes due to the Peltier effect. Due to the temperature gradient, ΔV_{dc} consists of an ohmic voltage drop and a thermoelectric voltage. Under adiabatic steady-state conditions, the heat fluxes due to the Peltier effect and due to the temperature gradient cancel each other.

J-8. THERMOELECTRIC POWER OF A SEMICONDUCTOR

Show that for a two-band model, the thermoelectric power is given by

$$S_s = \frac{S_n\sigma_n + S_p\sigma_p}{\sigma_n + \sigma_p},$$

where S_n and S_p are the thermoelectric powers, and σ_n and σ_p are the electrical conductivities due to electrons and holes, respectively.

J-9. HALL EFFECT OF A SEMICONDUCTOR

For a *p*-type semiconductor, the transition from impurity conduction to intrinsic conduction usually involves a sign change of the Hall coefficient R_H at the temperature T_H. Show that at T_H the electrical conductivity σ equals the intrinsic conductivity σ_i.

K. Magnetism

K-1. ADIABATIC DEMAGNETIZATION

For an ideal paramagnetic substance, calculate the temperature decrease due to the magnetocaloric effect. What is the ratio of the temperature decreases obtained at room temperature and at liquid helium temperature?

K-2. SATURATION MAGNETIZATION

In EuS, the saturation magnetization is obtained when the magnetic moments of all the Eu^{2+} ions are parallel. The magnetic moment of each Eu^{2+} ion is $7\mu_B$. Calculate the saturation magnetization of EuS. EuS crystallizes in the NaCl structure with a lattice constant of $a = 5.97$ Å.

K-3. LANGEVIN FUNCTION

Show that in the classical limit the magnetization of N identical atoms with permanent magnetic moments μ is given by

$$M_z = N\mu L(a_\infty)$$

with the Langevin function

$$L(a_\infty) = \coth a_\infty - \frac{1}{a_\infty}$$

and

$$a_\infty = \frac{\mu B_z}{kT}.$$

Remark: Each magnetic moment can rotate freely, and its potential energy under the influence of the magnetic induction $\boldsymbol{B}$ is $E_{\text{pot}} = -(\boldsymbol{\mu}\cdot\boldsymbol{B})$. The orienting action of the magnetic induction is counteracted by the thermal motions, such that the magnetization is

$$M_z = N\mu \overline{\cos\theta},$$

where θ is the angle between $\boldsymbol{\mu}$ and $\boldsymbol{B}$, and $\overline{\cos\theta}$ is the thermal average. This average value is obtained on the basis of Boltzmann statistics by integrating over all solid angles the relative probabilities of finding a magnetic moment in an element $d\Omega$ of solid angle.

K-4. LANDAU LEVELS AND CYCLOTRON ORBIT

A free electron moves in the x—y plane with velocity $\boldsymbol{v}_\perp = 10^4$ m/sec. Calculate the change of the absolute value $|\boldsymbol{v}_\perp|$, when the magnetic induction $B_z = 1$ V sec/m² perpendicular to the x—y plane is generated. How much energy does the electron gain or lose? Which Landau quantum number characterizes the state of the electron? Calculate the relative decrease of the radius of the cyclotron orbit if the magnetic induction is doubled.

K-5. TEMPERATURE DEPENDENCE OF THE SPIN PARAMAGNETISM

(a) Calculate the relative change $d\varkappa/\varkappa\, dT$ of the paramagnetic susceptibility for a typical metal ($\zeta = 5$ eV) at room temperature.

(b) Calculate the relative change of the paramagnetic susceptibility of a metal cube due to its thermal expansion. The linear thermal expansion coefficient is assumed to be $\beta = 1.6 \times 10^{-5}$ (°K)$^{-1}$.

K-6. SPIN POLARIZATION OF ELECTRONS IN A MAGNETIC FIELD

The electrons of a metal are spin-polarized under the influence of a magnetic induction $\boldsymbol{B}$. The degree P of polarization is defined as the difference of the relative concentrations of electrons with spins parallel and antiparallel to $\boldsymbol{B}$, i.e., $P = (N_\uparrow - N_\downarrow)/(N_\uparrow + N_\downarrow)$. Show that for free electrons at $T = 0$°K the spin polarization is given by $P = \mu_B B/2\zeta$, provided that $\mu_B B \ll \zeta$. Calculate P for $B = 1$ V sec/cm² and $\zeta = 5$ eV.

Suggestion: Derive first the quantity $P_D = (D_\uparrow - D_\downarrow)/(D_\uparrow + D_\downarrow)$, where $D_\uparrow$ and $D_\downarrow$ are the eigenvalue densities pertaining to electrons with spins parallel and antiparallel to $\boldsymbol{B}$, respectively. Then show that $P = 3P_D$.

K-7. ANTIFERROMAGNETISM

An antiferromagnetic material is assumed to consist of two sublattices A and B with sublattice magnetizations $\boldsymbol{M}_A$ and $\boldsymbol{M}_B$. Show, on the basis of the molecular field approximation, that the temperature dependence of the magnetic susceptibility $\varkappa$ in the paramagnetic range is given by

$$\varkappa = \frac{C}{T+\theta}.$$

Calculate the Curie constant C, the paramagnetic Curie temperature θ and the Néel temperature T_N.

Remark: The internal magnetic field in the sublattice A is $\boldsymbol{H}_A = \boldsymbol{H} - W_{AB}\boldsymbol{M}_B$, and that in the sublattice B is $\boldsymbol{H}_B = \boldsymbol{H} - W_{AB}\boldsymbol{M}_A$. Note that in each sublattice only half of the number of magnetic atoms in the antiferromagnetic crystal contribute to the sublattice magnetization.

K-8. HELIMAGNETISM

The magnetic energy density of a linear chain of atoms with spin operators $\boldsymbol{S}_i$ is given by the Hamiltonian

$$\mathcal{H}_{\text{exchange}} = -2\frac{A_1}{\hbar^2}\sum_{i=1}^{N}(\boldsymbol{S}_i\cdot\boldsymbol{S}_{i+1}) - 2\frac{A_2}{\hbar^2}\sum_{i=1}^{N}(\boldsymbol{S}_i\cdot\boldsymbol{S}_{i+2}),$$

where N is the number of spins per unit length, and A_1 and A_2 are parameters which characterize the exchange interaction between the nearest and next-nearest neighbour spins. The exchange interactions between more distant spins are neglected. Derive the conditions for ferromagnetic, antiferromagnetic, and helimagnetic (helical array of magnetic moments) order, and show that helimagnetic order occurs under the conditions

$$|A_1| < 4\,|A_2|, \quad \text{and} \quad A_2 < 0,$$

where the angle δ between neighbour spins is given by $\cos\delta = -A_1/4A_2$.

Suggestion: Consider the spin operators $\boldsymbol{S}_i$ as classical vectors and minimize the magnetic energy density

$$E_{\text{exchange}} = -\frac{2N}{\hbar^2}S^2(A_1\cos\delta + A_2\cos 2\delta).$$

Index

OTHER TITLES IN THE SERIES IN NATURAL PHILOSOPHY

Vol. 1. Davydov—Quantum Mechanics
Vol. 2. Fokker—Time and Space, Weight and Inertia
Vol. 3. Kaplan—Interstellar Gas Dynamics
Vol. 4. Abrikosov, Gor'kov and Dzyaloshinskii—Quantum Field Theoretical Methods in Statistical Physics
Vol. 5. Okun'—Weak Interaction of Elementary Particles
Vol. 6. Shklovskii—Physics of the Solar Corona
Vol. 7. Akhiezer *et al.*—Collective Oscillations in a Plasma
Vol. 8. Kirzhnits—Field Theoretical Methods in Many-body Systems
Vol. 9. Klimontovich—The Statistical Theory of Non-equilibrium Processes in a Plasma
Vol. 10. Kurth—Introduction to Stellar Statistics
Vol. 11. Chalmers—Atmospheric Electricity (2nd Edition)
Vol. 12. Renner—Current Algebras and their Applications
Vol. 13. Fain and Khanin—Quantum Electronics, Vol. 1—Basic Theory
Vol. 14. Fain and Khanin—Quantum Electronics, Vol. 2—Maser Amplifiers and Oscillators
Vol. 15. March—Liquid Metals
Vol. 16. Hori—Spectral Properties of Disordered Chains and Lattices
Vol. 17. Saint James, Thomas and Sarma—Type II Superconductivity
Vol. 18. Margenau and Kestner—Theory of Intermolecular Forces (2nd Edition)
Vol. 19. Jancel—Foundations of Classical and Quantum Statistical Mechanics
Vol. 20. Takahashi—An Introduction to Field Quantization
Vol. 21. Yvon—Correlations and Entropy in Classical Statistical Mechanics
Vol. 22. Penrose—Foundations of Statistical Mechanics
Vol. 23. Visconti—Quantum Field Theory, Volume 1
Vol. 24. Furth—Fundamental Principles of Modern Theoretical Physics
Vol. 25. Zheleznyakov—Radioemission of the Sun and Planets
Vol. 26. Grindlay—An Introduction to the Phenomenological Theory of Ferroelectricity
Vol. 27. Unger—Introduction to Quantum Electronics
Vol. 28. Koga—Introduction to Kinetic Theory: Stochastic Processes in Gaseous Systems
Vol. 29. Galasiewicz—Superconductivity and Quantum Fluids
Vol. 30. Constantinescu and Magyari—Problems in Quantum Mechanics
Vol. 31. Kotkin and Serbo—Collection of Problems in Classical Mechanics
Vol. 32. Panchev—Random Functions and Turbulence
Vol. 33. Talpe—Theory of Experiments in Paramagnetic Resonance
Vol. 34. ter Haar—Elements of Hamiltonian Mechanics (2nd Edition)
Vol. 35. Clarke and Grainger—Polarized Light and Optical Measurement
Vol. 36. Haug—Theoretical Solid State Physics, Volume 1
Vol. 37. Jordan and Beer—The Expanding Earth
Vol. 38. Todorov—Analytical Properties of Feynman Diagrams in Quantum Field Theory
Vol. 39. Sitenko—Lectures in Scattering Theory
Vol. 40. Sopel'man—Introduction to the Theory of Atomic Spectra
Vol. 41. Armstrong and Nicholls—Emission, Absorption and Transfer of Radiation in Heated Atmospheres

Vol. 42. BRUSH—Kinetic Theory, Volume 3
Vol. 43. BOGOLYUBOV—A Method for Studying Model Hamiltonians
Vol. 44. TSYTOVICH—An Introduction to the Theory of Plasma Turbulence
Vol. 45. PATHRIA—Statistical Mechanics
Vol. 46. HAUG—Theoretical Solid State Physics, Volume 2
Vol. 47. NIETO—The Titius–Bode Law of Planetary Distances: Its History and Theory
Vol. 48. WAGNER—Introduction to the Theory of Magnetism
Vol. 49. IRVINE—Nuclear Structure Theory
Vol. 50. STROHMEIER—Variable Stars
Vol. 51. BATTEN—Binary and Multiple Systems of Stars
Vol. 52. ROUSSEAU and MATHIEU — Problems in Optics
Vol. 53. BOWLER—Nuclear Physics
Vol. 54. POMRANING—The Equations of Radiation Hydrodynamics
Vol. 55. BELINFANTE—A Survey of Hidden-variables Theories
Vol. 56. SCHEIBE—The Logical Analysis of Quantum Mechanics
Vol. 57. ROBINSON—Macroscopic Electromagnetism
Vol. 58. GOMBÁS and KISDI—Wave Mechanics and Its Applications
Vol. 59. KAPLAN and TSYTOVICH—Plasma Astrophysics
Vol. 60. KOVÁCS and ZSOLDOS—Dislocations and Plastic Deformation
Vol. 61. AUVRAY and FOURRIER—Problems in Electronics
Vol. 62. MATHIEU—Optics
Vol. 63. ATWATER—Introduction to General Relativity
Vol. 64. MÜLLER—Quantum Mechanics: A Physical World Picture
Vol. 65. BILENKY—Introduction to Feynman Diagrams
Vol. 66. VODAR and ROMAND—Some Aspects of Vacuum Ultraviolet Radiation Physics
Vol. 67. WILLETT—Gas Lasers: Population Inversion Mechanisms
Vol. 68. AKHIEZER—Plasma Electrodynamics, Volume 1
Vol. 69. GLASBY—The Nebular Variables
Vol. 70. BILANICKY-BIRULA—Quantum Electrodynamics
Vol. 71. KARPMAN—Non-linear Waves in Dispersive Media
Vol. 72. CRACKNELL—Magnetism in Crystalline Materials
Vol. 73. PATHRIA—The Theory of Relativity
Vol. 74. SITENKO and TARTAKOVSKII—Lectures on the Theory of the Nucleus
Vol. 75. BELINFANTE—Measurements and Time Reversal in Objective Quantum Theory
Vol. 76. SOBOLEV—Light Scattering in Planetary Atmospheres
Vol. 77. NOVAKOVIC—The Pseudo-spin Method in Magnetism and Ferroelectricity
Vol. 78. NOVOZHILOV—Introduction to Elementary Particle Theory